Benchmark Papers
in Microbiology

Series Editor: Wayne W. Umbreit
Rutgers—The State University

PUBLISHED VOLUMES

MICROBIAL PERMEABILITY / *John P. Reeves*
CHEMICAL STERILIZATION / *Paul M. Borick*
MICROBIAL GENETICS / *Morad Abou-Sabé*
MICROBIAL PHOTOSYNTHESIS / *June Lascelles*
MICROBIAL METABOLISM / *H. W. Doelle*
ANIMAL CELL CULTURE AND VIROLOGY / *Robert J. Kuchler*
PHAGE / *Sewell P. Champe*
MICROBIAL GROWTH / *P. S. S. Dawson*
MICROBIAL INTERACTION WITH THE PHYSICAL ENVIRONMENT /
    D. W. Thayer
MOLECULAR BIOLOGY AND PROTEIN SYNTHESIS / *Robert A.
    Niederman*

Additional volumes in preparation

**Benchmark Papers
in Microbiology / 10**

A BENCHMARK® Books Series

MOLECULAR BIOLOGY
AND PROTEIN SYNTHESIS

Edited by

ROBERT A. NIEDERMAN
Rutgers—The State University

Dowden, Hutchinson & Ross, Inc.
STROUDSBURG, PENNSYLVANIA

Distributed by

HALSTED
PRESS

A Division of
John Wiley & Sons, Inc.

Exclusive Distributor: **Halsted Press**
A Division of John Wiley & Sons, Inc.

ACKNOWLEDGMENTS AND PERMISSIONS

ACKNOWLEDGMENTS

NATIONAL ACADEMY OF SCIENCES—*Proceedings of the National Academy of Sciences of the United States of America*

Analysis of DNA Polymerases II and III in Mutants of *Escherichia coli* Thermosensitive for DNA Synthesis

Mechanism of Activation of Catabolite-Sensitive Genes: A Positive Control System

N-Formylmethionyl-sRNA as the Initiator of Protein Synthesis

Replication and Repair of DNA in Cells of *Escherichia coli* Treated with Toluene

The Replication of DNA in *Escherichia coli*

RNA-Linked Nascent DNA Fragments in *Escherichia coli*

Structure and Function of *E. coli* Ribosomes: V. Reconstitution of Functionally Active 30S Ribosomal Particles from RNA and Proteins

Synthesis of a Large Precursor to Ribosomal RNA in a Mutant of *Escherichia coli*

PERMISSIONS

The following papers have been reprinted or translated with the permission of the authors and copyright holders.

ACADEMIC PRESS, INC.

Archives of Biochemistry and Biophysics

Studies on the Role of Factor Ts in Polypeptide Synthesis

Translation of the Genetic Message: VII. Role of Initiation Factors in Formation of the Chain Initiation Complex with *Escherichia coli* Ribosomes

Biochemical and Biophysical Research Communications

The Enzymic Incorporation of Ribonucleotides into Polyribonucleotides and the Effect of DNA

ACADEMIC PRESS, INC. (LONDON) LTD.—*Journal of Molecular Biology*

Codon-Anticodon Pairing: The Wobble Hypothesis

Ribonucleoprotein Particles from *Escherichia coli*

Translation and Translocation of Defined RNA Messengers

AMERICAN ASSOCIATION FOR THE ADVANCEMENT OF SCIENCE—*Science*

Active Center DNA Polymerase

Assembly of Bacterial Ribosomes

Structure of a Ribonucleic Acid

Three-Dimensional Structure of Yeast Phenylalanine Transfer RNA: Folding of the Polynucleotide Chain

Acknowledgments and Permissions

AMERICAN CHEMICAL SOCIETY—*Biochemistry*
    Effect of Aminoacyl Transfer Ribonucleic Acid on Competition Between Guanosine
        5'-Triphosphate and Guanosine 5'-Diphosphate for Binding to a Polypepetide Chain
        Elongation Factor from *Escherichia coli*
    Isolation and Physical Properties of the Ribosomal Ribonucleic Acid of *Escherichia coli*
    Stoichemistry of the 30S Ribosomal Proteins of *Escherichia coli*

AMERICAN SOCIETY OF BIOLOGICAL CHEMISTS—*The Journal of Biological Chemistry*
    The Enzymic Synthesis of Amino Acyl Derivatives of Ribonucleic Acid: I. The
        Mechanism of Leucyl-, Valyl-, Isoleucyl-, and Methionyl Ribonucleic Acid Formation

COLD SPRING HARBOR LABORATORY—*Cold Spring Harbor Symposia on Quantitative
        Biology*
    On the Catalytic Center of Peptidyl Transfer: A Part of the 50S Ribosome Structure
    Peptide Chain Termination, Codon, Protein Factor, and Ribosomal Requirements
    Polynucleotide Synthesis and the Genetic Code
    The RNA Code and Protein Synthesis
    In Vivo Mechanism of DNA Chain Growth

ELSEVIER SCIENTIFIC PUBLISHING CO.—*Biochimica et Biophysica Acta*
    On the Presence of Low-Molecular-Weight Ribonucleic Acid in the Ribosomes of
        Escherichia coli

MACMILLAN (JOURNALS) LTD.—*Nature*
    Conformation of the Anticodon Loop in tRNA
    Factor Stimulating Transcription by RNA Polymerase
    Genetical Implications of the Structure of Deoxyribonucleic Acid
    Isolation of an *E. coli* Strain with a Mutation Affecting DNA Polymerase
    Nucleotide Sequence of 5S-Ribosomal RNA from *Escherichia coli*
    Polypeptide Chain Initiation: Nucleotide Sequences of the Three Ribosomal Binding
        Sites in Bacteriophage R17 RNA
    Termination Factor for RNA Synthesis
    Two Compounds Implicated in the Function of the *RC* Gene of *Escherichia coli*
    Unstable Ribonucleic Acid Revealed by Pulse Labelling of *Escherichia coli*

SERIES EDITOR'S PREFACE

In the past decade there has clearly been no more exciting and advancing field than molecular biology; yet, to the novice, it may appear excessively complex. Dr. Niederman has prepared this volume with the needs of the novice in mind. So much information and such a wealth of literature exist that it has been important to select those papers which provide not merely superficial understanding but real knowledge of how and why and on what evidence we know how DNA is replicated, how proteins are synthesized, and how the complex, yet exquisitely tuned, machinery is controlled. Previous collections were designed for workers in the field, to whom the methods and the language were familiar. This volume has as its purpose to provide a knowing and useful guide to the fundamental papers upon which modern molecular biology is based. In this task, Dr. Niederman has brought not only his experience as an operator of a research laboratory actively involved in this field but also his experience as a teacher of undergraduates (and graduates) in a course that is a model of its kind. As such, we have here a volume useful to the novice—as well as to the advanced investigator whenever the latter wishes to refer to the major contributions to the field.

The volume is composed of five parts. In the first, eight papers cover the field of DNA structure and replication and introduce the novice to polymerase III and Okazaki pieces. Part II provides six papers on messenger RNA formation which introduce the novice to core enzyme, sigma factor, and other transcription factors (M, PSI, and rho), as well as information regarding the action of the promising antibiotic rifampicin and the beginnings of an explanation for catabolite repression. There is also a discussion of ppGpp and pppGpp, originally known as "magic spots."

The third part contains eight papers on the nature and the assembly of the ribosome, the organelle that translates messenger RNA into protein. The proteins, especially, seem to be exceedingly numerous and complex, and the RNA that they contain may well come from cleavage of a much larger molecule. Part IV, with eight papers, covers the actual mechanism of peptide-bond formation, and for persons brought up in an earlier day, the number of external, or at least dissociable, factors seems tremendous. Actually it appears that peptide-bond formation is considerably more complicated than originally supposed. In the fifth part, seven outstanding papers deal with the genetic code and transfer RNA, the latter known in such detail that one can begin to visualize three-dimensional structures.

Although a glance at the contents will reveal this arrangement, it seems sensible to outline it here so as to point out the unusual and illuminating findings that these sections report.

WAYNE W. UMBREIT

CONTENTS

Contents

PART II: TRANSCRIPTION

PART III: TRANSLATION: THE BACTERIAL RIBOSOME

PART IV: TRANSLATION: MECHANISM OF PEPTIDE-BOND FORMATION

Contents

PART V: TRANSLATION: THE GENETIC CODE AND TRANSFER RNA

CONTENTS BY AUTHOR

Contents by Author

MOLECULAR BIOLOGY
AND PROTEIN SYNTHESIS

INTRODUCTION

An assessment of the explosive growth in molecular biology that has occurred during the last two decades reveals many significant contributions from a variety of disciplines. Indeed, biochemistry, biophysics, crystallography, genetics, and microbiology have each provided major contributions. However, the demonstration of deoxyribonucleic acid (DNA) rather than protein as the carrier of the cell's genetic information by Avery et al. (1944) marks the beginning of this scientific revolution. These investigators identified highly polymerized DNA as the constituent that caused the transformation of one strain of bacterium to another. This startling finding stimulated a number of attempts at elucidating the structure of DNA in the hope that deciphering its molecular architecture would provide a basis for understanding how this molecule functions as the primary hereditary material within the cell. This hope was finally realized with the advent of the double-helical model of Watson and Crick (1953) for the structure of DNA. The double helix represented a series of brilliant insights by the latter investigators that were largely based upon X-ray diffraction analysis (Wilkins et al., 1953; Franklin and Gosling, 1953) and the purine and pyrimidine contents of a variety of DNA molecules (described in Part I). The complementarity of the two DNA chains suggested that, upon strand separation, each serves as a template for the synthesis of a new complementary strand. This provided a plausible model for replication of the chromosome, thus maintaining continuity in the transfer of genetic information from one generation to another. The area of molecular biology encompassed within this volume, that is the biochemical aspects of DNA replication and the processing of this genetic information into protein gene products, has largely stemmed from the funda-

1

mental discoveries made in these early investigations. The mechanism of DNA replication is the subject of Part I and has recently been extensively reviewed (Klein and Bonhoeffer, 1972; Kornberg, 1974; Schekman et al., 1974).

Another crucial aspect of the Watson–Crick double helix was the lack of restriction on the sequence of bases within the interior of the DNA helix. This suggested a coding mechanism for the expression of hereditary information. The manner in which this code, embodied within the base sequence of the DNA, is transcribed and translated into a specific linear array of amino acids in the protein gene products thus became the central question of molecular biology. These studies, in little more than a decade, led to the complete elucidation of the genetic code. In addition, many components of the transcription and translation apparatus have now been elucidated. However, very little is presently known about the manner in which the cell controls the selective transcription of specific regions of DNA templates into the various ribonucleic acid (RNA) species. (For a recent review on the selectivity of transcription see Chamberlin, 1974.) Part II deals with the transcription process and several of its regulatory aspects.

The overall process in which the specific information contained within the sequence of bases in the RNA messenger species is translated on the ribosome into a linear protein sequence can now be looked upon as a series of partial reactions involving initiation, elongation, and termination of peptide chains. The mechanisms of the individual steps and the details of the role that the ribosome plays in the translation sequence are presently the object of extensive investigations (Nomura et al., 1974). Several aspects of translation, for instance, the structure and function of the ribosome, the mechanism of peptide bond formation, the genetic code, and the role of transfer RNA, form the subject matter of Parts III, IV, and V.

This Benchmark volume has been designed primarily as an introduction for the beginning student of molecular biology. Accordingly, the comments at the beginning of each chapter and their accompanying bibliographies are mainly intended as a guide for the advanced undergraduate and beginning graduate student. These introductory comments are by no means exhaustive reviews of the literature (references to appropriate reviews are provided in each section). Instead, it is hoped that they will aid in bringing the collected papers within each section into a more coherent framework. In addition, they should serve both as an overview and a critical guide for the student. Where appropriate, they also function in updating the selected papers.

In that the papers in this collection represent diverse areas of molecular biology, it is also hoped that they will prove useful to investigators in the field who might appreciate having them compiled in a single volume. In addition to many of the classical papers of molecular biology, several recent papers of unusual impact have been included. Where essentially the same findings have been reported simultaneously by more than one research group, the task of choosing a single benchmark paper has been difficult. In these cases, largely because of space limitations, the papers not reprinted have been cited in the introductory comments preceding each section. Since this volume forms part of a series of classical papers in microbiology, it has been confined to studies that broadly fit into this overall discipline, and, in general, to those with prokaryotic microorganisms. In many cases, these studies have provided a basis for understanding the various aspects of molecular biology in higher forms.

It is a pleasure to thank Sewell P. Champe, Sam Kaplan, K. Sieglinde Neuhauser, John P. Reeves, and W. W. Umbreit (series editor) for their critical reading of the manuscript and for their many valuable suggestions. I also wish to express my appreciation to M. L. P. Collins, John A. Distasio, and Lawrence C. Parks, who also served in this capacity. I am especially grateful to K. S. Neuhauser for her excellent English translation of Paper 16, which appeared originally in French.

REFERENCES

Avery, O. T., C. M. MacLeod, and M. McCarty (1944). Studies on the chemical nature of the substance inducing transformation of pneumococcal types: I. Induction of transformation by a desoxyribonucleic acid fraction isolated from pneumococcus type III. *J. Exptl. Med., 79, 137-158.*

Chamberlin, M. J. (1974). The selectivity of transcription. *Ann. Rev. Biochem., 43,* 721-775.

Franklin, R. E., and R. G. Gosling (1953). Molecular configuration in sodium thymonucleate. *Nature (London), 171,* 740-741.

Klein, A., and F. Bonhoeffer (1972). DNA replication. *Ann. Rev. Biochem., 41,* 301-332.

Kornberg, A. (1974). *DNA Synthesis.* W. H. Freeman, San Francisco.

Nomura, M., A. Tissières, and P. Lengyel (1974). *Ribosomes.* Cold Spring Harbor Monograph Series, Cold Spring Harbor Laboratory, Cold Spring Harbor, N.Y.

Schekman, R., A. Weiner, and A. Kornberg (1974). Multienzyme systems of DNA replication. *Science, 186,* 987-993.

Watson, J. D., and F. H. C. Crick (1953). The structure of DNA. *Cold Spring Harbor Symp. Quant. Biol., 18,* 123-131.

Wilkins, M. F. H., A. R. Stokes, and H. R. Wilson (1953). Molecular structure of deoxypentose nucleic acids. *Nature (London), 171,* 738-739.

Part I
DNA REPLICATION

Editor's Comments
on Papers 1 Through 8

Much of today's molecular biology emanates largely from the Watson–Crick model for the structure of DNA (Watson and Crick, 1953). In this model, DNA is depicted as a double-helical molecule in which the two antiparallel chains follow right-handed helices coiled about a single axis. The backbone of the structure consists of deoxyribose residues in $3'$-, $5'$-phosphodiester linkage with the purine and pyrimidine bases located in the interior of the helix roughly perpendicular to the sugar residues. The bases are situated 3.4 Å apart with one repeating turn of the helix containing 10 residues on each chain, thus extending 34 Å per helical turn. The double helix is held together by hydrogen bonding between purine and pyrimidine base pairs. From tautomeric considerations as well as experimental determinations of base ratios (Zamenhof et al., 1952; Wyatt, 1952), it was proposed that adenine pairs with thymine and guanine with cytosine. Thus, the sequence of bases in one strand determines that of the complementary strand.

An important consequence of the Watson–Crick structure was that the complementarity of base pairing between DNA strands through hydrogen bonding suggested a copying mechanism for the genetic material (Watson and Crick, Paper 1). This mechanism assumes that the parental DNA duplex unwinds and each chain serves as a template for the synthesis of a new complementary chain. Each daughter duplex would therefore contain one new strand and one conserved parental strand. Although this semiconservative mechanism is a plausible outgrowth of the Watson–Crick model, Delbrück and Stent (1957) suggested conservative and dispersive mechanisms as alternative hypotheses for DNA replication. In the conservative model, parental molecules do not unwind; instead, they serve as a template for an entirely new daughter duplex. In the dispersive case, the parental duplex unravels by chain fragmentation at intervals and reunites to form two daughter strands that are composed of both old and new segments.

That the semiconservative model was the correct mechanism for DNA replication was elegantly settled in the landmark experiment of Meselson and Stahl (Paper 2). These authors developed an ingenious technique for detecting the distribution of material derived from parental DNA molecules in the progeny molecules. Their method of analysis was based upon the use of a density label (^{15}N, the heavy isotope of nitrogen) that imparts an increased buoyant density to the DNA of *Escherichia coli* and the examination of density distributions by sedimentation in the analytical ultracentrifuge. The sedimentation analysis was facilitated by equilibrium buoyant density centrifugation (Meselson et al., 1957). In this technique, a stable density gradient of cesium chloride is first established as a result of opposing forces of sedimentation and diffusion. When equilibrium is achieved, the DNA bands at its buoyant density within the gradient. With several species of DNA whose densities differ, each bands at a position where the density of the cesium chloride solution is equal to the buoyant density of the respective species. It was thus possible to separate ^{15}N DNA (density = 1.724 g/cc) from unlabeled DNA (density = 1.710 g/cc).

To detect the distribution of parental material within the progeny

DNA, *E. coli* was grown for several generations with $^{15}NH_4Cl$ as the sole source of nitrogen so that the DNA was heavily ^{15}N-labeled. The cells were shifted to an unlabeled medium and the DNA was isolated at intervals in the growth cycle. At 0 generations, a single band of ^{15}N DNA was observed, whereas after 1.0 generation, a single band with a hybrid density (half heavy, half light) was seen. At 1.9 generations, nearly equal quantities of a new band (fully light) and the conserved hybrid band appeared. These results agreed exactly with the Watson–Crick model for DNA duplication. However, to establish that the molecular subunits were indeed single polynucleotide chains, the DNA of hybrid density was separated into its component strands by heat denaturation. Upon buoyant density ultracentrifugation, two discrete bands were seen, thus establishing that the subunits of the DNA molecules resulting from a replicative cycle are indeed single, continuous structures.

Although the Meselson–Stahl experiment clearly established the distribution of parental atoms among progeny DNA molecules, much still remained to be learned about the mechanism of DNA duplication. Foremost among the questions left unanswered was the molecular structure of the DNA subunits from which the daughter duplexes are derived and the mechanism of synthesis and dissociation of the subunits. Despite numerous investigations (for reviews, see Kornberg, Paper 3; Richardson, 1969; Lark, 1969; Goulian, 1971), many of the molecular details of the copying mechanism and the individual enzymatic steps involved remain largely unknown.

The first major finding related to these questions was the demonstration by Kornberg and his associates (Kornberg et al., 1956; Lehman et al., 1958; summarized by Kornberg, Paper 3) that an enzyme in extracts of *E. coli* catalyzed the *in vitro* incorporation of deoxyribonucleotides into DNA in the presence of a single-stranded DNA template. This enzyme, which has now been designated DNA polymerase I, uses the four deoxyribonucleotide triphosphates as substrates and catalyzes a net synthesis of DNA by polymerization in a $5' \rightarrow 3'$ direction through covalent addition to the $3'$-end of a primer strand. This suggested that DNA contains all the specificity needed for its own replication. These results, together with the radioautographic demonstration of the circularity of the replicating bacterial chromosome (Cairns, 1963) and evidence for the association of replicating DNA with the cell membrane (Ryter and Jacob, 1963; Sueoka and Quinn, 1968; Earhart et al., 1968), form the basis of several models for the molecular structure of duplicating DNA (Gilbert and Dressler, 1968; Lark, 1969).

In these models, the replicating genome is depicted as a rolling circle in which the parental strands unravel and become duplicated. In Lark's model, replication proceeds sequentially and unidirectionally,* with both nascent strands of the new duplexes elongating simultaneously. The replicative origin is attached to the membrane and the inner parental strand (the negative strand) remains circular during the copying process. In contrast, the outer strand (the positive strand) is linearized during replication and its 5'-end is joined to a separate membrane site. Such attachments assure proper segregation of the completed DNA duplexes during cell division. In the Gilbert and Dressler model, the replicative origin is not attached to the membrane and only the free 5'-ends of the positive strands are so anchored. Instead, polymerization is controlled by a repressor at the replicative origin that must be removed before the duplication process can be initiated. However, the genetic evidence of Jacob et al. (1963) suggests that initiation is under the control of a diffusable gene product which serves as a positive regulatory element and interacts with a unique chromosome region.

A puzzling aspect of DNA replication is that simultaneous copying of each strand requires polymerization in both the $5' \rightarrow 3'$ and $3' \rightarrow 5'$ directions, whereas the known DNA polymerases elongate chains exclusively by the $5' \rightarrow 3'$ addition of nucleotides. This paradox has apparently been resolved by Okazaki et al. (Paper 4). (See also Okazaki et al., 1968.) These investigators showed that nascent DNA chains are first synthesized in *E. coli* and other bacteria as polynucleotide segments approximately 1,000 residues in length and then joined together to form the intact strands. Thus, short stretches could be synthesized along the strand with a free 3'-end in the $5' \rightarrow 3'$ direction. This discontinuous polymerization proceeds to the point at which the parental chains have unraveled. The polymerase then switches over to the other strand and copies it in the direction of the free 5'-end by the same discontinuous mechanism. In this manner, both strands would be copied by $5' \rightarrow 3'$ polymerization. These investigators also showed that the reaction in which the segments are joined is catalyzed by polynucleotide ligase (the ligase mechanism has been reviewed by Richardson, 1969).

Although the Okazaki mechanism provides a solution to the unidirectional polymerization problem of polymerase I, other results strongly suggest that this activity is not an obligatory component of the replicative machinery. Such evidence was first obtained with temperature-sensitive mutants of *E. coli* in which no alterations were found in polymerase I even though DNA synthesis ceased at nonpermissive temperatures (these studies are detailed by Hirota et al., 1972).

---

*The unidirectional nature of the replication process must be reevaluated in light of the evidence for bidirectional replication recently presented by Masters and Broda (1971).

These results, however, had the alternative explanation that the temperature-sensitive components were additional obligatory factors. More convincing evidence was that of DeLucia and Cairns (Paper 5), who isolated an *amber* mutant of *E. coli*, designated *polAl*, that grew normally and replicated its DNA but contained less than 1 percent of the usual polymerase I level. The *polAl* mutant was also more sensitive to ultraviolet irradiation, which suggests a repair function for polymerase I that is consistent with the known $5' \to 3'$ exonuclease activity of this enzyme.

Two additional enzymes, DNA polymerases II and III have been found in *E. coli polAl*. Polymerase II was first detected in 1970 by several groups (Knippers, 1970; T. Kornberg and Gefter, 1970; Moses and Richardson, 1970). This activity is chromatographically distinct from polymerase I and has different template requirements. Like polymerase I, polymerase II repairs single-stranded templates, adding nucleotides to the free 3'-hydroxyl group of the primer. Polymerase II also possesses exonuclease activity, but degrades single-stranded DNA only in the $3' \to 5'$ direction (R. B. Wickner et al., 1972).

DNA polymerase III was first detected as a chromatographically distinct entity in extracts of *E. Coli polAl* during purification of polymerase II (T. Kornberg and Gefter, 1971). Polymerase III can be distinguished from polymerase I and II by its different sensitivity to ionic strength and from polymerase II by its increased lability to heat (Moses et al., 1972). Both polymerases II and III are apparently present in wild-type *E. coli*, which suggests that they do not result from the *polAl* mutation.

The isolation of other mutants of *E. coli* defective in DNA polymerase activity has suggested that polymerase III is the only DNA polymerase activity indispensable to DNA replication. Richardson and his co-workers (cited in Moses et al., 1972) isolated a mutant of *E. coli polAl* that has no detectable polymerase II activity. It grows normally and has the same sensitivity to ultraviolet irradiation as the *polAl* mutant. This suggests that activities other than polymerases I and II are responsible for the replication of its DNA. More direct evidence for the role of polymerase III has been provided by Gefter et al. (Paper 6), who used *polAl* strains containing a second mutation that maps at the *dnaE* locus. These mutants replicate their DNA normally at 30°C, but not at 42°C, and contain polymerase II activity that functions normally at both temperatures. On the other hand, their *in vitro* polymerase III activity has essentially the same thermosensitive character as the overall replicative process *in vivo*. This suggests that polymerase III is the product of the *dnaE* gene and that it is required for DNA replication. In addition, the *dnaE* gene product has been purified by Nüsslein et al. (1971) and shown to possess the properties of polymerase III. DNA synthesis with polymerase III, however, is not semi-conservative and it requires gapped DNA as a primer (T. Kornberg and Gefter, 1972).* Thus, DNA duplication apparently requires protein factors in addition to polymerase III.

In this regard, Hirota et al. (1972) have summarized studies with thermosensitive mutants of *E. coli* that suggest the involvement of other components in the replicative mechanism. These studies involve mutants that map at *dna* loci other than *dnaE* and contain normal polymerase III activity (Gefter et al., Paper 6). Mutants of two such classes (*dnaA* and *C*) are defective in initiation, whereas those of the *dnaB, D, F,* and *G* loci are unable to elongate their DNA at nonpermissive temperatures. This supports the notion that multiple enzymes and structural components are required for copying the bacterial chromosome.

Unlike many other studies in which appropriate enzymes purified from cell-free extracts have resulted in the duplication of *in vivo* processes, the entire mechanism for copying the bacterial genome has so far not been reproduced with isolated components. An alternative *in vitro* approach has, however, yielded extended semiconservative replication at rates comparable to those in the intact cell. In this procedure, *E. coli* is treated with toluene; viability is lost, but the cells become permeable to nucleoside triphosphates. Moses and Richardson (Paper 7) have thereby established that the four deoxyribonucleoside triphosphates are the precursors of DNA and that the replicative process requires adenosine 5'-triphosphate (ATP)† and magnesium and is sensitive to *N*-ethylmaleimide (NEM), a sulfhydryl blocking reagent. They were also able to distinguish replicative synthesis from DNA polymerase I catalyzed repair, in that the latter process is endonuclease dependent, unaffected by ATP or NEM, and greatly reduced in toluene-treated *polAl* cells. Other *in vitro* systems for DNA replication have utilized ether-treated *E. coli* (Vosberg and Hoffmann-Berling, 1971), cell envelope–DNA complexes (Knippers and Strätling, 1970), and gently lysed cells embedded in agar (Smith et al., 1970). Each of these systems permits careful control of precursor pools.

A final aspect of DNA replication that has recently been clarified is the primer requirement for enzymes catalyzing DNA polymerization.

More recent evidence presented by W. Wickner et al. (1973) indicates that a new form of DNA polymerase III termed polymerase III carries out semiconservative replication of single-stranded templates in the presence of a second protein factor (copolymerase III*) and a primer fragment.
†This ATP requirement has been reconciled with polymerase III* as the essential replicative enzyme by the more recent report of W. Wickner and A. Kornberg (1973) in which ATP was shown to be indispensable for the copying of a single-stranded DNA template by polymerase III*.

This characteristic is not shared by RNA polymerase, which catalyzes the initiation and elongation of entirely new polyribonucleotide chains. This suggested a role for RNA polymerase in initiating the synthesis of DNA chains, with short segments of RNA serving as the primer for DNA polymerization. W. Wickner et al. (1972) have presented evidence for RNA-primed DNA synthesis during the formation of the double-stranded intermediate in the replication of the single-stranded circular chromosome of bacteriophage M13. Sugino et al. (Paper 8) have shown a requirement for RNA during discontinuous DNA replication in *E. coli*. The short RNA segments (50–100 nucleotides) are apparently located at the 5'-end of the Okazaki fragments containing approximately 400–2,000 nucleotides (Hirose et al., 1973). It is interesting to note that Sugino and Okazaki (1973) have also demonstrated that RNA-linked DNA fragments are intermediates in DNA replication in toluene-treated cells. Roychoudhury (1973) has recently presented evidence suggesting that the 5' → 3' exonuclease activity of DNA polymerase I is capable of the *in vivo* removal of the priming RNA from the final replication product.

REFERENCES

Cairns, J. (1963). The chromosome of *Escherichia coli. Cold Spring Harbor Symp. Quant. Biol., 28,* 43–46.

Delbrück, M., and G. S. Stent (1957). On the mechanism of DNA replication. In W. D. McElroy and B. Glass (eds.), *The Chemical Basis of Heredity,* Johns Hopkins Press, Baltimore, pp. 699–736.

Earhart, C. F., G. Y. Tremblay, M. J. Daniels, and M. Schaechter (1968). DNA replication studied by a new method for the isolation of cell membrane–DNA complexes. *Cold Spring Harbor Symp. Quant. Biol., 33,* 707–710.

Gilbert, W., and D. Dressler (1968). DNA replication. The rolling circle model. *Cold Spring Harbor Symp. Quant. Biol., 33,* 473–484.

Goulian, M. (1971). Biosynthesis of DNA. *Ann. Rev. Biochem., 40,* 855–898.

Hirose, S., R. Okazaki, and F. Tamanoi (1973). Mechanism of DNA chain growth: XI. Structure of RNA-linked DNA fragments of *Escherichia coli. J. Mol. Biol., 77,* 501–517.

Hirota, Y., J. Mordoh, I. Scheffler, and F. Jacob (1972). Genetic approach to DNA replication and its control in *Escherichia coli. Federation Proc., 31,* 1422–1427.

Jacob, F., S. Brenner, and F. Cuzin (1963). On the regulation of DNA replication in bacteria. *Cold Spring Harbor Symp. Quant. Biol., 28,* 329–348.

Knippers, R. (1970). DNA polymerase II. *Nature (London), 228,* 1050–1053.

——, and W. Strätling (1970). The DNA replicating capacity of isolated *E. coli* cell wall–membrane complexes. *Nature (London), 226,* 713–717.

Kornberg, A., I. R. Lehman, M. J. Bessman, and E. S. Simms (1956). Enzymic synthesis of deoxyribonucleic acid. *Biochim. Biophys. Acta, 21,* 197–198.

Kornberg, T., and M. L. Gefter (1970). DNA synthesis in cell-free extracts of a

DNA polymerase-defective mutant. *Biochem. Biophys. Res. Commun., 40,* 1348-1355.

—— (1971). Purification and DNA synthesis in cell-free extracts. Properties of DNA polymerase II. *Proc. Natl. Acad. Sci. U.S.A., 68,* 761-764.

—— (1972). Deoxyribonucleic acid synthesis in cell-free extracts: IV. Purification and catalytic properties of deoxyribonucleic acid polymerase III. *J. Biol. Chem., 247,* 5369-5375.

Lark, K. G. (1969). Initiation and control of DNA synthesis. *Ann. Rev. Biochem., 38,* 569-604.

Lehman, I. R., M. J. Bessman, E. S. Simms, and A. Kornberg (1958). Enzymatic synthesis of deoxyribonucleic acid: I. Preparation of substrates and partial purification of an enzyme from *Escherichia coli. J. Biol. Chem., 233,* 163-170.

Masters, M., and P. Broda (1971). Evidence for the bidirectional replication of the *Escherichia coli* chromosome. *Nature New Biol., 232,* 137-140.

Meselson, M., F. W. Stahl, J. Vinograd (1957). Equilibrium sedimentation of macromolecules in density gradients. *Proc. Natl. Acad. Sci. U.S.A., 43,* 581-588.

Moses, R. E., and C. C. Richardson (1970). A new DNA polymerase activity of *Escherichia coli:* I. Purification and properties of the activity present in *E. coli PolAl. Biochem. Biophys. Res. Commun., 41,* 1557-1564.

——, J. L. Campbell, R. A. Fleischman, G. D. Frenkel, H. L. Mulcahy, H. Shizuya, and C. C. Richardson (1972). Enzymatic mechanisms of DNA replication in *Escherichia coli. Federation Proc., 31,* 1415-1421.

Nüsslein, V., B. Otto, F. Bonhoeffer, and H. Schaller (1971). Function of DNA polymerase III in DNA replication. *Nature New Biol., 234,* 285-286.

Okazaki, R., T. Okazaki, K. Sakabe, K. Sugimoto, and A. Sugino (1968). Mechanism of DNA chain growth: I. Possible discontinuity and unusual secondary structure of newly synthesized chains. *Proc. Natl. Acad. Sci. U.S.A., 59,* 598-605.

Richardson, C. C. (1969). Enzymes in DNA metabolism. *Ann. Rev. Biochem., 38,* 795-840.

Roychoudhury, R. (1973). Transcriptional role in deoxyribonucleic acid replication. Nature of primer function of newly synthesized ribonucleic acid *in vitro. J. Biol. Chem., 248,* 8465-8473.

Ryter, A., and F. Jacob (1963). Étude au microscope électronique des relations entre mésosomes et noyaux chez *Bacillus subtilis. Compt. Rend. Acad. Sci., 257,* 3060-3063.

Smith, D. W., H. W. Schaller, and F. J. Bonhoeffer (1970). DNA synthesis *in vitro. Nature (London), 226,* 711-713.

Sueoka, N., and W. G. Quinn (1968). Membrane attachment of the chromosome replication origin in *Bacillus subtilis. Cold Spring Harbor Symp. Quant. Biol., 33,* 695-705.

Sugino, A., and R. Okazaki (1973). RNA-linked DNA fragments *in vitro. Proc. Natl. Acad. Sci. U.S.A., 70,* 88-92.

Vosberg, H., and H. Hoffmann-Berling (1971). DNA synthesis in nucleotide-permeable *Escherichia coli* cells: I. Preparation and properties of ether-treated cells. *J. Mol. Biol., 58,* 739-753.

Watson, J. D., and F. H. C. Crick (1953). Molecular structure of nucleic acids. A structure for deoxyribose nucleic acid. *Nature (London), 171,* 737-738.

Wickner, R. B., B. Ginsberg, I. Berkower, and J. Hurwitz (1972). Deoxyribonucleic acid polymerase II of *Escherichia coli:* I. The purification and characterization of the enzyme. *J. Biol. Chem., 247,* 489-497.

Wickner, W., and A. Kornberg (1973). DNA polymerase III star requires ATP to start synthesis on a primed DNA. *Proc. Natl. Acad. Sci. U.S.A., 70*, 3679–3683.

——, D. Brutlag, R. Schekman, and A. Kornberg (1972). RNA synthesis initiates *in vitro* conversion of M13 DNA to its replicative form. *Proc. Natl. Acad. Sci. U.S.A., 69*, 965–969.

——, R. Schekman, K. Geider, and A. Kornberg (1973). A new form of DNA polymerase III and a copolymerase replicate a long, single-stranded primer-template. *Proc. Natl. Acad. Sci. U.S.A., 70*, 1764–1767.

Wyatt, G. R. (1952). The nucleic acids of some insect viruses. *J. Gen. Physiol., 36*, 201–205.

Zamenhof, S., G. Brawerman, and E. Chargaff (1952). On the desoxypentose nucleic acid from several microorganisms. *Biochim. Biophys. Acta, 9*, 402–405.

Reprinted from *Nature*, 171(4361), 964–967 (1953)

GENETICAL IMPLICATIONS OF THE STRUCTURE OF DEOXYRIBONUCLEIC ACID

By J. D. WATSON and F. H. C. CRICK

Medical Research Council Unit for the Study of the
Molecular Structure of Biological Systems, Cavendish
Laboratory, Cambridge

THE importance of deoxyribonucleic acid (DNA) within living cells is undisputed. It is found in all dividing cells, largely if not entirely in the nucleus, where it is an essential constituent of the chromosomes. Many lines of evidence indicate that it is the carrier of a part of (if not all) the genetic specificity of the chromosomes and thus of the gene itself.

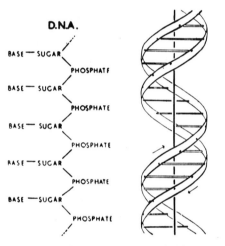

Fig. 1. Chemical formula of a single chain of deoxyribonucleic acid

Fig. 2. This figure is purely diagrammatic. The two ribbons symbolize the two phosphate-sugar chains, and the horizontal rods the pairs of bases holding the chains together. The vertical line marks the fibre axis

Until now, however, no evidence has been presented to show how it might carry out the essential operation required of a genetic material, that of exact self-duplication.

We have recently proposed a structure[1] for the salt of deoxyribonucleic acid which, if correct, immediately suggests a mechanism for its self-duplication. X-ray evidence obtained by the workers at King's College, London[2], and presented at the same time, gives qualitative support to our structure and is incompatible with all previously proposed structures[3]. Though the structure will not be completely proved until a more extensive comparison has been made with the X-ray data, we now feel sufficient confidence in its general correctness to discuss its genetical implications. In doing so we are assuming that fibres of the salt of deoxyribonucleic acid are not artefacts arising in the method of preparation,

since it has been shown by Wilkins and his co-workers that similar X-ray patterns are obtained from both the isolated fibres and certain intact biological materials such as sperm head and bacteriophage particles[3,4].

The chemical formula of deoxyribonucleic acid is now well established. The molecule is a very long chain, the backbone of which consists of a regular alternation of sugar and phosphate groups, as shown in Fig. 1. To each sugar is attached a nitrogenous base, which can be of four different types. (We have considered 5-methyl cytosine to be equivalent to cytosine, since either can fit equally well into our structure.) Two of the possible bases—adenine and guanine—are purines, and the other two—thymine and cytosine—are pyrimidines. So far as is known, the sequence of bases along the chain is irregular. The monomer unit, consisting of phosphate, sugar and base, is known as a nucleotide.

The first feature of our structure which is of biological interest is that it consists not of one chain, but of two. These two chains are both coiled around a common fibre axis, as is shown diagrammatically in Fig. 2. It has often been assumed that since there was only one chain in the chemical formula there would only be one in the structural unit. However, the density, taken with the X-ray evidence[2], suggests very strongly that there are two.

The other biologically important feature is the manner in which the two chains are held together. This is done by hydrogen bonds between the bases, as shown schematically in Fig. 3. The bases are joined together in pairs, a single base from one chain being hydrogen-bonded to a single base from the other. The important point is that only certain pairs of bases will fit into the structure. One member of a pair must be a purine and the other a pyrimidine in order to bridge between the two chains. If a pair consisted of two purines, for example, there would not be room for it.

We believe that the bases will be present almost entirely in their most probable tautomeric forms. If this is true, the conditions for forming hydrogen bonds are more restrictive, and the only pairs of bases possible are :

adenine with thymine ;
guanine with cytosine.

The way in which these are joined together is shown in Figs. 4 and 5. A given pair can be either way round. Adenine, for example, can occur on either chain ; but when it does, its partner on the other chain must always be thymine.

This pairing is strongly supported by the recent analytical results[5], which show that for all sources of deoxyribonucleic acid examined the amount of adenine is close to the amount of thymine, and the amount of guanine close to the amount of cytosine, although the cross-ratio (the ratio of adenine to guanine) can vary from one source to another. Indeed, if the sequence of bases on one chain is irregular, it is difficult to explain these analytical results except by the sort of pairing we have suggested.

The phosphate-sugar backbone of our model is completely regular, but any sequence of the pairs of bases can fit into the structure. It follows that in a long molecule many different permutations are possible, and it therefore seems likely that the precise sequence of the bases is the code which carries the genetical information. If the actual order of the

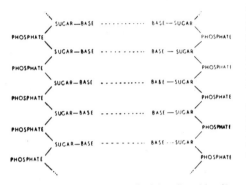

Fig. 3. Chemical formula of a pair of deoxyribonucleic acid chains. The hydrogen bonding is symbolized by dotted lines

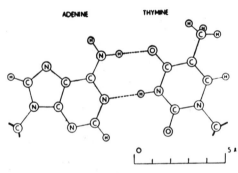

Fig. 4. Pairing of adenine and thymine. Hydrogen bonds are shown dotted. One carbon atom of each sugar is shown

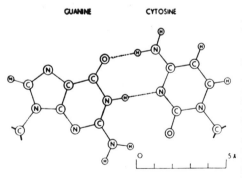

Fig. 5. Pairing of guanine and cytosine. Hydrogen bonds are shown dotted. One carbon atom of each sugar is shown

bases on one of the pair of chains were given, one could write down the exact order of the bases on the other one, because of the specific pairing. Thus one chain is, as it were, the complement of the other, and it is this feature which suggests how the deoxyribonucleic acid molecule might duplicate itself.

Previous discussions of self-duplication have usually involved the concept of a template, or mould. Either the template was supposed to copy itself directly or it was to produce a 'negative', which in its turn was to act as a template and produce the original 'positive' once again. In no case has it been explained in detail how it would do this in terms of atoms and molecules.

Now our model for deoxyribonucleic acid is, in effect, a *pair* of templates, each of which is complementary to the other. We imagine that prior to duplication the hydrogen bonds are broken, and the two chains unwind and separate. Each chain then acts as a template for the formation on to itself of a new companion chain, so that eventually we shall have *two* pairs of chains, where we only had one before. Moreover, the sequence of the pairs of bases will have been duplicated exactly.

A study of our model suggests that this duplication could be done most simply if the single chain (or the relevant portion of it) takes up the helical configuration. We imagine that at this stage in the life of the cell, free nucleotides, strictly polynucleotide precursors, are available in quantity. From time to time the base of a free nucleotide will join up by hydrogen bonds to one of the bases on the chain already formed. We now postulate that the polymerization of these monomers to form a new chain is only possible if the resulting chain can form the proposed structure. This is plausible, because steric reasons would not allow nucleotides 'crystallized' on to the first chain to approach one another in such a way that they could be joined together into a new chain, unless they were those nucleotides which were necessary to form our structure. Whether a special enzyme is required to carry out the polymerization, or whether the single helical chain already formed acts effectively as an enzyme, remains to be seen.

Since the two chains in our model are intertwined, it is essential for them to untwist if they are to separate. As they make one complete turn around each other in 34 A., there will be about 150 turns per million molecular weight, so that whatever the precise structure of the chromosome a considerable amount of uncoiling would be necessary. It is well known from microscopic observation that much coiling and uncoiling occurs during mitosis, and though this is on a much larger scale it probably reflects similar processes on a molecular level. Although it is difficult at the moment to see how these processes occur without everything getting tangled, we do not feel that this objection will be insuperable.

Our structure, as described[1], is an open one. There is room between the pair of polynucleotide chains (see Fig. 2) for a polypeptide chain to wind around the same helical axis. It may be significant that the distance between adjacent phosphorus atoms, 7·1 A., is close to the repeat of a fully extended polypeptide chain. We think it probable that in the sperm head, and in artificial nucleoproteins, the polypeptide chain occupies this position. The relative weakness of the second layer-line in the published X-ray pictures[3a,4] is crudely compatible with such an idea. The function of the protein might well be to control the coiling and uncoiling, to assist in holding a single polynucleotide chain in a helical configuration, or some other non-specific function.

Our model suggests possible explanations for a number of other phenomena. For example, spontaneous mutation may be due to a base occasionally occurring in one of its less likely tautomeric forms. Again, the pairing between homologous chromosomes at meiosis may depend on pairing between specific bases. We shall discuss these ideas in detail elsewhere.

For the moment, the general scheme we have proposed for the reproduction of deoxyribonucleic acid must be regarded as speculative. Even if it is correct, it is clear from what we have said that much remains to be discovered before the picture of genetic duplication can be described in detail. What are the polynucleotide precursors ? What makes the pair of chains unwind and separate ? What is the precise role of the protein ? Is the chromosome one long pair of deoxyribonucleic acid chains, or does it consist of patches of the acid joined together by protein ?

Despite these uncertainties we feel that our pro-posed structure for deoxyribonucleic acid may help to solve one of the fundamental biological problems—the molecular basis of the template needed for genetic replication. The hypothesis we are suggesting is that the template is the pattern of bases formed by one chain of the deoxyribonucleic acid and that the gene contains a complementary pair of such templates.

One of us (J.D.W.) has been aided by a fellowship from the National Foundation for Infantile Paralysis (U.S.A.).

[1] Watson, J. D., and Crick, F. H. C., *Nature*, **171**, 737 (1953).

[2] Wilkins, M. H. F., Stokes, A. R., and Wilson, H. R., *Nature*, **171**, 738 (1953). Franklin, R. E., and Gosling, R. G., *Nature*, **171**, 740 (1953).

[3] (a) Astbury, W. T., Symp. No. 1 Soc. Exp. Biol., 66 (1947). (b) Furberg, S., *Acta Chem. Scand.*, **6**, 634 (1952). (c) Pauling, L., and Corey, R. B., *Nature*, **171**, 346 (1953); *Proc. U.S. Nat. Acad. Sci.*, **39**, 84 (1953). (d) Fraser, R. D. B. (in preparation).

[4] Wilkins, M. H. F., and Randall, J. T., *Biochim. et Biophys. Acta*, **10**, 192 (1953).

[5] Chargaff, E., for references see Zamenhof, S., Brawerman, G., and Chargaff, E., *Biochim. et Biophys. Acta*, **9**, 402 (1952). Wyatt, G. R., *J. Gen. Physiol.*, **36**, 201 (1952).

2

Reprinted from *Proc. Natl. Acad. Sci.*, 44, 671–682 (1958)

THE REPLICATION OF DNA IN ESCHERICHIA COLI*

By Matthew Meselson and Franklin W. Stahl

GATES AND CRELLIN LABORATORIES OF CHEMISTRY,† AND NORMAN W. CHURCH LABORATORY OF CHEMICAL BIOLOGY, CALIFORNIA INSTITUTE OF TECHNOLOGY, PASADENA, CALIFORNIA

Communicated by Max Delbrück, May 14, 1958

Introduction.—Studies of bacterial transformation and bacteriophage infection[1–5] strongly indicate that deoxyribonucleic acid (DNA) can carry and transmit hereditary information and can direct its own replication. Hypotheses for the mechanism of DNA replication differ in the predictions they make concerning the distribution among progeny molecules of atoms derived from parental molecules.[6]

Radioisotopic labels have been employed in experiments bearing on the distribution of parental atoms among progeny molecules in several organisms.[6–9] We anticipated that a label which imparts to the DNA molecule an increased density might permit an analysis of this distribution by sedimentation techniques. To this end, a method was developed for the detection of small density differences among

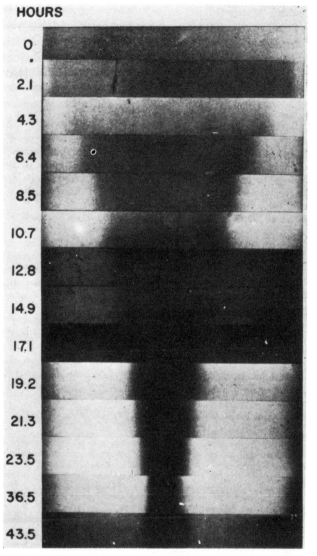

HOURS

0

2.1

4.3

6.4

8.5

10.7

12.8

14.9

17.1

19.2

21.3

23.5

36.5

43.5

Fig. 1.—Ultraviolet absorption photographs showing successive stages in the banding of DNA from *E. coli*. An aliquot of bacterial lysate containing approximately 10^8 lysed cells was centrifuged at 31,410 rpm in a CsCl solution as described in the text. Distance from the axis of rotation increases toward the right. The number beside each photograph gives the time elapsed after reaching 31,410 rpm.

macromolecules.[10]   By use of this method, we have observed the distribution of the heavy nitrogen isotope N^{15} among molecules of DNA following the transfer of a uniformly N^{15}-labeled, exponentially growing bacterial population to a growth medium containing the ordinary nitrogen isotope N^{14}.

Density-Gradient Centrifugation.—A small amount of DNA in a concentrated solution of cesium chloride is centrifuged until equilibrium is closely approached.

19

The opposing processes of sedimentation and diffusion have then produced a stable concentration gradient of the cesium chloride. The concentration and pressure gradients result in a continuous increase of density along the direction of centrifugal force. The macromolecules of DNA present in this density gradient are driven by the centrifugal field into the region where the solution density is equal to their own buoyant density.[11] This concentrating tendency is opposed by diffusion, with the result that at equilibrium a single species of DNA is distributed over a band whose width is inversely related to the molecular weight of that species (Fig. 1).

If several different density species of DNA are present, each will form a band at the position where the density of the CsCl solution is equal to the buoyant density of that species. In this way DNA labeled with heavy nitrogen (N^{15}) may be

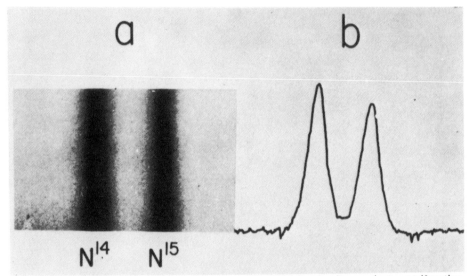

Fig. 2—*a:* The resolution of N^{14} DNA from N^{15} DNA by density-gradient centrifugation. A mixture of N^{14} and N^{15} bacterial lysates, each containing about 10^8 lysed cells, was centrifuged in CsCl solution as described in the text. The photograph was taken after 24 hours of centrifugation at 44,770 rpm. *b:* A microdensitometer tracing showing the DNA distribution in the region of the two bands of Fig. 2a. The separation between the peaks corresponds to a difference in buoyant density of 0.014 gm. cm.$^{-3}$

resolved from unlabeled DNA. Figure 2 shows the two bands formed as a result of centrifuging a mixture of approximately equal amounts of N^{14} and N^{15} *Escherichia coli* DNA.

In this paper reference will be made to the apparent molecular weight of DNA samples determined by means of density-gradient centrifugation. A discussion has been given[10] of the considerations upon which such determinations are based, as well as of several possible sources of error.[12]

Experimental.—*Escherichia coli* B was grown at 36° C. with aeration in a glucose salts medium containing ammonium chloride as the sole nitrogen source.[13] The growth of the bacterial population was followed by microscopic cell counts and by colony assays (Fig. 3).

Bacteria uniformly labeled with N^{15} were prepared by growing washed cells for

14 generations (to a titer of 2×10^8/ml) in medium containing 100 μg/ml of $N^{15}H_4Cl$ of 96.5 per cent isotopic purity. An abrupt change to N^{14} medium was then accomplished by adding to the growing culture a tenfold excess of $N^{14}H_4Cl$, along with ribosides of adenine and uracil in experiment 1 and ribosides of adenine, guanine, uracil, and cytosine in experiment 2, to give a concentration of 10 μg/ml of each riboside. During subsequent growth the bacterial titer was kept between

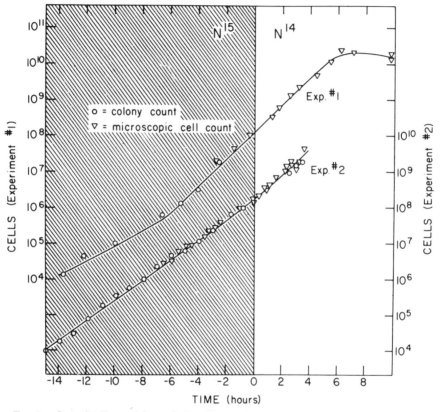

Fig. 3.—Growth of bacterial populations first in N^{15} and then in N^{14} medium. The values on the ordinates give the actual titers of the cultures up to the time of addition of N^{14}. Thereafter, during the period when samples were being withdrawn for density-gradient centrifugation, the actual titer was kept between 1 and 2×10^8 by additions of fresh medium. The values on the ordinates during this later period have been corrected for the withdrawals and additions. During the period of sampling for density-gradient centrifugation, the generation time was 0.81 hours in Experiment 1 and 0.85 hours in Experiment 2.

1 and 2×10^8/ml by appropriate additions of fresh N^{14} medium containing ribosides.

Samples containing about 4×10^9 bacteria were withdrawn from the culture just before the addition of N^{14} and afterward at intervals for several generations. Each sample was immediately chilled and centrifuged in the cold for 5 minutes at $1,800 \times g$. After resuspension in 0.40 ml. of a cold solution 0.01 M in NaCl and 0.01 M in ethylenediaminetetra-acetate (EDTA) at pH 6, the cells were lysed by the addition of 0.10 ml. of 15 per cent sodium dodecyl sulfate and stored in the cold.

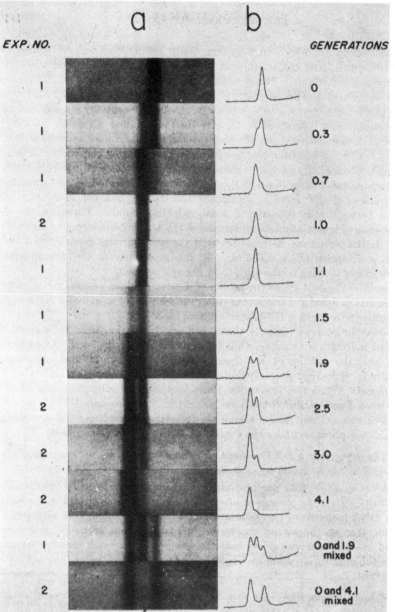

Fig. 4—*a:* Ultraviolet absorption photographs showing DNA bands resulting from density-gradient centrifugation of lysates of bacteria sampled at various times after the addition of an excess of N^{14} substrates to a growing N^{15}-labeled culture. Each photograph was taken after 20 hours of centrifugation at 44,770 rpm under the conditions described in the text. The density of the CsCl solution increases to the right. Regions of equal density occupy the same horizontal position on each photograph. The time of sampling is measured from the time of the addition of N^{14} in units of the generation time. The generation times for Experiments 1 and 2 were estimated from the measurements of bacterial growth presented in Fig. 3. *b:* Microdensitometer tracings of the DNA bands shown in the adjacent photographs. The microdensitometer pen displacement above the base line is directly proportional to the concentration of DNA. The degree of labeling of a species of DNA corresponds to the relative position of its band between the bands of fully labeled and unlabeled DNA shown in the lowermost frame, which serves as a density reference. A test of the conclusion that the DNA in the band of intermediate density is just half-labeled is provided by the frame showing the mixture of generations 0 and 1.9. When allowance is made for the relative amounts of DNA in the three peaks, the peak of intermediate density is found to be centered at 50 ± 2 per cent of the distance between the N^{14} and N^{15} peaks.

For density-gradient centrifugation, 0.010 ml. of the dodecyl sulfate lysate was added to 0.70 ml. of CsCl solution buffered at pH 8.5 with 0.01 M tris(hydroxymethyl)aminomethane. The density of the resulting solution was 1.71 gm. cm.$^{-3}$ This was centrifuged at 140,000$\times$ g. (44,770 rpm) in a Spinco model E ultracentrifuge at 25° for 20 hours, at which time the DNA had essentially attained sedimentation equilibrium. Bands of DNA were then found in the region of density 1.71 gm. cm.$^{-3}$, well isolated from all other macromolecular components of the bacterial lysate. Ultraviolet absorption photographs taken during the course of each centrifugation were scanned with a recording microdensitometer (Fig. 4).

The buoyant density of a DNA molecule may be expected to vary directly with the fraction of N^{15} label it contains. The density gradient is constant in the region between fully labeled and unlabeled DNA bands. Therefore, the degree of labeling of a partially labeled species of DNA may be determined directly from the relative position of its band between the band of fully labeled DNA and the band of unlabeled DNA. The error in this procedure for the determination of the degree of labeling is estimated to be about 2 per cent.

Results.—Figure 4 shows the results of density-gradient centrifugation of lysates of bacteria sampled at various times after the addition of an excess of N^{14}-containing substrates to a growing N^{15}-labeled culture.

It may be seen in Figure 4 that, until one generation time has elapsed, half-labeled molecules accumulate, while fully labeled DNA is depleted. One generation time after the addition of N^{14}, these half-labeled or "hybrid" molecules alone are observed. Subsequently, only half-labeled DNA and completely unlabeled DNA are found. When two generation times have elapsed after the addition of N^{14}, half-labeled and unlabeled DNA are present in equal amounts.

Discussion.—These results permit the following conclusions to be drawn regarding DNA replication under the conditions of the present experiment.

1. *The nitrogen of a DNA molecule is divided equally between two subunits which remain intact through many generations.*

The observation that parental nitrogen is found only in half-labeled molecules at all times after the passage of one generation time demonstrates the existence in each DNA molecule of two subunits containing equal amounts of nitrogen. The finding that at the second generation half-labeled and unlabeled molecules are found in equal amounts shows that the number of surviving parental subunits is twice the number of parent molecules initially present. That is, the subunits are conserved.

2. *Following replication, each daughter molecule has received one parental subunit.*

The finding that all DNA molecules are half-labeled one generation time after the addition of N^{14} shows that each daughter molecule receives one parental subunit.[14] If the parental subunits had segregated in any other way among the daughter molecules, there would have been found at the first generation some fully labeled and some unlabeled DNA molecules, representing those daughters which received two or no parental subunits, respectively.

3. *The replicative act results in a molecular doubling.*

This statement is a corollary of conclusions 1 and 2 above, according to which each parent molecule passes on two subunits to progeny molecules and each progeny molecule receives just one parental subunit.   It follows that each single molecular reproductive act results in a doubling of the number of molecules entering into that act.

The above conclusions are represented schematically in Figure 5.

The Watson-Crick Model.—A molecular structure for DNA has been proposed by Watson and Crick.[15]   It has undergone preliminary refinement[16] without alteration of its main features and is supported by physical and chemical studies.[17] The structure consists of two polynucleotide chains wound helically about a common axis.   The nitrogen base (adenine, guanine, thymine, or cytosine) at each level

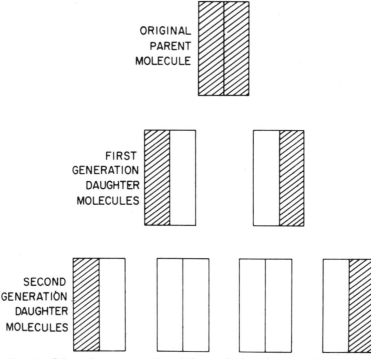

Fig. 5.—Schematic representation of the conclusions drawn in the text from the data presented in Fig. 4.   The nitrogen of each DNA molecule is divided equally between two subunits.   Following duplication, each daughter molecule receives one of these.   The subunits are conserved through successive duplications.

on one chain is hydrogen-bonded to the base at the same level on the other chain. Structural requirements allow the occurrence of only the hydrogen-bonded base pairs adenine-thymine and guanine-cytosine, resulting in a detailed complementariness between the two chains.   This suggested to Watson and Crick[18] a definite and structurally plausible hypothesis for the duplication of the DNA molecule. According to this idea, the two chains separate, exposing the hydrogen-bonding sites of the bases.   Then, in accord with the base-pairing restrictions, each chain serves as a template for the synthesis of its complement.   Accordingly, each daughter molecule contains one of the parental chains paired with a newly synthesized chain (Fig. 6).

The results of the present experiment are in exact accord with the expectations of the Watson-Crick model for DNA duplication.   However, it must be emphasized that it has not been shown that the molecular subunits found in the present experiment are single polynucleotide chains or even that the DNA molecules studied here correspond to single DNA molecules possessing the structure proposed by Watson and Crick.   However, some information has been obtained about the molecules and their subunits; it is summarized below.

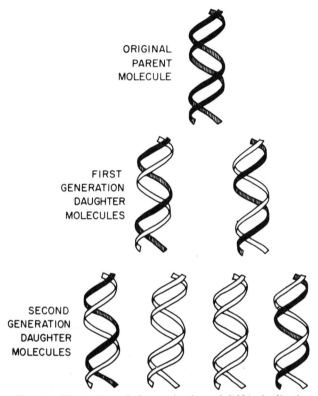

ORIGINAL
PARENT
MOLECULE

FIRST
GENERATION
DAUGHTER
MOLECULES

SECOND
GENERATION
DAUGHTER
MOLECULES

Fig. 6.—Illustration of the mechanism of DNA duplication proposed by Watson and Crick. Each daughter molecule contains one of the parental chains (*black*) paired with one new chain (*white*). Upon continued duplication, the two original parent chains remain intact, so that there will always be found two molecules each with one parental chain.

The DNA molecules derived from *E. coli* by detergent-induced lysis have a buoyant density in CsCl of 1.71 gm. cm.$^{-3}$, in the region of densities found for T2 and T4 bacteriophage DNA, and for purified calf-thymus and salmon-sperm DNA.   A highly viscous and elastic solution of N^{14} DNA was prepared from a dodecyl sulfate lysate of *E. coli* by the method of Simmons[19] followed by deproteinization with chloroform.   Further purification was accomplished by two cycles of preparative density-gradient centrifugation in CsCl solution.   This purified bacterial DNA was found to have the same buoyant density and apparent molecular weight, 7×10^6, as the DNA of the whole bacterial lysates (Figs. 7, 8).

Heat Denaturation.—It has been found that DNA from *E. coli* differs importantly from purified salmon-sperm DNA in its behavior upon heat denaturation.

Exposure to elevated temperatures is known to bring about an abrupt collapse of the relatively rigid and extended native DNA molecule and to make available for acid-base titration a large fraction of the functional groups presumed to be blocked by hydrogen-bond formation in the native structure.[19, 20, 21, 22] Rice and Doty[22] have reported that this collapse is not accompanied by a reduction in molecular weight as determined from light-scattering. These findings are corroborated by density-gradient centrifugation of salmon-sperm DNA.[23] When this material is

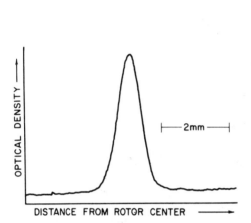

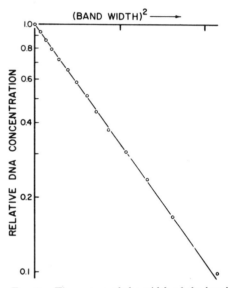

Fig. 7.—Microdensitometer tracing of an ultraviolet absorption photograph showing the optical density in the region of a band of N^{14} *E. coli* DNA at equilibrium. About 2 μg. of DNA purified as described in the text was centrifuged at 31,410 rpm at 25° in 7.75 molal CsCl at pH 8.4. The density gradient is essentially constant over the region of the band and is 0.057 gm./cm.[4] The position of the maximum indicates a buoyant density of 1.71 gm. cm.$^{-3}$ In this tracing the optical density above the base line is directly proportional to the concentration of DNA in the rotating centrifuge cell. The concentration of DNA at the maximum is about 50 μg./ml.

Fig. 8.—The square of the width of the band of Fig. 7 plotted against the logarithm of the relative concentration of DNA. The divisions along the abscissa set off intervals of 1 mm.² In the absence of density heterogeneity, the slope at any point of such a plot is directly proportional to the weight average molecular weight of the DNA located at the corresponding position in the band. Linearity of this plot indicates monodispersity of the banded DNA. The value of the the slope corresponds to an apparent molecular weight for the Cs·DNA salt of 9.4 × 10⁶, corresponding to a molecular weight of 7.1 × 10⁶ for the sodium salt.

kept at 100° for 30 minutes either under the conditions employed by Rice and Doty or in the CsCl centrifuging medium, there results a density increase of 0.014 gm. cm.$^{-3}$ with no change in apparent molecular weight. The same results are obtained if the salmon-sperm DNA is pre-treated at pH 6 with EDTA and sodium dodecyl sulfate. Along with the density increase, heating brings about a sharp reduction in the time required for band formation in the CsCl gradient. In the absence of an increase in molecular weight, the decrease in banding time must be ascribed[10] to an increase in the diffusion coefficient, indicating an extensive collapse of the native structure.

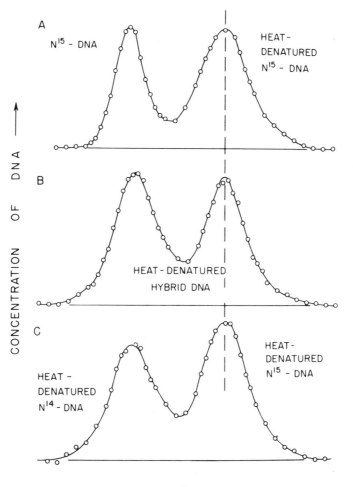

DENSITY ⟶

Fig. 9.—The dissociation of the subunits of *E. coli* DNA upon heat denaturation. Each smooth curve connects points obtained by micro-densitometry of an ultraviolet absorption photograph taken after 20 hours of centrifugation in CsCl solution at 44,770 rpm. The baseline density has been removed by subtraction. *A:* A mixture of heated and unheated N^{15} bacterial lysates. Heated lysate alone gives one band in the position indicated. Unheated lysate was added to this experiment for comparison. Heating has brought about a density increase of 0.016 gm. cm.$^{-3}$ and a reduction of about half in the apparent molecular weight of the DNA. *B:* Heated lysate of N^{15} bacteria grown for one generation in N^{14} growth medium. Before heat denaturation, the hybrid DNA contained in this lysate forms only one band, as may be seen in Fig. 4. *C:* A mixture of heated N^{14} and heated N^{15} bacterial lysates. The density difference is 0.015 gm. cm.$^{-3}$

The decrease in banding time and a density increase close to that found upon heating salmon-sperm DNA are observed (Fig. 9, *A*) when a bacterial lysate containing uniformly labeled N^{15} or N^{14} *E. coli* DNA is kept at 100° C. for 30 minutes in the CsCl centrifuging medium; but the apparent molecular weight of the heated bacterial DNA is reduced to approximately half that of the unheated material.

Half-labeled DNA contained in a detergent lysate of N^{15} *E. coli* cells grown for one generation in N^{14} medium was heated at 100° C. for 30 minutes in the CsCl centrifuging medium. This treatment results in the loss of the original half-labeled material and in the appearance in equal amounts of two new density species, each with approximately half the initial apparent molecular weight (Fig. 9, *B*). The density difference between the two species is 0.015 gm. cm. $^{-3}$, close to the increment produced by the N^{15} labeling of the unheated DNA.

This behavior suggests that heating the hybrid molecule brings about the dissociation of the N^{15}-containing subunit from the N^{14} subunit. This possibility was tested by a density-gradient examination of a mixture of heated N^{15} DNA and heated N^{14} DNA (Fig. 9, *C*). The close resemblance between the products of heating hybrid DNA (Fig. 9 *B*) and the mixture of products obtained from heating N^{14} and N^{15} DNA separately (Fig. 9, *C*) leads to the conclusion that the two molecular subunits have indeed dissociated upon heating. Since the apparent molecular weight of the subunits so obtained is found to be close to half that of the intact molecule, it may be further concluded that the subunits of the DNA molecule which are conserved at duplication are single, continuous structures. The scheme for DNA duplication proposed by Delbrück[24] is thereby ruled out.

To recapitulate, both salmon-sperm and *E. coli* DNA heated under similar conditions collapse and undergo a similar density increase, but the salmon DNA retains its initial molecular weight, while the bacterial DNA dissociates into the two subunits which are conserved during duplication. These findings allow two different interpretations. On the one hand, if we assume that salmon DNA contains subunits analogous to those found in *E. coli* DNA, then we must suppose that the subunits of salmon DNA are bound together more tightly than those of the bacterial DNA. On the other hand, if we assume that the molecules of salmon DNA do not contain these subunits, then we must concede that the bacterial DNA molecule is a more complex structure than is the molecule of salmon DNA. The latter interpretation challenges the sufficiency of the Watson-Crick DNA model to explain the observed distribution of parental nitrogen atoms among progeny molecules.

Conclusion.—The structure for DNA proposed by Watson and Crick brought forth a number of proposals as to how such a molecule might replicate. These proposals[6] make specific predictions concerning the distribution of parental atoms among progeny molecules. The results presented here give a detailed answer to the question of this distribution and simultaneously direct our attention to other problems whose solution must be the next step in progress toward a complete understanding of the molecular basis of DNA duplication. What are the molecular structures of the subunits of *E. coli* DNA which are passed on intact to each daughter molecule? What is the relationship of these subunits to each other in a DNA molecule? What is the mechanism of the synthesis and dissociation of the subunits in vivo?

Summary.—By means of density-gradient centrifugation, we have observed the distribution of N^{15} among molecules of bacterial DNA following the transfer of a uniformly N^{15}-substituted exponentially growing *E. coli* population to N^{14} medium.

We find that the nitrogen of a DNA molecule is divided equally between two physically continuous subunits; that, following duplication, each daughter molecule receives one of these; and that the subunits are conserved through many duplications.

* Aided by grants from the National Foundation for Infantile Paralysis and the National Institutes of Health.

† Contribution No. 2344.

[1] R. D. Hotchkiss, in *The Nucleic Acids*, ed. E. Chargaff and J. N. Davidson (New York: Academic Press, 1955), p. 435; and in *Enzymes: Units of Biological Structure and Function*, ed. O. H. Gaebler (New York: Academic Press, 1956), p. 119.

[2] S. H. Goodgal and R. M. Herriott, in *The Chemical Basis of Heredity*, ed. W. D. McElroy and B. Glass (Baltimore: Johns Hopkins Press, 1957), p. 336.

[3] S. Zamenhof, in *The Chemical Basis of Heredity*, ed. W. D. McElroy and B. Glass (Baltimore: Johns Hopkins Press, 1957), p. 351.

[4] A. D. Hershey and M. Chase, *J. Gen. Physiol.*, **36**, 39, 1952.

[5] A. D. Hershey, *Virology*, **1**, 108, 1955; **4**, 237, 1957.

[6] M. Delbrück and G. S. Stent, in *The Chemical Basis of Heredity*, ed. W. D. McElroy and B. Glass (Baltimore: Johns Hopkins Press, 1957), p. 699.

[7] C. Levinthal, these Proceedings, **42**, 394, 1956.

[8] J. H. Taylor, P. S. Woods, and W. L. Huges, these Proceedings, **43**, 122, 1957.

[9] R. B. Painter, F. Forro, Jr., and W. L. Hughes, *Nature*, **181**, 328, 1958.

[10] M. S. Meselson, F. W. Stahl, and J. Vinograd, these Proceedings, **43**, 581, 1957.

[11] The buoyant density of a molecule is the density of the solution at the position in the centrifuge cell where the sum of the forces acting on the molecule is zero.

[12] Our attention has been called by Professor H. K. Schachman to a source of error in apparent molecular weights determined by density-gradient centrifugation which was not discussed by Meselson, Stahl, and Vinograd. In evaluating the dependence of the free energy of the DNA component upon the concentration of CsCl, the effect of solvation was neglected. It can be shown that solvation may introduce an error into the apparent molecular weight if either CsCl or water is bound preferentially. A method for estimating the error due to such selective solvation will be presented elsewhere.

[13] In addition to NH_4Cl, this medium consists of 0.049 M Na_2HPO_4, 0.022 M KH_2PO_4, 0.05 M NaCl, 0.01 M glucose, 10^{-3} M $MgSO_4$, and 3×10^{-6} M $FeCl_3$.

[14] This result also shows that the generation time is very nearly the same for all DNA molecules in the population. This raises the questions of whether in any one nucleus all DNA molecules are controlled by the same clock and, if so, whether this clock regulates nuclear and cellular division as well.

[15] F. H. C. Crick and J. D. Watson, *Proc. Roy. Soc. London*, A, **223**, 80, 1954.

[16] R. Langridge, W. E. Seeds, H. R. Wilson, C. W. Hooper, M. H. F. Wilkins, and L. D. Hamilton, *J. Biophys. and Biochem. Cytol.*, **3**, 767, 1957.

[17] For reviews see D. O. Jordan, in *The Nucleic Acids*, ed. E. Chargaff and J. D. Davidson (New York: Academic Press, 1955), **1**, 447; and F. H. C. Crick, in *The Chemical Basis of Heredity*, ed. W. D. McElroy and B. Glass (Baltimore: Johns Hopkins Press, 1957), p. 532.

[18] J. D. Watson and F. H. C. Crick, *Nature*, **171**, 964, 1953.

[19] C. E. Hall and M. Litt, *J. Biophys. and Biochem. Cytol.*, **4**, 1, 1958.

[20] R. Thomas, *Biochim. et Biophys. Acta*, **14**, 231, 1954.

[21] P. D. Lawley, *Biochim. et Biophys. Acta*, **21**, 481, 1956.

[22] S. A. Rice and P. Doty, *J. Am. Chem. Soc.*, **79**, 3937, 1957.

[23] Kindly supplied by Dr. Michael Litt. The preparation of this DNA is described by Hall and Litt (*J. Biophys. and Biochem. Cytol.*, **4**, 1, 1958).

[24] M. Delbrück, these Proceedings, **40**, 783, 1954.

Active Center of DNA Polymerase

The operations are localized and arranged in
multiple sites within a single area of the molecule.

Arthur Kornberg

DNA polymerases have now been isolated from a variety of bacterial and animal cells. These enzymes, including those produced specifically in response to virus infection, catalyze the addition of mononucleotide units to the 3'-hydroxyl terminus of a primer DNA chain. Synthesis therefore proceeds in the direction of 5' to 3' (Fig. 1) (*1*). There is an absolute requirement for a DNA template, and errors in copying the template are very infrequent. The synthesis of DNA proceeds rapidly, at rates near 1000 nucleotides per minute per molecule of enzyme, with the production of chains several million in molecular weight.

The polymerases are remarkable enzymes. A polymerase takes instructions as it goes along to build a chain according to specifications by a template. Bacterial DNA polymerase will make animal DNA and animal polymerase will make bacterial DNA. The DNA polymerase from *Escherichia coli* has additional catalytic properties. It may degrade DNA progressively from either end (5' or 3') of the chain by hydrolysis to produce deoxyribonucleoside monophosphates. Or it may

degrade a chain by pyrophosphorolysis with inorganic pyrophosphate to produce deoxyribonucleoside triphosphates.

Until recently we understood little about how this enzyme works because we did not know enough about its physicochemical properties. We knew very little because our supplies of homogeneous enzyme were meager. Now with a simpler method for purification (*2*) and the invaluable use of the large-scale facilities of the New England Enzyme Center, we have had available 600 milligrams of homogeneous DNA polymerase obtained from 200 pounds (90 kilograms) of *E. coli.* The purpose of this article is twofold: (i) to assemble in a brief form the new physicochemical and functional observations concerning the pure enzyme; and (ii) to attempt an interpretation of these data in a model for the active center of the enzyme. This model is of course speculative, but it has helped us reconcile many hitherto confusing details and continues to suggest useful experiments.

Physicochemical Properties

The molecular weight of the homogeneous polymerase, determined by sed-

imentation equilibrium, is 109,000 (*2*). This large molecular weight and the presence of both polymerase and multiple nuclease activities suggest a subunit structure. However, the molecular weight measured by sedimentation equilibrium under denaturing and reducing conditions (6*M* guanidine hydrochloride and 0.3*M* mercaptoethanol) was found to be the same as that of the native protein. Optical rotatory dispersion and velocity sedimentation studies showed that polymerase loses ordered structure in 6*M* guanidine hydrochloride, and would therefore be expected to be fully dissociated in this solvent.

More than 95 percent of the protein migrated as a single band on polyacrylamide-gel electrophoresis at *p*H 3.5, *p*H 8, and *p*H 11 (with or without 7*M* urea at *p*H 3.5 and *p*H 8). This result is most consistent with a structure composed of either a single polypeptide chain or of two or more identical subunits. However, the possibility of multiple, identical subunits is ruled out by the fact that polymerase contains, per 109,000 molecular weight, a single sulfhydryl group and a single disulfide group. The sulfhydryl group is probably not part of the active site

The author is professor and chairman of the department of biochemistry, Stanford University School of Medicine, Stanford, California 94305.

because it can be modified either with iodoacetate or mercuric ion to give derivatives with full polymerase and exonuclease activity (3). The reaction with mercuric ion will give either a polymerase monomer, with one mercury atom per protein molecule, or, in the presence of a molar excess of enzyme, a dimer, with two protein molecules linked through a mercury atom. The dimer also has full activity. (The reaction of polymerase with a single atom of ^{203}Hg provides a convenient way of incorporating a radioactive label of about 10,000 counts per minute per microgram of enzyme, without affecting enzymatic activity. This label has served as a marker in DNA binding studies which will be described below.)

Amino acids (approximately 1000 per enzyme molecule) account for the dry weight. There is no evidence for any prosthetic group. The amino terminal residue, as determined by both the cyanate and the fluorodinitrobenzene procedures, is methionine.

Although these experiments indicate that DNA polymerase is a single polypeptide chain, the possibility still exists that there are subunits joined by nonpeptide linkages that resist disruption by guanidine hydrochloride–mercaptoethanol, or by urea, or by extremes of

Table 1. Influence of DNA structure on binding of DNA to DNA polymerase.

Conformation	Per DNA molecule	
	Nicks or ends	Polymerase molecules bound
d(AT)$_n$ oligomer		
Hairpin	1	1
ØX174 DNA		
Circular, single strand	0	20
Closed circular, duplex	0	< 0.1
Plasmid DNA		
Irreversibly denatured	0	21
3′-Hydroxyl nick	1	1
3′-Phosphate nick	5	6
T7 DNA		
Linear duplex	2.5	2.6
Single strand	2	240

pH. The presence of blocked amino termini has also not been ruled out. There have been two recent reports of *E. coli* polymerase preparations with molecular weights in the range of 30,000 and 50,000, and these have been designated as possible subunits by Cavalieri (4) and by Lezius (5). However, their preparations are of relatively low specific activity and have not yet been fully characterized. If DNA polymerase is assumed to be roughly spherical, its diameter is calculated to be near 65 angstroms. The diameter of a DNA helix is about 20 angstroms.

DNA Binding Site

DNA binding to DNA polymerase (6) was studied with a variety of DNA structures (Fig. 2). The alternating copolymer of deoxyadenylate and deoxythymidylate (dAT) was partially digested by deoxyribonuclease. Oligomers were isolated from these digests either by gel filtration or by polyacrylamide-gel electrophoresis. Preparations were obtained with chain lengths of approximately 24 and 40 nucleotide residues [d(AT)$_{12}$ and d(AT)$_{20}$]; they were induced to assume "hairpin" conformations by melting and quick cooling at low ionic strength.

Binding was measured by sucrose density-gradient centrifugation of mixtures containing DNA labeled with ^{3}H or ^{32}P and polymerase labeled with ^{203}Hg. The mixtures were layered on top of the gradients, and the enzyme-DNA complexes were identified after sedimentation.

With excess enzyme, d(AT)$_{12}$ sedimented almost quantitatively with the enzyme, an indication of a very high binding affinity. The polymerase-dAT complex sedimented at 7.7S, compared to 6.1S for the free enzyme. With the dAT oligomer present in excess, all the enzyme sedimented as a complex

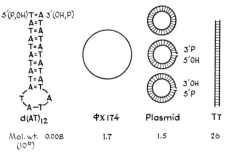

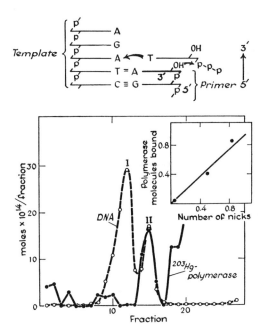

Fig. 1 (top left). Direction of chain growth catalyzed by DNA polymerase. Abbreviations are: A, adenine; C, cytosine; G, guanine; and T, thymine.

Fig. 2 (above). Various DNA structures used in studies of enzyme binding.

Fig. 3 (left). Binding of nicked plasmid DNA to the enzyme. Peak I contains intact, duplex, circular DNA in the supercoiled form; peak II contains the nicked form which is bound by polymerase. The inserted graph gives the number of polymerase molecules bound as a function of the number of nicks introduced by pancreatic deoxyribonuclease.

at 7.7S with an equimolar amount of dAT, an indication that DNA polymerase contains a single binding site for the dAT oligomer. Results identical to these were obtained with $d(AT)_{20}$. No difference was detected in the binding of oligomers terminated with a 3'-hydroxyl as compared with a 3'-phosphate group.

Binding of single-stranded circular ϕX174 DNA by polymerase resulted in about 20 enzyme molecules per molecule of DNA in the complex. A double-stranded, closed, circular plasmid DNA (7) was not bound at all by polymerase. However, when these duplex forms were denatured to make them single-stranded, they were bound by polymerase in proportion to their length and to the same extent as the single-stranded viral ϕX174 DNA.

We have introduced nicks into the plasmid DNA or the ϕX174 replicative form with pancreatic deoxyribonuclease. Such nicks have 3'-OH and 5'-P termini and are active points for replication (8). Nicks introduced with micrococcal nuclease produce 3'-P and 5'-OH termini and are not active points for replication; they inhibit replication (9). Regardless of the kind of nick, polymerase molecules formed complexes in numbers exactly equivalent to the number of nicks (Fig. 3, insert). The sedimentation pattern (Fig. 3) indicates that the polymerase molecules bound the nicked forms (II) and not the intact double circular forms of the plasmid (I).

The binding of DNA structures to polymerase is summarized in Table 1. There is no binding at all at helical regions. There is binding along single-stranded chains and to nicks and ends. In the case of the linear duplex DNA of bacteriophage T7, our preparation contained, on the average, one nick per two molecules; this explains why 2.6 molecules of the enzyme are bound per T7 DNA molecule.

Is binding of DNA at ends or nicks simply a consequence of fraying and single-strandedness in these regions, or is there more specific binding directed to the nucleotide termini at these points? We will assess this question in the formulation of a model for the active center of the enzyme.

Deoxyribonucleoside
Triphosphate Binding Site

Is one, or more than one, molecule of deoxyribonucleoside triphosphate bound to a polymerase molecule? This and related questions are crucial to understanding the nature of the active center and the mechanism of polymerase action.

Triphosphate binding was studied by equilibrium dialysis (10). Scatchard plots for triphosphate binding showed that there is one binding site for each triphosphate and that the dissociation constants for the enzyme-triphosphate complexes were 12, 33, 81, and 147 micromoles per liter for deoxyguano-

sine, deoxyadenosine, deoxythymidine, and deoxycytidine triphosphates, respectively. Although the four triphosphates differ in the affinity of their binding, the interpretation of these particular values is necessarily limited. DNA template and primer were not present, and their influence on binding, which is likely to be profound, has yet to be assessed.

Is there a separate site for each of the four triphosphates, or does the enzyme have a single site for which all four triphosphates compete? Equilibrium dialyses were run with each of the six possible combinations of two triphosphates, and each of the pair was labeled distinctively. These competition experiments established that there is a single binding site on the enzyme for which all four triphosphates compete. Further explorations (10) of the specificity of binding in this site emphasize the primary importance of the triphosphate moiety and the only secondary importance of the sugar and base components.

The effects of DNA template and primer on the binding of triphosphates are difficult to study experimentally. An active template and primer invariably promote polymerization or are degraded, and binding measurements are thus complicated. Among the questions to be answered are whether a template confers specificity on triphosphate binding and what influence the primer terminus exerts on the entry and orientation of the triphosphate in the site.

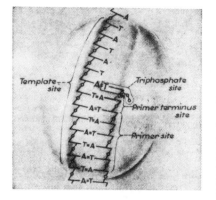

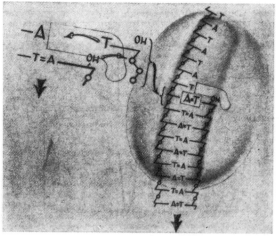

Fig. 4 (above). Sites in the active center of DNA polymerase. Fig. 5 (right). Mechanism of the polymerization step.

Multiple Functions of the Enzyme

At this point I would like to relate the results of the binding experiments to what we have learned about the catalytic properties of the enzyme in order to begin constructing a model of the many operations that may take place in the active center of the enzyme. These operations include: (i) the 5′ → 3′ growth of a DNA chain by the polymerization of nucleotides; (ii) hydrolysis of a DNA chain from the 3′-OH end (3′ → 5′ direction); (iii) hydrolysis of a DNA chain from the 5′ end (5′ → 3′ direction); (iv) pyrophosphorolysis of a DNA chain from the 3′ end; and (v) exchange of inorganic pyrophosphate (PP$_i$) with the terminal pyrophosphate group of a deoxyribonucleoside triphosphate.

We picture the active center of the enzyme as some specially adapted polypeptide surface that recognizes and accommodates several nucleotide structures (Fig. 4). We will present evidence that, within the active center, there are at least five major sites. (i) There is a site for a portion of the template chain. This area binds the chain where a base pair is formed and for a distance of several nucleotides on either side of it. The chain is oriented with a particular polarity. It seems likely that this is the site where circular, single-stranded DNA is bound, but we are uncertain whether this site recognizes an extended or a tightly stacked, helical conformation. (ii) There is a site for the growing chain, the primer. The primer is oriented with a polarity opposite that of the template. (iii) There is a site with special recognition for the 3′-OH group of the terminal nucleotide of the primer, the primer terminus. We shall discuss this region later as a site for the hydrolytic and pyrophosphorolytic cleavage of the 3′-OH–terminated primer chain (3′ → 5′ direction). (iv) There is a site for a triphosphate, and (v) there is an additional site, to be considered later, which provides for hydrolytic cleavage of the 5′-P–terminated chain (5′ → 3′ direction).

The Polymerization Step

When a linear duplex is partially degraded from each 3′-OH end, as for example by the action of certain exonucleases, these denuded portions are repaired with great facility by all DNA polymerases (11). How is this accomplished?

The triphosphate is bound adjacent to the 3′-OH group on the terminal nucleotide of the primer and oriented so that it can be brought into direct contact and form a base pair with the template (Fig. 5). When the correct base pair is formed, a nucleophilic attack by the 3′-OH of the primer terminus on the innermost phosphate of the triphosphate takes place. A plausible model, for reasons to be mentioned below, assumes that movement or translation of the chain relative to the enzyme is concurrent with diester bond formation. As the primer terminus loses its 3′-OH group during transformation into a diester bond, it is no longer held in the primer terminus site. Through movement of the entire chain, the old primer terminus is replaced by the newly added nucleotide, which has a terminal 3′-OH group and is therefore held in the primer terminus site. (The new primer terminus is now ready to attack another triphosphate and add the next nucleotide.) Inorganic pyrophosphate is displaced only as formation of the diester bond is being completed, and the chain movement is translating the newly added nucleotide into the primer terminus site.

The possibility has been raised that the interaction between template and triphosphate is not direct, but rather allosteric in nature. However, since there is only one triphosphate binding site, it is difficult to imagine this site assuming four conformations, each absolutely specific for one of the triphosphates.

The basis for specificity of DNA polymerase very likely is not in the recognition by the enzyme of an incoming triphosphate, but rather in its demand for one of the four base pairs. All of the Watson-Crick base pairs contain regions of identical dimensions and geometry and are symmetrical. When the correct base pair is within the active site, the enzyme may respond, possibly by a change in conformation, so that the subsequent catalytic steps can then proceed. If an

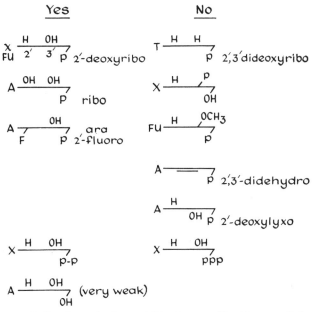

Fig. 6. Binding of monophosphates to DNA polymerase. Abbreviations are: X, the bases A, T, G, and C, as in the case of the 2′-deoxyribonucleoside 5′-monophosphate (X $\overset{H \quad OH}{\underset{P}{———}}$); FU, fluorouracil; F, fluorine; ara, arabinosyl; and lyxo, lyxosyl.

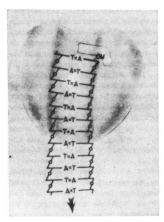

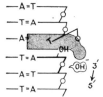

Fig. 7 (left). Binding of the end of a linear helix to the enzyme.

Fig. 8 (above). Binding of a nicked region of DNA to the enzyme.

Fig. 9 (right). Scheme for $5' \rightarrow 3'$ exonucleolytic degradation by the enzyme.

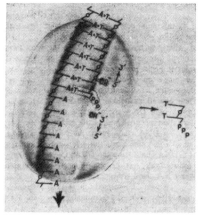

incorrect triphosphate were to bind to the enzyme, the correct base pair could not be formed, there would be no conformational change, and the triphosphate would be rejected.

Primer Terminus Site

Evidence for the specificity of the primer terminus site comes from studies of the binding and functional behavior of several nucleotide analogs. Among the analogs studied are those that lack a 3'-OH group. Such an analog, if added to a chain, would prevent further chain growth (12, 13). Dideoxythymidine triphosphate is one example. It binds to the triphosphate site. One nucleotide is added per chain. The chain terminated with such a dideoxynucleotide is inert to further elongation; it is relatively inert to exonuclease action at the primer terminus end of the chain. In such a chain, as will be mentioned presently, attack by PP_i is inhibited. However, the chain can be degraded from the 5' end ($5' \rightarrow 3'$ direction).

We interpret these results to mean that without a 3'-OH group, the chain cannot bind properly in the primer site and is therefore not an effective substrate for exonuclease action. In keeping with this interpretation are the results of the following studies of the binding of monophosphate in what may prove to be the primer terminus site (14).

Each of the four common deoxyribonucleoside monophosphates binds to and competes for a single site on the enzyme. This site is, however, entirely distinct from the triphosphate site. Replacement of the 3'-OH group—by hydrogen as in dideoxythymidine monophosphate, by esterification with phosphate, or by O-methylation—prevents binding to this site. However, many other alterations in the nucleotide can be tolerated, provided that the 3'-OH group is in the "ribo" configuration (Fig. 6). For example, a certain arabinosyl nucleotide binds to the monophosphate site, whereas lyxosyl nucleotide does not.

In support of the interpretation that the 3'-OH monophosphates bind at the primer terminus site is the finding that these monophosphates inhibit hydrolysis of polynucleotides from the 3' end of the chain. Other lines of evidence (15) have made it clear that there is only one primer terminus site serving for either polymerization or $3' \rightarrow 5'$ hydrolysis.

Completion of a Linear or Circular Duplex

Picture a duplex in which template copying has been completed. What potential does such an intact, linear, double-stranded helix have for further replication? Consider a polymerase molecule which binds the end of such a linear duplex (Fig. 7). The primer strand is in its site with the 3'-OH group oriented in the primer terminus site; the strand is hydrogen-bonded to its complementary strand which is in the template site. But the template strand extends only as far as the primer terminus (Fig. 7). When a triphosphate enters the triphosphate site, there is no purine or pyrimidine base to serve as a template and thus no replication can take place. An intact linear duplex must therefore be inert.

Thus it appears that the replication of a linear duplex from its terminus, as pictured in the original Watson and Crick model, should not apply to DNA polymerase action in vitro, even for one of the strands. In recent studies (16) the DNA of phage T7 was prepared with care to avoid any internal breaks in this linear duplex and this DNA was essentially inert in supporting replication.

A special case of template copying is the replication of single-stranded circular DNA, such as the viral DNA of ϕX174 (17, 18). The circle provides no primer terminus and initiation of new strands by the enzyme does not take place readily (18). Therefore addition of an oligonucleotide which can anneal to the circle promotes replication by furnishing the necessary primer terminus (19). Copying then proceeds rapidly around the circle. The product is an incomplete circle. However, if a joining enzyme, called ligase, is present, the diester bond between the 3'-OH and 5'-P termini is made, and a fully covalent, double-stranded circle is produced (20). This synthetic molecule, as well as the double, circular molecules isolated from nature (Replicative Form I) do not bind to polymerase and are inert for replication.

Nicked Helix as the Functional Template Primer

Circular duplexes serve in vivo as chromosomes in bacteria (*E. coli* and *Bacillus subtilis*) and in viruses (polyoma), as replicative intermediates for other viruses (ϕX174, M13, and λ), and as bacterial episomes (*21*). Studies of ϕX174 replication indicate that a nicked form is the active replicative intermediate in vivo (*22*). Whereas the closed circular form is not replicated by DNA polymerase in vitro, the introduction by pancreatic deoxyribonuclease of one single-strand nick enables this DNA to bind to a polymerase molecule and converts the DNA to a favorable template and primer for replication (*16*). After replication has been initiated by introduction of a nick, the product, early or later in replication, is associated with the nicked form and is covalently linked to it (*16*).

As indicated earlier, an intact linear duplex, such as the DNA of phage T7, does not support replication. Upon the introduction of nicks by pancreatic deoxyribonuclease, the binding of these nicked duplexes to polymerase molecules and the appearance of productive sites for replication increase in direct proportion to the number of nicks introduced. In this instance, too, at least 90 percent of the DNA product is covalently attached to the primer (*16*).

How can the binding of a nicked region be visualized in the active center of the enzyme? The template and primer sites are filled. But the triphosphate site may not be vacant, and growth of a chain from the primer terminus is obstructed by the presence of the 5′-P–terminated strand, hydrogen-bonded to the template strand (Fig. 8). This dilemma might be resolved temporarily by the 3′ → 5′ exonuclease activity of the polymerase. Hydrolytic removal of the primer terminus nucleotide, accompanied by movement of the chain upward one nucleotide would open the triphosphate site for insertion and addition of a triphosphate. However, were this succession of events to take place nothing more would be achieved than the restoration of the original nicked region.

For progressive replication of the template to take place, the 5′-P–terminated strand must be displaced. Our evidence indicates that during the first phase of replication there is in fact a burst of hydrolysis of the template-primer, roughly matching the extent of replication (*23*). This hydrolysis is predominantly from the 5′ end of the DNA and entails degrading the chain from 5′ → 3′. The locus of this hydrolytic function appears to be distinct from that responsible for 3′ → 5′ hydrolysis.

Distinctive Sites for 3′ → 5′ and 5′ → 3′ Hydrolysis

Klett, Cerami, and Reich (*24*) were the first to recognize that polymerase preparations contained an exonuclease activity which degraded from the 5′ end of a chain. Their conclusions were based on studies with a synthetic block polymer resistant to hydrolysis from the 3′ end of the chain. A similar discovery was made independently when we tried to explain how DNA with 3′-P termini (introduced by micrococcal nuclease) and presumably insensitive to 3′ → 5′ exonuclease action was nevertheless extensively degraded (*25*). It became clear that such a 3′-P–terminated chain is degraded exclusively from the 5′ end. The principal products of extensive hydrolysis proved to be mononucleotides and an oligonucleotide which bore the 3′-P terminus.

With the recognition of this new property of DNA polymerase, an interesting possibility was raised. If a DNA chain were initiated *de novo* by DNA polymerase, its starting 5′ terminal should, as in the case of RNA polymerase, be marked by the initiating triphosphate. Yet attempts to identify such a triphosphate initiation point have not succeeded. It seemed possible that the 5′ → 3′ exonuclease activity of polymerase might act also on a chain terminated in a 5′-triphosphate and would therefore have erased a terminal triphosphate group even if it had been present initially.

In order to test this possibility, a polydeoxythymidylate of about 300 residues was synthesized containing a ^{32}P-labeled triphosphate group at the 5′ terminus and ^{3}H in the thymidine residues (*26*). This polynucleotide was degraded by polymerase from the 3′ end but *not* from the 5′ end. However, when annealed to form a helix wtih a polydeoxyadenylate chain, it was degraded rapidly from both ends. After an incubation period limited to only 10 seconds, 90 percent of the ^{32}P was liberated from the polymer, whereas less than 5 percent of the ^{3}H was released. Most remarkably, the principal ^{32}P product proved to be not deoxythymidine triphosphate as expected, but instead a dinucleoside tetraphosphate (Fig. 9).

We interpret these results as follows. There is a hydrolytic site for progressive 5′ → 3′ release of mononucleotides from the 5′ end of a strand in a location just above the triphosphate site. When the chain is terminated in a triphosphate, the close resemblance of this terminus to a deoxyribonucleoside triphosphate directs its binding in the triphosphate site. As a consequence, the initial product is not a mononucleotide but rather a dinucleotide, the dinucleoside tetraphosphate. Subsequently as the chain moves downward one nucleotide at a time the products are principally mononucleotide residues.

The fact that the 5′ → 3′ degradation

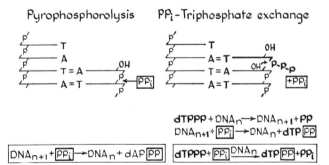

Fig. 10. Formulation of the pyrophosphorolysis and PP$_i$-triphosphate exchange reactions.

requires a helical structure indicates that this site cannot properly orient a single-stranded oligonucleotide chain. Because the site is occupied when such a chain is annealed to a complementary strand, we infer that the enzyme accommodates this complementary strand in the upper region of the template site.

The failure of the enzyme to degrade a 5'-terminated chain (from 5' → 3') when it is single stranded may help explain the stage in replication of helical DNA when synthesis proceeds without concomitant 5' → 3' hydrolysis. As the 3'-OH chain advances by growth along the template, the 5' chain may be displaced from the active site for a stretch of several nucleotides and thus be rendered insusceptible to hydrolysis. The 5' chain may be peeled back until some point when, for obscure reasons, it competes successfully for the template function by attracting the growing chain to switch templates (as in Fig. 13). Such a sequence of events would produce a covalent fork in the growing chain and has been suggested (27) as part of a mechanism to account for the branched structure and readily renaturable character of DNA synthesized on a helical template primer.

Pyrophosphorolysis and Pyrophosphate Exchange

Pyrophosphorolysis, the capacity of DNA polymerase to degrade DNA chains with PP_i, reaches a steady state when the accumulation of triphosphates supports synthesis at a rate that balances their removal (15). The enzyme also supports the exchange of PP_i

into the β, γ groups of a triphosphate (15). This reaction can occur with only a single triphosphate present, but otherwise requires all the primer and template conditions demanded of replication (Fig. 10). All the evidence fails to suggest the formation of a nucleotidyl-enzyme intermediate.

The behavior of chains terminated with a dideoxynucleotide suggests a mechanism of pyrophosphorolysis and PP_i exchange (12, 15). Such chains are relatively insusceptible to degradation by nucleophilic attack of PP_i, just as they are to that of OH^- (nuclease). Furthermore, a triphosphate analog lacking a 3'-OH group supports little, if any, PP_i exchange. Inasmuch as such an analog can be added to a chain, it would seem that PP_i exchange with the triphosphate does not occur entirely in the triphosphate site. Rather, the lack of a 3'-OH group prevents binding of the transition state formed by attack of the primer terminus on the triphosphate, and the terminal nucleotide in this transition state cannot be displaced efficiently by PP_i or by OH^-.

Pyrophosphorolysis is therefore taken to be a reversal of the polymerization step, including the concerted chain movement. Inorganic pyrophosphate exchange appears to be the result of a sequence of a polymerization step and a pyrophosphorolytic step repeated many times over (Fig. 10). Because the rate of PP_i exchange is considerably faster than that of pyrophosphorolysis (15), we suggest that the attack by PP_i occurs at a transition state short of completion of the polymerization step, and that this transition state is attained more readily from the direction of polymerization.

Insights from an Enzyme Modified by Acylation

Chemical alterations of DNA polymerase are beginning to provide important clues about its structure and function (3). Inasmuch as alkylation or Hg substitution of the single sulfhydryl group of DNA polymerase does not alter any of its activities, we assume that this group is relatively remote from the active center. However, acylation of the enzyme with N-carboxymethylisatoic anhydride results in a highly fluorescent derivative with markedly altered functional properties.

A derivative with 11 N-carboxymethylisatoyl groups had only 0.2 percent of the original polymerase activity but had 920 percent of the exonuclease activity when measured with DNA as primer or substrate at pH 7.4. Binding measurements and kinetic studies indicated that a major effect of this modification was a marked reduction in the affinity of deoxyribonucleoside triphosphate substrates for their site. Possibly, one or a few of the 61 lysine residues in the enzyme are located in this site, and their acylation had severely damaged the function of the site.

The remarkable increase in exonuclease activity implies that concomitant changes in the interaction of the enzyme with DNA are probably also involved. It remains to be determined whether one or both of the nuclease activities have been modified and how closer studies of the altered enzyme with various DNA's as substrates may elucidate the nature of the active center.

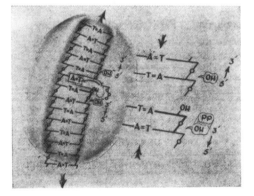

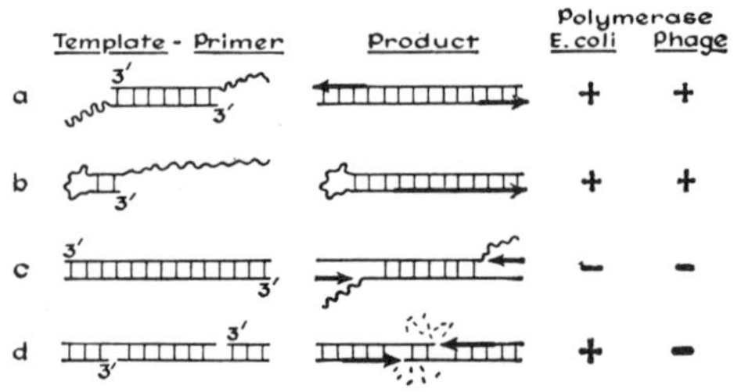

Fig. 11 (left). Model to account for the multiple functions of DNA polymerase within a single active center.

Fig. 12 (above). Template-primer requirements of the E. coli and phage-induced DNA polymerases.

36

Multiple Sites in One Active Center

A recapitulation of the several functional and physicochemical features of DNA polymerase is provided in a model proposed in Fig. 11. The template and primer functions in polymerization, the distinctive $3' \to 5'$ and $5' \to 3'$ exonuclease activities, and the various binding capacities of DNA are oriented in one active center. While alternative models have been considered (28–31), and some of these still have attractive features, the one we offer here goes farthest in explaining available data. A model by Beyersmann and Schramm (28), based exclusively on kinetic data, proposes an identical site for pyrophosphorolysis and hydrolysis and one active center for the degradative and polymerization functions of the enzyme. However, their model also proposes the pyrophosphorolysis of $3'$-P–terminated chains and roles for specific inhibitors that are inconsistent with our formulation.

Yet, our model, even if proven correct in basic outline, is still far too simple to account for many of the enzyme's features. For example, the capacity of the enzyme to discriminate secondary structure both in polymerization and in nuclease activities will require a more striking three-dimensional recognition of template and primer strands within the active center. To accommodate the screwlike translation of an essentially helical structure through the active center would demand a surface adapted to one or more turns of the helical duplex. The important influences of specific metal cations, ionic strength, and temperature on the enzyme functions have not even been considered. Finally, it seems likely that enzymes and factors related to the cellular functions of DNA polymerase will interact with it, at or near the active center, in the regulation of these functions.

Comparison of Various
DNA Polymerases

Studies analogous to those described here for the *E. coli* enzyme have not yet been carried out on DNA polymerase from other sources. However, a comparison of the template-primer requirements of the *E. coli* enzyme with those of enzymes induced by infection with phages T2 (8), T4 (32), and T5 (33) is of considerable interest

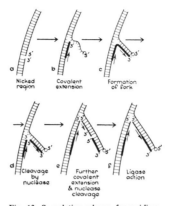

Fig. 13. Speculative scheme for unidirectional replication of a duplex chain.

(Fig. 12). Both the bacterial and the phage polymerases can utilize a double-stranded DNA helix, partially denuded from each $3'$-OH end by exonuclease III (Fig. 12a). The helix is restored by replication to its original length by extension of the $3'$ strands at each end. Similarly, both types of enzyme replicate single-stranded DNA. The latter, upon forming a looped structure with a short $3'$-OH end, is converted by replication to a hairpin-like product (Fig. 12b).

The bacterial enzyme was regarded as different from the phage-induced polymerases in its apparent ability to utilize a linear segment of double-stranded helical DNA. This difference had been ascribed to the unique capacity of the *E. coli* enzyme to initiate a DNA strand *de novo* with the $3'$-OH strand of the DNA duplex as template (Fig. 12c). It seems likely from the foregoing discussion that this earlier interpretation (34) was mistaken and that neither kind of polymerase can employ an intact linear duplex.

What seems clearer now is that a segment of duplex DNA which serves as template-primer for the *E. coli* enzyme does so by virtue of the nicked region that it contains and not by supporting new chain synthesis at its ends (Fig. 12d). It follows therefore that the phage-induced enzymes cannot exploit such nicked regions in the DNA. In support of this formulation are the long-standing observations that "activation" of duplex DNA by pancreatic deoxyribonuclease (that is, introduction of $3'$-OH nicks) increases its template-primer capacity 10- to 20-fold for the

E. coli enzyme while providing no significant improvement for the phage-induced enzymes (8).

How does one explain the inability of phage-induced enzymes to replicate at a nicked region of a duplex? Let us assume that the $5' \to 3'$ degradation by the $5' \to 3'$ nuclease function of the *E. coli* enzyme in the nicked region clears a path for the advancing synthetic chain, and that this is an essential step in the initiation of replication in vitro. Although the phage-induced polymerases are known to degrade polynucleotides from a $3' \to 5'$ direction, they may not possess the $5' \to 3'$ nuclease activity. Recent tests (26) show that there is, in fact, a striking absence of this $5' \to 3'$ activity in the phage T4-induced enzyme. Perhaps such a nuclease activity is present in the phage-infected cell, but has been eliminated on purification of the polymerase.

Inasmuch as the phage-induced DNA polymerase, as well as that from *E. coli*, has the $3' \to 5'$ nuclease activity, is there some physiological purpose attributable to this function? Again we can only conjecture. It appears that both nuclease activities are favored by some destabilization of the tight helical structure (23). Removal of DNA at or near disordered regions such as those produced by irradiation might then be performed by one of the nuclease activities. There is also the intriguing possibility that $3' \to 5'$ hydrolysis provides an opportunity for double-checking and editing to eliminate any newly polymerized nucleotide member of a poorly stacked or faulty base pair.

A Hypothetical Scheme for
Replication Process in vivo

How might DNA polymerase serve a physiological role in replication? What follows is speculative. Bacterial chromosomes are duplex circular structures and, when intact, are inert in replication. Introduction of a nick, possibly at a specific site, starts replication. DNA polymerase binds at the nick and replication proceeds by covalent extension of the $3'$-OH end (Fig. 13, a and b). The $5'$ end may be degraded to some extent by $5' \to 3'$ nuclease action, or, if displaced, is freed from further attack. Fixation to some membrane site (35, 36) may facilitate displacement and preservation of the $5'$ strand.

Replication proceeds for some distance and then switches to the complementary strand as template to form a fork; the fork is then cleaved by an endonuclease (Fig. 13, c and d). A repetition of this sequence leads to interruptions or small pieces of DNA near the replicating fork (Fig. 13e). Such pieces have been isolated by Okazaki and co-workers (37) at or near the nascent replicating region. These interruptions are sealed by ligase (Fig. 13f).

If this scheme were essentially correct, it would explain how one enzyme, DNA polymerase, replicating exclusively in a $5' \to 3'$ direction, would copy, almost simultaneously, the two maternal strands of opposite polarity. Examination of dividing bacteria at a gross level by autoradiography (38) or by gene duplication (39) makes it appear that there is a simultaneous, sequential replication of both strands. However, at the nucleotide level, as proposed in this scheme, the replicative action is staggered, alternating from one strand to the other.

Summary

DNA polymerase, a homogeneous protein of molecular weight 109,000, appears to be a single polypeptide chain. The enzyme contains one triphosphate substrate binding site and one site for binding a nicked region of duplex DNA. A model of the active center of the enzyme has been proposed (Fig. 11) in which there are distinctive sites for the template strand, the primer strand, the 3'-hydroxyl primer strand terminus, the triphosphate substrate, and the 5'-phosphate-terminated strand beyond the nick (point of scission). The model attempts to account for the various synthetic and degradative functions within the closely related sites in the active center of the enzyme.

Initiation of replication is favored at a nicked region of a duplex. In the first phase of replication, extension of the 3'-hydroxyl primer strand appears to be related to the $5' \to 3'$ hydrolytic removal of the 5'-phosphate-terminated strand. The failure of phage-induced DNA polymerases to initiate replication at a nicked region may be due to the lack of a $5' \to 3'$ nuclease function in the phage enzyme. A speculative model for helix replication, in vivo (Fig. 13), suggests how DNA polymerase, in conjunction with endonuclease and ligase, may achieve the sequential and almost simultaneous replication of both strands of a helix.

References and Notes

1. P. T. Englund, M. P. Deutscher, T. M. Jovin, R. B. Kelly, N. R. Cozzarelli, A. Kornberg, *Cold Spring Harbor Symp. Quant. Biol.*, in press.
2. T. M. Jovin, P. T. Englund, L. L. Bertsch, *J. Biol. Chem.*, in press.
3. T. M. Jovin, P. T. Englund, A. Kornberg, *ibid.*, in press.
4. L. F. Cavalieri and E. Carroll, *Proc. Nat. Acad. Sci. U.S.* 59, 951 (1968).
5. A. G. Lezius, S. B. Hennig, C. Menzel, E. Metz, *Eur. J. Biochem.* 2, 90 (1967).
6. P. T. Englund, R. B. Kelly, A. Kornberg, *J. Biol. Chem.*, in press.
7. N. R. Cozzarelli, R. B. Kelly, A. Kornberg, *Proc. Nat. Acad. Sci. U.S.* 60, 992 (1968).
8. H. V. Aposhian and A. Kornberg, *J. Biol. Chem.* 237, 519 (1962).
9. C. C. Richardson, C. L. Schildkraut, A. Kornberg, *Cold Spring Harbor Symp. Quant. Biol.* 28, 9 (1963).
10. P. T. Englund, J. A. Huberman, T. M. Jovin, A. Kornberg, *J. Biol. Chem.*, in press.
11. C. C. Richardson, R. B. Inman, A. Kornberg, *J. Mol. Biol.* 9, 46 (1964).
12. M. R. Atkinson, M. P. Deutscher, A. Kornberg, A. Russell, J. Moffatt, in preparation; M. R. Atkinson, J. A. Huberman, R. B. Kelly, A. Kornberg, *Fed. Proc.*, in press.
13. L. H. Toji and S. S. Cohen, personal communication.
14. J. A. Huberman, M. R. Atkinson, A. Kornberg, unpublished results.
15. M. P. Deutscher and A. Kornberg, *J. Biol. Chem.*, in press.
16. R. B. Kelly, N. R. Cozzarelli, A. Kornberg, unpublished results.
17. S. Mitra, P. Reichard, R. B. Inman, L. L. Bertsch, A. Kornberg, *J. Mol. Biol.* 24, 429 (1967).
18. M. Goulian and A. Kornberg, *Proc. Nat. Acad. Sci. U.S.* 58, 1723 (1967).
19. M. Goulian, *ibid.* 61, 284 (1968); *Cold Spring Harbor Symp. Quant. Biol.*, in press.
20. ———, A. Kornberg, R. L. Sinsheimer, *Proc. Nat. Acad. Sci. U.S.* 58, 2351 (1967).
21. J. Vinograd and J. Lebowitz, *J. Gen. Physiol.* 49 (6), 103 (1966).
22. R. Knippers, T. Komano, R. L. Sinsheimer, *Proc. Nat. Acad. Sci. U.S.* 59, 577 (1968); T. Komano, R. Knippers, R. L. Sinsheimer, *ibid.*, p. 911.
23. M. P. Deutscher, R. B. Kelly, A. Kornberg, unpublished results.
24. R. P. Klett, A. Cerami, E. Reich, *Proc. Nat. Acad. Sci. U.S.* 60, 943 (1968).
25. M. P. Deutscher and A. Kornberg, *J. Biol. Chem.*, in press.
26. N. R. Cozzarelli, R. B. Kelly, A. Kornberg, in preparation.
27. C. L. Schildkraut, C. C. Richardson, A. Kornberg, *J. Mol. Biol.* 9, 24 (1964).
28. D. Beyersmann and G. Schramm, *Biochim. Biophys. Acta* 159, 64 (1968).
29. H. Jehle, *Proc. Nat. Acad. Sci. U.S.* 53, 1451 (1965).
30. H. E. Kubitschek and T. R. Henderson, *ibid.* 55, 512 (1966).
31. A. Kornberg, in *Regulation of Nucleic Acid and Protein Biosynthesis*, V. V. Koningsberger and L. Bosch, Eds. (Elsevier, Amsterdam, 1967), p. 22.
32. M. Goulian, Z. J. Lucas A. Kornberg, *J. Biol. Chem.* 243, 627 (1968).
33. C. D. Steuart, S. R. Anand, M. J. Bessman, *ibid.*, p. 5319.
34. S. Mitra and A. Kornberg, *J. Gen. Physiol.* 49 (6), 59 (1966).
35. F. Jacob, S. Brenner, F. Cuzin, *Cold Spring Harbor Symp. Quant. Biol.* 28, 329 (1963).
36. W. Gilbert and D. Dressler, *ibid.*, in press.
37. R. Okazaki, T. Okazaki, K. Sakabe, K. Sugimoto, A. Sugino, *Proc. Nat. Acad. Sci. U.S.* 59, 598 (1968).
38. J. Cairns and C. I. Davern, *J. Cell. Physiol.* 70 (Suppl.), 65 (1967).
39. N. Sueoka, *Mol. Genet.* 2, 1 (1967).
40. The article is adapted from the inaugural Stanhope Bayne-Jones Lecture deliverd at Johns Hopkins University School of Medicine, 19 November 1968. For the recent work which forms the basis of this lecture, I want to acknowledge the contributions of M. R. Atkinson, N. R. Cozzarelli, M. P. Deutscher, P. T. Englund, J. A. Huberman, T. M. Jovin, and R. B. Kelly. I want particularly to cite Jovin's initiative and skill in the physicochemical aspects of the work. I am also appreciative of the skillful efforts of Mrs. L. M. Follett, director of medical illustration at Stanford.

4

Reprinted from *Cold Spring Harbor Symp. Quant. Biol.*, **33**, 129–143 (1968)

In Vivo Mechanism of DNA Chain Growth

Reiji Okazaki,* Tuneko Okazaki,* Kiwako Sakabe,† Kazunori Sugimoto,*
Ritsu Kainuma,* Akio Sugino,* and Norio Iwatsuki*

*Institute of Molecular Biology and Department of Chemistry, Faculty of Science, Nagoya
University, Chikusa-ku, Nagoya, Japan, and Division of Biology, Kansas State University,
Manhattan, Kansas*

Replication of DNA by the base-pairing mechanism proposed by Watson and Crick (1953a,b) has been supported by both in vivo (Meselson and Stahl, 1958) and in vitro experiments (Bessman et al., 1958; Lehman et al., 1958; Josse et al., 1961).

The Meselson and Stahl experiment and subsequent in vivo studies (Maaløe and Hanawalt, 1961; Hanawalt et al., 1961; Cairns, 1963a,b; Nagata, 1963; Yoshikawa and Sueoka, 1963a,b; Bonhoeffer and Gierer, 1963; Lark et al., 1963) indicated that replication of the bacterial chromosome proceeds sequentially. This led to the inference that both daughter strands of chromosomal DNA grow continuously, the direction of synthesis being 5′ to 3′ on one strand and 3′ to 5′ on the other (Fig. 1A). This mechanism of continuous chain growth, however, conflicts with the fact that the in vitro DNA synthesis by DNA polymerase proceeds in the 5′ to 3′ direction (Kornberg, 1960; Mitra and Kornberg, 1966) and no enzymatic reaction for the 3′ to 5′ synthesis of the DNA chain has been found. The difficulty could be avoided if DNA were synthesized in vivo by a discontinuous mechanism as shown in Figs. 1 B-D. These models assume that short stretches are synthesized, in either one or both strands, by a 5′ to 3′ reaction at the replicating point and are subsequently connected to the growing strands.

The following predictions are made from the models of discontinuous replication: (1) The most recently replicated portion of one or both daughter strands can be isolated after denaturation as short DNA chains distinct from large DNA molecules derived from the rest of the chromosome; (2) Selective and temporal inhibition of the enzyme for the formation of phosphodiester linkages between DNA chains will result in a marked accumulation of the nascent short chains. Both of these predictions are confirmed by the experiments described in this paper. Preliminary results also suggest that the direction of synthesis of DNA

* Present address: Institute of Molecular Biology, Faculty of Science, Nagoya University, Chikusa-ku, Nagoya, Japan.
† Present address: Department of Chemistry, Faculty of Science, Nagoya University, Chikusa-ku, Nagoya, Japan.

chains in vivo is 5′ to 3′. In addition, observations made during the course of these studies have revealed instability (or lack) of the double-stranded structure in the replicating portion. Possible implications of these findings will be discussed in relation to the replication mechanism.

DISCONTINUITY OF THE NEWLY SYNTHESIZED CHAINS

In order to test the first prediction, cells synthesizing chromosomal or phage DNA were exposed to ³H-thymidine (14–15 mc/µmole) at 20° (or 25°) for a short period and the pulse labeling was terminated by addition of KCN (0.02 M) and ice. In some experiments, cells were grown for several generations in ¹⁴C-thymidine (or ¹⁴C-thymine) to label the bulk of DNA prior to pulse labeling, or cells uniformly labeled with ¹⁴C-thymidine (or ¹⁴C-thymine) were added to the pulse-labeled cells prior to DNA extraction. Labeled DNA was obtained in the denatured form either by extraction by the procedure of Thomas, Berns, and Kelly (1966) followed by treatment with 0.1 N NaOH at room temperature or by direct extraction with 0.1 N NaOH containing 10 mM EDTA at 37°C for

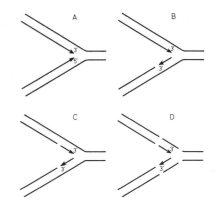

Figure 1. Models for the possible structure and reaction in the replication region of DNA.

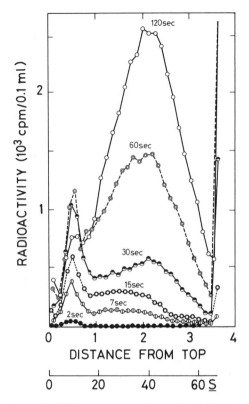

formed with *E. coli* 15T⁻ and were reported by Sakabe and Okazaki (1966).

Figure 2 shows a result obtained with T4D (wild-type)-infected *E. coli*. The pulse labeling was done at 70 min after infection at 20°, when phage DNA is being actively synthesized. After a two-second pulse, the radioactive label incorporated was found almost exclusively in a slowly sedimenting component with a sedimentation coefficient of 9 S. The radioactivity in this component increased and reached a maximum in about 30

FIGURE 2. Alkaline sucrose gradient sedimentation of pulse-labeled DNA from T4 phage-infected *E. coli* B. Cells were grown in M9 synthetic medium supplemented with 0.5% casamino acids (Medium C) at 37° to 5 × 10⁸ cells/ml, suspended in M9 medium containing no glucose at 10⁹ cells/ml, and incubated for 15 min at 37°. Following addition of DL-tryptophan (40 μg/ml), the cells were infected with T4 phage (m. o. i. = 10). After 5 min at 37°, the culture was cooled to 20°, and an equal volume of M9 medium containing twice as much glucose and casamino acids as Medium C added. After incubation with stirring at 20° for 70 min, the culture was pulse-labeled with 10⁻⁷ M ³H-thymidine (14 mc/μmole) for the indicated time. DNA was extracted by NaOH-EDTA treatment and sedimented in a Spinco SW 25.1 rotor for 15 hr at 20,500 rpm and 8°. The distance from the top is relative to that of infective DNA from phage δA (19 S, reference) (Okazaki et al., 1967, 1968).

20 min. Analysis of labeled DNA in alkaline sucrose gradients (5 to 20% linear sucrose gradients containing 0.1 N NaOH, 0.9 M NaCl and 1 mM EDTA) was carried out using infectious circular DNA of phage δA as an internal marker. In most experiments, recovery of labeled DNA in extraction and alkaline sucrose gradient sedimentation was more than 80%. The first experiments were per-

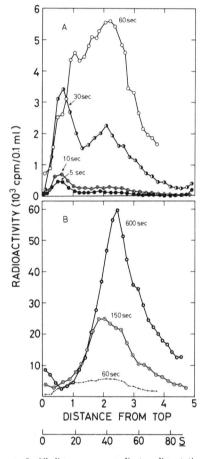

FIGURE 3. Alkaline sucrose gradient sedimentation of pulse-labeled DNA from *E. coli* B. Cells were grown in a glucose-salt medium (Medium A of Okazaki et al., 1968) at 37° to a titer of 3 × 10⁸ cells/ml and then at 20° to 5 × 10⁸ cells/ml. The culture was pulse-labeled with 10⁻⁷ M ³H-thymidine at 20° for the indicated time. DNA was extracted by NaOH-EDTA treatment and sedimented in a SW 25.3 rotor for 10 hr at 22,500 rpm and 4° (Okazaki et al., 1968).

seconds. At 2 seconds there was virtually no radio-activity in the fast sedimenting material, but the label then appeared in the fast sedimenting fraction and increased quickly. After two minutes of labeling, 90% of the label incorporated was present in the fast sedimenting DNA. The average sedimentation rate of the fast component increased gradually. The average rate was about 40 S after the two-minute pulse. In other experiments average sedimentation coefficients of 45 and 50 S were obtained for five- and ten-minute pulse DNA, respectively.

Figures 3 to 6 show examples of similar results obtained with various growing bacteria. The result with normal *E. coli* B (Fig. 3) indicates that ^{3}H-thymidine first appears in the slowly sedimenting short DNA chains before transition to large molecules as in T4-infected cells. The sedimentation coefficient of the initially labeled material was 10 to 11 S compared with the value of 8 to 9 S in the T4 system. Figure 4 shows that nascent short DNA chains with an average sedimentation rate of about 10 S are found also with an endonuclease I-deficient *E. coli* strain 1100 (Dürwald and Hoffmann-Berling, 1968) as well as a wild-type strain W3110. With *Bacillus subtilis* the result was similar but some difference was noted (Fig. 5). ^{3}H-thymidine was first incorporated into short chains (sedimentation coefficient = 8 S) but the increase of the sedimentation coefficient was slow and gradual. This is also evident from the pulse-chase experiment shown in Fig. 6. These results suggest the presence of a large number of short chains and many chains with intermediate sizes in *B. subtilis*.

These results conform to the first prediction of discontinuous replication. The fact that virtually all the radioactive label was found in the short chains after an extremely short pulse favors a two-strand discontinuous mechanism (Fig. 1 C or D), rather than a one-strand discontinuous mechanism (Fig. 1B), although more direct tests by annealing the short chains with the separated complementary strands would be needed for a definitive conclusion. The average sedimentation coefficient of the initially labeled material, which may consist of the growing and completed short chains, is 7 to 11 S, suggesting that the length of the 'unit' of the discontinuous synthesis may be 1000 to 2000 nucleotides. This corresponds to the estimated dimension of the cistron.

Figure 7 illustrates a possible structure of the daughter strands in the vicinity of the growing end. Units synthesized at the growing points would be joined together by phosphodiester bonds to form longer strands located in the nonterminal position. The number of the units and of chains with intermediate lengths would be determined by the relative rates of synthesis and of joining. *B. subtilis* may represent a system in which the joining activity is low relative to the synthesis.

An alternative possibility which could explain the above results is that artificial breaks may be introduced selectively in the newly replicated portion by the procedures used in our experiment because of selective weakness of that portion. To test this possibility, the effect of the following modifications of the experimental procedures were examined: (a) Omission of the phenol step from the Thomas procedure; (b) Extraction of denatured DNA with NaOH-EDTA after a lysozyme treatment; (c) Direct addition of NaOH-EDTA to the culture with or without prior addition of KCN and ice; (d) Denaturation with formamide after extraction by the Thomas procedure; (e) Extraction of DNA by the method of Nomura et al. (1962) followed by alkali denaturation. None of these changes of the procedure altered the essential feature of the result. Furthermore, as will be shown below, the nascent short DNA chains can be demonstrated without denaturation treatment by sedimentation in neutral sucrose gradients, where the DNA is not exposed to alkali at any stage. These facts argue against the above alternative, though they do not exclude it.

ACCUMULATION OF NASCENT SHORT DNA CHAINS UPON TEMPORARY INHIBITION OF POLYNUCLEOTIDE LIGASE

A test of the second prediction of the hypothesis of discontinuous replication was made possible by the discovery of polynucleotide ligase (DNA-joining enzyme), an enzyme whose existence is predicted by the hypothesis, in normal and phage-infected *E. coli* (Gellert, 1967; Olivera and Lehman,

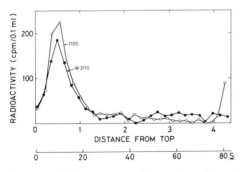

FIGURE 4. Alkaline sucrose gradient sedimentation of 10-sec pulse DNA of *E. coli* W3110 and 1100. The experiments were carried out as in Fig. 3 (Okazaki et al., 1968).

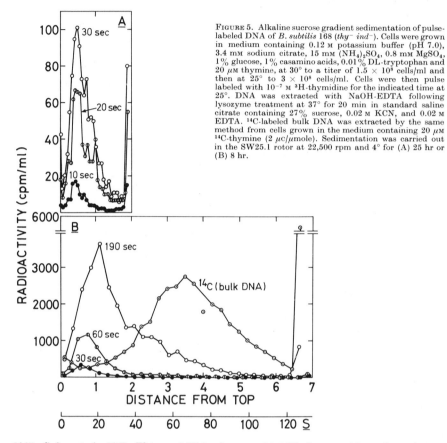

FIGURE 5. Alkaline sucrose gradient sedimentation of pulse-labeled DNA of *B. subtilis* 168 (*thy⁻ ind⁻*). Cells were grown in medium containing 0.12 M potassium buffer (pH 7.0), 3.4 mM sodium citrate, 15 mM $(NH_4)_2SO_4$, 0.8 mM $MgSO_4$, 1% glucose, 1% casamino acids, 0.01% DL-tryptophan and 20 μM thymine, at 30° to a titer of 1.5×10^8 cells/ml and then at 25° to 3×10^8 cells/ml. Cells were then pulse labeled with 10^{-7} M ³H-thymidine for the indicated time at 25°. DNA was extracted with NaOH-EDTA following lysozyme treatment at 37° for 20 min in standard saline citrate containing 27% sucrose, 0.02 M KCN, and 0.02 M EDTA. ¹⁴C-labeled bulk DNA was extracted by the same method from cells grown in the medium containing 20 μM ¹⁴C-thymine (2 μc/μmole). Sedimentation was carried out in the SW25.1 rotor at 22,500 rpm and 4° for (A) 25 hr or (B) 8 hr.

1967; Gefter et al., 1967; Weiss and Richardson 1967; Becker et al., 1967; Cozzarelli et al., 1967). Fareed and Richardson (1967) identified gene 30 of T4 as the structural gene of ligase and showed that temperature-sensitive mutants of this gene, e.g. T4 *ts* A80 and *ts* B20, produce thermosensitive ligase. As has been shown before (Fig. 2), ³H-thymidine that has been incorporated into T4 (wild-type)-infected cells during the period of active phage DNA synthesis first appears in short DNA chains having an average sedimentation rate of 8–9 S before transition to large chains (30–60 S). If these pulse-labeled short chains consist of the growing and completed units that are about to be joined in vivo by ligase, temporary impairment of the thermosensitive mutant ligase by high temperature should cause accumulation of the nascent short chains because of selective inhibition of the joining step. It is expected that the average size of the nascent short chains which accumulate under these conditions should be slightly larger than that of the short chains found in the normal

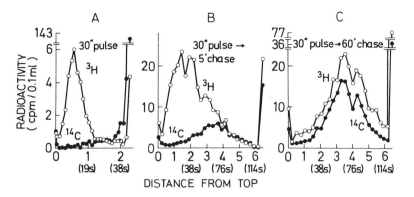

FIGURE 6. Effect of chase on the sedimentation of pulse-labeled DNA of *B. subtilis* 168 (*thy⁻ ind⁻*). Cells were pulse labeled with ³H-thymidine as in Fig. 5 for 30 sec at 25°. The chase was performed by adding a 10⁴-fold amount of unlabeled thymidine and incubating at 25° for the indicated time. Alkaline sucrose gradient sedimentation was carried out in the SW25.3 rotor at 22,500 rpm and 4° for (A) 24 hr or (B and C) 8 hr.

steady-state system, since most of the accumulated chains should be *completed* units.

Cells of *E. coli* B/5 infected with T4 *ts* A80, *ts* B20, or T4D (wild-type) were incubated for 70 min at 20° to reach the stage where phage DNA is being actively synthesized. The temperature was then raised to 43° or 44°, and one or two minutes later cells were pulse-labeled with ³H-thymidine. As shown in Fig. 8, the rate of incorporation of ³H-thymidine into DNA in the mutant-infected cells was comparable to the rate found with T4D-infected cells.

With T4D-infected cells (Fig. 9C), a small portion of the radioactivity was found, after a 20-sec pulse, in short chains with an average sedimentation coefficient of about 8 S, most of the radioactivity incorporated being found in the larger DNA chains. After 40- or 60-sec labeling, virtually all the radioactivity was found in the large DNA chains with a sedimentation coefficient of more than 30 S. This agrees with the expectation from the previous experiment at 20°.

In contrast to these results with wild-type T4, when cells infected with either T4 *ts* A80 or T4 *ts* B20 were pulse labeled with ³H-thymidine after the temperature shift-up, virtually all the radioactivity incorporated into DNA was found after a 10- or 20-sec pulse in the short DNA chains with an average sedimentation rate of about 9 S (Figs. 9A, B and 10). With an increase of pulse time (to 30, 40 or 60 seconds), the label in this region continued to increase and little radioactivity appeared in the

material sedimenting at faster rates. Thus, under these conditions the short DNA chains are produced but their joining is inhibited almost completely. Occurrence of some joining reaction is suggested by a slow increase of the sedimentation coefficient of the labeled DNA. This may be due to the residual activity of phage ligase or to the action of host ligase, which is not thermosensitive

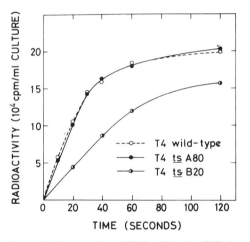

FIGURE 8. Incorporation of ³H-thymidine into DNA by gene 30 *ts* mutant and wild-type T4-infected cells at 2 min after temperature shift from 20° to 43°. *E. coli* B/5 was infected with the indicated strains of T4 and incubated at 20° for 70 min. The temperature was then raised to 43°, and 2 min later ³H-thymidine (15 mc/μmole) was added to 10⁻⁷ M. Samples were taken at the indicated time after the addition of ³H-thymidine and cold trichloroacetic acid-insoluble radioactivity was measured (Sugimoto et al., 1968).

FIGURE 7. Schematic illustration of a possible structure of the daughter strand in the vicinity of the growing end (Okazaki et al., 1968).

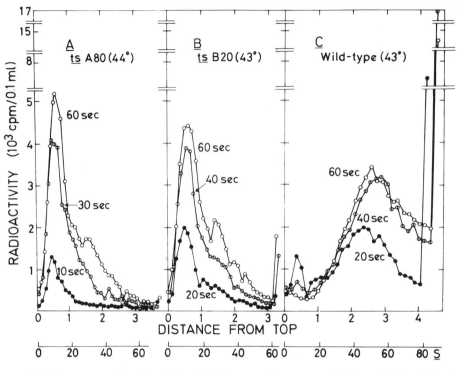

FIGURE 9. Alkaline sucrose gradient sedimentation of DNA from gene 30 *ts* mutant and wild-type T4-infected cells pulse labeled at high temperature. Cells were infected with the indicated strain of T4 and incubated at 20° for 70 min before the temperature was elevated to 43 or 44°. The cells were pulse labeled with 10^{-7} M ^{3}H-thymidine (15 mc/μmole) for the indicated time beginning at 2 min after the temperature shift. DNA was extracted by NaOH-EDTA treatment and sedimented in the SW25.3 rotor for 13–16 hr at 22,500 rpm and 4° (Sugimoto et al., 1968).

but shows a much lower level of activity than the phage-induced enzyme in T4-infected cells (Weiss and Richardson, 1967; Becker et al., 1967).

Such marked accumulation of radioactive short DNA chains, observed at 43–44°, was not found when the mutant-infected cells were pulse labeled at 20° or 30° (Fig. 11). At these temperatures, the results obtained with the mutant and wild-type phages were indistinguishable.

In the experiment shown in Fig. 12, T4 *ts* B20-infected cells were pulse labeled for 60 sec at 43° and then incubated at 30°. After the temperature shift-down, the size of the labeled DNA chains increased gradually and the peak in the 10 S region disappeared. Although a slow increase of the sedimentation rate of the accumulated short DNA chains was also found at 43°, as noted before, the rate was much faster at 30° than at 43°. Thus, it is evident that the effect of high temperature can be reversed in this system by subsequent cooling, and

the accumulated short chains were joined. The reversal may be due to relief from the inhibition of ligase or to the synthesis of active ligase after the shift-down of temperature.

All the above results are in complete accordance with the idea that DNA replicates in vivo by a discontinuous mechanism involving the synthesis and joining of short stretches of both strands and that polynucleotide ligase is involved in the joining step. It may be argued, however, that the accumulation of the nascent short chains might result from in vivo degradation of long DNA chains under the particular conditions of ligase inhibition. To explore this possibility, T4 *ts* A80- or T4 *ts* B20-infected cells were pulse labeled with ^{3}H-thymidine at 20° for 5 min and then incubated at 43°. It is to be noted that 5 min at 20° may correspond to 30–60 sec at 43°, and that most of ^{3}H-thymidine added to the medium is incorporated into the large DNA chains during the 5-min labeling at 20°. As shown in Fig. 13, the average sedimentation coefficient of the prelabeled DNA did not change during the 2-min incubation at 43°. A decrease from about 50 S to about 45 S was found between the second and third minutes after the temperature shift-up. This small decrease of the sedimentation coefficient, which is observed following a long incubation at 43°C, may reflect breakage and joining of phage DNA normally taking place in the cell. This change in the molecular size of the labeled long chains is too small to account for the accumulation of the large amount of the pulse-labeled short chains with a sedimentation coefficient of about 10 S. Moreover, the marked accumulation of the nascent short DNA chains is observed during the first two-minute period following the temperature shift-up, when no decrease of the size of the prelabeled DNA is detected. Therefore it is unlikely that the nascent short DNA chains are in vivo degradation products, unless there is selective breakage taking place in the restricted region near the growing end. This conclusion, of course, is also supported by the existence of a smaller amount of similar nascent short chains in the normal system (Fig. 2).

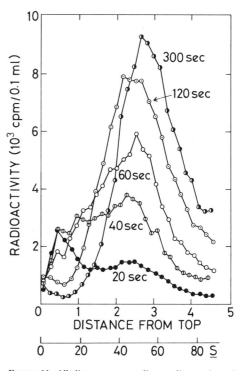

FIGURE 11. Alkaline sucrose gradient sedimentation of radioactive DNA from T4 *ts* B20-infected cells pulse labeled at low temperature (30°). The experiment was carried out as in Fig. 9, except that the temperature shift was from 20° to 30° (Sugimoto et al., 1968).

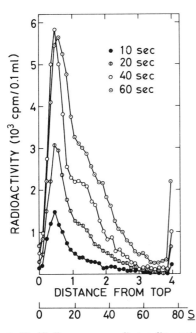

FIGURE 10. Alkaline sucrose gradient sedimentation of radioactive DNA from T4 *ts* B20-infected cells pulse labeled at 1 min after temperature shift-up. The experiment was carried out as in Fig. 9 except that the pulse labeling was begun at 1 min after the temperature shift from 20° to 43° (Sugimoto et al., 1968).

DIRECTION OF SYNTHESIS OF DNA CHAINS

The fact that both predictions were confirmed experimentally strongly supports the hypothesis of discontinuous DNA replication. However, there is still an alternative possibility that the DNA chains are synthesized by a continuous mechanism but selective nicks are introduced in vivo in the newly replicated portion, and that these nicks are subsequently sealed by the ligase reaction. This possibility would also be eliminated if one could prove that the short DNA chains are synthesized only in the 5′ to 3′ direction as assumed in the original hypothesis.

Study of the direction of synthesis of DNA chains in vivo was begun in this laboratory in 1964 before the first evidence of the discontinuity of the newly replicated chains was obtained early in 1966. One approach to this problem is to label the growing ends of the growing DNA strands and to identify

45

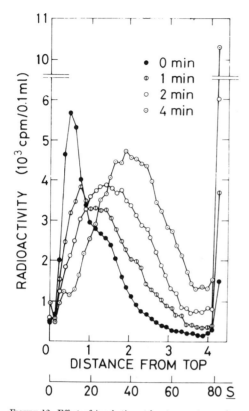

FIGURE 12. Effect of incubation at low temperature subsequent to pulse labeling at high temperature on the sedimentation pattern of radioactive DNA from T4 *ts* B20-infected cells. Cells were pulse labeled for 60 sec with ³H-thymidine at 43° as in Fig. 9 and then incubated at 30° for the indicated time. DNA extraction and alkaline sucrose gradient sedimentation were carried out as in Fig. 9 (Sugimoto et al., 1968).

the labeled growing end by stepwise degradation from the 3′ or 5′ end by specific exonucleases. The usefulness of *E. coli* exonuclease I (Lehman, 1960) and *B. subtilis* nuclease (Okazaki et al., 1966) for this purpose was demonstrated by model experiments which compared the kinetics of release of ³H and ¹⁴C from the ¹⁴C-labeled DNA whose 3′-end was labeled with ³H by the DNA polymerase reaction (Fig. 14). It is evident from the results that *E. coli* exonuclease degrades single-stranded DNA in a stepwise manner beginning from the 3′ end, whereas *B. subtilis* nuclease initiates an exonucleolytic attack at the 5′ end of the single-stranded DNA. This was also indicated by the kinetics of release of ³H or ³²P from the DNA chains

whose 3′ and 5′ ends were labeled with ³H and ³²P by the DNA polymerase and polynucleotide kinase reactions, respectively. These model experiments, however, also indicate that the size of the terminal label should be very small (less than 1 % portion of the whole chain) in order to obtain clear-cut results.

In the experiment shown in Fig. 15, *E. coli* 15T⁻ grown in a medium containing unlabeled thymidine was starved of thymidine for 10 min at 37° to deplete the intracellular pool of thymidine nucleotides. The cells were then labeled with ³H-thymidine at 0° for 5 min. (It is estimated that 20 to 40 nucleotides are added to each strand of chromosomal DNA during the 5-min period at 0°.) DNA was extracted by the method described by Nomura et al. (1962), denatured with formamide and subjected to the enzymatic digestion. Figure 16 shows a similar experiment, in which cells prelabeled with ¹⁴C-thymidine were exposed to ³H-thymidine at 0° for 5 min; DNA was extracted by the Thomas

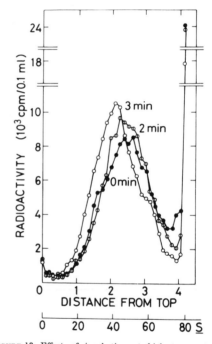

FIGURE 13. Effect of incubation at high temperature subsequent to pulse labeling at low temperature on the sedimentation pattern of radioactive DNA from T4 *ts*A 80-infected cells. Cells were pulse labeled with ³H-thymidine at 20° for 5 min from 65 to 70 min after infection and then incubated at 43° for the indicated time. DNA extraction and alkaline sucrose gradient sedimentation were carried out as in Fig. 9 (Sugimoto et al., 1968).

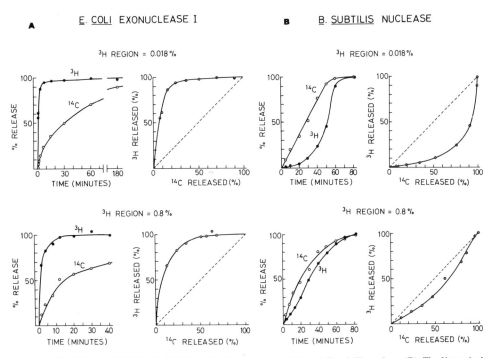

FIGURE 14. Hydrolysis of ³H¹⁴C-labeled DNA by *E. coli* exonuclease I (A) and *B. subtilis* nuclease (B). The 3′ terminal portion of ¹⁴C-labeled T7 DNA corresponding to 0.018% or 0.8% of the whole molecule was labeled with ³H-dTMP by the 'limited reaction' or 'repair' of *B. subtilis* DNA polymerase (Okazaki and Kornberg, 1964) respectively. DNA was heat-denatured before use. Digestion with exonuclease I of doubly labeled DNA prepared by the 'limited reaction' was carried out in the reaction mixture (70 μl) containing: 5 μmoles of glycine-KOH, pH 9.2; 0.5 μmole of MgCl₂; 50 mμmoles 2-mercaptoethanol; 4 mμmoles of DNA; and 0.6 unit of enzyme (a crystalline preparation generously provided by Dr. I. R. Lehman).

Digestion with exonuclease I of doubly labeled DNA prepared by 'repair' was carried out in the reaction mixture (50 μl) containing: 3.5 μmoles of glycine-KOH, pH 9.2; 0.35 μmole of MgCl₂; 35 mμmoles of 2-mercaptoethanol; 1 mμmole of DNA and 0.3 unit of enzyme (crystalline). Digestion with *B. subtilis* nuclease of doubly labeled DNA prepared by the 'limited reaction' was carried out in the reaction mixture (50 μl) containing: 2.5 μmoles of Tris-HCl, pH 8.8; 2 mμmoles of DNA and 0.11 unit of enzyme (Fraction I-A). Digestion with *B. subtilis* nuclease of doubly labeled DNA prepared by 'repair' was carried out in the reaction mixture (50 μl) containing: 2.5 μmoles of Tris-HCl, pH 8.8; 1 mμmole of DNA; and 0.15 unit of enzyme (Fraction I-A). Incubation was at 37° and acid-soluble radioactivities were measured (data on *B. subtilis* nuclease from Okazaki, Okazaki, and Sakabe, 1966).

procedure, and the doubly labeled nascent short DNA chains with a sedimentation coefficient of 10 S were isolated by repeated centrifugation in alkaline sucrose and hydrolyzed with the exonucleases. In both experiments, the release of the terminal ³H label preceded that of the bulk of nucleotides when the DNA was degraded by *E. coli* exonuclease I and followed it when *B. subtilis* nuclease was used. These results indicate that under these conditions most, if not all of the DNA chains are synthesized in the 5′ to 3′ direction.

Another way to determine the direction of synthesis of the DNA chains is to elucidate the structure of the 5′ terminus of the growing short chains and to follow the proportion of the radioactivity found in the 5′ termini of the growing chains of a certain size after various lengths of pulse labeling. The *B. subtilis* nuclease is also useful for these purposes because it cleaves deoxynucleoside 3′5′-diphosphates from the 5′ phosphoryl terminus of the DNA chain (Kanamori, Sakabe, Okazaki and Okazaki, unpubl.). Current study along this line suggests that the growing short chains of T4 DNA are terminated at the 5′ end with monophosphate and not with triphosphate. This is inconsistent with a hypothetical mechanism for the 3′ to 5′ chain elongation, in which deoxynucleoside 5′-triphosphates are added successively to the 5′-triphosphate end of the DNA chain by formation of covalent phosphodiester bonds.

OKAZAKI et al.

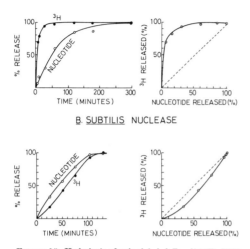

FIGURE 15. Hydrolysis of pulse-labeled *E. coli* 15T⁻ DNA with exonucleases. Cells were grown in Medium B (Okazaki et al., 1968) at 37° to a titer of 5×10^8 cells/ml washed with 0.85% NaCl and resuspended in Medium B containing no thymidine. After incubation at 37° for 10 min, the culture was chilled to 0° and 15 min later ³H-thymidine was added to 1.6×10^{-8} M. After 5 min of labeling at 0° with stirring, KCN (final concentration: 0.02 M) and ice were added and cells were collected by centrifugation. DNA was extracted by the method described by Nomura et al. (1962) and denatured by incubation in formamide at room temperature for 4.5 hr. For digestion with *E. coli* exonuclease I, the 250 μliter reaction mixture, containing 15 μmoles of glycine-KOH, pH 9.2, 1.5 μmoles of MgCl₂, 0.5 μmole of 2-mercaptoethanol, 140 mμmoles of DNA and 20 units of enzyme (DEAE-cellulose fraction, provided by Dr. I. R. Lehman), was incubated at 37°. For digestion with *B. subtilis* nuclease, the 250-μliter reaction mixture contained 15 μmoles of Tris-HCl, pH 8.8, 0.75 μmole of CaCl₂, 140 mμmoles of DNA and 4.2 units of enzyme (Fraction I-A of Okazaki, Okazaki, and Sakabe, 1966), and incubated at 37°. Release of acid-soluble radioactivity and diphenylamine-reactive material was measured (Okazaki, Sakabe, and Okazaki, 1966).

SECONDARY STRUCTURE OF THE REPLICATING REGION

Neutral sucrose gradient sedimentation of pulse-labeled DNA, isolated by the Thomas procedure but not subjected to denaturation treatment, gave rather unexpected results. An experiment with *E. coli* B is shown in Fig. 17. While most of the DNA isolated from the cells labeled with ³H-thymidine for 10 min sedimented at a rate faster than δA DNA, having a sedimentation coefficient of 29 S in 0.5 M NaCl, pH 7.0, a considerable fraction of 15-sec pulse DNA was recovered in a band sedimenting at a much slower rate. Other experiments indicated that the fraction of the radioactivity found in the slowly sedimenting band decreases with increasing pulse time.

At first sight this result seemed to suggest the mechanism represented by Fig. 1D, which assumes the occurrence of breaks also in the parental strands. However, further studies revealed that it is due to the extreme instability (or absence) of the double-stranded structure of the newly replicated portion. As shown in Table 1, nearly half the 5-sec-pulse DNA was susceptible to *E. coli* exonuclease I, which is specific for single-stranded linear DNA. The percentage of radioactive DNA sensitive to this enzyme decreased with increasing pulse time. Approximately the same portions of the labeled DNA were eluted from hydroxylapatite at the relatively low phosphate concentration expected for single-stranded DNA and were found to be completely susceptible to the action of exonuclease I (Fig. 18, Table 1). The single-strand nature of the pulse-labeled DNA was also demonstrated by equilibrium density gradient sedimentation in the cesium sulfate-HgCl₂ system (Fig. 19). On the other hand, the component which sediments slowly in the neutral sucrose gradient (Fraction I) was found to be highly susceptible to exonuclease I, while the fast sedimenting material had a low

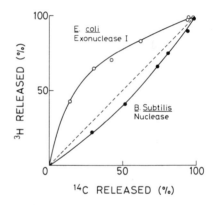

FIGURE 16. Hydrolysis of pulse-labeled short DNA chains with exonucleases. Pulse labeling was carried out as in Fig. 15 except that cells were grown in medium containing ¹⁴C-thymidine before starvation. DNA was extracted by the Thomas procedure and denatured in 0.1 N NaOH. Doubly labeled "10 S DNA" was obtained by sedimentation in alkaline sucrose gradients and purified by a second alkaline sucrose gradient sedimentation. For digestion with *E. coli* exonuclease I the 100-μliter reaction mixture contained: 10 μmoles of glycine-KOH, pH 9.3, 1.5 μmoles of MgCl₂ 0.5 μmole of 2-mercaptoethanol, 28 $\mu\mu$moles of DNA and 0.5 unit of enzyme (DEAE-cellulose fraction). For digestion with *B. subtilis* nuclease the 100-μliter reaction mixture contained: 10 μmoles of Tris-HCl, pH 8.7, 28 $\mu\mu$moles of DNA and 0.06 unit of enzyme (Fraction I-A). Incubation was at 37° and release of acid-soluble radioactivities was measured.

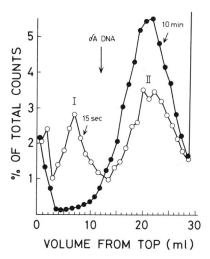

FIGURE 17. Neutral sucrose gradient sedimentation of pulse-labeled DNA from *E. coli* B. Cells were pulse labeled at 20° as in Fig. 3. Native DNA samples prepared by the Thomas procedure were sedimented in the SW 25.1 rotor for 15 hr at 20,500 rpm and 8° (Okazaki et al., 1968).

susceptibility to this enzyme (Table 1). Therefore, an appreciable fraction of the newly synthesized chains as isolated by the Thomas procedure is single-stranded, and this fraction sediments slowly in the neutral sucrose gradient. Essentially the same situation as in normal *E. coli* was found in the T4 systems. As shown in Table 2, appreciable portions of both the nascent short chains found in normal steady-state system and those accumulated at high temperature in cells infected with T4 *ts* A80 were susceptible to exonuclease I after extraction by the Thomas method.

DISCUSSION

Alkaline sucrose gradient sedimentation of pulse-labeled DNA supports the first prediction from the hypothesis of discontinuous DNA replication, that is, the discontinuity of the newly synthesized DNA chains. An alternative possibility that artificial breaks may be introduced selectively in the newly replicated region by some operation involved in the experiments is diminished by the fact that similar results are obtained using different procedures in a number of different systems (including an endonuclease-deficient *E. coli* strain).

The hypothesis is further supported and the above alternative possibility further diminished by a test of the second prediction that temporary inhibition of the enzyme for the formation of phosphodiester linkages between DNA chains should result in a marked accumulation of the nascent short chains. Experiments have been carried out using T4 phage mutants, which produce thermosensitive ligase. The result agrees perfectly with the prediction. It is highly unlikely, though not impossible, that the observed accumulation of the nascent short chains results from in vivo degradation of large DNA chains synthesized by a continuous mechanism, because the observation is made under conditions where little or no degradation of the prelabeled DNA is detectable. Furthermore, the average size of the accumulated nascent short chains is almost the same as, but slightly larger than the average size of the pulse-labeled short chains found in the normal steady-state system of T4. This is exactly what one would expect from the hypothesis.

Our experiments suggest that the direction of synthesis of the DNA chains in vivo is 5′ to 3′, in agreement with the assumption in the original

TABLE 1. SUSCEPTIBILITY OF PULSE-LABELED DNA TO *E. coli* EXONUCLEASE I PRIOR TO DENATURATION TREATMENT

	Extent of degradation by exonuclease I (%)				
		Hydroxylapatite fraction		Neutral sucrose gradient fraction	
Pulse time	Unfractionated	I	II	I	II
5 sec	45				
10 sec	32*			77*	24*
15 sec	30	96	24		
30 sec	24, 24*	96, 88*	18, 15*		
10 min	4, 0*	78	2		

E. coli B was pulse labeled as in Fig. 3. Extraction and fractionation of labeled DNA were carried out as in Fig. 17 and 18. SDS treatment for DNA extraction was at 37° (*) or at 60°. Hydroxylapatite Fractions I and II are shown in Fig. 18, and neutral sucrose gradient fractions I and II in Fig. 17.

For susceptibility to exonuclease the 150-μl reaction mixture, containing 10 μmoles glycine-KOH buffer, pH 9.2, 1 μmole MgCl₂, 0.15 μmole 2-mercaptoethanol, 60 μl DNA sample (300–12,000 count/min) and 3 units of *E. coli* exonuclease I (DEAE-cellulose fraction), was incubated at 37°. After 60 min, 3 units of enzyme were added to the mixture and the incubation was continued for another 60 min. Acid-soluble and insoluble counts were determined at 0, 60, and 120 min. More than 85% of the radioactive DNA degraded during the 120-min period was already acid soluble at 60 min (Okazaki et al., 1968).

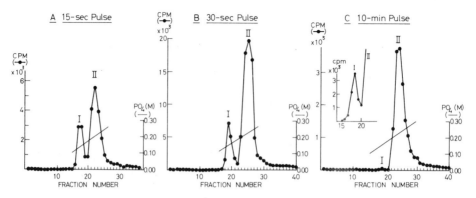

FIGURE 18. Hydroxylapatite chromatography of pulse-labeled DNA of *E. coli* B. The native DNA samples of Fig. 17 were dialyzed against 0.01 M potassium phosphate buffer, pH 6.8, and added to a column (1.4 × 10 cm) of hydroxylapatite, which had been equilibrated with 0.005 M potassium phosphate buffer, pH 6.8. Elution was achieved with a linear gradient of the same buffer (total volume of 140 ml) ranging from 0.01 to 0.7 M. 2.5 ml fractions were collected at 30-min intervals.

hypothesis. However, further thorough study is needed for a definitive conclusion on this crucial point.

The fact that an appreciable portion of the nascent short chains is isolated in the single-stranded form (also observed by Oishi [1968] and by Nagata and Meselson [pers. commun.]) is puzzling. One possibility is that the parental strands wind back to break the association between the nascent short DNA chains and the parental strands when cellular elements (replication apparatus), which attach to the replicating point and serve to stabilize its structure, are removed (Fig. 20). This would be possible, particularly if the replication along one of the parental strands is ahead of the replication along the other strand as in Fig. 20B.

Assuming that DNA in fact replicates by a discontinuous mechanism in vivo, one of the important questions to ask is how the discontinuity of the chain growth is controlled, that is, what is the mechanism of initiation and termination

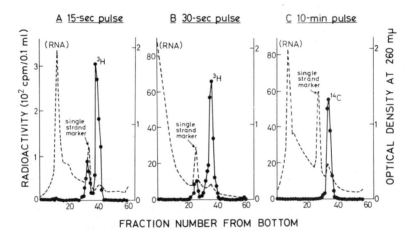

FIGURE 19. Equilibrium density gradient sedimentation of pulse-labeled *E. coli* DNA in the cesium sulfate-Hg²⁺ system. Pulse labeling and DNA extraction were carried out as in Fig. 17, except that ^{14}C-thymidine was used for the 10-min pulse. Heat denatured unlabeled *E. coli* DNA was added to the labeled DNA samples as a marker. Calculated quantities of Cs_2SO_4 solution (ρ = 1.99 g/cm³), borate buffer (0.1 M $Na_2B_4O_7$, pH 9.2), DNA solution (in 0.1 M Na_2SO_4, pH 9.2), $HgCl_2$ solution (10^{-3} M) and water were mixed, in this order, to give a final borate concentration of 5 mM, a molar ratio of Hg²⁺ to DNA phosphorus of (A and B) 0.19 or (C) 0.173 and an initial density of 1.52 g/cm³. Samples were centrifuged in a Spinco SW50 rotor at 45,000 rpm and 15° for 36 hr. - - - - -: Optical density at 260 mμ. ————: Radioactivity.

of synthesis of the 'replication unit'. An intriguing possibility is that there are initiation and/or termination signals on the template strands. The signals could be special base sequences, unusual bases, interruptions of phosphodiester linkages or some other structure. Our results suggest that the size of the unit of discontinuous synthesis corresponds to the estimated size of the cistron. This suggests that the replication unit might be the cistron and that the ends of the cistron might be the initiation and termination points in the discontinuous replication. Alternatively, the size of the segment synthesized by one continuous act of polymerase might be determined primarily by the nature of the replication apparatus. How such mechanisms can be reconciled with the properties of DNA polymerase remains to be elucidated in future in vivo and in vitro studies.

One merit of having a discontinuous mechanism of replication is that there could be multiple, simultaneous points of synthesis along one template strand. This would increase the overall rate of the chain growth. If discontinuous replication involves the production of breaks in the parental strands, it would also circumvent the difficulty of unwinding and winding of large chromosomal DNA.

The discontinuous mechanism may be a device evolved for replication of genomes which exceed a certain size. A small circular DNA such as ϕX 174 phage DNA may be synthesized by a continuous mechanism, although its replication is similar to the discontinuous replication of a large genome in that the covalent joining by polynucleotide ligase may be involved as suggested by in vitro experiments by Goulian and Kornberg (1967) and Goulian et al. (1967). RNA genomes are generally small in size (Fraenkel-Conrat, 1965), and no

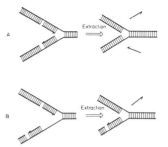

FIGURE 20. Possible models for 'peeling off' of the nascent short chains from the replicating point.

closed circular RNA genome is known. Some double-stranded RNA genomes appear to consist of several small fragments (Dunnebacke and Kleinschmidt, 1967; Watanabe and Graham, 1967; Bellamy et al., 1967; Miura et al., 1967). These facts might be correlated to the possible absence of 'RNA-joining enzyme'.

ACKNOWLEDGMENTS

This work was aided by the Research Funds of the Ministry of Education of Japan and grants from the Jane Coffin Childs Memorial Fund for Medical Research and the National Science Foundation.

REFERENCES

BECKER, A., G. LYN, M. GEFTER, and J. HURWITZ. 1967. The enzymatic repair of DNA. II. Characterization of phage-induced sealase. Proc. Nat. Acad. Sci. 58: 1996.

BELLAMY, A. R., L. SHAPIRO, J. T. AUGUST, and W. K. JOKLIK. 1967. Studies on reovirus RNA. I. Characterization of reovirus genome RNA. J. Mol. Biol. 29: 1.

BESSMAN, M. J., I. R. LEHMAN, J. ADLER, S. B. ZIMMERMAN, E. S. SIMMS, and A. KORNBERG. 1958. Enzymatic synthesis of deoxyribonucleic acid. III. The incorporation of pyrimidine and purine analogues into deoxyribonucleic acid. Proc. Nat. Acad. Sci. 44: 633.

BONHOEFFER, F., and A. GIERER. 1963. On the growth mechanism of the bacterial chromosome. J. Mol. Biol. 7: 534.

CAIRNS, J. 1963a. The bacterial chromosome and its manner of replication as seen by autoradiography. J. Mol. Biol. 6: 208.

———. 1963b. The chromosome of Escherichia coli. Cold Spring Harbor Symp. Quant. Biol. 28: 43.

COZZARELLI, N. R., N. E. MELECHEN, T. M. JOVIN, and A. KORNBERG. 1967. Polynucleotide cellulose as a substrate for a polynucleotide ligase induced by phage T4. Biochem. Biophys. Res. Commun. 28: 578.

DUNNEBACKE, T. H., and A. KLEINSCHMIDT. 1967. Ribonucleic acid from reovirus as seen in protein monolayers by electron microscopy. Z. Naturforsch. 22B: 159.

DÜRWALD, H., and H. HOFFMANN-BERLING. 1968. Endonuclease I-deficient and ribonuclease I-deficient Escherichia coli mutants. J. Mol. Biol. 34: 331.

TABLE 2. SUSCEPTIBILITY OF PULSE-LABELED T4 DNA TO EXONUCLEASE I

A. T4D(Wild-type): 20°

Pulse time	Exonuclease I-sensitive ³H-DNA % of total	count/min/ml culture
7 sec	46	2.7×10^3
15 sec	36	4.1×10^3
30 sec	28	6.6×10^3
60 sec	18	9.4×10^3

B. T4 ts A80 (Gene 30): 44°

Pulse time	Exonuclease I-sensitive ³H-DNA % of total	count/min/ml culture
10 sec	22	$1,1 \times 10^4$
20 sec	15	1.8×10^4
30 sec	13	1.9×10^4
60 sec	9	1.8×10^4

Pulse labeling was done (A) as in Fig. 2 or (B) as in Fig. 9. DNA was extracted by the Thomas procedure, and susceptibility to exonuclease I was measured as in Table 1.

FAREED, G. C. and C. C. RICHARDSON. 1967. Enzymatic breakage and joining of deoxyribonucleic acid. II. The structural gene for polynucleotide ligase in bacteriophage T4. Proc. Nat. Acad. Sci. *58:* 665.

FRAENKEL-CONRAT, H. 1965. Structure and function of virus proteins and of viral nucleic acid, p. 99–151. *In* H. Neurath [ed.] The proteins, v. *3.* Academic Press, New York.

GEFTER, M. L., A. BECKER, and J. HURWITZ. 1967. The enzymatic repair of DNA. I. Formation of circular DNA. Proc. Nat. Acad. Sci. *58:* 240.

GELLERT. M., 1967. Formation of covalent circles of lambda DNA by *E. coli* extracts. Proc. Nat. Acad. Sci. *57:* 148.

GOULIAN M. and A. KORNBERG. 1967. Enzymatic synthesis of DNA. XXIII. Synthesis of circular replicative form of phage ϕX174 DNA. Proc. Nat. Acad. Sci. *58:* 1723.

GOULIAN, M., A. KORNBERG, and R. L. SINSHEIMER. 1967. Enzymatic synthesis of DNA. XXIV. Synthesis of infectious phage ϕX174 DNA. Proc. Nat. Acad. Sci. *58:* 2321.

HANAWALT, P. C., O. MAALØE, D. J. CUMMINGS and M. SCHAECHTER. 1961. The normal DNA replication cycle. II. J. Mol. Biol. *3:* 156.

JOSSE, J., A. D. KAISER and A. KORNBERG. 1961. Enzymatic synthesis of deoxyribonucleic acid. VIII. Frequencies of nearest neighbor base sequences in deoxyribonucleic acid. J. Biol. Chem. *234:* 864.

KORNBERG, A. 1960. Biologic synthesis of deoxyribonucleic acid. Science *131:* 1503.

LARK, K. G., T. REPKO and E. J. HOFFMAN. 1963. The effect of amino acid deprivation on subsequent deoxyribonucleic acid replication. Biochem. Biophys. Acta *76:* 9.

LEHMAN, I. R. 1960. The deoxyribonuclease of *Escherichia coli*. I. Purification and properties of a phosphodiesterase. J. Biol. Chem. *235:* 1479.

LEHMAN, I. R., S. B. ZIMMERMAN, J. ADLER, M. J. BESSMAN, E. S. SIMMS, and A. KORNBERG. 1958. Enzymatic synthesis of deoxyribonucleic acid. V. Chemical composition of enzymatically synthesized deoxyribonucleic acid. Proc. Nat. Acad. Sci. *44:* 1191.

MAALØE, O., and P. C. HANAWALT. 1961. Thymine deficiency and the normal DNA replication cycle. I. J. Mol. Biol. *3:* 144.

MESELSON, M., and F. W. STAHL. 1958. The replication of DNA in *Escherichia coli*. Proc. Nat. Acad. Sci. *44:* 671.

MITRA, S., and A. KORNBERG. 1966. Enzymatic mechanisms of DNA replication. J. Gen. Physiol. *49:* 59.

MIURA, K., I. FUJII, T. SAKAKI, M. FUKE, and S. KAWASE. 1968. Double-stranded RNA from cytoplasmic-polyhedrosis virus of silk-worm. J. Virol. *2:* 1211.

NAGATA, T. 1963. The molecular synchrony and sequential replication of DNA in *Escherichia coli*. Proc. Nat. Acad. Sci. *49:* 551.

NOMURA, M., K. MATSUBARA, K. OKAMOTO and R. FUJIMURA. 1962. Inhibition of host nucleic acid and protein synthesis by bacteriophage T4: Its relation to the physical and functional integrity of host chromosome. J. Mol. Biol. *5:* 535.

OISHI, M. 1968. Studies of DNA replication in vivo. I. Isolation of the first intermediate of DNA replication in bacteria as single-stranded DNA. Proc. Nat. Acad. Sci. *60:* 329.

OKAZAKI, R., T. OKAZAKI and K. SAKABE. 1966. An extracellular nuclease of *Bacillus subtilis:* Some novel properties as a DNA exonuclease. Biochem. Biophys. Res. Commun. *22:* 611.

OKAZAKI, R., T. OKAZAKI, K. SAKABE, and K. SUGIMOTO. 1967. Mechanism of DNA replication: Possible discontinuity of DNA chain growth. Jap. J. Med. Sci. Biol. *20:* 255.

OKAZAKI, R., T. OKAZAKI, K. SAKABE, K. SUGIMOTO, and A. SUGINO. 1968. Mechanism of DNA chain growth. I. Possible discontinuity and unusual secondary structure of newly synthesized chains. Proc. Nat. Acad. Sci. *59:* 598.

OKAZAKI, R., K. SAKABE, and T. OKAZAKI. 1966. Replication of chromosomal DNA. Paper presented at 13th Symp. on Nucleic Acids. Tanpakushitsu kakusan koso: *11:* 610–611 (Abstract).

OKAZAKI, T., and A. KORNBERG. 1964. Enzymatic synthesis of deoxyribonucleic acid. XV. Purification and properties of a polymerase from *Bacillus subtilis*. J. Biol. Chem. *239:* 259.

OLIVERA, B. M., and I. R. LEHMAN. 1967. Linkage of polynucleotides through phosphodiester bonds by an enzyme from *Escherichia coli*. Proc. Nat. Acad. Sci. *57:* 1426.

SAKABE, K., and R. OKAZAKI. 1966. A unique property of the replicating region of chromosomal DNA. Biochim. Biophys. Acta *129:* 651.

SUGIMOTO, K., T. OKAZAKI and R. OKAZAKI. 1968. Mechanism of DNA chain growth. II. Accumulation of newly synthesized short chains in *Escherichia coli* infected with ligase-defective T4 phages. Proc. Nat. Acad. Sci. *60:* in press.

THOMAS, C. A., Jr., K. I. BERNS, and T. J. KELLY, Jr. 1966. Isolation of high molecular weight DNA from bacteria and cell nuclei, p. 535–540. *In* G. L. Cantoni and D. R. Davies [ed.] Procedures in nucleic acid research. Harper and Row, New York and London.

WATANABE, Y. and A. GRAHAM. 1967. Structural units of reovirus RNA and their possible functional significance. J. Virology *1:* 665.

WATSON, J. D., and F. H. C. CRICK. 1953a. Genetical implications of the structure of deoxyribonucleic acid. Nature *171:* 964.

——, ——. 1953b. The structure of DNA. Cold Spring Harbor Symp. Quant. Biol. *18:* 123.

WEISS, B. and C. C. RICHARDSON. 1967. Enzymatic breakage and joining of deoxyribonucleic acid. I. Repair of single-strand breaks in DNA by an enzyme system from *Escherichia coli* infected with T4 bacteriophage. Proc. Nat. Acad. Sci. *57:* 1021.

YOSHIKAWA, H. and N. SUEOKA. 1963a. Sequential replication of *Bacillus subtilis* chromosome. I. Comparison of marker frequencies in exponential and stationary growth phases. Proc. Nat. Acad. Sci. *49:* 559.

——, ——. 1963b. Sequential replication of the *Bacillus subtilis* chromosome. II. Isotopic transfer experiments. Proc. Nat. Acad. Sci. *49:* 806.

DISCUSSION

(Presented with reference to all the papers on Okazaki's model—Englund et al., Okazaki et al., Kidson, Richardson et al., Newman and Hanawalt, and Sadowski et al.)

W. R. GUILD: My question is whether any of today's speakers have critically examined their data in terms of a model which is slightly different from those discussed so far, and yet which seems very plausible, particularly in view of Dr. Goulian's report that *E. coli* polymerase does *not* self-initiate on a single strand without a primer to build on. The model is simply that the growing 3'OH end turns the corner, as reported for in vitro synthesis by Dr. Kornberg's group about four years ago, and

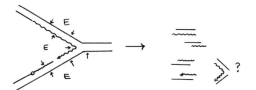

FIGURE 2. Postulated effect of excess cuts by endonuclease (→) and a few of the possible products if extraction occurred before ligase sealed the cuts.

thought then to be an artifact of the in vitro system. All that is needed for this system to work in vivo and to generate much of today's data is to have ligase and an endonuclease (see figures). The result is in a sense a combination of Okazaki's Models B and C, but with a quite different connotation; that is, it requires no free initiation without a 3'OH to build on.

Note a few of the predictions to be tested:

(a) If only one endonuclease cut is made, in the apex of the Y, then half of the new DNA would be of high molecular weight, but probably very heterogeneous, and therefore perhaps hard to detect reliably above background for very short pulses. How good are the data?

(b) If, to insure sufficient cuts to maintain propagation of the Y joint, the endonuclease were in excess, making a number of cuts near the Y joint, then *after* the fact of synthesis, both new strands would be present in the low molecular weight region.

(c) If (b) is true, hybridization tests of the new material are definitive only if the result is asymmetric.

(d) If (b) is true, we may expect cuts in the parental strands also, generating an apparent result like Okazaki's Model D. There seems to be disagreement about the single vs. double strand nature of the pieces. Has there been a critical test for presence of parental DNA in these small pieces, and if they are double, is not one of the strands likely to be parental?

If there are arguments against this model, we need to hear them discussed.

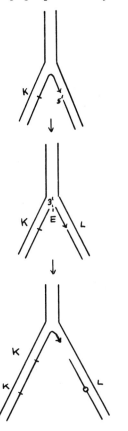

FIGURE 1. Model for propagation of Y-joint without requiring initiation except on 3' OH end.
K, L, E- polymerase, ligase, endonuclease, respectively.
— site of past initiation by polymerase on 3'OH end.
○ site joined by ligase.

Isolation of an *E. coli* Strain with a Mutation affecting DNA Polymerase

by

PAULA DE LUCIA
JOHN CAIRNS

Cold Spring Harbor Laboratory,
Cold Spring Harbor,
New York 11724

By testing indiscriminately several thousand colonies of mutagenized *E. coli*, a mutant has been isolated that on extraction proves to have less than 1 per cent of the normal level of DNA polymerase. The mutant multiplies normally but has acquired an increased sensitivity to ultraviolet light.

KORNBERG's discovery of an enzyme that could faithfully copy DNA *in vitro*[1] was a crucial step in the history of molecular biology because it firmly established the fact that only a small part of a cell's DNA is needed to code for a mechanism that can duplicate the whole. Whether this is the enzyme responsible for DNA duplication *in vivo* was rightly thought, at that time, to be of secondary importance. Since then, however, circumstantial evidence has accumulated suggesting that, at least in bacteria, this particular enzyme is used for the repair of DNA rather than for its duplication. The various mutants of *Escherichia coli* and *Bacillus subtilis* that are unable to duplicate their DNA at high temperature have all been shown to contain normal polymerase and normal deoxyribonucleoside triphosphate pools at the non-permissive temperature[2-6], and at least one of them has been shown to carry out repair synthesis at high temperature[7]. Repair replication and the process of DNA duplication apparently differ in the extent to which they discriminate against 5-bromouracil as an acceptable substitute for thymine, suggesting that the two reactions involve different polymerases[8]. Finally, the 5'-exonucleolytic activity, recently shown to be an intrinsic property of the *E. coli* polymerase[9], is clearly a desirable attribute for an enzyme responsible for excision and repair but is of no obvious advantage for an enzyme carrying out semiconservative replication.

These and other less persuasive arguments prompted us to look for mutants of the polymerase, in the hope that they would either establish a role for the polymerase in DNA duplication or exclude it and, at the same time, provide convenient strains in which to search for the right enzyme. Although we have not succeeded in these more distant objectives, we have isolated such a mutant and here describe the method of isolation and some of its properties. The accompanying article describes a genetic study of the mutation.

The Selective Procedure

The successful isolation of mutants of *E. coli* lacking ribonuclease I[10] demonstrated that it is possible to find the mutant one wants simply by testing individually several hundred colonies grown from a heavily mutagenized stock. Because we wished to avoid having to guess what symptoms, if any, would result from a lack of DNA polymerase, we decided to follow that example and assay the polymerase in clones of a mutagenized stock until we found what we were looking for. We had to allow for the possibility that the mutation we sought might be a conditional lethal, so we began by assaying at 45° extracts made from clones grown at 25° or 30°; later we tested clones grown at 37°, thinking that temperature-sensitive mutants of the polymerase might be more readily detectable if the enzyme had been assembled at a higher temperature. As it turned out, the mutant we eventually isolated would have been found whatever approach had been adopted, and we shall therefore simply give the history of the mutant when we describe its isolation and properties.

Extraction of Polymerase

Because we expected to have to test many hundred colonies, we required a very simple method for preparing extracts. In addition, we needed a procedure which made the bacteria incapable of incorporating deoxyribonucleosides, to ensure that labelled triphosphates could not enter DNA by way of breakdown to nucleosides and incorporation by those few cells that might have survived the extraction procedure. These two requirements were satisfied by the slight modification of a method devised for extracting polysomes, using the non-ionic detergent Brij-58 (ref. 11). *E. coli* is suspended at a concentration of about 3×10^9/ml. in ice cold 10 per cent sucrose 0·1 M Tris (pH 8·5); lysozyme and EDTA are added to final concentrations of 50 μg/ml. and 0·005 M, respectively, and the mixture is kept on ice for 30 min; addition of a mixture of Brij and MgSO$_4$ (at room temperature) to give final concentrations of 5 per cent and 0·05 M, respectively, results in partial clearing; following centrifugation (1,500g for 30 min), the deposit contains 99·9 per cent of the DNA and the supernatant contains the polymerase, which may then be assayed simply by adding sonicated calf thymus DNA (to 50 μg/ml.) and the four deoxyribonucleoside triphosphates (to a final concentration of 4 nmoles/ml. dATP, dGTP, dCTP and 2 nmoles/ml. ³H-TTP).

This extraction procedure demonstrates one point of interest: any method of lysis that liberates fragmented DNA will automatically create sites for the attachment of polymerase and therefore cannot give a true picture of the location of the polymerase *in vivo*[12]. Extraction with Brij yields cells which still contain their DNA but, on resuspension, have little if any ability to incorporate deoxyribonucleoside triphosphates. Because Brij apparently does not dissociate polymerase from its template (the polymerase being assayable in the presence of Brij), we can conclude that most of the polymerase in *E. coli* is normally not attached to DNA but lies free within the cell—as might befit an enzyme awaiting the summons to repair synthesis. This conclusion is supported by the observation that when *E. coli* segregates daughter cells which lack DNA these cells nevertheless retain their full quota of DNA polymerase[13,14].

Isolation of the Mutant

E. coli W3110 *thy⁻*, growing in minimal medium, was washed and suspended in 0·15 M acetate (pH 5·5), treated with N-methyl-N′-nitro-N-nitrosoguanidine (1 mg/ml.) for 30 min, and then centrifuged and suspended in Penassay broth[15]. Following growth at 25° C for 18 h, the culture was plated; after incubation overnight at 37° C, the colonies were picked into 1 ml. lots of Penassay broth which were incubated overnight at 37° C and then centrifuged and extracted with lysozyme and Brij.

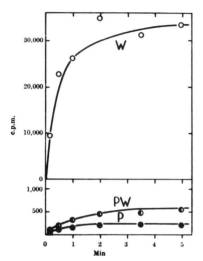

Fig. 1. Triphosphate incorporation by extracts of the parent strain (W), the mutant strain (P) and a mixture of 99 per cent mutant and 1 per cent parent (PW). *E. coli* W3110 *thy⁻* and the mutant derivative, p3478, were grown with aeration in Penassay broth at 37° to about 5 × 10⁸/ml. Each culture was then chilled, centrifuged and suspended in 0·1 M Tris–0·01 M MgSO₄ (pH 7·4) at a concentration of 6 × 10⁹/ml. A mixture of 1 per cent parent strain, 99 per cent mutant strain was prepared. This and the two unmixed suspensions were centrifuged and suspended in Tris-Mg²⁺ at a concentration of 1 × 10¹⁰/ml. The three suspensions were disrupted by sonic vibration and mildly centrifuged (1,000*g* for 10 min). To 0·9 ml. of each supernatant at 25° 0·1 ml. sonicated calf thymus DNA was added (final concentration 20 µg/ml.) and, 5 min later, 0·3 ml. triphosphate solution (final concentrations 100 nmoles/ml. dGTP, dATP, dCTP, and 0·6 nmoles/ml., 2·5 µCi/ml., ³H-TTP). Samples of 0·2 ml. were taken from each reaction mixture into 5 ml. 5 per cent trichloroacetic acid–1 per cent sodium pyrophosphate[16]. These samples were then washed on Whatman GFA filters with 5 per cent trichloroacetic acid and with 5 per cent acetic acid, dried and counted in a scintillation counter.

After testing a few thousand colonies in this way we found a clone, p3478, that appeared to lack polymerase activity. It was therefore tested again using a more conventional method for extracting the enzyme. According to this test (Fig. 1), extracts of the mutant have 0·5–1·0 per cent of the normal activity. This decrease in activity does not seem to arise from the presence of an inhibitor.

Some Properties of the Mutant

As far as we can determine, the mutant multiplies at the same rate as the parent strain, in minimal and complete media, and at temperatures from 25° to 42° C. On plating, it forms slightly smaller colonies than those of the parent strain, and occasionally it seems to have difficulty in getting out of stationary phase, but we have not investigated further either of these phenomena.

Parent and mutant are equally susceptible to infection with T4, T5, T7 and λ bacteriophages. When converted to spheroplasts, they are equally susceptible to φX174 DNA and produce equal yields of phage (personal communication from David Dressler). This finding was somewhat surprising, but it should be remembered that all stages in the replication of φX174 DNA are temperature sensitive in a mutant that is temperature sensitive for normal DNA replication[17] but not for repair synthesis[7].

With regard to host cell reactivation, there is no detectable increase in the rate of inactivation of T7 by ultraviolet light, when the survivors are assayed on the mutant rather than the parent. Thus the mutant is *hcr⁺*.

The mutant has a marked increase in sensitivity to ultraviolet light. For convenience, this effect will be documented in the following article[18], where the sensitivities of various derivative strains are compared.

The parent strain will form colonies normally in the presence of 0·04 per cent methylmethanesulphonate, whereas the mutant plates with an efficiency of about 10⁻⁷. We assume that these rare methylmethane-sulphonate-resistant cells are revertants that have either arisen spontaneously or been created by the methylmethanesulphonate. Because every one of twenty such independently arising revertants exhibited normal sensitivity to ultraviolet light and had normal or near-normal levels of polymerase, it is clear that the three basic properties of the mutant (UVs, MMSs and lack of polymerase) are the result of a single mutational step.

Repair or Replication

The accompanying article[18] demonstrates that we are dealing with an amber mutation which is recessive in partial diploids. We assume that it is in the gene coding for DNA polymerase, although proof will require the demonstration that it—or other similar mutations—results in changes in the polymerase protein. Because the mutation produces an increased sensitivity to ultraviolet light, it seems likely that recovery from the effects of ultraviolet light is partly the responsibility of this polymerase.

Unfortunately, it is not going to be easy, by a study of this or other such mutants, to show that this polymerase plays no part in normal DNA duplication. Because *E. coli* contains several hundred polymerase molecules per bacterium[19], the residual activity found in extracts of our mutant could represent 5–10 molecules per cell—a number that could well be sufficient for normal duplication. Even if we could somehow prove that the residual activity were entirely that of another enzyme (in other words, that this amber mutation is not measurably leaky), we should still not have proved that duplication is carried out by some other enzyme, for it could readily be argued that those few polymerase molecules concerned with duplication are necessarily incorporated into some larger enzyme complex the activity of which is not assayable *in vitro*. It could even be argued that more of the polymerase molecule must be intact for it to serve as a repair enzyme (and, incidentally, to survive extraction) than for it to act when part of the replicating machinery. We therefore believe that the question will be resolved either by engineering a total deletion of the polymerase gene or by determining, in some direct manner, which enzymes and what precursors are used for normal DNA duplication. It is our hope that each of these exercises will have been made easier now that the polymerase gene has probably been located[18] and a mutant is generally available.

We thank Dr Raymond Gesteland (who pioneered this kind of mutant hunt) for encouragement; Dr David Dressler for testing our mutant with φX174 and for permission to cite his results; and Drs Julian and Marilyn Gross for arranging to stay on at Cold Spring Harbor to conduct most of the experiments reported in the next article.

The work was supported by a grant from the US National Science Foundation.

Received November 26, 1969.

[1] Lehman, I. R., Bessman, M. J., Simms, E. S., and Kornberg, A., *J. Biol. Chem.*, **233**, 163 (1958).

[2] Bonhoeffer, F., *Z. Vererbungslehre*, **98**, 141 (1966).

[3] Buttin, G., and Wright, M., *Cold Spring Harbor Symp. Quant. Biol.*, **33**, 259 (1968).

[4] Fangman, W. L., and Novick, A., *Genetics*, **60**, 1 (1968).

[5] Gross, J. D., Karamata, D., and Hempstead, P. G., *Cold Spring Harbor Symp. Quant. Biol.*, **33**, 307 (1968).

[6] Hirota, Y., Ryter, A., and Jacob, F., *Cold Spring Harbor Symp. Quant. Biol.*, **33**, 677 (1968).

[7] Couch, J., and Hanawalt, P. C., *Biochem. Biophys. Res. Commun.*, **29**, 779 (1967).

[8] Kanner, L., and Hanawalt, P. C., *Biochim. Biophys. Acta*, **157**, 532 (1968).

[9] Kornberg, A., *Science*, **163**, 1410 (1969).

[10] Gesteland, R. F., *J. Mol. Biol.*, **16**, 67 (1966).

[11] Godson, G. N., and Sinsheimer, R. L., *Biochim. Biophys. Acta*, **149**, 476 (1967).

[12] Billen, D., *Biochim. Biophys. Acta*, **68**, 342 (1963).

[13] Cohen, A., Fisher, W. D., Curtiss, R., and Adler, H. I., *Cold Spring Harbor Symp. Quant. Biol.*, **33**, 635 (1968).

[14] Hirota, Y., Jacob, F., Ryter, A., Buttin, G., and Nakai, T., *J. Mol. Biol.*, **35**, 175 (1968).

[15] Adelberg, E. A., Mandel, M., and Chien Ching Chen, G., *Biochem. Biophys. Res. Commun.*, **18**, 788 (1965).

[16] Hurwitz, J., Gold, M., and Anders, M., *J. Biol. Chem.*, **239**, 3462 (1964).

[17] Steinberg, R. A., and Denhardt, D. T., *J. Mol. Biol.*, **37**, 525 (1968).

[18] Gross, J. D., and Gross, M., *Nature*, **224**, 1166 (1969).

[19] Richardson, C. C., Schildkraut, C. L., Aposhian, H. V., and Kornberg, A., *J. Biol. Chem.*, **239**, 222 (1964).

6

Reprinted from *Proc. Natl. Acad. Sci.*, **68**(12), 3150–3153 (1971)

Analysis of DNA Polymerases II and III in Mutants of *Escherichia coli* Thermosensitive for DNA Synthesis

(*pol*A1 mutants/phosphocellulose chromatography/*dna*E locus)

MALCOLM L. GEFTER, YUKINORI HIROTA*, THOMAS KORNBERG, JAMES A. WECHSLER, AND C. BARNOUX*

Department of Biological Sciences, Columbia University, New York, N.Y. 10027; and * Service de Génétique Cellulaire de l'Institut Pasteur, Paris

Communicated by Cyrus Levinthal, October 18, 1971

ABSTRACT A series of double mutants carrying one of the thermosensitive mutations for DNA synthesis (*dna*A, B, C, D, E, F, and G) and the *pol*A1 mutation of DeLucia and Cairns, were constructed. Enzyme activities of DNA Polymerases II and III were measured in each mutant. DNA Polymerase II activity was normal in all strains tested. DNA Polymerase III activity is thermosensitive specifically in those strains having thermosensitive mutations at the *dna*E locus. From these results we conclude that DNA Polymerases II and III are independent enzymes and that DNA Polymerase III is an enzyme required for DNA replication in *Escherichia coli*.

The isolation by DeLucia and Cairns (1) of an *Escherichia coli* mutant that lacks DNA Polymerase I activity (*pol*A1) has prompted many investigations into the nature of the DNA synthetic capacity of such strains. The purification and characterization of DNA Polymerase II has been reported by ourselves (2) and others (3, 4). In addition, we have reported the existence of a third DNA polymerase in *E. coli* (DNA Polymerase III) (2). A physiological function for these enzymes has not been determined.

The viability of cells devoid of measurable DNA Polymerase I activity suggests that this enzyme is not an obligatory component of the DNA replication machinery of *E. coli*. To determine whether polymerases II and III are essential for replication, we examined the DNA polymerases of *E. coli* mutants that were temperature-sensitive for DNA replication in an attempt to correlate the genetic lesions with altered DNA polymerase activity *in vitro*. We will present evidence indicating that DNA Polymerase III is the product of an essential gene mapping at the *dna*E locus.

MATERIALS AND METHODS

The following bacterial strains were used†:

(*1*) CRT4637: F⁻ *thr⁻ leu⁻ his⁻ str*ʳ *malA mtl⁻ thi⁻ dna*AT46

(*2*) CRT2667: F⁻ *his⁻ str*ʳ *malA thi⁻ pol*A1 *sup⁻ dna*BT266

(*3*) BT1029: H560 *thy⁻ endo*I⁻ *pol*A1 *dna*B

(*4*) PC22: F⁻ *his⁻ str*ʳ *malA xyl⁻ arg⁻ mtl⁻ thi⁻ pol*A1 *sup⁻ dna*C2

† In the text, these strains will be referred to by their number in the above list, followed by the *dna* mutation designation in parenthesis.

(*5*) PC79: F⁻ *his⁻ str*ʳ *malA xyl⁻ mtl⁻ thi⁻ pol*A1 *sup⁻ dna*D7

(*6*) E5111: F⁻ *his⁻ str*ʳ *malA xyl⁻ mtl⁻ arg⁻ thi⁻ sup⁻ pol*A1 *dna*E511

(*7*) E4860: F⁻ *his⁻ str*ʳ *malA xyl⁻ mtl⁻ arg⁻ thi⁻ sup⁻ dna*E486

(*8*) E4868: F⁻ *his⁻ str*ʳ *malA xyl⁻ mtl⁻ arg⁻ thi⁻ sup⁻ pol*A1 *dna*E486

(*9*) BT1026: H560 *thy⁻ endo*I⁻ *pol*A1 *dna*E

(*10*) BT1040: H560 *thy⁻ endo*I⁻ *pol*A1 *dna*E

(*11*) E1011: F⁻ *his⁻ str*ʳ *malA xyl⁻ mtl⁻ arg⁻ thi⁻ sup⁻ pol*A1 *dna*F101

(*12*) JW207: *thy⁻ rha⁻ str*ʳ *pol*A1 *dna*F101

(*13*) NY73: *leu⁻ thy⁻ metE rif*ʳ *str*ʳ *pol*A1 *dna*G3

(*14*) CRT2668: F⁻ B1⁻ *his⁻ malA str*ʳ *sup⁻ pol*A1 *dna⁺*

(*15*) JG112: W3110 *thy⁻ rha⁻ lac⁻ sup⁻ pol*A1 *dna⁺*

The isolation of the double mutant *dna*B *pol*A1, was described (5). A further series of *dna–pol*A1 double mutants, *dna*C, D, E, and F, with *pol*A1, were constructed through two successive steps. (*i*) Each thermosensitive mutation was introduced into an Hfr strain (HfrP4x8: an Hfr that injects its chromosome in the order, O-*proA-leu—lac*-F, or Hfr-Cavalli: an Hfr that injects its chromosome in the order, O-*lac-leu—gal*-F) by crossing the Hfr with an F-strain carrying a thermosensitive mutation affecting DNA synthesis. *lac*y (The site of F integration of HfrP4x8 is near *lac*y) (6) and *gal* (the site of F integration of Hfr Cavalli is near *gal*) (6) are used for selection. (*ii*) Each thermosensitive Hfr strain isolated was then crossed with an F⁻ strain, PA33612; F⁻ *arg⁻ his⁻ thi⁻ leu⁻ malA⁻ xyl⁻ mtl⁻ pol*A1⁻ *sup⁻*, using a closely linked marker (*leu* for *dna*E, C, and D and *his* for *dna*F) for selection. Recombinants were then tested for both *pol*A1, *dna*, and *sup*.

As controls, thermoresistant *dna⁺pol*A1⁻ were constructed by selection at a high temperature, either after P1-transduction of the thermosensitive allele, *dna⁺*, or spontaneous occurrence of revertants from the double mutants. Strain JW 207 was isolated after bacteriophage P1 transduction of *dna*F101 from strain E101 (7) into W3110 *thy⁻ rha⁻ pol*A1 *pur*F, by selection for *pur*F⁺. Strain NY73 was isolated by introduction of *pol*A1 into PC3 with JG78 (Hfr *R1*, *metE rha⁺ pol*A1 Rifʳ) (Peacey, M., and J. D. Gross, unpublished data).

The isolation and mapping of thermosensitive mutants have been reported by others (5, 7–14). Strains PC2:*dna*C2 and PCF *dna*D7 were a gift from Dr. P. Carl. Strains 1026, 1040 (*dna*E), and 1029 (*dna*B) were a gift from Dr. F. Bonhoeffer (classification of *dna* lesion was by co-transduction, ref. 11).

The materials used for purification and assay of DNA Polymerases II and III were described. [³H]TTP (2 × 10⁵ cpm/nmol) was used throughout to assay DNA polymerase activity.

Cells were grown in three-times concentrated L.B. broth (Difco) (15) at 25°C with aeration, and harvested in mid-log phase at 3–4 × 10⁹ cells/ml. Cell-free extracts (10 g of cells) and the S100 fraction were prepared as described (2). Separation of DNA Polymerases II and III (see Fig. 1*A*) was also described, except that all volumes and column dimensions were scaled down 10-fold and preliminary dialysis and batch elution from phosphocellulose (step II) were omitted. For the addition of large amounts of DNA Polymerase III to reaction mixtures, the enzyme activity that eluted from phosphocellulose was concentrated 10-fold by precipitation with ammonium sulfate.

Assays of rates of reaction at 30 and 45°C were done by first equilibrating the assay mixture (0.9 ml) at the appropriate temperature. The reaction was begun by the addition of enzyme. 0.2-ml aliquots were withdrawn at various times and pipetted into 1.0 ml of 5% trichloroacetic acid. The acid-insoluble material was collected on a Millipore filter and the radioactivity was determined in a liquid scintillation counter.

RESULTS

The results of a typical isolation of DNA Polymerase II and III are shown in Fig. 1*A*. Polymerases II and III are distinguished on the basis of chromatographic behavior; Polymerase III elutes at 0.1 M PO₄³⁻ (fraction 17) and Polymerase II at 0.2 M PO₄³⁻ (fraction 37). The two enzymes can further be distinguished on the basis of their response to ionic strength (2). In two instances (strains 1 and 7), Polymerases II and III were isolated from cells containing the normal amount of DNA Polymerase I. The result of phosphocellulose chromatography of extracts from such cells is shown in Fig. 1*B*. Although DNA Polymerase III is not completely resolved from Polymerase I activity (fraction 20), the activity of Polymerase III can be uniquely determined by assay of column fractions in the presence of either *N*-ethylmaleimide (dotted line, Fig. 1*B*) or antiserum to DNA Polymerase I. Since Polymerase III activity is completely abolished in the presence of *N*-ethylmaleimide, and is unaffected by antiserum directed against DNA Polymerase I (2), it is possible to obtain preparations active only due to Polymerase III despite the presence of Polymerase I. DNA Polymerase II is obtained in normal yield from *pol*⁺ strains; it is completely resolved from Polymerase I by phosphocellulose chromatography.

The peak fraction of each polymerase activity was used directly for measurements of the rate of synthesis at 30 and 45°C. The rate of reaction catalyzed by Polymerase II was 1.8-times faster at 45°C than at 30°C; the rate of the Polymerase III reaction was 1.5-times faster at 45°C than at 30°C. DNA Polymerase III activity is not linear with time after 5 min at 45°C and, therefore, relative rates were calculated only from the initial slopes.

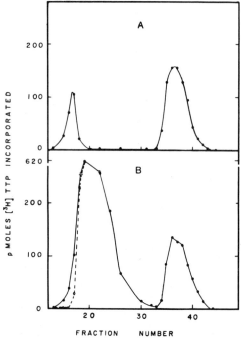

FIG. 1.   Separation of DNA Polymerases II and III by phosphocellulose chromatography. Cell-free extracts from *pol*A1 strains (*A*) and *pol*⁺ strains (*B*) were prepared and subjected to phosphocellulose chromatography. Each fraction was assayed for DNA polymerase activity. Results obtained by assay of fractions (12–19) in the presence of *N*-ethylmaleimide are shown by the *dotted line*. Polymerase II and III activities elute at fractions 35–45 and 12–20, respectively.

The results of a typical analysis of DNA Polymerase II activity at 30 and 45°C are shown in Fig. 2. Fig. 2*A* represents the Polymerase II activity isolated from strain 13 (*dna*G) and 2*B* the results from strain 9 (*dna*E). On the basis of these analyses, DNA Polymerase II activity appeared normal in all strains tested. The results of these relative rate measurements are summarized in Table 1.

The results of a typical analysis of wild-type DNA Polymerase III activity are shown in Fig. 3*A*. These results were obtained with DNA Polymerase III isolated from strain 13 (*dna*G) and are representative of all enzyme preparations tested except for those isolated from strains 6, 9, and 10 (carrying *dna*E mutations). These results are also summarized in Table 1.

The results obtained for DNA Polymerase III activity isolated from strain 9 (*dna*E) are shown in Fig. 3*B*. (Polymerase III activity isolated from strain 10 (*dna*E) behaves essentially the same way.) In contrast to the rate observed with a normal enzyme, the rate of synthesis at 45°C with enzyme preparations from the *dna*E mutants 9 and 10 was undetectable. To rule out the possibility of a temperature-dependent inhibitor present in these preparations, concentrated Polymerase III preparations (see *Methods*) from strains 9 (*dna*E) and 13 (*dna*G) were mixed and the rates at

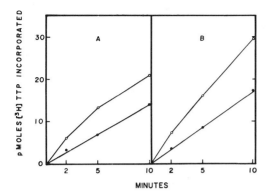

FIG. 2. Effect of temperature on Polymerase II-catalyzed synthesis. The rate of polymerization at 30°C (—●—) and 45°C (—○—) was measured. The results obtained for enzymes isolated from strain 13 (*dna*G) and strain 9 (*dna*E) are shown in parts *A* and *B*, respectively.

TABLE 1.   *Effect of temperature on the rate of reaction*

	Strain	Reaction rate $\dfrac{45°C}{30°C}$	
		DNA Polymerase II	DNA Polymerase III
1	(*dna*A)	1.85	1.45
2	(*dna*B)	1.73	1.74
3	(*dna*B)	1.95	1.55
4	(*dna*C)	1.86	1.20
5	(*dna*D)	1.99	1.47
6	(*dna*E)	1.87	1.0
7	(*dna*E)	1.70	—*
8	(*dna*E)	2.00	—*
9	(*dna*E)	1.95	<0.1
10	(*dna*E)	1.70	<0.1
11	(*dna*F)	1.55	1.33
12	(*dna*F)	1.77	1.43
13	(*dna*G)	1.75	1.57
14	(*dna*+)	1.92	1.70
15	(*dna*+)	1.90	1.55

* A dash indicates that the enzyme activity was not detectable at 30°C.

30 and 45°C were determined. The presence of Polymerase III from strain 9 does not render the wild-type enzyme temperature sensitive (Fig. 3*C*).

In order to obtain further evidence that the temperature-sensitive character of Polymerase III, derived from strains 9 or 10, was due to a specific alteration of the enzyme, Polymerase III from strain 9 was further purified. The procedures used (T. K. and M. G., manuscript in preparation) are sufficient to purify the wild-type enzyme 2000-fold with respect to the S100 fraction. The properties of such an enzyme preparation are identical to those described for the enzyme activity eluted from the phosphocellulose column.

The specific activity (assayed at 30°C) of preparations from strains 9 and 10 are 10% that of the wild-type enzyme throughout the purification procedure. In all cases "mutant" Polymerase III is totally inactive when assayed at 45°C.

Polymerase III activity could not be detected in extracts from strains 7 and 8 (*dna*E486). The DNA Polymerase III

isolated from strain 6 (*dna*E511) was only marginally temperature sensitive.

DISCUSSION

The enzymes DNA Polymerase II and III have been analyzed in several mutants thermosensitive for DNA synthesis. DNA Polymerase II activity appears to be normal in all strains tested, including a strain carrying a thermosensitive *rec*A gene (data not presented). The failure to associate Polymerase II activity with any *dna* locus tested does not prove that this enzyme is dispensable; isolation of a mutant defective in Polymerase II will be required to resolve this question. That Polymerase II is not thermolabile in *dna*E mutants indicates that this enzyme is not related to Polymerase III, as was previously suggested (2).

DNA Polymerase III activity appears to be normal in strains carrying mutations at the *dna*A, B, C, D, F, and G loci. Of four independently isolated *dna*E mutants, all had altered DNA Polymerase III activity *in vitro*. DNA Polymerase III activity could not be detected in extracts of *dna*E486 mutants, either in the presence or absence of DNA Polymerase I. We believe that this result is due to instability *in vitro* of the Polymerase III in these strains. DNA Polymerase III in the *dna*E511 mutant is only slightly temperature sensitive, which perhaps reflects the slow cessation of *in vivo* DNA synthesis that this mutant displays at the restrictive temperature (11). Polymerase III activity in strains 9 and 10 (*dna*E) is thermolabile.

Our results suggest that the structural gene for DNA Polymerase III is located at the *dna*E locus. Furthermore, since *dna*E mutants fail to replicate their DNA at 42°C *in vivo* (10) and contain a thermolabile Polymerase III *in vitro*, we conclude that this enzyme is required for DNA replication in *E. coli*.

NOTE ADDED IN PROOF

Drs. H. Schaller, B. Otto, V. Nüsslein, J. Huf, R. Herrmann, and F. Bonhoeffer have isolated 50 independent

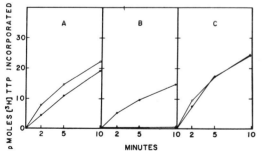

FIG. 3. Effect of temperature on Polymerase III-catalyzed synthesis. The rate of polymerization at 30°C (—●—) and 45°C (—○—) was measured. The results obtained for enzymes isolated from strain 13 (*dna*G) and strain 9 (*dna*E) are shown in Parts *A* and *B*, respectively. The results obtained with a mixture of equal amounts of enzyme from strain 13 and strain 9 are shown n part *C*.

dna^{ts} mutants of *E. coli*. Of these, 20 were shown to be temperature-sensitive in highly concentrated lysates designed to measure DNA replication *in vitro*. Four such mutants that include strains 9 and 10, were shown not to complement each other but were complemented by a soluble factor isolated from *dna*⁺ cells. Drs. B. Otto and V. Nüsslein have purified the soluble factor (E-protein) 1,000-fold using a complementation assay, and have independently shown that its properties (polymerase activity, stability, response to ionic conditions, and molecular weight) are in keeping with those determined for DNA polymerase 111. According to the above criteria, DNA polymerase 111 and the E-protein appear to be the same.

We thank Miss C. Ganier, Mrs. S. Yang, and Mr. G. Grandusky for excellent technical assistance and Dr. M. Oishi for helpful discussions. Dr. H. Ogawa performed enzyme isolations from the *recA*^{ts} strain. We are grateful to Dr. F. Bonhoeffer for making strains 3, 9, and 10 available before publication of their isolation. This work was supported by a grant from the Délégation Générale a la Recherche Scientifique et Technique to Y. H.; by Public Health Service Research Grant CA12590-01 to J. A. W., and by Grant no. NP-6A from the American Cancer Society to M. G.

1. DeLucia, P., and J. Cairns, *Nature*, **224**, 1164 (1969).
2. Kornberg, T., and M. L. Gefter, *Proc. Nat. Acad. Sci. USA*, **68**, 761 (1971).
3. Knippers, R., *Nature*, **228**, 1050 (1970).
4. Moses, R., and C. C. Richardson, *Biochem. Biophys. Res. Commun.*, **41**, 1565 (1970).
5. Mordoh, J., Y. Hirota, and F. Jacob, *Proc. Nat. Acad. Sci. USA*, **67**, 773 (1970).
6. Jacob, F., and E. Wollman, *Sexuality and Genetics of Bacteria* (Academic Press, Inc., New York, 1961).
7. Carl, P., *Mol. Gen. Genet.*, **109**, 107 (1970).
8. Ricard, M., and Y. Hirota, *C.R.-Acad. Sci.*, **268**, 1335, (1969).
9. Bonhoeffer, F., *Z. Vererbungslehre.*, **98**, 141 (1966).
10. Gross, J. D., *Current Topics in Microbiology and Immunology* (Berlin, Springer-Verlag, 1971).
11. Wechsler, J. A., and J. D. Gross, *Mol. Gen. Genet.*, in press.
12. Kohiyama, M., D. Cousin, A. Ryter, and F. Jacob, *Ann. Inst. Pasteur*, **110**, 465 (1966).
13. Hirota, Y., A. Ryter, and F. Jacob, *Cold Spring Harbor Symp. Quant. Biol*, **33**, 677 (1968).
14. Hirota, Y., J. Mordoh, and F. Jacob, *J. Mol. Biol*, **53**, 369 (1970).
15. Bertani, G., *J. Bacteriol.* **62**, 293 (1951).

7

Reprinted from *Proc. Natl. Acad. Sci.*, **67**(2), 674–681 (1970)

Replication and Repair of DNA in Cells of *Escherichia coli* Treated with Toluene*

Robb E. Moses† and Charles C. Richardson‡

DEPARTMENT OF BIOLOGICAL CHEMISTRY, HARVARD MEDICAL SCHOOL, BOSTON, MASS. 02115

Communicated by Eugene P. Kennedy, July 13, 1970

Abstract. DNA synthesis has been studied in *Escherichia coli* cells made permeable to nucleotides by treatment with toluene. Replicative synthesis, as distinguished from repair synthesis, occurs at a rate comparable to that observed *in vivo;* it is dependent on the presence of all four deoxyribonucleoside triphosphates, but does not require exogenous DNA; and it is stimulated by ATP. Furthermore, replicative synthesis can be abolished at the restrictive temperature in DNA temperature-sensitive mutants. *N*-ethylmaleimide completely inhibits this type of synthesis, whereas it does not inhibit repair synthesis. Repair synthesis further differs from replicative synthesis in the following points: it does not require ATP; it persists at the restrictive temperature in DNA temperature-sensitive mutants; it can be induced by endogenous or exogenous nuclease activity; and its demonstration requires a Pol+ strain.

The bacterial chromosome replicates semiconservatively. Synthesis starts at a fixed point and proceeds in a linear, sequential fashion.[1-3] Attempts to duplicate this process *in vitro* using purified DNA polymerases have failed thus far. While DNA polymerase can accurately repair regions of single-stranded DNA, when it is presented with a duplex DNA the product is biologically inactive and contains structural aberrations.[4-6]

Several attempts have been made to isolate the replicating complex while maintaining its integrity. On the hypothesis that this complex may involve the cell membrane, some investigators have attempted to isolate membrane fractions capable of synthesizing DNA.[7-9] In another approach, Smith *et al.*[10] have utilized gentle lysis of cells imbedded in agar. Both approaches have yielded a system capable of carrying out semiconservative DNA synthesis for short periods of time at rates comparable to those observed *in vivo.*[9,10] Buttin and Kornberg[11] have described a technique for measuring intracellular DNA synthesis. Cells treated with EDTA–Tris become permeable to deoxynucleoside triphosphates, but the synthesis observed is mainly of the repair type.[12]

This report describes DNA synthesis in *Escherichia coli* cells treated with toluene. Such cells maintain many of their physiological functions,[13] but have become permeable to compounds of low molecular weight, including deoxynucleoside triphosphates. Although these cells are no longer viable, it has been possible to obtain extended semiconservative replication, and to distinguish this process from a repair type synthesis. In this report the DNA synthesis induced by nuclease action is termed "repair synthesis." Ether-treated cells have been used to study intracellular ϕX174 DNA replication.[14]

Materials. Unlabeled deoxynucleotides, ATP, [³H]dATP, and [¹⁴C]dTDP were obtained from Schwarz BioResearch. [α-³²P]dTTP and [α-³²P]dATP were obtained from International Chemical and Nuclear Corporation, and their purity was established enzymatically and chromatographically. Deoxybromouridine triphosphate (BrdUTP) was a gift from Dr. A. Kornberg. Crystalline pancreatic DNase and snake venom phosphodiesterase were purchased from Worthington. Antibody to purified *E. coli* DNA polymerase was the gift of Dr. I. R. Lehman.

Bacterial strains: *E. coli* W 3110 (*pol⁺endI⁺*), a *K12* derivative, was provided by Dr. J. Cairns, as was *P3478* (*pol⁻endI⁺*), a derivative of *W3110* lacking DNA polymerase activity in extracts. In the original description[15,16] *P3478* is indicated as *polA1*, but since it is the only polymerase-negative mutant used in this study, *pol⁻* will here indicate a mutation at the *polA1* locus. Both strains require thymine. *E. coli* ER22 is a B derivative lacking endonuclease I (*pol⁺endI⁻*). *D110* is an endonuclease I negative strain (*pol⁻endI⁻*) derived by us from *P3478* using the method of Dürwald and Hoffmann-Berling.[17] Extracts of this strain contain less than 2% of wild type endonuclease I activity. *E. coli* strains *CR266-26* and *CRT26-43* were provided by Dr. G. Buttin. *recB₂₁*, a strain lacking Rec B exonuclease, was obtained from Dr. C. A. Thomas. Strains were routinely grown in tryptone broth supplemented with thymine (10 μg/ml) or in M-9 medium[18] supplemented with casamino acids (2 mg/ml) and thymine (10 μg/ml).

Toluene treatment was by a modification of the method of Levin *et al.*[19] Cells in the log phase of growth were harvested by centrifugation at 4°C, resuspended in 0.05 M potassium phosphate buffer (pH 7.4), the suspension was made 1% in toluene and shaken at 25° or 37°C for 10 min unless otherwise noted. After toluene treatment, survival was less than 10^{-8}, as measured by ability to grow and form colonies on agar plates. Toluene-treated cells, washed once with buffer and frozen at -60°C, maintain their activity for at least 1 month.

Results. Characteristics of the reaction: (*a*) *Requirements:* Synthesis of DNA in toluene-treated cells has been followed by incorporation of [α-³²P]dTTP, [α-³²P]dATP, [³H]dATP, or [³H]dCTP into acid-precipitable material. The reaction requires the presence of all four deoxynucleoside triphosphates and Mg⁺⁺ (Table 1). Mn⁺⁺ is 50% as effective as Mg⁺⁺.

(*b*) *Stimulation by ATP:* The rate of reaction is markedly stimulated by ATP; Mg⁺⁺ is present in the reaction mixture at ten times the concentration of ATP (Table 1). The optimal concentration of ATP for freshly prepared cell

TABLE 1. *Characteristics of DNA synthesis in toluene-treated* E. coli.

System	Activity (%)	System	Activity (%)
Complete	100	$-$dTTP	8
+DNA	104	$-$dCTP	11
$-$ATP	9	$-$4 dXTP	0.3
$-$Mg⁺⁺	5	$-$4 dXTP, + 4 dXMP	1
$-$dGTP	0.4	$-$4 dXTP, + 4 dX	1

W3110 cells were grown to 7×10^8 cells/ml, concentrated 5-fold in 0.05 M potassium phosphate buffer (pH 7.4), and agitated 10 min at 37°C with 1% toluene. The reaction mixture (0.3 ml) contained 70 mM potassium phosphate buffer (pH 7.4), 13 mM Mg⁺⁺, 1.3 mM ATP, 33 μM [³H]-dATP, dGTP, dTTP, dCTP, and 1.5×10^8 toluene-treated cells. Salmon sperm DNA was present at 20 μg/0.3 ml reaction where indicated. After incubation for 30 min at 37°C, the reaction was stopped by the addition of cold 10% TCA–0.1 M PP$_i$. After mixing, each sample was filtered through a Whatman GF/C glass filter (2.4 cm) and washed three times with 3 ml of cold TCA–PP$_i$, followed by three washes of 3 ml each with cold 0.01 M HCl. The filters were dried and the radioactivity was measured. dX, dXMP, and dXTP stand for deoxynucleoside, 5′-deoxynucleoside monophosphate, and 5′-deoxynucleoside triphosphate, respectively.

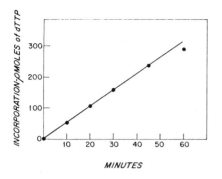

FIG. 1. Rate of DNA synthesis in toluene-treated cells. *D110* cells were grown to a concentration of 6 × 10⁸ cells/ml, harvested by centrifugation at 4°C, concentrated 50-fold in 0.05 M potassium phosphate buffer (pH 7.4), and shaken for 10 min in 1% toluene at 37°C before freezing at −60°C. The reaction mixture was as in Table 1, except that $[\alpha\text{-}^{32}P]dTTP$ was used in place of $[^{3}H]dATP$. The reaction was initiated by the addition of 0.14 ml of thawed cell suspension to 2.1 ml of reaction mixture. At each time point 0.3 ml of the reaction mixture was withdrawn, the reaction stopped, and the product assayed as in Table 1.

suspensions is 1.0–2.0 mM. The stimulation does not appear to be due to the *recB* gene product, an ATP-requiring exonuclease,[20] since it is also observed in toluene-treated *recB₂₁* cells.

(c) *Time course of reaction:* The incorporation of labeled dTTP into DNA in toluene-treated cells is linear for 1 hr (Fig. 1). The extent of the reaction is directly proportional to the concentration of the cell suspension from 10⁷ to 10⁹ cells per reaction mixture. With freshly prepared cell suspensions of *pol⁻ endI⁻* cells and with optimal ATP concentrations, a rate of 1.5 × 10³ nucleotides cells⁻¹ sec⁻¹ can be achieved at 35°C, comparable to that observed *in vivo*.

This ATP-stimulated synthesis represents semiconservative replication (see below). However, nuclease action can induce a repair-type synthesis that can achieve a rate several times that observed during replicative synthesis.

Toluene treatment: (a) *Time course:* The optimal time of exposure of the cells to toluene for assays performed in the presence of ATP is 10 min (Fig. 2, left column); the stimulation of synthesis produced by ATP can be as much as 20-fold. After 20 min the amount of ATP-stimulated synthesis decreases. Polymerase-negative strains (*P3478* and *D110*) show normal levels of ATP-stimulated activity.

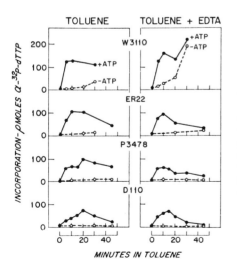

FIG. 2. Time course of toluene treatment. Cells were grown to a concentration of 7–9 × 10⁸ cells/ml, collected, concentrated 5-fold in 0.05 M potassium phosphate buffer (pH 7.4), exposed to 1% toluene or to 1% toluene–2 mM EDTA, and placed on a rotary shaker at 37°C. At various times, aliquots were withdrawn and assayed directly for 30 min at 37°C as in Table 1, with or without ATP. $[\alpha\text{-}^{32}P]$-dTTP was present in the reaction.

(b) *Effect of EDTA on toluene treatment:* If EDTA is present during toluene treatment, a stimulation of DNA synthesis is observed in *pol⁺endI⁺* (*W3110*) cells between 20 and 30 min of treatment (Fig. 2, right column). The enhanced synthesis does not depend on ATP.

A possible explanation, which might be applicable to Tris–EDTA-treated cells as well,[14] is that EDTA produces ribosomal breakdown, release of RNase I, and unmasking of endonuclease I activity. In support of this hypothesis is the observation that *pol⁺endI⁻* (*ER22*) cells lacking endonuclease I do not show increased activity when EDTA is present during toluene treatment (Fig. 2.) Neither *endI⁺* nor *endI⁻* strains of *pol⁻* (*P3478, D110*) show increased synthesis under these conditions. In *E. coli pol⁻endI⁺* this suggests a defect in the ability to repair damage induced by endonuclease I.

Characterization of the DNA product: (a) *Product of hydrolysis:* The radioactive product synthesized in toluene-treated cells is acid-precipitable and alkali-resistant. It can be degraded to acid-soluble material by pancreatic DNase and snake venom phosphodiesterase. When the product of this enzymatic digestion, after incorporation of [α-³²P]dTTP, is analyzed by chromatography, the label is associated with dTMP. We conclude that the isolated product is DNA synthesized from deoxynucleoside triphosphate precursors.

(b) *Sedimentation analysis:* *E. coli* was grown for several generations in the presence of [³H]dT. The cells were then treated with toluene and incubated in the standard DNA-synthesizing system containing [α-³²P]dTTP. As shown in Fig. 3, the newly synthesized DNA is distributed over the same size range as the prelabeled DNA. This is true for both *pol⁺* (*W3110*) and *pol⁻* (*P3478*), but *pol⁻* shows more pieces of smaller size. These findings are in contrast to those observed with Tris–EDTA-treated cells, where newly synthesized product was found in pieces smaller than the DNA synthesized before treatment.

(c) *Pycnographic analysis:* Pycnographic analysis (Fig. 4) of newly syn-

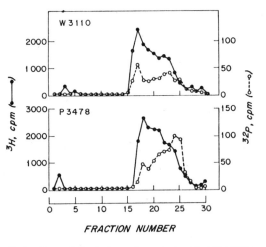

FIG. 3. Alkaline sedimentation analyis of DNA synthesized in toluene-treated cells. Cells were grown in [³H]thymidine for two and one-half generations, harvested, concentrated, and treated with toluene as in Fig. 4. Cells were added to a reaction mixture as in Table 1, with [α-³²P]dTTP. Reaction was continued for 40 min at 37°C. The cells were then chilled to 4°C, harvested by centrifugation, resuspended in 0.5 M KOH, and incubated 20 min at 37°C for lysis. The sample was centrifuged to remove debris. An aliquot of the supernatant was placed on a linear 0.7 M NaCl–0.3 M NaOH–1 mM EDTA 30–70% sucrose gradient, and centrifuged for 3.5 hr at 35,000 rpm in the Spinco type SW50.1 rotor at 4°C. 3-drop fractions were collected from the bottom of the tube and treated with 10% TCA in the presence of 25 μg of carrier DNA. The precipitates were washed 3 times with 3 ml of 0.01 M HCl over Whatman GF/C filters (2.4 cm), dried, and counted in a toluene scintillator.

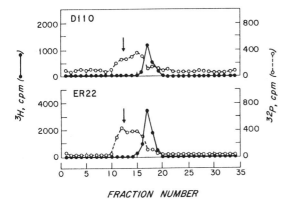

FIG. 4. Equilibrium sedimentation of BrdUTP-containing DNA synthesized in toluene-treated cells. Ten ml of cells were grown to a concentration of 8×10^8 cells/ml in [³H]thymidine, harvested, and suspended in 0.5 ml of 0.05 M potassium phosphate buffer (pH 7.4). After shaking for 10 min in 1% toluene at 37°C, the cells were added to a reaction mixture as in Table 1, but containing in addition 0.13 mM NAD, α[³²P]dATP (specific activity of 1.0 Ci/mmole), and 33 μM BrdUTP in place of dTTP. The reaction mixture was incubated for 20 min at 37°C. Cells were lysed and DNA was extracted.[21] DNA was sedimented in neutral CsCl[22] for 50 hr at 22°C at 30,000 rpm in the IEC SB-405 rotor. Two-drop fractions were collected, precipitated with acid, and the precipitates were collected on Whatman GF/C filters, dried, and counted in toluene-based scintillator. The arrows indicate the position of hybrid DNA.

thesized density-labeled DNA demonstrates that the synthesis occurring in toluene-treated cells is semi-conservative. *E. coli* cells, grown in the presence of [³H]dT, were treated with toluene and incubated in the presence of BrdUTP, [α-³²P]dATP, dGTP, and dCMP. After sedimentation to equilibrium in CsCl, a major portion of the newly synthesized DNA is found at a position characteristic of molecules of hybrid density, thereby indicating semiconservative synthesis. When this DNA was sedimented to equilibrium in alkaline CsCl, all of the newly synthesized DNA was found at the position of heavy single strands.

Effect of DNase on the reaction: Repair synthesis can be stimulated not only by activation of endogenous endonuclease as described above, but by exogenous DNase as well. Hoffmann-Berling[14] has demonstrated an increased rate of DNA synthesis in ether-treated cells exposed to pancreatic DNase. As shown in Table 2, addition of pancreatic DNase to toluene-treated *pol+endI-* (*ER22*) cells leads to a 3-fold stimulation of DNA synthesis. DNA synthesis in toluene-treated *pol- endI-* (*D110*) cells (lacking DNA polymerase) is inhibited rather than enhanced by DNase treatment (Table 2), suggesting that the stimulation in *pol+* cells is due to the repair activity of polymerase at sites created by the DNase. This conclusion is further supported by the finding that the DNase stimulated synthesis does not depend on ATP. Furthermore, when pycnographic analysis, as described in Fig. 4, was performed on the product of the DNase-stimulated reaction in *pol+* cells, the newly synthesized DNA was found at a position corresponding to molecules of light density. With *pol-* cells the

TABLE 2.   *Effect of pancreatic DNase on synthesis.*

Strain	—Activity—	
	—DNase (nmol)	+DNase (nmol)
ER22 (pol+endI-)	0.11	0.27
D110 (pol-endI-)	0.06	0.004

The reaction was as in Table 1, with 0.3 μg/ml of pancreatic DNase present where indicated. ATP was omitted from the reactions with DNase present.

newly synthesized DNA, although reduced in amount, appears at hybrid density. This result suggests a defect in repair synthesis in *pol⁻* cells.

DNA synthesis in DNA thermosensitive mutants (DNA$_{ts}$): Replicative and repair synthesis can be differentiated in toluene-treated mutant cells which are unable to replicate DNA at elevated temperatures *in vivo* (Table 3). Two such mutants, *CR266-26* and *CRT26-43*, rapidly cease DNA synthesis at 41°C, although extracts of these cells possess normal DNA polymerase activity, even at the restrictive temperature.[12] Replicative synthesis (ATP-stimulated DNA synthesis) after toluene treatment in these mutants is 10 times lower at 43°C as compared to only 2 times lower in wild-type cells. However, normal levels of repair synthesis (DNase-stimulated activity) are present in the temperature-sensitive strains at the restrictive temperature.

Differential inhibition: DNA synthesis in toluene-treated cells is not inhibited by 5 mM cyanide or 10 mM azide. ATP-stimulated synthesis is inhibited by sulfhydryl compounds or sulfhydryl-blocking agents (Table 4). However, DNase-stimulated activity is not inhibited by sulfhydryl blockers, but is inhibited by antibody to *E. coli* DNA polymerase.

Discussion. Two types of DNA synthesis have been identified *in vivo:* that responsible for replication of the chromosome, and that occurring during repair of DNA.[6] In fact, mutants of *E. coli* have been isolated which can carry out

TABLE 3. *Effect of temperature on DNA synthesis in toluene-treated DNA$_{ts}$ mutants.*

	DNA synthesis			
	Assayed at 33°C		Assayed at 43°C	
Mutant	+ATP (%)	+DNase (%)	+ATP (%)	+DNase (%)
ER22	100	150	51	230
D110	100	...	52	...
CR266-26	100	300	12	300
CRT26-43	100	94	8	170

Cells were prepared as in Table 1, but with growth at 28°C and toluene treatment at 29°C. 50-μl aliquots of treated cells were added directly to warmed reaction mixtures, and incubation continued for 30 min, then terminated as in Table 1. Pancreatic DNase was present at 0.3 μg/ml in the reaction. ATP was omitted from reactions with DNase. Activity for each strain is expressed as percent of activity at 33°C with ATP.

TABLE 4. *Inhibition of DNA synthesis.*

	Activity	
Additions	*D110* (*pol⁻endI⁻*) %	*ER22* (*pol⁺endI⁻*) %
Control	100	100
+2-Mercaptoethanol	45	52
+NEM	3	2
+Antibody	95	97
+DNase, +NEM	...	135
+DNase, +Antibody	...	33

Cells were prepared and the reaction was performed as in Table 1, but with [α-³²P]dTTP in place of [³H]dATP. 2-Mercaptoethanol was present at 13 mM. When *N*-ethylmaleimide (NEM) was present (1.5 mM), the cells were first incubated for 2 min at 0°C with 1 mM 2-mercaptoethanol. Antibody was present at 20 μl of a 1:10 dilution in 0.15 M NaCl/0.3 ml of reaction mixture. The antibody was heated at 70°C for 20 min before use to destroy nucleases. ATP was omitted from the reactions with DNase present. Pancreatic DNase was present at 0.3 μg/ml when indicated.

DNA repair but not replication under nonpermissive conditions.   Both repair and replication of DNA can be observed in toluene-treated *E. coli* cells.   The permeability of toluene-treated cells to compounds of low molecular weight permits further distinction between the two types of DNA synthesis.

DNA synthesis in toluene-treated cells resembles replication *in vivo* in its rate, the size of the product, a requirement for all four deoxyribonucleotides, and its semiconservative nature.   The phase of DNA synthesis we believe to represent replication occurs even in toluene-treated *pol⁻* cells at normal levels, as replication must.   Toluene-treated DNA$_{ts}$ mutants lack the capacity for replicative synthesis at the restrictive temperature, as do untreated mutants *in vivo*.   Freedom from a requirement for an exogenous template in the toluene-treated cells implicates the bacterial chromosome as the site of synthetic activity.

Our studies with toluene-treated cells significantly add to observations made *in vivo* in revealing that replicative synthesis depends, not only on the presence of all four deoxyribonucleoside triphosphates, but also on the presence of ATP. The ATP stimulation of DNA synthesis is not highly specific, since other nucleoside triphosphates can partially substitute for ATP.   The effect of ATP could be due to protection of the precursors against a phosphatase, or it may serve either as an energy source or as an enzyme cofactor.   ATP appears to be required only for replicative synthesis, without effect on repair synthesis.

Toluene-treated cells furnish a new tool for distinguishing repair from replication with endogenous nuclease and polymerase activities.   *PolAl* strains show a normal capacity for replicative synthesis, but a markedly reduced capacity for repair synthesis as evidenced by the results in *endI⁻* cells of *pol⁻* compared to *endI⁻pol⁺*.   These observations agree with current concepts of a role for DNA polymerase in repair synthesis.[23-25]   Although replicative synthesis appears to proceed normally in the toluene-treated *pol⁻* mutant, sedimentation analysis suggests a large proportion of low molecular DNA, as would be expected if repair activity were impaired and/or degradation of DNA were taking place.

DNase-stimulated synthesis in toluene-treated cells furnishes readily controllable, easily induced levels of repair synthesis in cells which are *endI⁻*.   We believe that the DNase-stimulated synthesis is similar to that observed during repair of ultraviolet damage of DNA *in vivo*, but this has not yet been determined. *Pol⁻* cells do not respond to stimulation by DNase, a finding compatible with the lack of DNA polymerase activity in extracts of this mutant.   Treatment of the toluene-treated cells with antibody to DNA polymerase inhibits the DNase-stimulated synthesis, thus further implicating the polymerase in repair synthesis. The results with DNA$_{ts}$ mutants extend the differentiation between replicative and repair synthesis further: repair synthesis can be demonstrated in DNA$_{ts}$ strains under conditions where replicative synthesis, as well as replication *in vivo* has ceased.   Studies with sulfhydryl blocking agents provide separate documentation for the duality of DNA synthetic processes; they demonstrate inhibition of replicative synthesis under conditions not affecting repair synthesis as demonstrated here, or DNA polymerase *in vitro*.[26]

It is clear that toluene-treated cells differ from the Tris–EDTA cells described by Buttin and Kornberg.[11]   These appear to be dependent upon the presence of an endogenous nuclease or on the activation of a nucleolytic activity. In the Tris–EDTA system pol^- strains show lower synthetic activity than do pol^+ strains, supporting this interpretation.

Our results demonstrate a dichotomy of events in DNA synthesis: repair and replication. Repair synthesis seems to require an activity similar to, or identical with, the known functions of DNA polymerase. Our data do not exclude DNA polymerase from functioning in replication. However, the observed differences in the effect of inhibitors and ATP on repair and replicative synthesis imply a necessity for steps or structures in addition to those already known.

Abbreviations: DNA_{ts}, a mutant with a block in DNA synthesis at elevated temperatures; NEM, N-ethylmaleimide; TCA, trichloroacetic acid; BrdUTP, bromodeoxyuridine triphosphate.

* The work described in this paper was sponsored in part by research grants from the National Institutes of Health, U.S. Public Health Service, No. AI-06045, and from the American Cancer Society, No. P-486.

† Supported by Public Health Service Fellowship, 5-FO3-GM42, 968-02 from General Medical Sciences.

‡ Recipient of a Public Health Service Research Career Program Award, GM-13,634.

[1] Cairns, J., *J. Mol. Biol.*, **6**, 208 (1963).

[2] Nagata, T., *Proc. Nat. Acad. Sci. USA*, **49**, 551 (1963).

[3] Yoshikawa, H., and N. Sueoka, *Proc. Nat. Acad. Sci. USA*, **49**, 559 (1963).

[4] Schildkraut, C. L., C. C. Richardson, and A. Kornberg, *J. Mol. Biol.*, **9**, 24 (1964).

[5] Richardson, C. C., R. B. Inman, and A. Kornberg, *J. Mol. Biol.*, **9**, 46 (1964).

[6] Richardson, C. C., *Ann. Rev. Biochem.*, **38**, 795 (1969).

[7] Frankel, F. R., C. Majumdar, S. Weintraub, and D. M. Frankel, *Cold Spring Harbor Symp. Quant. Biol.*, **33**, 495 (1968).

[8] Ganesan, A. T., *Cold Spring Harbor Symp. Quant. Biol.*, **33**, 45 (1968).

[9] Knippers, R., and W. Strätling, *Nature*, **226**, 713 (1970).

[10] Smith, D. W., H. Schaller, and F. J. Bonhoeffer, *Nature*, **226**, 711 (1970).

[11] Buttin, G., and A. Kornberg, *J. Biol. Chem.*, **241**, 5419 (1965).

[12] Buttin, G., and M. Wright, *Cold Spring Harbor Symp. Quant. Biol.*, **33**, 259 (1968).

[13] Jackson, R. W., and J. A. DeMoss, *J. Bacteriol.*, **90**, 1420 (1965).

[14] Hoffmann-Berling, H., in *Molecular Genetics*, eds. Wittmann, H. C., and M. Schuster (Berlin: Springer, 1968), p. 38.

[15] DeLucia, R., and J. Cairns, *Nature*, **224**, 1164 (1969).

[16] Gross, J., and M. Gross, *Nature*, **224**, 1166 (1969).

[17] Dürwald, H., and H. Hoffmann-Berling, *J. Mol. Biol.*, **34**, 331 (1968).

[18] Anderson, E. H., *Proc. Nat. Acad. Sci. USA*, **32**, 120 (1946).

[19] Levin, D. H., M. N. Thang, and M. Grunberg-Manago, *Biochim. Biophys. Acta*, **76**, 558 (1963).

[20] Oishi, M., *Proc. Nat. Acad. Sci. USA*, **64**, 1292 (1969).

[21] Thomas, C. A., Jr., K. I. Berns, and T. J. Kelley, Jr., in *Procedures in Nucleic Acid Research*, eds. Cantoni, G. L., and D. R. Davies, (New York: Harper and Row, 1966), p. 535.

[22] Pettijohn, D. E., and P. C. Hanawalt, *J. Mol. Biol.*, **9**, 395 (1964).

[23] Kelly, R. B., N. R. Cozzarelli, M. P. Deuscher, I. R. Lehman, and A. Kornberg, *J. Biol. Chem.*, **245**, 39 (1970).

[24] Kanner, L., and P. Hanawalt, *Biochem. Biophys. Res. Commun.*, **39**, 149 (1970).

[25] Boyle, J. M., M. C. Patterson, and R. B. Setlow, *Nature*, **226**, 708 (1970).

[26] Englund, P. T., J. A. Huberman, T. M. Jovin, and A. Kornberg, *J. Biol. Chem.*, **244**, 3038 (1969).

8

Reprinted from *Proc. Natl. Acad. Sci.*, **69**(7), 1863–1867 (1972)

RNA-Linked Nascent DNA Fragments in *Escherichia coli**

(DNA biosynthesis/discontinuous replication/pulse labeling/density gradient centrifugation/
RNA biosynthesis/rifampicin)

AKIO SUGINO, SUSUMU HIROSE, AND REIJI OKAZAKI

Institute of Molecular Biology, Faculty of Science, Nagoya University, Nagoya, Japan

Communicated by Arthur Kornberg, May 8, 1972

ABSTRACT   Nucleic acid that is extracted from *E. coli* labeled by a brief pulse of [³H]dT and denatured by treatment with heat, formamide, or formaldehyde bands in a region with a density higher than that of single-stranded *E. coli* DNA in a Cs_2SO_4 equilibrium density gradient. If treated with alkali or RNase, it then exhibits the density of single-stranded DNA. These results suggest the presence of a short strand of RNA covalently linked to the nascent DNA. Evidence for the presence of covalently linked RNA–DNA molecules is also obtained by pulse labeling with [³H]U. Analyses of nascent nucleic acids from cells pulse labeled for various times, and of the molecules with different sizes, support the hypothesis that the short DNA fragments are formed by extension of even shorter RNA chains, which are synthesized on the parental DNA strands and are removed before ligation of the DNA fragments. The synthesis of the RNA segment of the RNA–DNA molecule is much less sensitive to rifampicin than is the synthesis of bulk RNA.

Evidence has accumulated that DNA is replicated by a discontinuous mechanism involving systhesis and joining of short DNA units (1–8). Little is known, however, about the mechanism of creation of the short DNA chains.

We postulated (2) that short DNA units might be synthesized by a mechanism in which specific structures serve as initiation or termination signals. Although short DNA similar to that found *in vivo* can be produced and joined in crude *in vitro* systems (refs. 9–13, Ogawa and Okazaki, to be published), synthesis of such DNA has not been demonstrated with purified DNA polymerases. Unlike RNA polymerases, the known DNA polymerases do not seem to initiate new chains along parental templates, but rather catalyze the extention of a preexisting chain (14–18). Furthermore, DNA polymerases have relatively poor template specificities (18, 19), in contrast to those of RNA polymerases and other proteins involved in transcription (20, 21). These facts, as well as the recent findings of Kornberg and coworkers (22, 23) indicating a requirement for RNA synthesis to produce a primer for initiation of replication of phage M13 DNA, suggest that RNA synthesis (transcription) might be involved in the initiation of synthesis of the short DNA chains at specific sites on the template in the discontinuous mode of replication. In accord with this possibility, we have obtained evidence that a short stretch of RNA is covalently linked to the nascent DNA fragments.

Abbreviations: SDS, sodium dodecyl sulfate; SSC, standard saline citrate (0.15 M NaCl–0.015 M sodium citrate).
* This paper is No. IX in a series, "Mechanism of DNA Chain Growth." The preceding paper is ref. 6.

MATERIALS AND METHODS

The following *Escherichia coli* strains were used: AB301 (*Hfr met⁻ λ⁺*); D10 (*met⁻ λ⁺ rnsI⁻*); A19 (*Hfr met⁻ tyr⁻ λ⁺ rnsI⁻*); Q13 (*Hfr met⁻ tyr⁻ rnsI⁻ pnp⁻*); N464 (*Hfr met⁻ tyr⁻ rnsI⁻ pnp⁻ rnsII^{ts}*); NK2 (F⁺ *leu⁻ his⁻ λˢ strʳ pyr⁻*); NK3 (F⁺ *leu⁻ his⁻ λˢ strʳ pyr⁻ rnsI⁻*); 1000 (F⁻ *λˢ supᴱ*); 1200 (F⁻ *λˢ supᴱ rnsI⁻ endI⁻*). AB301 and D10 were provided by Dr. R. F. Gesteland, A19 and Q13 by Dr. M. Takanami, N464 by Dr. D. Schlessinger, and NK2, NK3, 1000, and 1200 by Dr. H. Hoffmann-Berling.

[*Methyl*-³H]thymidine (dT) (20 Ci/mmol) was obtained from New England Nuclear Corp. and [5-³H]uridine (U) (29 Ci/mmol) and [*U*-¹⁴C]dT (520 Ci/mol) from the Radiochemical Centre. *E. coli* DNA was extracted by a modified Thomas method (24), followed by RNase IA treatment and ethanol precipitation. Single-stranded *E. coli* DNA labeled with ¹⁴C, with a sedimentation coefficient of 7–10 S, was prepared as described (5). ¹⁴C-labeled *E. coli* RNA was a gift of Dr. Y. Hayashi. *E. coli* exonuclease I was a gift of Dr. I. R. Lehman. *Neurospora* nuclease was purified by a modification of the procedure described by Rabin *et al.* (25). RNase IA was obtained from Worthington Biochemical Corp. RNase T₁ was provided by Dr. T. Uchida. The RNases were used after heating at 90° for 10 min in 10 mM acetate buffer (pH 5.0). RNase-free sucrose was purchased from Nakarai Chemicals, Ltd. (Kyoto) and autoclaved before use. Rifampicin was obtained from Calbiochem.

To pulse label bacteria with [³H]dT or [³H]U, a culture (20–100 ml), grown at 37° to 8 × 10⁸ cells/ml in Medium A (3) supplemented with 1% Casamino Acids (and 40 μg/ml uracil in the case of *E. coli* NK2 and NK3), was transferred to 14°, at which temperature the cells grow with a doubling time of 400–600 min. After 90 min of shaking at 14°, the radioactive compound was added to a final concentration of 0.1 or 0.2 μM. The reaction was stopped by the addition of an equal volume of 75% ethanol–21 mM acetate buffer (pH 5.3)–2 mM EDTA–2% phenol ("ethanol–phenol") (5), and cells were collected by centrifugation. Nucleic acid was extracted by a modified Thomas procedure (24), except that Pronase P, which had been autodigested at 37° for 1 hr, was used and, in the experiments presented in Figs. 5–8, samples were concentrated by means of ethanol precipitation.

For Cs_2SO_4 equilibrium centrifugation, nucleic acid samples, 2.7 g of Cs_2SO_4, 0.45 ml of 1 M Tris·HCl (pH 8.0)–0.1 M EDTA were mixed, and then the solution was adjusted to 4.5 ml and pH 7.6. The mixture was placed in a nitrocellulose tube, overlaid with liquid paraffin, and centrifuged in a

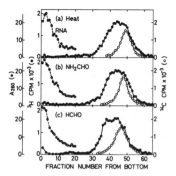

FIG. 1.   Cs$_2$SO$_4$ equilibrium centrifugation of nascent *E. coli* Q13 DNA after denaturation by various methods. Nucleic acid was extracted from cells pulse labeled with [^{3}H]dT for 15 sec at 14° and denatured by heating in 0.1 × SSC at 100° for 5 min (*a*), by incubating in 90% formamide at 37° for 30 min (*b*), or by heating in 12% formaldehyde at 80° for 10 min (*c*). Each sample was subjected to equilibrium centrifugation in a 50 Ti rotor, together with ^{14}C-labeled denatured *E. coli* DNA.

Beckman SW50L or 50 Ti rotor for 42–48 hr at 36,000 rpm and 15°; 65- μl fractions were collected.

For neutral sucrose gradient sedimentation, DNA of phage δA (labeled with ^{14}C or unlabeled) was used as a molecular weight reference marker, and the distance of sedimentation was shown relative to the distance from the top to the band of δA DNA (24).

To measure [^{3}H]dT- or [^{14}C]dT-labeled DNA and [^{3}H]U-labeled RNA, radioactivity in material insoluble in cold trichloroacetic acid that was not or was labile in alkali was counted, respectively (24).

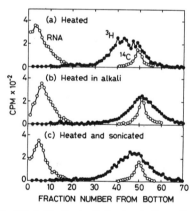

FIG. 2.   Effect of heating in alkali and sonication on the density of nascent DNA. Nucleic acid was extracted from *E. coli* Q13 pulse labeled with [^{3}H]dT for 15 sec at 14°, and heated in SSC at 100° for 5 min. It was subjected to Cs$_2$SO$_4$ equilibrium centrifugation in a 50 Ti rotor without further treatment (*a*), after heating in 0.3 M NaOH at 100° for 5 min (*b*), or after sonic treatment (for 30 sec at 0° with Branson Sonifier S75 at setting 1) (*c*). *E. coli* [^{14}C]RNA and denatured *E. coli* [^{14}C]DNA were included as markers.

RESULTS

To test the possibility that a short RNA strand might be linked to nascent DNA, nucleic acid was extracted from *E. coli* after a very brief pluse of [^{3}H]dT (15 sec at 14°, or 0.04% of the generation time) and analyzed, after various treatments, by density equilibrium centrifugation in Cs$_2$SO$_4$. Strain Q13, a mutant deficient in RNase I and polynucleotide phosphorylase, was used in most experiments. The pulse labeling was terminated by addition of "ethanol–phenol" (5) to stop the reaction instantaneously and to inactivate nucleases.

The results presented in Fig. 1 show that after denaturation by heat, by formamide, or by hot formaldehyde the pulse-labeled DNA banded heterogeneously, with an average density significantly higher than that of reference single-stranded DNA from *E. coli*. When the pulse-labeled DNA was heated in alkali, which would hydrolyze RNA and cause some fragmentation of DNA, the pulse-labeled DNA banded at the density of single-stranded DNA (Fig. 2*b*). Fragmentation by sonic treatment, however, did not cause the labeled material to shift to the position of the reference DNA in the density gradient (Fig. 2*c*). Incubation with RNase IA and T$_1$, or in 0.3 M NaOH at 37°, milder treatments for RNA hydrolysis, resulted in a shift of the nascent DNA to the density of reference DNA (Fig. 3).

After denaturation, the ^{3}H-labeled DNA was degraded to the same extent as control single-stranded DNA by *E. coli* exonuclease I and *Neurospora* nuclease, enzymes specific for single-stranded nucleic acids (Table 1).

These results indicate the following: (*a*) An appreciable fraction of the nascent DNA fragments are linked to short RNA strands by bonds resistant to the various treatments for denaturation; (*b*) both RNA and DNA are susceptible to nucleases specific for single-stranded structures. Therefore, it is likely that in the nascent molecule a short stretch of

TABLE 1.   *Susceptibility of pulse-labeled DNA to nucleases specific for single-stranded nucleic acid*

Prior treatment	Extent of degradation (%)	
	E. coli exonuclease I	*Neurospora* nuclease
SSC at 100°	88.1 (89.5)*	94.1 (94.7)
SSC at 100° and RNase	90.5 (89.7)	
HCHO at 80°	86.3 (90.3)	94.3 (95.8)

Nucleic acid was extracted from *E. coli* Q13 pulse labeled with [^{3}H]dT for 15 sec at 14°, treated as indicated as in Figs. 1–3, and dialyzed against 10 mM Tris·HCl (pH 8.0)–10 mM NaCl. To test the susceptibility to exonuclease I, a 60-μl sample (3400 dpm) was incubated at 37° for 90 min in a reaction mixture (100 μl) containing 60 mM glycine–KOH (pH 9.0), 10 mM 2-mercaptoethanol, 60 mM MgCl$_2$, 0.3 μM heated *E. coli* [^{14}C]DNA, and 1.5 μg of enzyme. For digestion with *Neurospora* nuclease, a 100-μl sample (6500 dpm) was incubated at 37° for 4 hr in a reaction mixture (300 μl), containing 0.1 M Tris·HCl (pH 8.0), 10 mM MgCl$_2$, 0.1 μM dTMP, 30 μg of bovine serum albumin, 0.1 μM heated *E. coli* [^{14}C]DNA, and 12 μg of enzyme. The percentage of radioactivity rendered acid soluble was measured (4).

* Numbers in parentheses indicate the percentage of [^{14}C]DNA (internal control) degraded.

RNA is covalently linked to DNA, and that there is no duplex structure over an extended region of the molecule.

That the high density is characteristic of very nascent DNA is indicated by the experiment of Fig. 4. Nucleic acid was extracted from cells labeled with [³H]dT for various times and subjected, after denaturation, to isopycnic and zone sedimentation analysis. The shift of labeled DNA to high density relative to the reference single-stranded DNA was evident only when the pulse time was 30 sec or less at 14° (≤0.08 generations).

Evidence for the presence of the linked RNA-DNA molecule was also obtained by a pulse label with either [³H]U or with [³H]U and [¹⁴C]dT. Fig. 5 shows such an experiment. After a pulse label with [³H]U and [¹⁴C]dT for 15 or 30 sec at 14°, a small portion (about 1%) of the [³H]U-labeled RNA, as well as a large portion of [¹⁴C]dT-labeled DNA, banded in the region with a density a little higher than reference single-stranded DNA (Fig. 5a and b). The ³H band was broad and had a higher average density than the ¹⁴C band; the ³H band was also shifted to lighter densities when the pulse time was increased from 15 to 30 sec. Administration of large amounts of unlabeled U and dT after the 15-sec pulse, which had an incomplete chase effect, greatly reduced the percentage of the ³H label in the "light RNA" (Fig. 5c). It seems likely from these results, and the results of Fig. 4, that RNA constitutes a relatively large portion of the very nascent molecule, and is linked to a short DNA chain, which may undergo elongation to a definite length [to the size of the "units" in discontinuous replication (2, 3)]. This process may be followed by the removal of the RNA segment from the molecule.

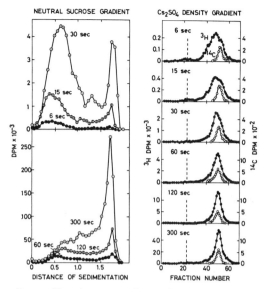

FIG. 4. Neutral sucrose gradient sedimentation and Cs₂SO₄ equilibrium centrifugation of *E. coli* Q13 DNA pulse labeled with [³H]dT for various times at 14°. Pulse-labeled DNA was heated in 0.1 × SSC–10 mM Tris·HCl (pH 7.6)–1 mM EDTA–0.1% SDS at 100° for 5 min and chilled quickly. A 1-ml sample was layered on 31 ml of a 5–20% sucrose gradient containing 0.1 × SSC–10 mM Tris·HCl (pH 7.6)–1 mM EDTA–0.1% SDS made on 80% sucrose (5 ml), and centrifuged in a Beckman SW27 rotor for 13 hr at 24,000 rpm and 20°. For density analysis, a 1-ml sample and denatured *E. coli* [¹⁴C]DNA were centrifuged in a Cs₂SO₄ solution in a SW50L rotor. The vertical *dotted lines* indicate the position of the peak of RNA.

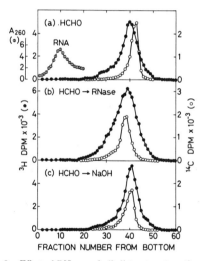

FIG. 3. Effect of RNase and alkali treatment on the density of nascent DNA. Nucleic acid, extracted from *E. coli* Q13 pulse labeled with [³H]dT for 15 sec at 14° and heated in 12% formaldehyde at 100° for 10 min, was subjected to Cs₂SO₄ equilibrium centrifugation in a SW50L rotor without further treatment (*a*), or after incubation for 1 hr at 37° in SSC containing 50 μg/ml of RNase IA and 10 μg/ml of RNase T₁ (*b*), or after incubation in 0.3 M NaOH at 37° for 15 hr (*c*). Denatured *E. coli* [¹⁴C]DNA was included as marker.

This idea was supported by analysis of nascent molecules of different sizes. Fig. 6 shows density analysis of [³H]dT pulse-labeled DNA that sedimented at different rates in neutral sucrose gradients. The material sedimenting very slowly in the sucrose gradient (Fractions A and B) had higher average densities than the material sedimenting more rapidly (Fraction C or D). Fraction D banded almost coincidentally with reference single-stranded DNA in the Cs₂SO₄ density gradient.

Although the initiation of a new round of chromosome replication in *E. coli* has been reported to be inhibited by rifampicin (26), the continuation of an ongoing round of chromosome replication is relatively insensitive to this antibiotic. Thus, treatment of *E. coli* Q13 with 200 μg/ml rifampicin at 14° for 30 min reduced the rate of incorporation of [³H]thymidine into DNA to about 80% of the control, while the rate of [³H]U incorporation into RNA was suppressed to 2.2% by the same treatment. The experiment shown in Fig. 7 demonstrated that the incorporation of [³H]U into the RNA–DNA molecule is much less sensitive to rifampicin than is the synthesis of the bulk of pulse-labeled RNA.

All the experiments described above were performed with *E. coli* Q13. To see whether the linked RNA–DNA molecules are also found in other *E. coli* strains, the experiment of Fig. 8 was performed. Density analysis of nucleic acid extracted

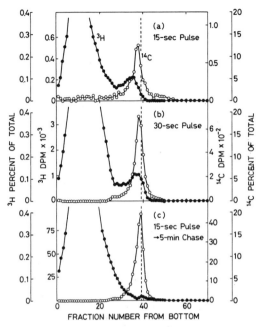

FIG. 5. Cs₂SO₄ equilibrium centrifugation of nucleic acid pulse labeled with [³H]U and [¹⁴C]dT. *E. coli* Q13 was pulse labeled with [³H]U and [¹⁴C]dT (0.1 μM each) at 14° for 15 sec (*a*), or for 30 sec (*b*), or pulse labeled for 15 sec and "chased" for 5 min by the addition of (unlabeled) 20 μM U and 100 μM dT (*c*). Nucleic acid was extracted, and rRNA was removed by precipitation with 1 M NaCl. The rest of the nucleic acid was precipitated with carrier tRNA by ethanol, dissolved in 0.1 × SSC, and heated at 100° for 5 min. Equilibrium centrifugation was performed as in Fig. 3, except that heat-denatured, unlabeled *E. coli* DNA was used as marker. The vertical *dotted lines* indicate the position of the peak of marker DNA.

from cells labeled by a 15-sec pulse of [³H]dT at 14° revealed the presence of the RNA–DNA molecules in all the strains examined. These include AB301, which has normal RNase I and polynucleotide phosphorylase and from which Q13 and other strains shown in the left column of Fig. 8 are derived, NK3 and 1200, other RNase I-deficient strains, and their parental strains.

DISCUSSION

The possibility that RNA serves as primer for the synthesis of DNA has recently been suggested by several groups (18, 22, 23, 27–29). Our work presents evidence that a short RNA strand is covalently linked to the nascent short DNA fragments of *E. coli*, suggesting that the synthesis of these fragments in the discontinuous mode of replication is primed by even shorter chains of RNA. The possibility that the molecule is an aggregate of RNA and DNA is unlikely due to the following facts. (*a*) Its presence is invariably indicated by procedures described here and by other procedures involving Sepharose gel filtration, after various treatments that would destroy hydrogen bonds and other weak linkages. (*b*) The RNA moiety of the molecule is metabolically distinguished from

the bulk of pulse-labeled RNA, as shown in Fig. 5, and by a different sensitivity of its synthesis to rifampicin (Fig. 7). The data presented here and those to be reported elsewhere indicate that the RNA of covalently linked RNA-DNA molecules may consist of 50-100 nucleotides. The RNA appears to be synthesized from low molecular weight precursors, used immediately for the formation of the RNA-DNA molecule, and quickly removed from it.

These results are consistent with the following mechanism for DNA replication. There are specific structures, possibly specific nucleotide sequences, on the parental strands of DNA that serve as signals for chain initiation; such signals occur repeatedly along the chromosome. In the replicating region, DNA-dependent RNA polymerase molecules specifically bind to these structures and initiate synthesis of short RNA chains, which terminate at certain points that may also be designated by specific structures on the template DNA. A DNA polymerase then extends the chains in the 5′ to 3′ direction, first forming a phosphodiester bond between the ribo- and deoxyribonucleotides. The elongation of the chains by the DNA polymerase is followed by removal of the RNA portion of the chains by nuclease action. The gaps thus created between the adjacent fragments are filled by the action of a DNA polymerase. Finally, the completed DNA fragments are covalently linked to each other by DNA ligase to form long DNA. These chemical processes may be preceded by unwinding of the parental strands of DNA, mediated either by an unwinding protein such as that reported by Alberts and Frey (30), or by the transcriptional operation that initiates the short chains.

The finding that the nascent DNA fragments are linked to RNA further diminishes the possibility that these fragments are products of endonucleolytic cleavage of DNA strands synthesized by a continuous mechanism, or some kind of artifact (31). The present results indicate that a large

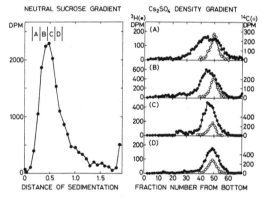

FIG. 6. Density analysis of pulse-labeled DNA of various sizes. DNA extracted from *E. coli* Q13 pulse labeled with [³H]dT for 15 sec at 14° was heated in 12% formaldehyde at 80° for 10 min. A 4-ml sample was layered on 53 ml of a 5–20% sucrose gradient containing 0.1 × SSC–1 mM EDTA–2% formaldehyde and centrifuged in a Beckman SW25.2 rotor for 18 hr at 25,000 rpm and 4°. The indicated fractions (*A*, *B*, *C*, and *D*) were dialyzed against 0.1 × SSC–10 mM Tris·HCl (pH 7.6)–0.1% SDS–1 mM EDTA, and were subjected to Cs₂SO₄ equilibrium centrifugation as in Fig. 3.

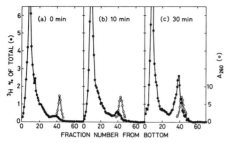

FIG. 7. Effect of rifampicin on the synthesis of bulk RNA and of RNA–DNA molecules. A 40-ml culture of *E. coli* Q13 was treated with 200 μg/ml of rifampicin at 14° for the times indicated, then pulse labeled for 15 sec with [³H]U. Nucleic acid was extracted, precipitated by ethanol, dissolved in 0.1 × SSC–10 mM Tris·HCl (pH 7.6)–0.1% SDS, dialyzed against the same solution, heated at 100°, and quickly chilled. Equilibrium density centrifugation was performed as in Fig. 5. Total radioactivity incorporated into RNA was: (*a*) 74,667 cpm, (*b*) 24,392 cpm, and (*c*) 1,620 cpm.

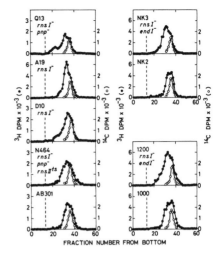

FIG. 8. Cs₂SO₄ equilibrium centrifugation of pulse-labeled DNA of various strains. A 20-ml culture of each strain was pulse labeled with [³H]dT for 15 sec at 14°. Nucleic acid was extracted, precipitated by ethanol, dissolved in 0.1 × SSC–10 mM Tris·HCl (pH 7.6)–0.1% SDS, dialyzed against the same solution, heated at 100°, and quickly chilled. Equilibrium centrifugation was performed as in Fig. 4. The *dotted lines* indicate the position of the peak of RNA.

fraction of the nascent DNA fragments of random cultures pulse labeled for 0.04% of the generation time are linked to RNA. Therefore, it is very unlikely that the majority of the labeled RNA-DNA molecules derive from the chromosome origin. Initiation, but not continuation, of a round of chromosome replication in *E. coli* is inhibited strongly by rifampicin, a specific inhibitor of the initiation of RNA synthesis by *E. coli* RNA polymerase. Our data (Fig. 7) indicate that the synthesis of the RNA segment of the covalently linked RNA–DNA molecule is much less sensitive to rifampicin than is the synthesis of other RNA of *E. coli*. Thus, an RNA polymerase different from the known *E. coli* RNA polymerase might be involved in DNA replication. Alternatively, it is possible that the RNA polymerase is resistant to rifampicin when it is a part of the replication complex. Still another possibility is that the synthesis of the RNA segment of the RNA-DNA molecule is primed by preformed oligonucleotides.

We thank Dr. Arthur Kornberg for valuable discussions and critical comments on the manuscript. This work was supported by grants from the Ministry of Education of Japan, the Jane Coffin Childs Memorial Fund for Medical Research, and Toray Science Foundation.

1. Sakabe, K. & Okazaki, R. (1966) *Biochim. Biophys. Acta* **129**, 651–654.
2. Okazaki, R., Okazaki, T., Sakabe, K., Sugimoto, K., Kainuma, R., Sugino, A. & Iwatsuki, N. (1968) *Cold Spring Harbor Symp. Quant. Biol.* **33**, 129–142.
3. Okazaki, R., Okazaki, T., Sakabe, K., Sugimoto, K. & Sugino, A. (1968) *Proc. Nat. Acad. Sci. USA* **59**, 598–605.
4. Okazaki, T. & Okazaki, R. (1969) *Proc. Nat. Acad. Sci. USA* **64**, 1242–1248.
5. Sugino, A. & Okazaki, R. (1972) *J. Mol. Biol.* **64**, 61–85.
6. Okazaki, R., Arisawa, M. & Sugino, A. (1971) *Proc. Nat. Acad. Sci. USA* **68**, 2954–2957.
7. Yudelevich, A., Ginsberg, B. & Hurwitz, J. (1968) *Proc. Nat. Acad. Sci. USA* **61**, 1129–1136.
8. Schandl, E. K. & Taylor, J. H. (1969) *Biochem. Biophys. Res. Commun.* **34**, 291–300.
9. Okazaki, R., Sugimoto, K., Okazaki, T., Imae, Y. & Sugino, A. (1970) *Nature* **228**, 223–226.
10. Schaller, H., Otto, B., Nüsslein, V., Huf, J., Herrmann, R. & Bonhoeffer, F. (1972) *J. Mol. Biol.* **63**, 183–200.
11. Strätling, W. & Knippers, R. (1971) *J. Mol. Biol.* **61**, 471–487.
12. Geider, K. & Hoffmann-Berling, H. (1971) *Eur. J. Biochem.* **21**, 374–384.
13. Moses, R. E., Campbell, J. L., Fleischman, R. A. & Richardson, C. C. (1971) *Miami Winter Symposia* (North-Holland, Amsterdam, Neth.), Vol. 2, pp. 48–66.
14. Goulian, M. (1968) *Proc. Nat. Acad. Sci. USA* **61**, 284–291.
15. Kornberg, A. (1969) *Science* **163**, 1410–1418.
16. Knippers, R. (1970) *Nature* **228**, 1050–1053.
17. Wickner, R. B., Ginsberg, B. & Hurwitz, J. (1972) *J. Biol. Chem.* **247**, 498–504.
18. Wells, R. D., Flügel, R. M., Larson, J. E., Schendel, P. F. & Sweet, R. W. (1972) *Biochemistry* **11**, 621–629.
19. Karkas, J. D., Stavrianopoulos, J. G. & Chargaff, E. (1972) *Proc. Nat. Acad. Sci. USA* **69**, 398–402.
20. Burgess, R. R. (1971) *Annu. Rev. Biochem.* **40**, 711–740.
21. Gilbert, W. & Müller-Hill, B. (1967) *Proc. Nat. Acad. Sci. USA* **58**, 2415–2421.
22. Brutlag, D., Schekman, R. & Kornberg, A. (1971) *Proc. Nat. Acad. Sci. USA* **68**, 2826–2829.
23. Wickner, W., Brutlag, D., Schekman, R. & Kornberg, A. (1972) *Proc. Nat. Acad. Sci. USA* **69**, 965–969.
24. Okazaki, R. (1971) *Methods Enzymol.* **21D**, 296–304.
25. Rabin, E. Z., Preiss, B. & Fraser, J., *Prep. Biochem.*, in press.
26. Lark, K. G. (1972) *J. Mol. Biol.*, **64**, 47–60.
27. Verma, I. M., Meuth, N. L., Bromfeld, E., Manly, K. F., & Baltimore, D. (1971) *Nature New Biol.* **233**, 131–134.
28. Leis, J. P. & Hurwitz, J. (1972) *J. Virol.* **9**, 130–142.
29. Chang, L. M. S. & Bollum, F. J. (1972) *Biochem. Biophys. Res. Commun.* **46**, 1354–1360.
30. Alberts, B. M. & Frey, L. (1970) *Nature* **227**, 1313–1318.
31. Werner, R. (1971) *Nature* **230**, 570–572.

Part II
TRANSCRIPTION

Editor's Comments
on Papers 9 Through 14

In Part II we detail the mechanism by which the DNA genome is transcribed into complementary RNA copies, as well as the manner in which this process is regulated. The transcription process is catalyzed by a DNA-dependent RNA polymerase. Although essentially all bacterial RNA is polymerized in this manner, messenger RNA (mRNA) is the transcription product that ultimately provides the code words for protein synthesis. Genetic information dictated by the base sequence of DNA is transcribed into mRNA. The mRNA is subsequently translated on ribosomes, which results in the sequential assembly of amino acids into the appropriate protein gene products. These messenger molecules thus function in transmitting specific instructions from DNA to the site of peptide bond formation on the ribosome.

The earliest evidence for a mRNA species was the observation by Hershey et al. (1953) that metabolically active RNA was produced during infection of *Escherichia coli* with bacteriophage T2. Later, Volkin and Astrachan (1956) demonstrated that the base composition of the phage-specific RNA reflected that of the phage DNA (assuming equivalency of thymine and uracil). Furthermore, Nomura et al. (1960) showed that the newly synthesized RNA was associated with ribosomes. The possibility that the unstable RNA was a component of the ribosome was eliminated when Brenner et al. (1961) showed that no new ribosomes were assembled after the onset of phage infection and that the short-lived RNA was a discrete entity reversibly associated with preexisting host ribosomes. Ultimate proof that this T2-RNA represented actual transcripts of the phage DNA was provided by Hall and Spiegelman (1961) who demonstrated specific RNA–DNA complex (hybrid) formation between the complementary nucleic acid chains. That transient RNA species were not limited to bacteriophage infection was convincingly shown by Gros et al. (Paper 9) in uninfected *E. coli* cells. They found that RNA molecules physically similar to the phage-specific RNA were associated with the ribosomes under suitable ionic conditions. Pulse labeling revealed that the unstable RNA component was distinct from ribosomal RNA (rRNA) and transfer RNA (tRNA).

Among the earliest reports on the nature of the transcription apparatus was the detection of a DNA-dependent RNA polymerase in bacteria (Hurwitz et al., Paper 10; for other early references, see Hurwitz et al., 1962). The properties of highly purified preparations from *E. coli* were later described by Hurwitz et al. (1962) and Chamberlin and Berg (1962). RNA synthesis was found to require all four ribonucleotide triphosphates, magnesium, and manganese, in addition to a DNA template. That RNA was synthesized in the reaction was indicated by the alkali lability of the product and its sensitivity to ribonuclease but not deoxyribonuclease. Salmon sperm DNA was the most efficient template, but significant rates of RNA synthesis were also obtained with DNA from *E. coli* and several bacteriophages. A deoxyadenylate–deoxythymidylate copolymer with the component bases adjacent to each other was employed as a template by Furth et al. (1961). The product consisted of polyadenylate–polyuridylate with 5'-adenylic and 5'-uridylic acids in alternating sequence, suggesting that complementarity with template was achieved.

With respect to the initiation of the RNA polymerase reaction, Maitra and Hurwitz (1965) demonstrated that RNA chains begin with purine ribonucleotide triphosphates, which become the 5' termini. Additional ribonucleotides are then added to the 3'-hydroxyl end of the growing chain. Thus, RNA, like DNA, is polymerized in the 5' → 3' direction; however, the RNA polymerase catalyzed reaction does not start by addition to short stretches of a preexisting polymer as does that of DNA polymerase.

Other features of the transcription machinery in *E. coli* have recently been elucidated (for reviews, see Burgess, 1971; Losick, 1972). The complete RNA polymerase holoenzyme consists of five polypeptide subunits in the ratios of $\alpha_2 \beta \beta' \sigma$. Burgess et al. (Paper 11) resolved holoenzyme into core enzyme $(\alpha_2 \beta \beta')$ and sigma factor (σ). The core enzyme initiates polymerization nonspecifically, whereas the addition of σ restores initiation at specific sites (Bautz et al., 1969). Other transcription factors that are apparently not a part of the holoenzyme include M (Davison et al., 1969), psi (ψ) (Travers et al., 1970), the catabolite gene-activating protein (Zubay et al., Paper 12; Emmer et al., 1970), and rho (ρ) (Roberts, Paper 13). A more complete list has recently been compiled (Ramakrishnan and Echols, 1973).

Although unequivocal evidence on the function of the individual subunits of the core enzyme is lacking, the α, β, and β' components appear to be required for polymerization. This was demonstrated by Heil and Zillig (1970), who reconstituted catalytic activity with the resolved subunits, which were inactive by themselves. That the β subunit is involved in both the initiation and elongation of RNA chains is suggested by studies with mutants in which this subunit was altered. One of these mutants (Rabussay and Zillig, 1969) was resistant to rifampicin, an antibiotic that normally binds to the β subunit and inhibits initiation, whereas a second mutant was partially resistant to streptolydigin, an antibiotic that inhibits the elongation process (Schleif, 1969). Heil and Zillig (1970) restored normal sensitivity to these antibiotics by resolving the mutant enzymes and replacing their β subunits with unaltered β from wild-type cells. With regard to a specific role for the β' subunit, it is apparently required for the interaction of DNA with the enzyme surface, since it was the only separated subunit to cause binding of DNA to membrane filters (Losick, 1972). An additional polypeptide, (ω), usually purifies with holoenzyme (Burgess, 1969), but it has no apparent catalytic role.

The role of the σ subunit in the transcription process has been extensively investigated. The binding studies of Hinkle and Chamberlin (1970) suggest that it is required for stable interactions between the polymerase and specific initiation sites within the operon known as promoters (Ippen et al., 1968). Results presented by Dunn and Bautz (1969) suggest that σ is also involved in the initiation of catalysis. Travers and Burgess (1969) have shown that, soon after initiation, σ is released from the holoenzyme–DNA complex. It may then be cyclically reutilized by other core complexes for the initiation of new RNA chains. Bautz et al. (1969) demonstrated that, during infection of *E. coli* with phage T4, the host σ apparently restricts initiation to the promoter regions of only those phage genes expressed immediately after infection. Thereafter, the host polymerase is modified and restricted to the transcription of genes utilized later in infection. Accordingly, host σ acts as a positive control element permitting selective transcription of a fraction of the T4 genome. An analogous situation also prevails for the earliest genes transcribed in phage T7 infection (Summers and Siegel, 1969). The positive control of late T7 transcription, however, results from an entirely new RNA polymerase, which is the product of a phage gene expressed early in infection (Chamberlin et al., 1970); a phage-specified polymerase of similar function is also synthesized in T3 infection (Dunn et al., 1971). Thus, the original hope voiced by Burgess and his colleagues (Paper 11) for the existence of several σ-like factors, each with a specificity for a different type of promoter, has not been realized. Instead, it appears that σ functions by somehow permitting the core enzyme to interact with a variety of specific DNA sequences.

Other protein factors that reportedly stimulate transcription include M, ψ, and the catabolite-gene activating protein. M factor was first detected by Davison et al. (1969) in a high-salt wash of *E. coli* ribosomes. Highly purified preparations resulted in a 17-fold stimulation of transcription with phage λ DNA and σ-saturated holoenzyme (Davison et al., 1970). M appears to interact with the polymerase and stimulate initiation at already specified promoter regions. The ψ factor was thought to be a positive control element that interacts with guanosine 5'-diphosphate, 3'-diphosphate (ppGpp) and specifically regulates transcription of rRNA genes (Travers et al., 1970); but, more recently, Haseltine (1972) and Pettijohn (1972) have shown that significant quantities of rRNA are synthesized with highly purified holoenzyme from *E. coli* in the absence of ψ. Similar results were also reported by Birnbaum and Kaplan (1973), who employed DNA templates enriched in rRNA genes.

The catabolite-gene activating protein (CAP) is involved in the transcription of genes subject to catabolite repression (Zubay et al., Paper 12; Emmer et al., 1970). During catabolite repression, the synthesis of enzymes involved in catabolic processes (e.g., β-galactosidase and other proteins specified by the structural genes of the lactose operon) is severely curtailed by the growth of susceptible strains on glucose and several related carbon sources (for a detailed discussion of this phenomenon, see Pastan and Perlman, 1970). This is accompanied by a decreased intracellular level of adenosine 3',5'-monophosphate (cyclic AMP), a cyclic nucleotide required for the expression of these genes. Using an *in vitro* protein-synthesizing system dependent upon added DNA and its subsequent transcription, Zubay et al. (Paper 12) demonstrated that the effect of cyclic AMP on β-galactosidase synthesis is mediated through CAP. They observed low β-galactosidase levels when DNA from CAP-deficient mutants was employed, even in the presence of cyclic AMP; however, the levels were greatly increased by the addition of a CAP-containing fraction. The stimulation by CAP was diminished with DNA containing a mutation in the *lac* promoter. It thus appears that CAP, which interacts directly with cyclic AMP (Emmer et al., 1970), exerts its effect upon transcription at the promoter region. Arditti et al. (1970) have demonstrated that the σ factor is necessary for stimulation of transcription by CAP, and Burgess (1971) has suggested that CAP may interact with a variety of promoters.

Insight into the mechanism by which the transcription process is terminated and the completed RNA chains are released has been provided by Roberts (Paper 13). He reported the isolation of a protein factor from *E. coli* that causes specific termination and release of transcripts of a discrete size from a phage λ template at low ionic strength. This factor, which was named rho (ρ), binds to the polymerase (Darlix et al., 1971) and brings about the completion of RNA chains at termination sites not recognized by holoenzyme alone. The transcripts obtained in the presence of ribonuclease-free ρ preparations are thus shorter than those terminated in its absence. At higher ionic strength, however, specific termination and release of discrete RNA chains occurs without ρ; under these conditions active polymerase is also released (Millette et al., 1970; Maitra et al., 1970; Richardson, 1970; Pettijohn et al., 1970). It thus appears that under appropriate conditions the polymerase is capable of recognizing both ρ-dependent and ρ-independent termination sequences. That these *in vitro* transcripts are apparently identical to natural messengers has been suggested in hybridization studies with DNA from phages λ (Roberts, Paper 13; Roberts, 1970), T4 (Richardson, 1970; Witmer, 1971), and T7 (Davis and Hyman, 1970). Recently, a role for ρ has been suggested in the transcription of bacterial operons (DeCrombrugghe et al., 1973). The assigning of any definitive *in vivo* role to ρ and other transcription factors, however, awaits the isolation of the appropriate conditional lethal mutants.

Another aspect of the transcription process that has been investigated extensively is the apparent change in template specificity which occurs during bacterial sporulation. Under these circumstances, the transcription of sporulation genes is activated while that of unneeded vegetative genes is arrested (Losick, 1972). In the early stages of sporulation in *Bacillus subtilis,* the ability to transcribe phage φe DNA *in vitro* is lost (Losick et al., 1970) along with vegetative σ activity (Linn et al., 1973); this suggests that template specificity is altered. Although modifications of the β subunit were thought to occur during sporulation (Losick et al., 1970; Leighton et al., 1972), Linn et al. (1973) have shown that these alterations are due to *in vitro* proteolysis during the purification of the enzyme. Recently, Greenleaf et al. (1973) have reported that a protein which binds RNA polymerase appears during the first hours of the sporulation process. This finding may provide a basis for understanding altered transcription processes that accompany cellular differentiation.

The manner in which the transcription process is regulated has also been extensively investigated. These studies have been largely conducted during step-down growth by transfer from a rich to a nutrient-poor medium (Neidhardt and Fraenkel, 1961) and in auxotrophic mutants of *E. coli* deprived of an essential amino acid (Schaechter, 1961). In each case, RNA accumulation is severely restricted. During amino acid deprival in these auxotrophs, protein synthesis ceases and RNA synthesis is reduced (stringent response). In other mutant strains, however, removal of an essential amino acid results in continued RNA synthesis (relaxed response). These responses are under the regulation of the RNA control (*rel*) locus (Edlin and Broda, 1968). In stringent strains (*rel*$^+$), the regulation of RNA and protein synthesis is coupled, whereas in relaxed strains (*rel*$^-$), RNA synthesis continues after protein synthesis has ceased. In contrast, both relaxed and stringent strains respond normally to step-down growth transitions, which indicates that the latter may be controlled by a distinct mechanism. Lazzarini and Winslow (1970) have shown that stable RNA (rRNA and tRNA) and not mRNA synthesis is curtailed during the stringent response. In addition, it has been shown by Cashel and Gallant (Paper 14) that substantial quantities of two unusual nucleotides (now identified as ppGpp and pppGpp) accumulate during the stringent response. Haseltine et al. (1972) reported that the synthesis of these guanosine nucleotides requires a factor washed from ribosomes of *rel*$^+$ strains, in addition to elongation factor G and the ribosome itself, thereby providing a mechanism for the regulatory coupling of transcription and translation. However, Sy et al. (1973) have recently demonstrated ppGpp synthesis with the *rel*$^+$ ribosomal wash in the absence of ribosomes. It thus appears that the *rel* gene product may have a role in the control of RNA synthesis, but the exact mechanism by which its effect is exerted remains to be established.

REFERENCES

Arditti, R., L. Eron, G. Zubay, G. Tocchini-Valenti, S. Connaway, and J. Beckwith (1970). *In vitro* transcription of the *lac* operon genes. *Cold Spring Harbor Symp. Quant. Biol., 35*, 437–442.

Bautz, E. K. F., F. A. Bautz, and J. J. Dunn (1969). *E. coli* σ factor: a positive control element in phage T4 development. *Nature (London), 223*, 1022–1024.

Birnbaum, L. S., and S. Kaplan (1973). *In vitro* synthesis of *Escherichia coli* ribosomal RNA. *J. Mol. Biol., 75*, 73–81.

Brenner, S., F. Jacob, and M. Meselson (1961). An unstable intermediate carrying information from genes to ribosomes for protein synthesis. *Nature (London)*, *190*, 675–581.

Burgess, R. R. (1969). Separation and characterization of the subunits of ribonucleic acid polymerase. *J. Biol. Chem., 244*, 6168–6176.

—— (1971). RNA polymerase. *Ann. Rev. Biochem., 40*, 711–740.

Chamberlin, M., and P. Berg (1962). Deoxyribonucleic acid-directed synthesis of ribonucleic acid by an enzyme from *Escherichia coli. Proc. Natl. Acad. Sci. U.S.A., 48*, 81–94.

——, J. McGrath, and L. Waskell (1970). Isolation and characterization of a new RNA polymerase from *Escherichia coli* infected with bacteriophage T7. *Nature (London), 228*, 227–231.

Darlix, J. L., A. Sentenac, and P. Fromageot (1971). Binding of termination factor rho to RNA polymerase and DNA. *FEBS Letter, 13*, 165–168.

Davis, R. W., and R. W. Hyman (1970). Physical locations of the *in vitro* RNA initiation site and termination sites of T7M DNA. *Cold Spring Harbor Symp. Quant. Biol., 35*, 269–281.

Davison, J., L. M. Pilarski, and H. Echols (1969). A factor that stimulates RNA synthesis by purified RNA polymerase. *Proc. Natl. Acad. Sci. U.S.A., 63*, 168–174.

——, K. Brookman, L. Pilarski, and H. Echols (1970). The stimulation of RNA synthesis by M factor. *Cold Spring Harbor Symp. Quant. Biol., 35*, 95–99.

DeCrombrugghe, B., S. Adhya, M. Gottesman, and I. Pastan (1973). Effect of rho on transcription of bacterial operons. *Nature New Biol., 241*, 260–264.

Dunn, J. J., and E. K. F. Bautz (1969). DNA-dependent RNA polymerase from *E. coli:* studies on the role of σ in chain initiation. *Biochem. Biophys. Res. Commun., 36*, 925–930.

——, F. A. Bautz, and E. K. F. Bautz (1971). Different template specificities of phage T3 and T7 RNA polymerases. *Nature New Biol., 230*, 94–96.

Edlin, G., and P. Broda (1968). Physiology and genetics of the "ribonucleic acid control" locus in *Escherichia coli. Bacteriol. Rev., 32*, 206–226.

Emmer, M., B. DeCrombrugghe, I. Pastan, and R. Perlman (1970). The cyclic AMP receptor of *E. coli:* its role in the synthesis of inducible enzymes. *Proc. Natl. Acad. Sci. U.S.A., 66*, 480–487.

Furth, J. J., J. Hurwitz, and M. Goldmann (1961). The directing role of DNA in RNA synthesis. Specificity of the deoxyadenylate deoxythymidylate copolymer as primer. *Biochem. Biophys. Res. Commun., 4*, 431–435.

Greenleaf, A. L., T. G. Linn, and R. Losick (1973). Isolation of a new RNA polymerase-binding protein from sporulating *Bacillus subtilis. Proc. Natl. Acad. Sci. U.S.A., 70*, 490–494.

Hall, B. D., and S. Spiegelman (1961). Sequence complementarity of T2-DNA and T2-specific RNA. *Proc. Natl. Acad. Sci. U.S.A., 47*, 137–146.

Haseltine, W. A. (1972). *In vitro* transcription of *Escherichia coli* ribosomal genes. *Nature (London), 235*, 329–333.

——, R. Block, W. Gilbert, and K. Weber (1972). MSI and MSII made on ribosome in idling step of protein synthesis. *Nature (London), 238*, 381–384.

Heil, A., and W. Zillig (1970). Reconstitution of bacterial DNA-dependent RNA-polymerase from isolated subunits as a tool for the elucidation of the role of the subunits in transcription. *FEBS Letter, 11*, 165–168.

Hershey, A. D., J. Dixon, and M. Chase (1953). Nucleic acid economy in bacteria infected with bacteriophage T2. *J. Gen. Physiol., 36*, 777–789.

Hinkle, D. C., and M. Chamberlin (1970). The role of sigma subunit in template site selection by *E. coli* RNA polymerase. *Cold Spring Harbor Symp. Quant. Biol., 35*, 65–72.

Hurwitz, J., J. J. Furth, M. Anders, and A. Evans (1962). The role of deoxyribonucleic acid in ribonucleic acid synthesis: II. The influence of deoxyribonucleic acid on the reaction. *J. Biol. Chem., 237,* 3752-3759.

Ippen, K., J. H. Miller, J. Scaife, and J. R. Beckwith (1968). New controlling element in the *lac* operon of *E. coli. Nature (London), 217,* 825-827.

Lazzarini, R. A., and R. M. Winslow (1970). The regulation of RNA synthesis during growth rate transitions and amino acid deprivation in *E. coli. Cold Spring Harbor Symp. Quant. Biol., 35,* 383-390.

Leighton, R. J., P. K. Freese, R. H. Doi, R. A. J. Warren, and R. A. Kelln (1972). Initiation of sporulation in *Bacillus subtilis:* requirement for serine protease activity and ribonucleic acid polymerase modification. In H. O. Halvorson, R. Hanson, L. L. Campbell (eds.), *Spores V.* American Society for Microbiology, Washington, D.C., pp. 238-246.

Linn, T. G., A. Greenleaf, R. Shorenstein, and R. Losick (1973). Loss of the sigma activity of RNA polymerase of *Bacillus subtilis* during sporulation. *Proc. Natl. Acad. Sci. U.S.A., 70,* 1865-1869.

Losick, R. (1972). *In vitro* transcription. *Ann. Rev. Biochem., 41,* 409-446.

———, A. L. Sonenshein, R. G. Shorenstein, and C. Hussey (1970). Role of RNA polymerase in sporulation. *Cold Spring Harbor Symp. Quant. Biol., 35,* 443-450.

Maitra, U., and J. Hurwitz (1965). The role of DNA in RNA synthesis: IX. Nucleoside triphosphate termini in RNA polymerase products. *Proc. Natl. Acad. Sci. U.S.A., 54,* 815-822.

———, A. H. Lockwood, J. S. Dubnoff, and A. Guha (1970). Termination, release, and reinitiation of RNA chains from DNA templates by *Escherichia coli* RNA polymerase. *Cold Spring Harbor Symp. Quant. Biol., 35,* 143-156.

Millette, R. L., C. D. Trotter, P. Herrlich, and M. Schweiger (1970). *In vitro* synthesis, termination, and release of active messenger RNA. *Cold Spring Harbor Symp. Quant. Biol., 35,* 135-142.

Neidhardt, F. C., and D. G. Fraenkel (1961). Metabolic regulation of RNA synthesis in bacteria. *Cold Spring Harbor Symp. Quant. Biol., 26,* 63-74.

Nomura, M., B. D. Hall, and S. Spiegelman (1960). Characterization of RNA synthesized in *Escherichia coli* after bacteriophage T2 infection. *J. Mol. Biol., 2,* 306-326.

Pastan, I., and R. Perlman (1970). Cyclic adenosine in bacteria. *Science, 169,* 339-344.

Pettijohn, D. E. (1972). Ordered and preferential initiation of ribosomal RNA synthesis *in vitro. Nature New Biol., 235,* 204-206.

———, O. G. Stonington, C. R. Kossman (1970). Chain termination of ribosomal RNA synthesis *in vitro. Nature (London), 228,* 235-239.

Rabussay, D., and W. Zillig (1969). A rifampicin resistant RNA-polymerase from *E. coli* altered in the β-subunit. *FEBS Letter, 5,* 104-106.

Ramakrishnan, T., and H. Echols (1973). Purification and properties of M protein: an accessory factor for RNA polymerase. *J. Mol. Biol., 78,* 675-686.

Richardson, J. P. (1970). Rho factor function in T4 RNA transcription. *Cold Spring Harbor Symp. Quant. Biol., 35,* 127-133.

Roberts, J. W. (1970). The ρ factor: termination and anti-termination in lambda. *Cold Spring Harbor Symp. Quant. Biol., 35,* 121-126.

Schaechter, M. (1961). Patterns of cellular control during unbalanced growth. *Cold Spring Harbor Symp. Quant. Biol., 26,* 53-62.

Schleif, R. (1969). Isolation and characterization of a streptolydigin resistant RNA polymerase. *Nature (London), 223,* 1068-1069.

Summers, W. C., and R. B. Siegel (1969). Control of template specificity of *E. coli*

RNA polymerase by a phage coded protein. *Nature (London), 223,* 1111–1113.

Sy, J., Y. Ogawa, and F. Lipmann (1973). Nonribosomal synthesis of guanosine 5′,3′-polyphosphates by the ribosomal wash of stringent *Escherichia coli. Proc. Natl. Acad. Sci. U.S.A., 70,* 2145–2148.

Travers, A. A., and R. R. Burgess (1969). Cyclic reuse of the RNA polymerase sigma factor. *Nature (London), 222,* 537–540.

——, R. Kamen, and M. Cashel (1970). The *in vitro* synthesis of ribosomal RNA. *Cold Spring Harbor Symp. Quant. Biol., 35,* 415–418.

Volkin, E., and L. Astrachan (1956). Phosphorus incorporation in *Escherichia coli* ribonucleic acid after infection with bacteriophage T2. *Virology, 2,* 149–161.

Witmer, H. J. (1971). *In vitro* transcription of T4 deoxyribonucleic acid by *Escherichia coli* ribonucleic acid polymerase. Sequential transcription of immediate early and delayed early cistrons in the absence of release factor, rho. *J. Biol. Chem., 246,* 5220–5227.

9

Reprinted from *Nature*, **190**(4776), 581–585 (1961)

UNSTABLE RIBONUCLEIC ACID REVEALED BY PULSE LABELLING OF *ESCHERICHIA COLI*

By Drs. FRANCOIS GROS and H. HIATT

The Institut Pasteur, Paris

Dr. WALTER GILBERT

Departments of Physics, Harvard University

AND

Dr. C. G. KURLAND, R. W. RISEBROUGH and Dr. J. D. WATSON

The Biological Laboratories, Harvard University

WHEN *Escherichia coli* cells are infected with *T* even bacteriophage particles, synthesis of host proteins stops[1], and much if not all new protein synthesis is phage-specific[2]. This system thus provides an ideal model for observing the synthesis of new proteins following the introduction of specific DNA. In particular, we should expect the appearance of phage-specific RNA, since it is generally assumed that DNA is not a direct template for protein synthesis but that its genetic information is transmitted to a specific sequence of bases in RNA. It was thus considered paradoxical when it was first noticed[3] that, following infection by the *T* even phages, net RNA synthesis stops even though protein synthesis continues at the rate of the uninfected bacterium. This could mean that DNA sometimes serves as a direct template for protein synthesis. Alternatively, net RNA synthesis may not be necessary so long as there exists the synthesis of a genetically specific RNA that turns over rapidly. This possibility was first suggested by experiments of Hershey[4], who, in 1953, reported that *T*2 infected cells contain a metabolically active RNA fraction comprising about 1 per cent of the total RNA. Several years later, Volkin and Astrachan[5] reported that this metabolic RNA possessed base ratios similar, if not identical (considering uracil formally equivalent to thymine), to those of the infecting *T*2 DNA. By 1958 they[6] extended their observation to *T*7 infected cells, where again the RNA synthesized after phage infection had base ratios similar to those of the phage DNA.

During these years, evidence[7] accumulated that the sites of much, if not all, protein synthesis are the ribosomal particles, and it was thought most likely that ribosomal RNA was genetically specific, with each ribosome possessing a base sequence which coded for a specific amino-acid sequence (one ribosome–one protein hypothesis). Direct verification of this hypothesis was lacking, and its proponents[8] were troubled by the fact that, except for phage-specific RNA, it was impossible to find any correlation within a given organism between the base ratios of DNA and RNA. Moreover, there was no evidence that phage-specific RNA was ribosomal RNA.

Nomura, Hall and Spiegelman[9] have recently discovered that following *T*2 infection there is no synthesis of typical (see below) ribosomal RNA and that the phage-specific RNA sediments at a slower rate (8*s*) than ribosomal RNA (16*s* and 23*s*). The genetic information for the synthesis of phage-specific proteins does not reside in the usual ribosomal RNA. Instead, if we assume that the synthesis of phage-specific proteins also occurs on ribosomes, then the phage-specific RNA might be viewed as a 'messenger' (to use the terminology of Monod and Jacob[10]) which carries the genetic information to the ribosomes. Furthermore, unless we postulate that there exist two different mechanisms for protein synthesis, there should also exist within uninfected normal cells RNA molecules physically similar to the phage-specific RNA and having base ratios similar to its specific DNA.

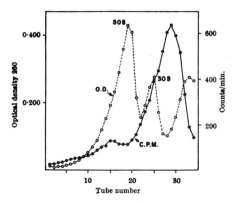

Fig. 1. Sedimentation of pulse-labelled 10^{-4} Mg²⁺ extract. 1 ml. of a crude extract labelled by a 20-sec. pulse of ¹⁴C-uracil and made in 10^{-4} M Mg²⁺ and 5×10^{-2} *tris* buffer (pH 7·4) was carefully layered on top of a 24-ml. sucrose gradient (5–20 per cent) in a Spinco $SW25$ swinging-bucket tube. It was spun 2 hr. 45 min. at 25,000 r.p.m. at 4° C. The tube bottom was then punctured and 10 drop samples taken to measure 2600 Å. absorption (2 drops) and the radioactivity in 5 per cent trichloracetic acid precipitates (8 drops). All points were counted to at least 1,000 counts

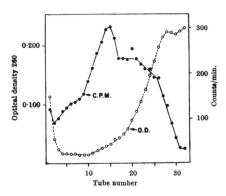

Fig. 2. Sedimentation of pulse-labelled 10^{-4} Mg²⁺ extract, long run. An extract as in Fig. 1 run for 11 hr. at 25,000 r.p.m., 15° C., on a sucrose gradient. Drop collection and radioactivity measurements were made as described in Fig. 1

Here we present evidence that RNA molecules physically similar to phage-specific RNA exist in normal *E. coli* cells and that, under suitable ionic conditions, they are associated with ribosomal particles.

Experimental plan. Most (80–85 per cent) RNA in actively growing *Escherichia coli* cells is found in ribosomal particles composed of 64 per cent RNA and 36 per cent protein[11]. There are two sizes[12] of ribosomal RNA, 16s (molecular weight = $5 \cdot 5 \times 10^5$) and 23s (molecular weight = $1 \cdot 1 \times 10^6$). The 16s RNA is derived from both 30s and 50s ribosomes, while 23s RNA is only found in 50s ribosomes. The other principal (10–15 per cent) form of RNA in *E. coli* is soluble RNA (now more appropriately called transfer RNA), which functions in the movement of activated amino-acids to the ribosomes. At least 20 (one for each amino-acid) different transfer RNA molecules exist[13], all of which have molecular weights about 25,000 and sedimentation constants of 4s.

Collectively, ribosomal and transfer RNA comprise at least 95 per cent of *E. coli* RNA. Thus messenger RNA, if present, can amount to at most only several per cent of the total RNA. Now if the messenger were stable, only a corresponding fraction of newly synthesized RNA could be messenger RNA; the great majority of new RNA being the metabolically stable ribosomal and transfer RNA's. If, however, messenger RNA is turning over (as is suggested by the original Hershey experiments) then a much larger fraction of newly made RNA must be messenger. For example, if the messenger functions only once for the synthesis of a single protein molecule, its lifetime might be only several seconds. In this event, if we look at the RNA synthesis occurring during a very short interval, then most of the newly synthesized RNA would be messenger even though this fraction may comprise only 1 per cent of the total RNA.

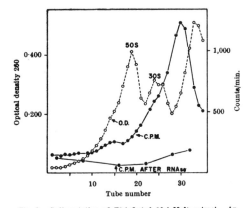

Fig. 3. Sedimentation of $T2$-infected 10^{-4} Mg²⁺ extract. An extract in 10^{-4} Mg²⁺ of cells infected with $T2$ at a multiplicity of 20 and labelled with phosphorus-32 between the second and fifth minutes after infection. Run as in Fig. 1. Several samples were treated with ribonuclease (10 γ/c.c.) to determine the background of label not in RNA

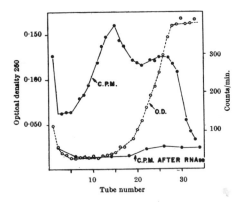

Fig. 4. Sedimentation of $T2$-infected 10^{-4} Mg²⁺ extract, long run. An extract as in Fig. 3 run for 11 hr. at 25,000 r.p.m. 20° C. Measured as in Fig. 1

We have therefore exposed *E. coli* cells to short pulses (10–20 sec. at 25° C. where the time of generation is about 90 min.) of radioactive RNA precursors (^{32}P or ^{14}C-uracil), rapidly chilled the cells with crushed ice and $M/100$ azide, prepared cell-free extracts by alumina grinding, added deoxyribonuclease at $5\gamma/\text{ml.}$, and examined the newly made RNA using the sucrose-gradient centrifugation technique[14]. In some experiments ^{14}C-uracil was given to cells of a pyrimidine-requiring mutant (*B*148) which was briefly starved (30 min. at 25° C.) for uracil. No difference has been seen between the properties of RNA labelled in these two ways. Starved cells incorporate more radioactivity, and they were used in most of the experiments reported below.

Results. Radioactive uracil (or phosphorus-32) is incorporated in RNA within several seconds after addition of the isotope. The RNA labelled by 10–20 sec. pulses is stable at 4° C. in cell-free extracts where ribonuclease and polynucleotide phosphorylase are

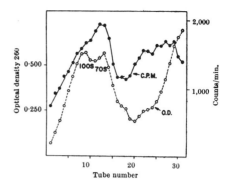

Fig. 7. Sedimentation of *T*2-infected 10^{-2} M Mg^{2+} extract. An extract in 10^{-2} M Mg^{2+} of cells infected with *T*2 and labelled with phosphorus-32 from the second to the fifth minutes after infection. Run on a sucrose gradient in 10^{-2} M Mg^{2+} for 2 hr. 45 min. at 25,000 r.p.m. Drop collection and radioactivity measurements as for Fig. 1

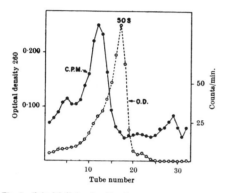

Fig. 5. Pulse-labelled active 70*s* ribosomes in 10^{-4} Mg^{2+}. An extract of cells labelled with a 20-sec. pulse of ^{14}C-uracil, made in 10^{-4} M Mg^{2+}, was purified by three 1 hr. centrifugations at 40,000 r.p.m. in 10^{-4} M Mg^{2+}, 5×10^{-3} M *tris* (pH 7·4). The resuspended pellet was run as for Fig. 1

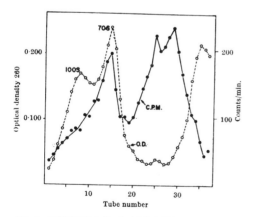

Fig. 6. Sedimentation of pulse-labelled 10^{-4} Mg^{2+} extract in 10^{-2} M Mg^{2+}. Magnesium was added to the 10^{-4} M Mg^{2+} extract of Fig. 1 to bring it to 10^{-2} M Mg^{2+}. The sample was then run on a sucrose gradient in 10^{-2} M Mg^{2+}, 5×10^{-3} M *tris*, for 2 hr. 45 min. The measurements were made as described for Fig. 1

not active. Traces ($1\gamma/\text{ml.}$) of ribonuclease degrade it under conditions where RNA in ribosomes is untouched. This suggests that it exists less protected than bound ribosomal RNA. Similarly, addition of phosphate permits the polynucleotide phosphorylase[15] in the cell extract to degrade preferentially pulse-labelled RNA. Our experiments thus as a matter of routine avoid the use of phosphate buffers.

Figs. 1 and 2 show how RNA labelled by a 20-sec. exposure of uninfected cells to ^{14}C-uracil sediments in a sucrose gradient. The cell-free extract contains 10^{-4} M magnesium ions in which 30*s* and 50*s* ribosomes predominate. The majority of the radioactivity is not associated with ribosomes but moves with a 14–16*s* peak. There is also RNA which moves more slowly, in addition to a faster forward fraction moving at 70*s*. The slower sedimenting fraction shows up more clearly in the longer centrifugation shown in Fig. 2. Here a sharp 14–16*s* component is seen together with material sedimenting at 4–8*s*. Figs. 3 and 4 illustrate experiments with extracts from *T*2 infected cells exposed to phosphorus-32 from 2 to 5 min. after infection. They are similar, if not identical, to Figs. 1 and 2. Both have a major 14–16*s* component, a slower trailing fraction, and about 10 per cent of the material moving at 70*s*. The 70*s* component can be purified by three 1-hr. centrifugations at 40,000 r.p.m. with 10^{-4} M magnesium ions to concentrate the faster-moving ribosomes of Fig. 1. Fig. 5 shows the purified ribosomes consisting largely of 50*s* particles together with a now visible 70*s* component. The radioactivity, however, sediments as 70*s* ribosomes.

Still more label is attached to 70*s* particles when extracts are made in 10^{-2} M magnesium ions or when additional magnesium ions are added to 10^{-4} M extracts. Figs. 6 and 7 again illustrate the parallel appearance of normal and phage-infected extracts. About 30 per cent of label sediments with the 70*s* and 100*s* ribosomes, with the specific activity of 70*s* ribosomes generally greater than that of 100*s* ribosomes. When slightly lower concentrations of magnesium ions are used to give similar amounts of 30*s*, 50*s* and 70*s* ribosomes, label is specifically associated only with 70*s* particles and no label is in 50*s* ribosomes. The radioactivity sedimenting about

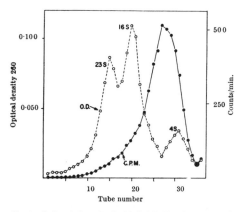

Fig. 8. Sedimentation of pulse-labelled RNA. An extract of cells labelled with a 20-sec. pulse of ¹⁴C-uracil was treated with 0·5 per cent sodium lauryl sulphate ('Duponol') and then extracted three times with phenol. The RNA was then precipitated three times with alcohol and finally resuspended in 10⁻¹ acetate buffer (pH 5·1). The RNA was run on a sucrose gradient, containing 10⁻¹ acetate (pH 5·1) and M/10 sodium chloride, for 10 hr. at 25,000 r.p.m., 4° C. The drop collection and radioactivity measurements were made as described for Fig. 1

30s shows a very broad peak (20–40s) and it is probable that high concentrations of magnesium ions cause the 14–16s material either to aggregate or to assume a more compact shape.

Purified RNA, prepared by sodium lauryl sulphate ('Duponol') and phenol treatment of 20-sec. pulse extracts, sediments as shown in Fig. 8. Again the pattern is identical to that of labelled RNA from $T2$ infected cells (Fig. 9). Most newly synthesized RNA has a sedimentation constant about 8s; there is no appreciable synthesis of typical 16s or 23s ribosomal RNA. Since much new RNA in crude extracts moves at 14–16s, we considered the possibility that pulse-RNA is more sensitive than ribosomal RNA to degradation during the phenol extractions. Sodium lauryl sulphate (0·5 per cent) alone was therefore added to labelled crude extracts. This detergent treatment[12] both separates RNA from protein and completely inhibits action of the latent ribonuclease[16] release upon ribosome breakdown. When such

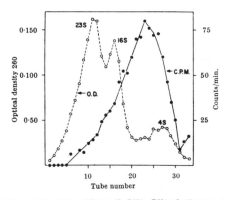

Fig. 9. Sedimentation of $T2$-specific RNA. RNA of cells infected with $T2$ and labelled with phosphorus-32 between the second and fifth minute, prepared and run as described for Fig. 8

treated extracts were run in sucrose gradients, the results were identical to those with phenol-prepared RNA.

The turnover of pulse-RNA is seen in chase experiments in which an excess of cold uracil is added after a 10-sec. ¹⁴C-uracil pulse. After 15 min. at 37° C., all the label in crude extracts leaves the 14–16s component and is incorporated in the metabolically stable ribosomal and soluble RNA. In our kinetics experiments, the 16s RNA molecules always become labelled before appreciable labelling of 23s RNA molecules; likewise the 30s ribosomes become labelled before the 50s ribosomes. Since the base ratios of 16s and 23s chains are identical[17], it is likely that the 23s molecules form from two 16s chains.

Discussion. Our pulse experiments show that uninfected cells contain unstable RNA with sedimentation constants and attachment properties similar to those of $T2$-specific RNA. It is tempting to believe that these unstable molecules convey genetic information and are 'messenger' RNA. Complete homology with phage-specific RNA, however, will be demonstrated only if the base ratios of pulse-RNA are those of DNA. Unfortunately, in *E. coli*, the DNA base ratios do not differ greatly from ribosomal RNA, and as yet the base ratios of *E. coli* pulse-RNA are not precisely known. But we do have preliminary results (Hayes, D., and Gros, F., unpublished results) from *Staphylococcus aureus* which indicate an RNA component rapidly turning over and possessing DNA-like base ratios. Moreover, in yeast, there is also reported an unstable RNA resembling DNA in base composition[18]. We thus believe that our current measurements with *E. coli* will extend this fact and rule out the possibility that pulse-RNA is a precursor sub-unit from which ribosomal RNA is built up.

A messenger role for pulse-RNA fits nicely with its specific attachment to 70s ribosomes, the sites of protein synthesis[19]. Our experiments reveal two types of attachment. In one, the labelled RNA moves reversibly on or off 70s ribosomes depending on the concentration of magnesium ions. The second type of attachment binds pulse-RNA irreversibly to 'active' 70s ribosomes. These are 70s ribosomes which do not break apart with 10⁻⁴ M magnesium ions and which Tissières *et al.*[19] have shown to be the principal, if not sole, site of *in vitro* protein synthesis. They can be washed several times with 10⁻⁴ M magnesium ions without losing their ability to act in protein synthesis, and so they must have their genetic RNA firmly attached. It is reassuring that they can contain pulse-RNA. At least two stages, one reversible, the other not, thus exist in active 70s formation.

So far as we can tell, pulse-RNA, when associated with ribosomes, sediments at exactly 70s or 100s. Many experiments show that the label sediments neither faster nor slower than ordinary ribosomes. We find it hard to believe that the 14–16s RNA component which exists in 10⁻⁴ M magnesium ion extracts can combine with 70s or 100s ribosomes without altering their rate of sedimentation. Another possibility is that with 10⁻² M magnesium ions, the 14–16s component assumes a more compact configuration similar to that of the RNA in a 30s particle. It could then associate with a free 50s ribosome to form a 70s ribosome which is transformed to an active 70s ribosome.

In conclusion, we state our findings: bacteria contain an RNA component turning over rapidly which is physically distinct from ribosomal or soluble (transfer) RNA. This fraction behaves, in its range of sedimentation constants and its attachment to ribosomes in high magnesium ion concentrations, exactly as does the phage-specific RNA made after T2 infection. Furthermore, it is associated with the active 70s ribosomes, the site of protein synthesis.

Our working hypothesis is that no fundamental difference exists between protein synthesis in phage-infected and uninfected bacteria. In both cases typical ribosomal RNA does not carry genetic information, but has another function, perhaps to provide a stable surface on which transfer RNA's can bring their specific amino-acids to the messenger RNA template.

These experiments were initiated when F. Gros was visiting the Biological Laboratories (May–August 1960). The pyrimidine-requiring strain B148 was kindly provided by Dr. Martin Lubin. The financial support of the National Science Foundation and the National Institutes of Health is gratefully appreciated.

[1] Cohen, S. S., *Bact. Rev.*, **13**, 1 (1949).

[1] Koch, G., and Hershey, A. D., *J. Mol. Biol.*, **1**, 260 (1959). Kornberg, A., Zimmerman, S. B., Kornberg, S. R., and Josse, J., *Proc. U.S. Nat. Acad. Sci.*, **45**, 772 (1959).

[2] Cohen, S. S., *J. Biol. Chem.*, **174**, 271 (1948).

[4] Hershey, A. D., Dixon, J., and Chase, M., *J. Gen. Physiol.*, **36**, 777 (1953).

[5] Volkin, E., and Astrachan, L., *Virology*, **2**, 149 (1956).

[6] Volkin, E., Astrachan, L., and Countryman, J. L., *Virology*, **6**, 545 (1958).

[7] Zamecnik, P. Z., *The Harvey Lectures* (1958–59), 256 (Academic Press, New York, 1960).

[8] Crick, F. H. C., *Brookhaven Symposia in Biology*, **12**, Structure and Function of Genetic Elements, 35 (1959).

[9] Nomura, M., Hall, B. D., and Spiegelman, S., *J. Mol. Biol.*, **2**, 306 (1960).

[10] Jacob, F., and Monod, J., *J. Mol. Biol.* (in the press).

[11] Tissières, A., Watson, J. D., Schlessinger, D., and Hollingworth, B. R., *J. Mol. Biol.*, **1**, 221 (1959).

[12] Kurland, C. G., *J. Mol. Biol.*, **2**, 83 (1960).

[13] Berg, P., and Ofengand, E. J., *Proc. U.S. Nat. Acad. Sci.*, **44**, 78 (1958). Tissières, A., *J. Mol. Biol.*, **1**, 365 (1959).

[14] McQuillen, K., Roberts, R. B., and Britten, R. J., *Proc. U.S. Nat. Acad. Sci.*, **45**, 1437 (1959).

[15] Grunberg-Manogo, M., Ortiz, P. J., and Ochoa, S., *Science*, **122**, 907 (1955).

[16] Elson, D., *Biochim. Biophys. Acta*, **27**, 217 (1958). Spahr, P. F., and Hollingworth, B., *J. Biol. Chem.* (in the press).

[17] Spahr, P. F., and Tissières, A., *J. Mol. Biol.*, **1**, 237 (1959).

[18] Yčas, M., and Vincent, W. S., *Proc. U.S. Nat. Acad. Sci.*, **46**, 804 (1960).

[19] Tissières, A., Schlessinger, D., and Gros, Françoise, *Proc. U.S. Nat. Acad. Sci.*, **46**, 1450 (1960).

10

Reprinted from *Biochem. Biophys. Res. Commun.*, 3(1), 15-19 (1960)

The Enzymic Incorporation of Ribonucleotides into Polyribonucleotides and the Effect of DNA

Jerard Hurwitz*, Ann Bresler** and Renata Diringer***

Department of Microbiology
New York University College of Medicine
New York, New York

Received July 12, 1960

Cell free extracts of <u>Escherichia coli</u> utilize the nucleoside triphosphates for the incorporation of UMP, AMP, CMP and GMP into an acid-insoluble product. Enzyme fractions which catalyze the incorporation of AMP and CMP (Preiss and Berg, 1960, Alexander et al, 1960) require the addition of RNA, ATP or CTP, and Mg^{++}. This fixation is unaffected by the addition of UTP, GTP, or DNase but is sensitive to RNase. In contrast to these requirements, it has been observed that the incorporation of radioactivity from UTP is dependent on the addition of the nucleoside triphosphates ATP, GTP and CTP. Similar observations have been reported by Weiss and Gladstone (1959) for CMP incorporation with liver nucleic preparations. This incorporation of UMP by <u>E. coli</u> preparations can be prevented if RNase or DNase is included in the reaction mixture. The latter observation suggested that the incorporation of the UMP might require the presence of DNA. A DNA-dependent incorporation of the ribonucleotide CMP has previously been reported with enzyme fractions from <u>E. coli</u> (Hurwitz, 1959) as well as thymus gland (Krakow and Kammen, 1960).

---

* Senior Postdoctoral Fellow of the National Institutes of Health

** Postdoctoral Fellow of the Dazian Foundation for Medical Research

*** Postdoctoral Fellow of the Jane Coffin Childs Memorial Fund for Medical Research

The enzyme fraction responsible for the incorporation of UMP was prepared from either glass bead extracts or from sonically disrupted cells. The enzyme activity has been purified approximately 30-fold by precipitation with protamine sulfate and elution with 0.005M phosphate buffer, pH 7.4, followed by ammonium sulfate fractionation (30-45% saturation). Such preparations exhibit the requirements summarized in Table I. Omission of any one of the nucleoside triphosphates results in a marked decrease in UMP incorporation. In the presence of RNase or DNase no detectable acid-insoluble product is formed. The nucleic acid preparation used in these experiments (Table I) was obtained from ribosome-free fractions of E. coli by lanthanum precipitation followed by phenol treatment (Kirby, 1956). Other DNA preparations from thymus gland (Kay et al, 1952), from the bacteriophage T2, or from rat liver

Table I

REQUIREMENTS FOR UMP INCORPORATION

	mμmoles incorporated
Complete System	0.81
-ATP	0.05
-GTP	0.07
-CTP	0.05
-ATP, GTP or CTP	<0.02
- Nucleic acid	0.17
Complete + DNase (1 μg)	<0.02
Complete + RNase (1 μg)	<0.02

The complete system (0.5 ml) contained uridine P*-P-P (30 mμmoles, 1 x 10^6 cpm/μmole), GTP, ATP and CTP (50 mμmoles) E. coli nucleic acid (2 optical density units at 260 mμ), MgCl$_2$ (2 μmoles), MnCl$_2$ (2 μmoles), mercaptoethanol (1 μmole), acetyl phosphate (4 μmoles), 0.04 units of acetokinase (Rose et al.), and 40 μg of an ammonium sulfate fraction obtained from E. coli W. After 20 minutes at 38°, the reaction was stopped with 0.2 ml of 7 per cent HClO$_4$ and 0.05 ml of 10 per cent albumin was added as carrier. The acid-insoluble material was washed 3 times with 1 per cent HClO$_4$ and dissolved with NH$_4$OH, plated and the radioactivity measured.

nuclei[1], similarly stimulate UMP incorporation (Table II).   These nucleic

Table II

NUCLEIC ACID REQUIREMENT

Exp.	Nucleic Acid Fraction	mμmoles incorporated
1	E. coli nucleic acid	0.32
2	T2-DNA	0.52
3	Thymus DNA	0.40
4	E. coli RNA	0.03
5	No addition	0.04
6	1 or 3 + DNase	<0.02
7	1 or 3 + RNase	<0.02

The additions were as in Table I with the exception that T2 DNA (1 optical density unit at 260 mμ) and thymus DNA (2 optical density units at 260 mμ) replaced the E. coli nucleic acid fraction where indicated and Mg^{++} (8 μmoles) was added instead of the Mg^{++} - Mn^{++} mixture. The enzyme preparation was pretreated with DNase prior to the assay (0.005 μg of DNase per ml of enzyme solution); this treatment reduces the amount of incorporation in the absence of added DNA.

acid fractions are inactivated by pretreatment with DNase.   RNA isolated from E. coli does not replace DNA in this system (Table II).

The acid-insoluble radioactive material produced in the reaction has the properties of RNA.   The reaction mixture was dialyzed against large volumes of a salt solution to remove the starting material and small oligonucleotides.   The non-dialyzable radioactive product was sensitive to treatment with NaOH or RNase, which rendered it completely acid soluble, but was not affected by treatment with DNase.   Following hydrolysis with NaOH, the mononucleotides were separated by paper electrophoresis; all four nucleotides were labelled.   The distribution of P^{32} in cpm was as follows: cytidylate 220, adenylate 420, guanylate 390 and uridylate 230.   This was not exclusively end-group addition since degradation of the product with snake venom phosphodiesterase (Razell

[1]   The authors are indebted to Dr. R. Rendi of the Department of Biochemistry for this preparation.

and Khorana, 1959, Singer et al, 1958) resulted in liberation of both radioactivity and ultraviolet absorbing material at parallel rates.

With respect to the requirement for all four nucleoside triphosphates (Table I) it has been established that these are also incorporated into an acid-insoluble product. The possibility that RNA is also involved has not been completely excluded since the enzyme preparations contain small amounts of RNA.

This enzyme system may be related to that previously described (Hurwitz, 1959) for the incorporation of CMP into DNA. In each case inactivation of the enzyme preparations during storage can be reversed by the addition of Mn^{++} to the incubation mixture. This effect of Mn^{++} can be related to inhibition of nucleases in the enzyme preparations. In the previous work a requirement for all four ribonucleotides for the incorporation of CMP was detected and the system was also dependent upon the presence of DNA.

While the present results with UMP incorporation are in accord with the formation of chains of ribonucleotides, the nature of the product, and the precise role of DNA must await further purification of the system.

This program was supported by grants from the National Institutes of Health.

References

Alexander, M., Bresler, A., Furth, J. and Hurwitz, J. (1960) Fed. Proc. 19, 318.

Hurwitz, J. (1959) J. Biol. Chem. 234, 2351.

Kay, E.R.M., Simmons, N.S. and Dounce, A.L.   (1952) J. Am. Chem. Soc. 74, 1724.

Kirby, K.S. (1956) Biochem. J. 64, 405.

Krakow, J.S. and Kammen, H.O. (1960) Fed. Proc. 19, 307.

Preiss, J. and Berg, P. (1960) Fed. Proc. 19, 317.

Razell, W.E. and Khorana, H.G. (1959) J. Biol. Chem. 234, 2114.

Rose, I.A., Grunberg-Manago, M., Korey, S.R. and Ochoa, S. (1954) J. Biol. Chem. 211, 737.

Singer, M.F., Hilmoe, R.J. and Heppel, L.A. (1958) Fed. Proc. 17, 312.

Weiss, S.B. and Gladstone, L. (1959) J. Am. Chem. Soc. 81, 4118.

11

Reprinted from *Nature*, **221**(5175), 43–46 (1969)

Factor Stimulating Transcription by RNA Polymerase

by
RICHARD R. BURGESS
ANDREW A. TRAVERS
Biological Laboratories,
Harvard University

JOHN J. DUNN
EKKEHARD K. F. BAUTZ
Institute of Microbiology,
Rutgers University

A protein component usually associated with RNA polymerase can be separated from the enzyme by chromatography on phosphocellulose. The polymerase is unable to transcribe T4 DNA unless this factor is added back.

IN *E. coli* the synthesis of all types of cellular RNA is thought to be mediated by a single enzyme, DNA-dependent RNA polymerase. The highly purified enzyme[1-6] can catalyse the synthesis of RNA *in vitro* in the presence of DNA and the ribonucleoside triphosphates. When an intact double helical DNA is used as template, transcription *in vitro* is asymmetric—at any given region along the DNA only one of the complementary DNA strands is transcribed[7-10]. This is also characteristic of transcription *in vivo*[11-13]. Moreover, the selective transcription *in vitro* of certain regions of T4 and λ DNA[10,14-17], coupled with studies on the binding of RNA polymerase to DNA[18-22], suggests that the polymerase initiates RNA synthesis at specific sites on the DNA. The state of aggregation of the enzyme is strongly influenced by ionic strength[19,21,23,24], substrate[25], and possibly by enzyme concentration and temperature. Most investigators agree that the molecular weight of the active enzyme is in the range of 350,000–700,000 daltons. This large size is consistent with the observation that the enzyme is composed of several different polypeptide chains[26,27]. Furthermore, it has been assumed that "highly purified" RNA polymerase is a protein entity from which nothing can be further removed without destroying its enzyme activity. We report here, however, the separation of polymerase into two components. One contains enzyme activity, but its ability to transcribe certain DNA templates is greatly reduced; the other is a factor able to stimulate RNA synthesis on these restrictive templates to normal levels.

It was independently observed at Harvard and Rutgers that when RNA polymerase was purified by chromatography on a phosphocellulose column, the enzyme obtained, although able to transcribe calf thymus DNA almost normally, was much less active when assayed with T4 DNA as template. Enzyme purified by an alternative procedure, however, was almost equally active on both templates. This suggested that some component necessary for the transcription of T4 DNA was separated from RNA polymerase by the phosphocellulose column. Furthermore, the activity of the phosphocellulose-purified enzyme on T4 DNA could be greatly enhanced by the addition of another fraction from the phosphocellulose column. This fraction lacked significant RNA polymerase activity of its own. We describe here the identification and some properties of this stimulating component.

Isolation of RNA Polymerase

The RNA polymerase we used was purified as outlined in Fig. 1 from *E. coli* K12. Three methods were used to achieve the final purification of the polymerase. Enzyme purified with phosphocellulose (PC enzyme) and with glycerol gradient (GG enzyme) was prepared by the method of Burgess (manuscript in preparation). Using

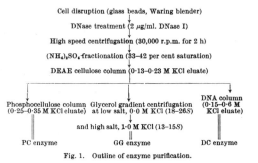

Cell disruption (glass beads, Waring blender)

DNase treatment (2 μg/ml. DNase I)

High speed centrifugation (30,000 r.p.m. for 2 h)

(NH₄)₂SO₄ fractionation (33–42 per cent saturation)

DEAE cellulose column (0·13–0·23 M KCl eluate)

Phosphocellulose column (0·25–0·35 M KCl eluate) → PC enzyme

Glycerol gradient centrifugation at low salt, 0·0 M KCl (18–26S) and high salt, 1·0 M KCl (13–15S) → GG enzyme

DNA column (0·15–0·6 M KCl eluate) → DC enzyme

Fig. 1. Outline of enzyme purification.

the same method through the DEAE cellulose step, J. J. D. and E. K. F. B., from Rutgers, purified the enzyme by binding it to a cellulose column containing immobilized T4 DNA and eluting the bound enzyme with high salt[28]. DNA column enzyme (DC enzyme) obtained in this way and GG enzyme are essentially identical and do not differ significantly from enzymes purified by the method of Chamberlin and Berg[1]. Unless otherwise stated, the work described here was carried out with GG enzyme, but similar results were obtained with DC enzyme.

Identification of the Component Stimulating Transcription of T4 DNA

To isolate the component necessary for the transcription of T4 DNA, GG enzyme was chromatographed on a

Table 1. ASSAY OF STIMULATING ACTIVITY OF PEAKS *A*, *B* AND *C*

Sample assayed for stimulation	μμmole AMP incorporated/min		Stimulation ratio
	No PC enzyme added	4 μg PC enzyme added	
PC enzyme (4·0)	3	9	2
GG enzyme (1·1)	57	170	38
Peak *A* (1·0)	13	114	34
Peak *B* (1·3)	66	203	46
Peak *C* (1·9)	2	8	2

The assay mixture (0·25 ml.) contained 0·04 M *tris*-HCl buffer, pH 7·9, at 25° C, 0·01 M MgCl₂, 0·0001 M EDTA, 0·0058 M 2-mercaptoethanol, 0·15 M KCl, 0·5 mg/ml. bovine serum albumin (Calbiochem, crystalline), 0·15 mM CTP, GTP and UTP, 0·15 mM ¹⁴C-ATP (1 mCi/mmole), 20 μg/ml. T4 phage DNA, and varying amounts of enzyme and stimulating factor. The mixture was incubated for 10 min at 37° C, chilled, precipitated with 5 per cent TCA, and filtered on 'Millipore' filters. Samples were counted on an end-window gas-flow counter. The amount of incorporation is expressed as μμmoles of AMP incorporated per min of incubation. To obtain a measure of the stimulating activity in a particular sample, that sample was assayed in the absence and presence of PC enzyme. The additional incorporation in the presence of PC enzyme was due to the stimulation of this added PC enzyme by the factor in the sample. When this additional incorporation is divided by the small amount of incorporation obtained with PC enzyme alone, the stimulation ratio is obtained. The values in parentheses indicate the amount of protein, in μg, added to the assay mixture.

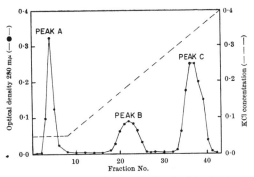

Fig. 2. **Phosphocellulose column profile.** Phosphocellulose (Whatman P11, 7·4 mequiv./g) was washed with base and acid and titrated to pH 7·9 at 25° C with KOH. This material was placed in a column (0·9 × 10 cm) and equilibrated extensively with a buffer containing 0·05 M *tris*-HCl, pH 7·9, at 25° C, 0·05 M KCl, 0·0001 M EDTA, 0·0001 M dithiothreitol and 5 per cent glycerol. The pH of the outflow was 8·1 ± 0·05 at 4° C. A sample containing 9 mg of GG enzyme in 4 ml. of this buffer was applied to the column. The column was then washed with 5 ml. of buffer and eluted with a 100 ml. gradient from 0·05 M to 0·4 M KCl. The flow rate was 0·5 ml./min and 2·2 ml. fractions were collected. Tubes 4, 22 and 36 were taken as representative of peaks *A*, *B* and *C* respectively. Assuming a value of $E^{1\%}_{280\,m\mu} = 6\cdot6$ for all fractions, these tubes contain 0·5, 0·14 and 0·4 mg of protein per ml. respectively. A small amount of polymerase activity was present in the flow-through (peak *A*). If the sample is applied to the column too rapidly (at a rate greater than one column volume per 30 min), then much more enzyme will be found in the flow-through.

phosphocellulose column. This column separated the material present in GG enzyme into three peaks, *A*, *B* and *C* (Fig. 2). Each peak was assayed for ability to promote the transcription of T4 DNA in the presence and absence of PC enzyme. Peak *A* contains stimulating activity, peak *B* is a mixture of PC enzyme and stimulating activity, while peak *C* is identical to normal PC enzyme. The results (Table 1) show that, in the absence of PC enzyme, peak *B* can use T4 DNA as a template for

RNA synthesis while peaks *A* and *C* have little activity. The addition of peak *A* material to PC enzyme, however, results in a stimulation of RNA synthesis which increases with increasing amounts of peak *A*, until a rate of RNA synthesis comparable with that obtained with GG enzyme is reached. Although peak *B* is able to stimulate PC enzyme, it can itself be stimulated by the addition of peak *A*, and thus is a mixture of PC enzyme and stimulating component.

Each of the fractions was analysed by polyacrylamide gel electrophoresis in various conditions to determine the species of protein present. These gels were run in the presence of either 8 M urea or 0·1 per cent sodium dodecyl sulphate (SDS), both of which can cause the dissociation of oligomeric proteins into single polypeptide chains. The protein bands observed in this analysis are shown in Fig. 3.

Previous studies[27] have shown that phosphocellulose enzyme is composed of two chief types of polypeptide chains, which we here designate α and β. These chains are present in equimolar amounts and have molecular weights of ~40,000 and ~160,000 respectively. α and β are present in all fractions but only in very small amounts in peak *A*. GG enzyme contains, in addition, two extra bands which we shall designate σ and τ. These bands are present in peak *A* in amounts greatly in excess of the amounts of α and β present. Peak *B* contains bands α, β and σ, but lacks τ. These patterns are observed on both the 8 M urea gels and the 0·1 per cent SDS gels. These data are consistent with band σ being the component responsible for the stimulating activity, but they do not exclude the possibility that band τ or even some other material might also stimulate.

The stimulating component was identified as σ by the zone sedimentation of peak *A* material on a glycerol density gradient (Fig. 4). The factor required for T4 transcription was found to sediment at ~5S. Analysis of the gradient fractions by electrophoresis on 8 M urea gels showed that band σ sedimented identically to the stimulating activity whereas band τ sedimented at about 8S in a region with no stimulating activity.

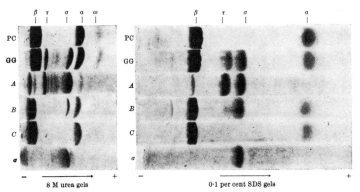

Fig. 3. Polyacrylamide gel electrophoresis patterns of different preparations of RNA polymerase and of purified factor. From top to bottom: PC enzyme (20 μg), GG enzyme (20 μg), peak *A* (10 μg), peak *B* (7 μg), peak *C* (8 μg) and purified factor (2 μg) from tube 17 of the glycerol gradient shown in Fig. 4. 8 M urea gels were prepared and run according to the general method described by Davis[34]. These gels contained 7·5 per cent acrylamide. 0·1 per cent SDS gels were run according to the procedure of Shapiro *et al.*[35] and contain 5 per cent acrylamide. In both cases the gels were stained by immersing them in a 0·25 per cent solution of Coomassie brilliant blue in methanol : acetic acid : water (5 : 1 : 5 v/v/v) for at least 2 h. The gels were then soaked in 7·5 per cent acetic acid, 5 per cent methanol for 0·5 h and finally destained electrophoretically in this same solvent. The bands observed are designated β, τ, σ, α and ω; α, β and ω are the bands normally seen in PC enzyme. The molecular weights of α and β are ~40,000 and ~160,000 respectively. The molecular weight of β was previously reported to be 110,000 (ref. 27), but recent measurements (Burgess, in preparation) indicate that 160,000 is a more accurate value. The band β as seen on 0·1 per cent SDS gels appears as two closely spaced bands of equal intensity which probably correspond to two different polypeptide chains, with molecular weights of about 155,000 and 165,000. For simplicity these are both called β in the text. In addition, a small polypeptide, ω, with a molecular weight of about 10,000 can be seen moving ahead of α on the 8 M urea gels. It is present in GG enzyme, PC enzyme and DC enzyme but it is not yet known whether it is a component of RNA polymerase or merely a tightly binding impurity. The amount of band τ observed in GG enzyme and also in peak *A* is variable. DC enzyme (not shown) contains only traces of τ and thus it is probably an impurity. From the SDS gels it is estimated that 45 per cent of peak *A* protein is σ. Purified factor is estimated to be about 80 per cent pure and is completely free of τ. The right hand two-thirds of the urea gels which contain no bands are not shown.

Table 2. HEAT INACTIVATION OF THE STIMULATING FACTOR

Treatment of factor before assay	$\mu\mu$mole AMP incorporated/min	Per cent inactivation of stimulating activity
5 min at 0° C	294	0
5 min at 37° C	295	0
5 min at 45° C	154	52
5 min at 50° C	31	98

Peak A protein (0·9 μg) was added to each assay tube which also contained assay solution lacking the four triphosphates and T4 DNA. Separate tubes were incubated at the indicated temperatures for 5 min and then chilled to 0° C. The triphosphates, T4 DNA, and PC enzyme (6 μg) were added and the assay performed as described in the legend to Table 1. With no added factor, PC enzyme incorporated 17 $\mu\mu$moles of AMP. Factor alone incorporated 9 $\mu\mu$moles of AMP. The percentage inactivation at a given temperature was calculated by subtracting 26 $\mu\mu$moles from the total incorporation observed after treatment at that temperature and then setting the incorporation of the unheated sample equal to 100 per cent activity.

Stimulating Factor is a Protein

The association of the stimulating activity with a specific band on polyacrylamide gels suggests that the factor is a protein. To confirm this, we determined the heat stability of the stimulating activity (Table 2). Incubation of peak A material for 5 min at 45° C and 50° C resulted in an inactivation of the stimulating activity of 52 per cent and 98 per cent respectively. Furthermore, the activity was sensitive to trypsin. We conclude therefore that the factor is a heat labile protein, although it is still possible that the factor is associated with a nucleic acid component of low molecular weight.

The molecular weight of a polypeptide chain can be estimated from its mobility on 0·1 per cent SDS polyacrylamide gels[29]. Using this procedure with β-galactosidase, bovine serum albumin and ovalbumin, with molecular weights of 130,000, 67,000 and 45,000 respectively, as molecular weight markers, the molecular weight of σ was calculated to be about 95,000 ± 5,000. This is consistent with the S value of about 5 obtained from the glycerol gradient.

Factor Requirements on Various DNA Templates

RNA synthesis in the presence and absence of factor was measured for several different DNA templates (Table 3). The greatest stimulation was observed with native T4 DNA, where the presence of factor increased the amount of synthesis seventy-five-fold. With all other templates tested the stimulation was considerably lower. This variation may be ascribed to two factors. First, the fully stimulated levels of RNA synthesis vary according to the template used: this has been observed by several investigators[1-3]. Second, in the absence of factor, different DNA templates direct the synthesis of differing amounts of RNA. From analytical gels we estimate that PC enzyme contains less than 2 per cent as much factor as GG enzyme. Thus it is possible that the very small amount of RNA synthesis off T4 DNA with PC enzyme is due to traces of remaining factor. With calf thymus DNA, however, this could not easily explain the results. A more likely possibility is that there are some sites at which initiation can occur in the absence of factor. It is clear that such sites are not merely fully denatured regions, for denaturation of T4 DNA does not result in increased synthesis with PC enzyme. Furthermore, φX-174 DNA is also a poor template. The behaviour of PC enzyme on

Table 3. STIMULATION ON VARIOUS DNA TEMPLATES

DNA template	mμmoles AMP incorporated/min/mg enzyme		
	PC enzyme	PC enzyme + factor	GG enzyme
T4—native	0·5	33·0	37·5
T4—denatured	0·5	6·1	3·0
Calf thymus—native	14·2	32·8	30·5
Calf thymus—denatured	3·3	14·5	10·7
φX-174	0·9	6·2	4·9

RNA synthesis in the presence and absence of factor was assayed as described in the legend to Table 1, except that the DNA concentration was 10 μg/ml. DNA was denatured by adding 1/10 volume of 2 N NaOH to a 20 μg/ml. DNA solution. After standing at 25° C for 10 min, the solution was neutralized. Almost identical results were obtained if the DNA was denatured by heating at 95° C for 10 min, and then rapidly chilling in ice. The concentrations of PC enzyme and GG enzyme in the reaction mixture were both 4 μg/ml. Peak A protein was present, where indicated, at a concentration of 4 μg/ml. This corresponds to 2 μg/ml. of factor. The ratio of factor to enzyme in the mixture of PC enzyme and factor was about twice that normally occurring in GG enzyme. Incorporation by factor alone was negligible for all types of DNA tested. In all cases saturating amounts of DNA were used. The φX-174 phage DNA was a gift from Dr D. T. Denhardt.

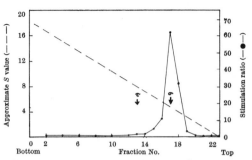

Fig. 4. Analysis of peak A by zone centrifugation. A 0·2 ml. sample containing 100 μg of peak A protein was layered on a 4·8 ml. linear 10–30 per cent glycerol density gradient containing 0·01 M *tris*-HCl buffer, pH 7·9, 0·01 M MgCl$_2$, 0·15 M KCl, 0·0001 M EDTA, and 0·0001 M dithiothreitol. The gradient was centrifuged for 11 h at 60,000 r.p.m. in a Spinco *SW*65 rotor at 4° C. 0·22 ml. fractions were collected. Each fraction was assayed for stimulating activity as described in the legend to Table 1. The proteins present in each fraction were analysed by electrophoresis on 8 M urea polyacrylamide gels. Band σ peaked in tube 17, band τ in tube 18. Molecular weight markers (*E. coli* β-galactosidase, human gamma-globulin, and egg lysozyme with sedimentation coefficients of 16S, 7S and 1·9S, respectively) were centrifuged on a parallel gradient. The sedimentation coefficients of σ and τ were estimated to be about 5S and 8S, respectively.

T4 DNA and calf thymus DNA is remarkably similar to that of the RNA polymerase isolated from T4 infected cells by Walter *et al.*[30].

Factor is not a Nuclease

It could be argued that σ is a type of nuclease which makes single-stranded breaks in the DNA, thus "activating" the DNA. One prediction of this hypothesis is that DNA which has been used as a template for transcription in the presence of both enzyme and factor should then be capable of supporting the initiation of RNA chains in the absence of factor. This is not the case. The experiment shown in Table 4 shows that DNA which has been used as a template for factor-stimulated transcription exhibits virtually the same unstimulated and stimulated levels of transcription on re-use as DNA which has not previously been incubated with factor. In addition, evidence obtained by several investigators[19-21,31] argues that DNA used for transcription by polymerase purified by the methods of Chamberlin[1] and Furth[2] remains intact. Because enzyme prepared in this way contains σ, it seems unlikely that the stimulation observed is due to nuclease action.

Formation of a Factor–Polymerase Complex

Several lines of evidence suggest that the factor can exist in a complex with RNA polymerase. First, the factor purifies with polymerase through steps involving protamine sulphate precipitation, ammonium sulphate fractionation and DEAE–cellulose chromatography. Furthermore, it remains with polymerase during low and high salt glycerol gradient centrifugation ($\Gamma/2 = 0.04$ and 1·0, respectively) where the polymerase sediments at 24S and 14S. The free factor sediments at about 5S, so it must be tightly bound to polymerase in all these conditions. Second, from the molecular weights of α, β and σ, and from the intensity of their stained bands on polyacrylamide gels, it is possible to make an estimate of the relative amounts of each band present in GG enzyme and also in peak B. In both, such an estimate yields a very approximate molar ratio of α : β : σ of 2 : 2 : 1. Third, complex formation between PC enzyme and factor can be demonstrated by running standard pH 8·7 polyacrylamide gels in the absence of dissociating agents (Fig. 5). In these conditions PC enzyme is resolved into several bands, which probably represent aggregates. Peak B, which contains the complex of α, β and σ, migrates as a single band which moves ahead of the PC enzyme bands. Purified σ moves

Table 4. **EFFECT OF PREINCUBATION WITH FACTOR AND PC ENZYME ON THE ABILITY OF T4 DNA TO DIRECT FACTOR-DEPENDENT TRANSCRIPTION**

Material present during preincubation of DNA	$\mu\mu$moles AMP incorporated/min	
	PC enzyme alone	PC enzyme + factor
No additions	8	264
PC enzyme (10 μg)	10	282
Factor (1·5 μg)	17	238
PC enzyme (10 μg) + factor (1·5 μg)	15	259
No preincubation	12	259

Four reaction mixtures (0·25 ml. each) for RNA synthesis were set up as described in the legend to Table 1, with the modification that BSA was omitted and non-radioactive ATP replaced ¹⁴C-ATP. PC enzyme and factor were added as indicated. Each reaction mixture was incubated for 10 min at 37° C. They were then diluted to 1 ml. with distilled water and extracted with 1 ml. of water-saturated phenol. The DNA was precipitated from the aqueous phase by the addition of 2 ml. ethanol. The precipitates were washed twice with 2 ml. of ethanol, collected by centrifugation and dried *in vacuo* to remove all traces of ethanol. Finally the DNA samples were redissolved in 0·2 ml. 0·01 M *tris* buffer pH 7·9. The ability of this DNA to direct RNA synthesis by PC enzyme in the presence and absence of factor was then assayed as described in the legend to Table 1 with the modification that the reaction volume was 0·10 ml. and BSA was omitted. Each tube contained 4 μg of PC enzyme. 0·6 μg factor (1·3 μg of peak *A* material) was also added where indicated.

faster than the complex. If σ and PC enzyme are mixed in approximately equivalent amounts and then subjected to gel electrophoresis, a single band corresponding to the complex is seen.

Even though PC enzyme and σ form a complex, for example as in peak *B*, the addition of either factor or PC enzyme to this complex results in stimulation. We are investigating whether the enzyme and factor are in rapid equilibrium with complex and whether factor is released for re-use in other complexes by the act of initiation.

Enzyme and complex elute at different ionic strengths from a phosphocellulose column, so it seems possible that they would elute from a DNA cellulose column at different ionic strengths. We found, however, that they behave identically on such a column. This provides some indication that the factor does not function merely by increasing the affinity of the polymerase for DNA. Furthermore, the free factor was not retained by the DNA cellulose column at 0·1 M KCl. This suggests that the free factor does not bind to DNA in conditions where the basic enzyme and the complex do so.

The Function of σ

The results clearly show that PC enzyme can by itself initiate the synthesis of RNA chains, and can catalyse chain elongation. Thus it is possible that this enzyme is the fundamental RNA polymerase. The presence of the stimulating factor, σ, greatly enhances the amount of RNA synthesis, the degree of enhancement being dependent on the DNA template used. Several possible modes of action of σ can be proposed. It could stimulate initiation, increase the rate of polymerization or prevent unusually early cessation of chain growth. Preliminary evidence (Travers and Burgess, manuscript in preparation) indicates that σ markedly increases the number of

RNA chains initiated. This suggests that σ acts at the level of initiation.

We can thus pose the question: is the additional initiation observed in the presence of σ merely due to an increase in the rate of initiation at sites poorly utilized in its absence, or does this initiation occur at sites which absolutely require σ for their expression? The first possibility implies that the PC enzyme determines the specificity of initiation and that σ may have some other function in the process of initiation. If σ itself determines the specificity of initiation, however, the interesting possibility arises that several similar factors could exist, each with a specificity for a different type of initiation site. This latter idea is attractive, for recent studies using the antibiotic rifamycin suggest that *in vivo* only one kind of RNA polymerase exists[32]. Yet there is also much evidence to indicate that *in vivo* the control of mRNA, tRNA and rRNA synthesis is not coordinate[33]. σ and similar factors could then act as positive control elements regulating the amount of synthesis of different classes of RNA, including the late RNA of certain bacteriophages.

We thank Professor J. D. Watson for his enthusiastic interest and support, Professor K. Weber, Mr J. Roberts and Dr J. Tkacz for helpful discussions and suggestions, and Miss Anne-Marie Piret and Mrs Christine Roberts for technical assistance. This work was supported by grants from the US National Science Foundation and the US Public Health Service. One of us (A. A. T.) is a postdoctoral fellow of the Damon Runyon Memorial Fund for Cancer Research, and R. R. B. is a National Science Foundation predoctoral fellow.

Received December 2, 1968.

[1] Chamberlin, M., and Berg, P., *Proc. US Nat. Acad. Sci.*, **48**, 81 (1962).
[2] Furth, J. J., Hurwitz, J., and Anders, M., *J. Biol. Chem.*, **237**, 2611 (1962).
[3] Stevens, A., and Henry, J., *J. Biol. Chem.*, **239**, 196 (1964).
[4] Fuchs, E., Zillig, W., Hofschneider, P. H., and Preuss, D., *J. Mol. Biol.*, **10**, 546 (1964).
[5] Richardson, J. P., *Proc. US Nat. Acad. Sci.*, **55**, 1616 (1966).
[6] Babinet, C., *Biochem. Biophys. Res. Commun.*, **26**, 639 (1967).
[7] Hayashi, M., Hayashi, M. N., and Spiegelman, S., *Proc. US Nat. Acad. Sci.*, **51**, 351 (1964).
[8] Geiduschek, E. P., Tocchini-Valentini, G. P., and Sarnat, M., *Proc. US Nat. Acad. Sci.*, **52**, 486 (1964).
[9] Green, M., *Proc. US Nat. Acad. Sci.*, **52**, 1388 (1964).
[10] Luria, S. E., *Biochem. Biophys. Res. Commun.*, **18**, 735 (1965).
[11] Hayashi, M., Hayashi, M. N., and Spiegelman, S., *Proc. US Nat. Acad. Sci.*, **50**, 664 (1963).
[12] Tocchini-Valentini, G. P., Stodolsky, M., Sarnat, M., Aurisicchio, A., Graziosi, F., Weiss, S. B., and Geiduschek, E. P., *Proc. US Nat. Acad. Sci.*, **50**, 935 (1963).
[13] Marmur, J., and Greenspan, C. M., *Science*, **142**, 387 (1963).
[14] Khesin, R. B., Shemyakin, M. F., Gorlenko, Zh. M., Bogdanova, S. L., and Afanaseva, T. P., *Biokhimiya*, **27**, 1092 (1962).
[15] Geiduschek, E. P., Snyder, L., Colvill, A. J. E., and Sarnat, M., *J. Mol. Biol.*, **19**, 541 (1966).
[16] Naono, S., and Gros, F., *Cold Spring Harbor Symp. Quant. Biol.*, **31**, 363 (1966).
[17] Cohen, S. N., Maitra, U., and Hurwitz, J., *J. Mol. Biol.*, **26**, 19 (1967).
[18] Crawford, L. V., Crawford, E. M., Richardson, J. P., and Slayter, H. S., *J. Mol. Biol.*, **14**, 593 (1965).
[19] Richardson, J. P., *J. Mol. Biol.*, **21**, 83 (1966).
[20] Jones, O. W., and Berg, P., *J. Mol. Biol.*, **22**, 199 (1966).
[21] Pettijohn, D., and Kamiya, T., *J. Mol. Biol.*, **29**, 275 (1967).
[22] Sentenac, A., Ruet, A., and Fromageot, P., *Europ. J. Biochem.*, **5**, 385 (1968).
[23] Priess, H., and Zillig, W., *Biochim. Biophys. Acta*, **140**, 540 (1967).
[24] Stevens, A., Emery, jun., A. J., and Sternberger, N., *Biochem. Biophys. Res. Commun.*, **24**, 929 (1966).
[25] Smith, D. A., Martinez, A. M., Ratliff, R. L., Williams, D. L., and Hayes, F. N., *Biochemistry*, **6**, 3057 (1967).
[26] Zillig, W., Fuchs, E., and Millette, R., in *Proc. Nucleic Acid Res.* (edit. by Cantoni, G. L., and Davies, D. R.), 323 (Academic Press, New York, 1966).
[27] Burgess, R. R., *Fed. Proc.*, **27**, 295 (1968).
[28] Alberts, B. M., Amodio, F. J., Jenkins, M., Gutman, E. D., and Ferris, F. L., *Cold Spring Harbor Symp. Quant. Biol.*, **33** (in the press).
[29] Shapiro, A., Vinuela, E., and Maizel, J. V., *Biochem. Biophys. Res. Commun.*, **28**, 815 (1967).
[30] Walter, G., Seifert, W., and Zillig, W., *Biochem. Biophys. Res. Commun.*, **30**, 240 (1968).
[31] Maitra, U., and Hurwitz, J., *J. Biol. Chem.*, **242**, 4897 (1967).
[32] Tocchini-Valentini, G. P., Marino, P., and Colvill, A. J., *Nature*, **220**, 275 (1968).
[33] Maaløe, O., and Kjeldgaard, N. O., *Control of Macromolecular Synthesis* (Benjamin, New York, 1966).
[34] Davis, B. J., *Ann. NY Acad. Sci.*, **121**, 406 (1964).

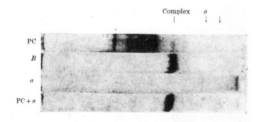

Fig. 5. Reconstitution of the enzyme–factor complex as demonstrated by electrophoresis on polyacrylamide gels. From top to bottom PC enzyme (4 μg), peak *B* (4 μg), purified factor (1 μg) from tube 17 of the glycerol gradient shown in Fig. 4, and a mixture of PC enzyme and purified factor. Polyacrylamide gels, pH 8·7, 4 per cent acrylamide, were prepared and run as described by Davis[34], and stained as described in the legend to Fig. 3. σ appears as two very faint bands which move about 20 and 30 per cent faster than the complex and are indicated by arrows. The marker dye, seen very near the right end of the third gel from the top, just ran off of the other three gels.

12

Reprinted from *Proc. Natl. Acad. Sci., 66*(1), 104–110 (1970)

Mechanism of Activation of Catabolite-Sensitive Genes: A Positive Control System*

Geoffrey Zubay,† Daniele Schwartz,‡ and Jon Beckwith‡

COLUMBIA UNIVERSITY, NEW YORK, NEW YORK, AND HARVARD MEDICAL SCHOOL,
BOSTON, MASSACHUSETTS

Communicated by Sol Spiegelman, February 26, 1970

Abstract. Catabolite repression is defined as the inhibition of enzyme induction by glucose or related substances. In the bacterium *E. coli*, the effect of glucose appears to be due to a lowering of the cyclic AMP level. A DNA-directed cell-free system for β-galactosidase synthesis has served as a model system for studying the mechanism of action of cyclic AMP. Previously, it was reported that in this system cyclic AMP is required for normal initiation of mRNA synthesis. A protein factor which acts in conjunction with the cyclic AMP has been partially purified. This protein factor has a high affinity for cyclic AMP. These and other results presented herein lead us to the conclusion that cyclic AMP and a protein factor called the catabolite gene activator protein are part of a positive control system for activating catabolite-sensitive genes.

Introduction. Catabolite repression involves the inhibition of enzyme induction by glucose or biochemically related compounds such as glucose 6-phosphate, fructose, or glycerol.[1] The extent of inhibition, which varies from a few per cent to more than 90%, is dependent on the bacterial strain, the source of the catabolite, and the growth conditions. Enzymes which show the repression effect include enzymes of glycerol regulation, the *gal* operon, the arabinose operon, the *lac* operon, tryptophanase, D-serine deaminase, and histidase. In fact, most enzymes classified as inducible show some effect. Although observations on catabolite repression date back to the turn of the century,[2] it is only recently that significant understanding of the mechanism of the reaction has been achieved.

For a long time the most serious impediment to progress was in determining, out of the multiplicity of effects of glucose, the ones which were directly linked to catabolite repression. A turning point in our understanding of this phenomenon was provided when Makman and Sutherland[3] showed that a rapid decrease in the intracellular level of cyclic 3':5'-adenosine monophosphate (cyclic AMP) occurs in the presence of glucose. From that time on it has become increasingly clear that the glucose effect is due to a lowering of the cyclic AMP level. Thus Perlman and Pastan[4] and Ullman and Monod[5] showed that the repressing effect of glucose on the synthesis of enzymes could be reversed by the addition of cyclic AMP. Having determined that cyclic AMP plays a key role in catabolite repression, furrher understanding of the phenomenon requires that

98

two questions be answered: (1) How does glucose control the cyclic AMP level and (2) How does cyclic AMP control the enzyme level? Our efforts are directed toward answering the second question.

Methods. (*a*) **Isolation of *E. coli* strain X7901:** The starting strain was CA-8000, an Hfr Hayes prototroph. After mutagenic treatment with nitrosoguanidine, approximately 500 bacteria are spread on tetrazolium agar containing arabinose and maltose. Mutants incapable of metabolizing the two carbon sources in the media give rise to red colonies with a frequency of about 0.5%. Most of the *ara⁻ mal⁻* are also *lac⁻*. Some of these show phenotypic reversion to *lac⁺* in the presence of cyclic AMP, others do not. One of the latter (CA-7900) was crossed with an F⁻ strain which was *arg⁻*, *met* B⁻, *pyr* F⁻, *trp⁻*, and *arg⁺ met⁺* recombinants selected. One recombinant (X-7900) which was *arg⁺ met⁺ trp⁺ pyr* F⁻ was selected for further crosses. An Hfr (CA-7033) carrying the *lac-pro* A, B deletion, X-111, was crossed with X-7900; *pyr* F⁺ *trp⁻* recombinants were selected and scored for the *pro⁻* character. A *pro⁻ trp⁻* recombinant, (X7901), carrying the *lac-pro* deletion was isolated. This strain is *ara⁻*, *mal⁻*, *lac⁻* and is unaffected by cyclic AMP in the growth medium. Work to be reported elsewhere has shown that these deficiencies are due to a single gene alteration at a distant point from the operons.

(*b*) **Procedures for β-galactosidase synthesis and assay:** Except for slight modifications described herein, all procedures used for synthesis, enzyme assay, and preparation of bacterial extracts and DNA have been described in detail elsewhere. The procedures for synthesis and assay will be reviewed here. The incubation mixture contains per ml: 44 μmoles Tris-acetate, pH 8.2; 1.37 μmoles dithiothreitol; 55 μmoles KAc; 27 μmoles NH₄Ac; 14.7 μmoles MgAc₂; 7.4 μmoles CaCl₂; 0.22 μmole amino acids; 2.2 μmoles ATP; 0.55 μmole each GTP, CTP, UTP; 21 μmoles phosphoenol-pyruvic acid; 100 μg tRNA; 27 μg pyridoxine HCl; 27 μg triphosphopyridine nucleotide; 27 μg flavine adenine dinucleotide; 11 μg *p*-amino-benzoic acid. The above ingredients are preincubated for 3 min at 37°C with 50 γ/ml DNA with shaking before 6.5 mg S-30 protein extract is added. When catabolite gene activator protein (CAP) (see (*c*) below) is present, addition is made by mixing the protein extract with S-30. When cyclic AMP is present, the concentration is 5×10^{-4} M; the cyclic AMP is added to the mixture of salts and cofactors described above. Incubations with shaking are allowed to continue for 60 min at 37°C. The enzyme assays are performed at 28°C in 0.1 M sodium phosphate buffer; pH 7.3, 0.14 M β-mercaptoethanol, and 0.35 mg *O*-nitrophenyl β-D-galactoside (ONPG)/ml. Samples (0.2 ml) of the incubation mixture are mixed with 1.5 ml of the ONPG solution. At the end of the incubation with the substrate ONPG, 1 drop of glacial acetic acid is added to each tube to precipitate the protein, thus decreasing the background absorption and preventing errors due to turbidity. The tubes are quickly stirred and chilled in ice, then centrifuged in the cold for 15 min at 2000 $\times$ *g*. The supernatant is transferred to a clean tube and an equal volume of 1 M Na₂CO₃ is added. The optical density is determined at 420 mμ.

(*c*) **Partial purification of CAP protein:** About 200 gm of frozen *E. coli* strain 514 is homogenized in 700 ml of buffer I (0.01 M Tris-acetate, pH 8.2, 0.01 M Mg(Ac)₂, 0.06 M KAc, 1.4 mM dithiothreitol) and centrifuged for 30 min at 16,000 $\times$ *g*. The sediment containing the bacteria is homogenized and recentrifuged. The final sediment is resuspended in 260 ml of buffer I. The suspension of cells is lysed in an Aminco pressure cell at pressures between 4000 and 8000 psi. The lysate is centrifuged for 30 min at 30,000 $\times$ *g* in a small Sorvall rotor. The resulting supernatant is dialyzed for 16 hr against buffer II (0.01 M KH₂PO₄—KOH, pH 7.7 + 0.01 M mercaptoethanol). The dialyzed extract is passed over a 5 $\times$ 20 cm DEAE-cellulose column previously equilibrated with buffer II. A linear gradient containing buffer II and increasing NaCl concentration is run to 0.25 M NaCl. Tubes around 0.1 M NaCl containing appreciable CAP activity are pooled and dialyzed for 16 hr against buffer III (0.01 M K₂HPO₄—HAc, pH 7.0, + 0.01 M mercaptoethanol). This material is passed over a 2.5 $\times$ 15 cm phosphocellulose column   A linear gradient is executed starting with buffer III + 0.15 M KCl and finishing with buffer III + 0.45 M KCl. Tubes around 0.3 M KCl containing the main peak of activity are dialyzed for 16 hr against buffer I, quick frozen and stored at −90°C until ready for use.

(*d*) **Dialysis-binding studies:** The partially purified CAP described above is concentrated to about 0.5% protein by burying a dialysis sac in G-200 Sephadex for 18 hr at 5°C. Concentrated extract (0.7 ml) is placed in another dialysis sac and dialyzed for 18 hr against buffer I and the appropriate concentration of ^{3}H-cyclic AMP. A 0.3-ml aliquot from inside and outside the dialysis sac are dissolved in 3 ml of formic acid; 1-ml aliquots are plated and counted in a windowless gas flow counter.

Results and Discussion. The direct involvement of cyclic AMP in triggering catabolite genes was made clearest by the finding that this compound is required in a DNA-directed cell-free system for normal initiation[6-8] leading to the synthesis of the enzymes of the *lac* operon.   Since normal initiation of mRNA synthesis occurs at the promoter locus of the operon, it seems likely that cyclic AMP or a closely related derivative interacts at this gene site.[9]   This stimulatory action of cyclic AMP is most probably mediated by a protein since a single mononucleotide would not be expected to interact strongly and specifically with a DNA molecule.   With this in mind it was suggested[6] that cyclic AMP triggers the synthesis of catabolite enzymes such as β-galactosidase by interacting with the RNA polymerase.   A more elaborate working hypothesis, currently favored by us, is that cyclic AMP binds to a protein subunit, which we shall call the catabolite gene activator protein (CAP), and thereby stimulates the binding of this protein subunit to RNA polymerase, producing a complex active in initiation at the promoter locus.

Detection of the CAP protein has been greatly aided by the finding of a mutant bacterial strain which appears to produce defective CAP.   This mutant was obtained by Schwartz and Beckwith[10] by isolation and analysis of a number of variant strains of *E. coli* that could not grow on either lactose or arabinose. These variants all produce low levels of most catabolite repressible enzymes tested.   Some are defective in cyclic AMP production as evidenced by their phenotypic reversion under the influence of this compound.   Others are not, and one of the latter mutants has been found to be lacking a protein factor required for "turning on" the *lac* operon.   The assignment of this defect to a protein factor has been made possible through use of the cell-free system for β-galactosidase synthesis described below.

A DNA-directed cell-free system for synthesis of *lac* operon proteins has been developed which contains DNA with the *lac* operon, a cell-free extract of *E. coli*, and all the cofactors and substrates essential for RNA and protein synthesis (see *Methods* for details).   In the cell-free synthesis studies presented here two strains, 514 and X7901, have been used to produce the cell-free bacterial extracts.   Both strains contain a deletion of the *lac* region including the repressor so that the results would not be complicated by the presence of enzymes of the *lac* operon or *lac* operon repressor at the beginning of synthesis.   Strain 514 is normal in other respects and strain X7901 is the defective strain isolated by Schwartz and Beckwith (described above).   In all cases the DNA used to stimulate synthesis was derived from ϕ80d*lac* virus containing a normal *lac* operon region—a region which includes promoter, operator, structural gene for β-galcato-sidase, and other structural genes of the *lac* operon, in that order.

In all our studies we have found that the amount of β-galactosidase produced by the cell-free system is directly related to the amount of active *lac* operon present.[7, 8] When the strain believed to contain the defective CAP factor is used as the source of the cell-free extract, only about 5% of the usual level of β-galactosidase is made (Table 1, compare lines 2 and 6). This level of activity is

TABLE 1. *Effect of cyclic AMP and CAP protein on DNA-directed cell-free synthesis of β-galactosidase.*

	Source of cell-free bacterial extract	Cyclic AMP	CAP protein	β-galactosidase activity (relative values*)
1	X7901	−	−	1
2	"	+	−	1
3	"	−	+	1
4	"	+	+	5
5	514	−	−	1
6	"	+	−	20
7	"	−	+	1
8	"	+	+	24

* A unit of 1 on this scale is equivalent to 5×10^{-4} International Units. One unit of β-galacto-sidase is defined as that amount of enzyme producing 1 μmole of O-nitrophenol/min at 28°C and pH 7.3. All measurements are the average of duplicate determinations. Conditions for cell-free synthesis are described in *Methods*.

When present during synthesis, the concentration of cyclic AMP is 5×10^{-4} M and the concentration of CAP-containing extract is 10 γ/ml. See text for explanation of results.

interpreted as resulting from abnormal initiations in transcription, since the presence of cyclic AMP, which usually stimulates synthesis (Table 1, compare lines 5 and 6), has no effect (Table 1, compare lines 1 and 2), and DNA with a defective *lac* promoter yields about the same level of activity. The addition of unfractionated extract from a normal strain stimulates the amount of β-galactosidase synthesized but only in the presence of cyclic AMP and normal DNA. An extract can be tested for the presence of CAP by this stimulation effect. With this assay, CAP protein has been purified about 200-fold relative to the total protein in a crude extract. The purification procedure (described in *Methods*) consists of subjecting the crude bacterial lysate to a high speed centrif-ugation, fractionation of the resulting supernatant by gradient elution on a diethylaminoethyl cellulose column (DEAE), and further fractionation of the eluate containing most of the activity by gradient elution on phosphocellulose. Most of the DNA and ribosomes are removed in the high-speed centrifugation and the remaining nucleic acid is removed in the DEAE step.

The stimulation effect of the partially purified CAP protein on β-galactosidase synthesis is most pronounced when the defective strain X7901 is used as the source of the cell-free extract; this stimulation effect requires the presence of cyclic AMP (Table 1, compare lines 1 and 2 with 3 and 4). Little stimulation by added CAP is seen when the normal strain is used to make the cell-free ex-tract (Table 1, compare lines 6 and 8). This is undoubtedly because the normal extract contains a large supply of CAP protein. Thus far, the CAP protein has been used to augment cell-free synthesis of β-galactosidase tenfold, with the upper limit yet to be determined. Over the range of concentration studied, the amount of stimulation is proportional to the amount of CAP protein added (Fig. 1). This linear response to CAP suggests that gene activation requires a single molecule of the CAP protein.

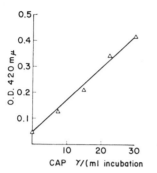

CAP γ/(ml incubation

Fig. 1.—By using the cell-free system described in *Methods* in conjunction with cell-free extract prepared from strain X7901 and otherwise standard conditions, the β-galactosidase synthesized as a function of varying levels of partially purified CAP protein was determined. The β-galactosidase activity is expressed in arbitrary units.

Some of the physicochemical properties of the partially purified factor have been studied. The total extract has an absorption maximum in the ultraviolet at 278 mμ. The activity is completely destroyed by heating for 5 min at 60°C even though longer exposures to 50°C have no effect. The size of the CAP protein has been estimated by its flow rate on G-100 Sephadex (Fig. 2). It elutes in a volume about 60% greater than the column exclusion volume, and bovine serum albumin of molecular weight 6.8×10^4 elutes in a volume about 43% greater than the exclusion volume. A molecular weight of 4.5×10^4 is estimated for CAP on the basis of its elution behavior;[11] we do not know whether the molecule is composed of one or more polypeptide chains.

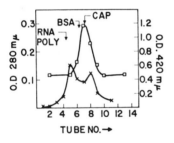

TUBE NO.→

Fig. 2.—Elution diagram of partially purified CAP protein on G-100 Sephadex. Column dimensions, 1.5 × 30 cm; flow rate, 18 ml/hr; 4 ml/fraction. Column and samples are equilibrated in a buffer containing 0.01 *M* Tris-acetate, pH 8.2, 0.04 *M* MgAc₂, 0.06 *M* KAc, and 1.4 mM dithiothreitol. The ultraviolet absorption profile (*lower curve*) of the eluted fractions is indicated on the left and the corresponding stimulation effect on β-galactosidase synthesis (in arbitrary units) is on the right. The elution pattern of RNA polymerase and bovine serum albumin was determined in separate runs. Arrows indicate the position of elution of *E. coli* RNA polymerase and bovine serum albumin.

The binding of cyclic AMP to the enriched CAP-containing extracts has been examined extensively. Whereas crude extracts of *E. coli* show no measurable affinity for cyclic AMP, the partially purified and concentrated preparations of CAP show a substantial affinity. With ³H-labeled cyclic AMP, the binding has been measured by equilibrium dialysis over a broad range of cyclic AMP concentrations (from about 10^{-9} to 10^{-5} *M*). The standard equation relating bound to free ligand, when it is assumed that there exists only one type of binding site, is $I_b/I_f = K_f n(\mathrm{CAP}) - K_f I_b$, where I_b is the concentration of bound ligand, I_f the concentration of free ligand, CAP the concentration of CAP protein, n the number of ligand binding sites per CAP protein, and K_f the formation constant for the CAP-cyclic AMP complex defined by the equation:

$$K_f = \frac{(CAP^* + \text{cyclic AMP})}{(CAP^*)(\text{cyclic AMP})}.$$

In this equation (CAP^*) is the concentration of ligand binding sites. If $n = 1$, then $(CAP^*) = (CAP)$; if $n = 2$, then $(CAP^*) = 2(CAP)$; and so on. All concentrations are expressed as molar amounts. When I_b/I_f is plotted against I_b (as in Fig. 3) the slope is equal to $-K_f$. The numerical value obtained for K_f is 0.6×10^5 liters moles^{-1}.

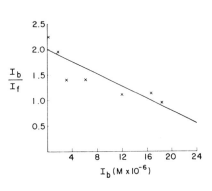

FIG. 3.—Binding curve for ^{3}H-labeled cyclic AMP to CAP-containing extract. The preparation of partially purified CAP and technique for dialysis binding are described in _Methods_. Calculations made in the text assume the concentration of free cyclic AMP, I_f, is proportional to the radioactivity found outside the dialysis sac and the concentration of bound cyclic AMP, I_b, is proportional to the difference in radioactivity found inside and outside the sac. Protein concentration is determined on an aliquot of material inside the sac by the Lowry method with a serum albumin standard. All plotted measurements have been normalized to a protein concentration of 1%. The experimental point plotted closest to the ordinate represents an average of 12 independent measurements between I_f values of 10^{-9} and 10^{-7} M. Other points each represent single experiments.

Since the CAP-containing extracts are impure, we have been concerned with the possibility that species other than the CAP are binding the cyclic AMP. Previously it had been found that cyclic GMP inhibited the cyclic AMP stimulatory effect on β-galactosidase synthesis at comparable concentrations.[7] In parallel binding studies, the level of ^{3}H-cyclic AMP was fixed at 5×10^{-6} M and the concentration of competing nucleotide was fixed at 10^{-4} M. Cyclic GMP inhibits 70% of the binding whereas either 3'-GMP or 5'-GMP have less than an 8% inhibitory effect. The parallelism between cyclic GMP in inhibiting the stimulatory action of cyclic AMP in cell-free synthesis and in inhibiting the binding of cyclic AMP to CAP-containing extracts supports the view that most of the cyclic AMP binding in the CAP-containing extracts is due to the CAP protein itself, but it does not eliminate other possibilities. The following discussion assumes that most of the cyclic AMP binding is to the CAP.

The formation constant for the CAP-cyclic AMP complex can also be used to estimate the concentration of CAP present from the amount of bound cyclic AMP. Only approximate calculations are possible since we do not know n, the number of cyclic AMP binding sites per CAP molecule. Assuming $n = 1$ we calculate that the molarity of CAP at the point of maximum stimulation in Figure 1 is 6×10^{-8} M. This is about 35 times the molarity of the _lac_ operon containing DNA and about twice the molarity of the RNA polymerase.[12]   It seems unlikely that there are many other catabolite-sensitive gene promoter sites on the ϕ80d*lac* DNA.   Therefore, the large excess of CAP over *lac* operon makes it most unlikely that CAP functions by forming a strong stoichiometric complex with the *lac* operon promoter.   The formation of a strong stoichiometric complex between CAP and RNA polymerase is also unlikely since such an event would cause an appreciable decrease in the pool of free polymerase, thereby causing departure from linearity over the range of CAP concentrations studied in Figure 1.   We conclude that CAP does not form a strong complex with either DNA or RNA polymerase.   In quantitative terms, the K_f's for the complexes must be less than 10^8.   Since the hypothesized function of CAP is to trigger initiation, a complex with more than transient existence may not be needed.   We are in the process of studying the interaction between CAP, RNA polymerase, and DNA by direct means; it remains to be demonstrated that CAP functions by forming such a complex.

We are indebted to Tetteh Blankson for technical assistance.

* This work was supported by grants from the National Institute of Health, 5-RO1-GM-16648-02, and the American Cancer Society, E-545.

† Department of Biological Sciences, Columbia University, New York, N.Y.
‡ Department of Bacteriology and Immunology, Harvard Medical School, Boston, Mass.

[1] For reviews of the subject of catabolite repression see Magasanik, B., in *Cold Spring Harbor Symposia on Quantitative Biology*, vol 26, (1961), p 249; Magasanik, B., in *The Lac Operon*, ed. D. Zipser and J. R. Beckwith, *Cold Spring Harbor Laboratory on Quantitative Biology* (1970).

[2] Dennert, F., *Ann. Inst. Pasteur*, **14**, 139 (1900).

[3] Makman, R. S., and E. W. Sutherland, *J. Biol. Chem.*, **240**, 1309 (1965).

[4] Perlman, R. L., and I. Pastan, *J. Biol. Chem.*, **243**, 5420 (1968).

[5] Ullman, A., and J. Monod, *FEBS Letters*, **2**, 57 (1968).

[6] Chambers, D. A., and G. Zubay, these PROCEEDINGS, **63**, 118 (1969).

[7] Zubay, G., D. Chambers, and L. Cheong, in *The Lac Operon*, ed. D. Zipser and J. Beckwith, *Cold Spring Harbor Laboratory on Quantitative Biology* (1970).

[8] Zubay, G., and D. Chambers, in *Cold Spring Harbor Symposia on Quantitative Biology* (1969), vol. 64, in press.

[9] Silverstone, A. E., B. Magasanik, W. S. Reznikoff, J. H. Miller, and J. Beckwith, *Nature*, **221**, 1012 (1969).

[10] Schwartz, D., and J. Beckwith, in *The Lac Operon*, ed. D. Zipser and J. Beckwith, *Cold Spring Harbor Laboratory on Quantitative Biology* (1970).

[11] For the method of molecular weight determination, see Murphy, W. H., G. Barrie Kitto, J. Everse, and N. O. Kaplan, *Biochemistry*, **6**, 603 (1967).

[12] The molarity of the operon is calculated from the molarity of ϕ80d*lac* DNA added with a molecular weight of 30×10^6.   The molarity of RNA polymerase is estimated from the total protein present in the cell free extract, 6500 γ/ml, and assuming RNA polymerase comprises 1 part per 1000 of the protein with a molecular weight of 4.95×10^6 as suggested by Burgess, A., *J. Biol. Chem.*, **244**, 6168 (1969).

Reprinted from *Nature*, **224**(5225), 1168–1174 (1969)

Termination Factor for RNA Synthesis

by
JEFFREY W. ROBERTS

The Biological Laboratories,
Harvard University

A new protein has been isolated from *E. coli* which causes specific termination and release of RNA during synthesis *in vitro*. It has been given the name ρ-factor.

THE synthesis of a messenger RNA molecule requires that the enzyme RNA polymerase initiate and terminate transcription at appropriate sites on the DNA template. RNA polymerase does frequently initiate RNA synthesis at the correct sites *in vitro*[1-5], a process which requires a protein factor (the sigma factor) that is normally bound in a tight complex to purified RNA polymerase[6,7]. The usual observation, however, has been that RNA synthesis *in vitro* does not end at unique sites, but instead terminates at random to produce RNA molecules of various, and often clearly excessive, lengths (refs. 8, 9 and personal communication from M. Green). Furthermore, these RNA molecules are not released from the enzyme, but remain complexed with DNA and RNA polymerase when synthesis is completed[8,10]. I now report the isolation and purification of a protein factor from *Escherichia coli* which causes the termination and release of RNA molecules in an *in vitro* reaction using bacteriophage λ DNA as a template. The product contains several discrete RNA species which may correspond to the natural early messenger RNA of lambda.

The termination protein, which is named the ρ-factor (ρ for release), was discovered during a search for factors in a crude extract of *E. coli* which might increase the overall accuracy of *in vitro* RNA synthesis. The criterion of accuracy was that RNA synthesis initiated at the c_{17} promoter of the DNA of bacteriophage λc_{17} should dominate the *in vitro* transcription from λc_{17} DNA; a discussion of the reasoning which suggested this approach is given at the end of this article, where the exact relation of the termination factor to the c_{17} mutation is explained. This criterion provided an assay which allowed a partial purification of the ρ-factor, after which a convenient activity of the ρ-factor became evident: it depresses net RNA synthesis in a standard *in vitro* synthesis reaction. This depression is the basis for a much simpler assay which has allowed purification of the ρ-factor to homogeneity; a summary of the purification procedure is given at the end of this text. The experiments presented below demonstrate that the ρ-factor depresses RNA synthesis from λ DNA template by terminating the synthesis of RNA molecules at specific sites on the template.

ρ-Factor affects Chain Propagation

Fig. 1 illustrates an experiment which shows two properties of the depression activity: only a limited fraction of *in vitro* transcription is depressed by the ρ-factor, and the propagation rather than initiation of RNA chains is affected. Both total RNA synthesis and the number of RNA chains initiated with GTP are measured with increasing concentrations of ρ-factor in a standard reaction mixture, which contains purified λ DNA, purified RNA polymerase, nucleotide triphosphate precursors and the appropriate ions. Total RNA synthesis is determined as the incorporation of ^{3}H-UTP into acid insoluble material; the number of RNA chains initiated with GTP is indicated by the incorporation of ^{32}P from GTP labelled in the γ position with ^{32}PO$_4$, because only the γ phosphate from the 5' terminal nucleotide is retained in the product after polymerization. Total RNA synthesis is depressed by ρ-factor to a plateau value, whereas the number of molecules initiated with GTP remains constant or increases slightly as a saturating amount of ρ-factor is added. An identical result is found if initiation with ATP is measured. Because essentially no initiation occurs with pyrimidine nucleotide triphos-

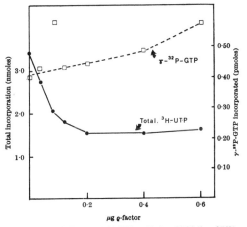

Fig. 1. Effect of ρ-factor on total RNA synthesis and initiation of RNA synthesis with GTP. Each reaction mixture contained in 100 μl.: 0·02 M Tris (pH 7·9); 0·10 M KCl; 0·010 M magnesium acetate; 0·0001 M dithiothreitol; 0·15 mM CTP; 0·05 mM ATP; 0·05 mM ^{3}H-UTP at 80 μCi/μmole; 0·05 mM γ-^{32}P-GTP at 2·1 mCi/μmole; 1·2 μg λ DNA; 3 μg RNA polymerase; 25 μg bovine serum albumin; and ρ-factor as indicated. The amount of ρ-factor was estimated by assuming an absorbance at 280 nm for the purified protein of 0·5/mg/ml. After incubation for 20 min at 37°, the reaction was stopped by addition of 3 ml. 6 per cent trichloroacetic acid (TCA) containing 0·01 M sodium pyrophosphate and the RNA was filtered on a Millipore filter. The filters were washed on the filtration apparatus with 20–30 ml. of 2 per cent TCA with pyrophosphate, followed by soaking for 1 h in 2 per cent TCA containing 1·0 M KCl and pyrophosphate. They were dried and counted in toluene-liquifluor on a liquid scintillation spectrometer. The growth and purification of phage and preparation of DNA were described previously[1]. *E. coli* RNA polymerase was purified to greater than 90 per cent purity by the glycerol gradient technique[11]. ^{3}H-UTP was purchased from Schwarz Bioresearch and γ-^{32}P-GTP was made by the procedure of Glynn and Chappell[12].

phates[4], this experiment shows that the ρ-factor does not interfere with initiation. The depression activity must therefore be the result of some later effect on chain growth. Depression of RNA synthesis by ρ-factor also occurs with DNA templates from bacteriophages T4 and T7, so that the effect does not depend on any peculiar property of phage λ DNA.

ρ-Factor produces Shorter RNA Molecules

The most direct demonstration that the ρ-factor terminates RNA synthesis and produces shorter RNA molecules is provided by examining the size of RNA synthesized in the presence and absence of the factor; in order to make such an experiment meaningful, it is necessary to establish rigorously that the ρ-factor preparation contains no ribonuclease activity which could produce fragments from longer RNA molecules. The following experiment achieves this, and also reveals that the ρ-factor does cause shorter RNA chains to be synthesized. RNA labelled with ^{3}H-UTP was synthesized in a standard reaction with no ρ-factor present; it was purified by phenol extraction and equal portions were added to two new RNA synthesis reactions incubated in the presence and absence of ρ-factor. The new RNA synthesized in the presence of ρ-factor was labelled with ^{14}C-ATP, and no further label was added to the second incubation mixture without ρ-factor. After the second incubation period each reaction was treated with sodium dodecyl sulphate (SDS) and the sizes of the ^{3}H-RNA and ^{14}C-RNA were estimated by sucrose gradient centrifugation. Fig. 2a reveals that ^{3}H-RNA synthesized in the absence of ρ-factor has a broad distribution of sizes from greater than 35S to less than 5S; this appearance is typical of RNA synthesized in the conditions used. By contrast, the ^{14}C-RNA synthesized in the presence of ρ-factor is restricted to sizes less than 15S (Fig. 2b). It will be demonstrated that this peak contains two RNA species which are not resolved here. Because the ^{3}H-RNA present in the synthesis reaction with ρ-factor has essentially the same size distribution as that in the reaction without ρ-factor, I conclude that the ρ-factor decreases the size of the product by directly interacting with the synthesizing complex, and that the factor preparation does not fragment RNA under the conditions of RNA synthesis.

Other less rigorous tests confirm the absence of ribonuclease activity. Incubation of bacteriophage R17 RNA with purified ρ-factor does not reduce its infectivity to spheroplasts, nor does incubation of factor with ^{3}H-poly U release acid soluble radioactivity.

The ρ-factor is free of deoxyribonuclease activity to the extent that it can be incubated with λ DNA in the conditions of the standard assay (including Mg[2+] and nucleotide triphosphates) without introducing detectable single strand breaks in the DNA. Thus the decrease in size of the RNA product cannot be attributed to fragmentation of the template. This absence of nuclease activity might be expected from the final purification step, for most known nucleases are considerably smaller than the ρ-factor.

ρ-Factor releases RNA Molecules

Besides restricting the size of RNA synthesized *in vitro*, the ρ-factor causes it to be released from the complex of DNA and RNA polymerase. After a normal *in vitro* synthesis of RNA with purified RNA polymerase, most of the RNA is found in a fast sedimenting complex with DNA and RNA polymerase[8,10]. This is revealed in the experiment of Fig. 3a, in which an *in vitro* reaction with λ DNA was stopped by addition of EDTA and chilling and centrifuged on a glycerol gradient without being subjected to conditions which denature proteins. Most of the RNA and template DNA sediment faster than free DNA, at approximately 50S. Fig. 3b presents a companion experiment for which ρ-factor was present during the

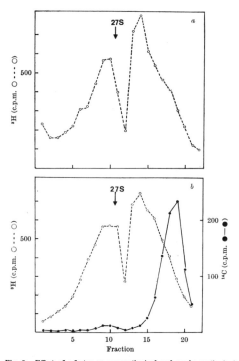

Fig. 2. Effect of ρ-factor on pre-synthesized and newly synthesized RNA. ^{3}H-RNA was synthesized without ρ-factor in a reaction mixture containing in 100 μl.; 0·02 M Tris (pH 7·9); 0·10 M KCl; 0·010 M magnesium acetate; 0·0001 M dithiothreitol; 0·15 mM ATP; 0·15 mM GTP; 0·15 mM CTP; 0·05 mM ^{3}H-UTP at 2·0 mCi/μmole; 1·2 μg λ DNA; and 2 μg RNA polymerase. After incubation for 20 min at 37°, the reaction mixture was chilled and extracted with 50 μl. phenol; the aqueous phase was dialysed against 0·01 M Tris (pH 7·4), 0·0001 M EDTA, and 1·0 M KCl, and finally against 0·01 M Tris (pH 7·4) and 0·0001 M EDTA. Equal portions of this ^{3}H-RNA were added to two further reaction mixtures: (a) as above, except with 1·0 μg RNA polymerase and 0·15 mM UTP (unlabelled), and (b) as above, except with 1·0 μg RNA polymerase, 0·15 mM UTP (unlabelled), 0·15 mM ^{14}C-ATP at 40 μCi/μmole, and 1·0 μg ρ-factor. These were incubated 20 min at 37°, followed by addition of 50 μl. 0·3 per cent sodium dodecyl sulphate (SDS); each was mixed and chilled. Precipitated SDS was removed by centrifugation, and each reaction mixture was layered on a 5 ml. 5–20 per cent sucrose gradient in SSC (0·15 M NaCl; 0·015 M sodium citrate), followed by centrifugation for 1·8 h at 50,000 r.p.m. and 4° in the Spinco SW50 rotor. Twenty-one fractions were collected of each gradient, and the fractions were precipitated with 6 per cent TCA and 0·01 M sodium pyrophosphate and filtered on Millipore filters. Distribution of ^{3}H and ^{14}C was determined by counting in toluene-liquifluor on a liquid scintillation spectrometer. R17 RNA was centrifuged in a separate gradient as a marker at 27S. ○ - - - ○, Radioactivity (c.p.m.) in ^{3}H; ●—● in [14]C.

incubation. In this case most of the DNA cosediments with free λ DNA (32S), and the peak of DNA is only slightly skewed toward the higher S values. Most of the RNA sediments well behind the DNA so that clearly it has been released from the DNA. Because the previous experiment ensures that the factor contains no nuclease activity which could cut RNA chains away from the complex, the ρ-factor must free RNA chains from the complex of DNA and RNA polymerase, an event which is probably part of the act of termination.

RNA Product consists of Discrete Sizes

If the ρ-factor terminates at specific sites on the DNA template, the resulting RNA molecules should be of discrete sizes. It will be shown that the RNA product from λ DNA does consist of two discrete sizes, each initiated at a single site on the DNA; thus the ρ-factor causes specific rather than random termination of RNA synthesis.

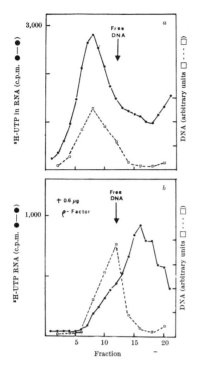

Fig. 3. Release of RNA from complex by ρ-factor. The conditions of synthesis were as in Fig. 2, except with 0·05 mM ³H-UTP at 1·0 mCi/μmole, 1·8 μg RNA polymerase, and 25 μg bovine serum albumin per 100 μl. reaction mixture. The reaction mixture of 3b contained in addition 0·6 μg ρ-factor. After incubation for 20 min at 37°, synthesis was stopped by chilling and addition of EDTA to 0·02 M, and each was layered on a 5 ml. 10–30 per cent glycerol gradient in 0·05 M Tris (pH 7·4), 0·1 M KCl, 0·01 M magnesium acetate, 0·0001 M EDTA and 0·0001 M dithiothreitol. The gradients were centrifuged 1·6 h at 64,000 r.p.m. in the Spinco *SW*65 rotor at 4°, and twenty-one fractions were collected. To determine the distribution of DNA in the gradients, 50 μl. portions of each fraction were precipitated with TCA, filtered on to Millipore filters, and counted as before. The distribution of DNA in the gradients was determined by assaying portions of each fraction for template activity, measured as incorporation of ¹⁴C-ATP into RNA under conditions of excess polymerase. The position of free DNA was determined in a parallel gradient. ●—●, Radioactivity (c.p.m.) in ³H-RNA; □- - -□, DNA, arbitrary units. *a*, No ρ-factor; *b*, with 0·6 μg ρ-factor.

What parts of the λ genome does one expect to be transcribed by RNA polymerase in the absence of phage specific functions? Fig. 4 is a map of λ which shows the positions of the promoters for early transcription and the directions of RNA synthesis. If phage repressor is absent or is removed, synthesis of messenger RNA for the *N* gene is initiated at a promoter close to the beginning of the *N* gene, and proceeds to the end of the gene. Unless the *N* protein is made from this messenger, there is no transcription of genes to the left of *N* (refs. 13 and 14). In an *in vitro* RNA synthesis reaction which lacks the *N* protein, one therefore would expect transcription to the left only of the *N* gene itself. (The function of the *N* gene product will be discussed below.) RNA synthesis to the right is initiated at a promoter which is likely to be in region *x* (refs. 15 and 16). This transcription also appears to be dependent on the *N* gene product, although some genes to the right are expressed at least slightly in its absence[17,18]. Thus if both initiation and termination occur at the correct sites *in vitro*, the RNA product should include discrete species initiated at the leftward and possibly the rightward promoter.

Fig. 5a shows the sedimentation profile of RNA synthesized by RNA polymerase *in vitro* from λb₂ template in the presence of ρ-factor. Two peaks of RNA are apparent, sedimenting at approximately 12S and 7S. If RNA is synthesized in the absence of ρ-factor and treated identically, there is little RNA in this area of the gradient (see also Fig. 2) and there is no trace of these peaks. The following experiments establish that these peaks represent distinct RNA species by showing (*a*) that they are synthesized from small regions of the genome adjacent to the promoters; and (*b*) that they are initiated at the rightward and leftward promoters.

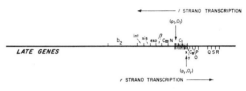

Fig. 4. A genetic and physical map of bacteriophage λ. The shaded zone is the immunity region, which includes the C₁ (repressor) gene and the promoters and operators of early transcription[19,20]. Messenger RNA synthesis toward the right from the immunity region is initiated at a promoter *p*ᵣ located in region *x*, and the RNA is homologous to the *r* strand of DNA[15,16]. Transcription toward the left of messenger for gene *N* (and probably genes *c*ᵢᵢᵢ–*int* as well) is initiated from a promoter *p*ₗ at the beginning of the *N* gene, and the RNA is homologous to the *l* strand[16,20].

To determine the regions of the template to which the 12S and 7S RNA is homologous, portions of each fraction of the gradient of Fig. 5a were hybridized to separated strands of λb₂ or wild type, λi⁴³⁴, and λb₂i²¹ (see Fig. 6). RNA homologous to *l* strands DNA is synthesized toward the left of the standard λ map (see Fig. 4), and RNA homologous to *r* strand DNA is synthesized toward the right[14,16]. The λi⁴³⁴ DNA effectively deletes the immunity region for hybridization purposes, and the λb₂i²¹ DNA effectively deletes the immunity region and the *N* gene. The presence of the *b*₂ region in the λi⁴³⁴ DNA is irrelevant because the template does not contain the *b*₂ region. Fig. 5b reveals that the 12S peak is homologous to the *l* strands of both λb₂ and λi⁴³⁴ DNA, but not to the *l* strand of λb₂i²¹ DNA; thus the 12S RNA is synthesized to the left and from a region in the vicinity of the *N* gene. The 7S RNA is homologous to the *r* strand of λb₂ DNA, but not to *r* strands of either λi⁴³⁴ or λb₂i²¹ DNA (Fig. 5c). The 7S RNA is therefore synthesized towards the right and from an area within the immunity region. Both RNA peaks arise from highly limited regions of the template DNA.

The 12S and 7S RNA combined constitute only approximately 50 per cent of the total product. The remaining background of RNA is transcribed from both strands and is not found in definite peaks, but instead exhibits a wide range of sizes. This RNA might result from incorrect initiation by RNA polymerase at unknown locations on the template, followed by termination events which produce RNA molecules of random length; such incorrect initiation is known to occur (see ref. 1). Substantially more of the background RNA is synthesized from λ DNA template which contains the *b*₂ region, so that λb₂ DNA (that is, DNA deleted for the *b*₂ region) has been used for these experiments whenever possible.

12S and 7S RNA initiated at Known λ Promoters

How can it be shown that each of the 7S and 12S RNA species is initiated at a unique locus on the template? Mutants of λ exist which abolish or greatly reduce transcription *in vivo* either toward the right or toward the left from the immunity region. These are good candidates to be mutations in the promoters which prevent the normal interaction between RNA polymerase and DNA to initiate RNA synthesis. The experiments below demonstrate that DNA which carries either of these mutations is not a template for *in vitro* transcription of the RNA species synthesized in the direction corresponding to the *in vivo* defect; thus the two RNA species are initiated at the sites of the mutations, and these sites are in fact the promoters.

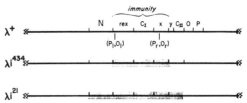

Fig. 6. Phage DNA used for hybridization. The shading indicates regions of non-homology with bacteriophage λ (refs. 19, 22).

The mutant defective in transcription toward the left is λ$_{sex}$. This mutant was isolated by M. Gottesman, who showed that it greatly reduces the expression of several gene products in the region to the left of gene *N*. The sex mutation diminishes *in vivo* RNA synthesis throughout the whole p_L-*N-int* region by approximately a factor of ten (H. J. J. Nijkamp and W. Szybalski, in preparation), and thus differs from N$^-$ mutations, which do not affect the transcription of the p_L-*N* region but abolish RNA synthesis in the region to the left of *N* (refs. 14, 23). Lambda$_{sex}$ maps as indicated in Fig. 9. These facts strongly suggest that it affects the promoter at the beginning of the *N* gene.

To test this mutant, RNA was synthesized from λb$_2$ and λb$_2$ sex DNA templates in the usual *in vitro* system in the presence of ρ-factor. RNA from λb$_2$ template was labelled with ^{3}H-ATP, and RNA from λb$_2$ sex template was labelled with ^{14}C-ATP; the RNA preparations were mixed and analysed by sucrose gradient centrifugation as before. Fig. 7 shows that the 7*S* r strand RNA is synthesized from both templates, whereas the 12*S* *l* strand peak is not synthesized in significant amount from the λ$_{sex}$ DNA. It might be argued that the 12*S* RNA is not seen because the sex mutation has created immediately adjacent to the promoter a new termination site which the ρ-factor recognizes; this is not true, however, because I have observed that the sex mutation also decreases net *l* strand RNA synthesis in an *in vitro* reaction without ρ-factor. The sex mutation must therefore affect RNA synthesis by impairing the interaction between RNA polymerase and DNA by which synthesis is initiated, and the 12*S* RNA must be initiated at the site affected in λ$_{sex}$. I conclude that the 12*S* RNA is initiated at the leftward promoter at the beginning of gene *N*.

The mutation which has been shown to abolish rightward transcription in λ is t_{11} (ref. 16). Lambda t_{11} maps in region *x* (see Fig. 9), as do numerous other mutants which are deficient in expression of phage genes on the right[15,16]. The x mutant examined here for its effect on *in vitro* RNA synthesis is λx$_{13}$, which was chosen because it could be grown in sufficient quantities to provide template DNA (see legend to Fig. 9); it is presumed to be similar to t_{11}. For this experiment, ^{3}H-RNA synthesized *in vitro* from wild type λ in the presence of ρ-factor was mixed with ^{14}C-RNA synthesized similarly from λx$_{13}$ template and the combined RNA was analysed by sucrose gradient. This experiment is complicated by the fact that neither template contains the b$_2$ deletion, and that, as mentioned above, there is RNA synthesized *in vitro* from the b$_2$ region which somewhat obscures the sucrose gradient pattern of RNA seen in the previous figures. Nevertheless, it is evident from Fig. 8*a* that considerably less RNA in the 7*S* region is synthesized from λx$_{13}$ DNA than from wild type λ DNA.

The 7*S* RNA can be examined without interference from *r* strand RNA synthesized on other regions of the template by measuring RNA which anneals to the *r* strand of φ80i^λ DNA; this phage is a hybrid of λ and φ80 which is homologous to λ only to the right of some point probably located in the exonuclease gene[24,25], so that RNA synthesized from the b$_2$ region or farther to the left will not

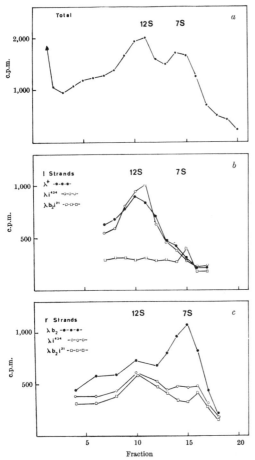

Fig. 5. Sucrose gradient analysis of RNA synthesized from λb$_2$ DNA template in the presence of ρ-factor. The conditions of synthesis were as in Fig. 2, except that the 200 μl. reaction mixture contained 2·4 μg λb$_2$ DNA, 4·0 μg RNA polymerase, 0·05 mM ^{3}H-UTP at 2·0 mCi/μmole, and 1·5 μg ρ-factor. After incubation for 20 min at 37°, the reaction mixture was made 0·1 per cent in SDS, mixed, chilled and centrifuged to remove precipitated SDS. It was layered on to a 5 ml. 5–20 per cent sucrose gradient in SSC and centrifuged 4 h at 50,000 r.p.m. and 4° in the Spinco *SW50* rotor. Twenty-one fractions were collected. The sedimentation constants were determined by centrifuging 16*S* ribosomal RNA and 4*S* tRNA as markers in a parallel gradient. *a*, Thirty μl. portions of each fraction were precipitated with TCA, filtered on to Millipore filters, and counted as before. *b*, Thirty μl. portions were annealed to *l* strand DNA from wild type λ, λi^{434}, and λb$_2$i^{21}. The growth of phage and preparation of separated strands were described previously[1]. For hybridization, each RNA portion was combined with 200 μl. 2 × SSC containing 0·4–0·5 μg DNA and phenol at 25 per cent saturation. The mixtures were annealed for 5 h at 67°, chilled, diluted to 1·0 ml. with 2 × SSC containing 10 μg ribonuclease A, incubated 30 min at 37°, and diluted to 4 ml. with 0·01 M Tris (pH 7·4) and 0·50 M KCl (ref. 21). They were filtered on to Schleicher and Schuell B6 filters, rinsed with 30–40 ml. of Tris-KCl, dried, and counted in toluene-liquifluor. It was previously determined that 0·4–0·5 μg DNA is a saturating amount. RNA homologous to wild type λ *l* strand DNA (●—●); to λi^{434} *l* strand DNA (○—○); and λb$_2$i^{21} *l* strand DNA, (□—□). *c*, Thirty μl. portions were hybridized to 0·3–0·5 μg *r* strand DNA of λb$_2$ (●—●); λi^{434} (○—○); and λb$_2$i^{21} (□—□).

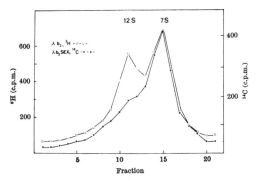

Fig. 7. Sucrose gradient analysis of RNA products from λb, and λb,sex DNA templates. Synthesis reactions were as in Fig. 2, except that 100 μl. reaction mixtures contained 0·05 mM UTP (unlabelled), 1·5 μg DNA template, 1·3 μg RNA polymerase, and 0·8 μg ρ-factor. RNA synthesized from λb, DNA was labelled with ³H-ATP (0·15 mM at 330 μCi/μmole), and RNA from λb,sex DNA was labelled with ¹⁴C-ATP (0·15 mM at 67 μCi/μmole). After a 20 min incubation at 37°, the reaction mixtures were made 0·1 per cent in SDS, chilled, combined and centrifuged to remove precipitated SDS. The RNA was layered on to a sucrose gradient as before and centrifuged 4 h at 65,000 r.p.m. and 5° in a Spinco SW 65 rotor. Twenty-one fractions were collected; each was precipitated with TCA, filtered on to Millipore filters, dried and counted in toluene-liquifluor on a liquid scintillation spectrometer. ○, Radioactivity (c.p.m.) in ³H; ●, in ¹⁴C.

Table 1. TRANSCRIPTION FROM WILD TYPE λ AND λc₁₇ TEMPLATES IN THE PRESENCE OF ρ-FACTOR

	Template	
Annealed to	λc₁₇	λ⁺
λ⁺ l	697	704
λ⁺ r	1,847	947
φ80iλ r	1,164	297

RNA synthesis in 200 μl. reaction mixtures was as in Fig. 2, except with 0·05 mM ³H-UTP at 1·0 mCi/μmole, 0·6 μg template DNA, 0·9 μg RNA polymerase, 50 μg bovine serum albumin, and 1 μg ρ-factor. Incubation was for 20 min at 37° C. Aliquots of each reaction mixture were hybridized to separated DNA strands by a method described previously[1]. Figures represent radioactive material (c.p.m.) annealed to a saturating amount of DNA, and are the averages of duplicate samples; a background of 307 c.p.m. was subtracted.

Conclusion and Discussion

Fig. 9 summarizes the effect of ρ-factor on *in vitro* transcription from bacteriophage λ DNA. This work has shown that a large fraction of *in vitro* RNA synthesis is initiated at promoters p_L and p_R, as defined experimentally by the mutations λsex and λx₁₃. When ρ-factor is present, RNA synthesis is terminated at points indicated by triangles to produce a 12S RNA species from the *l* strand and a 7S RNA species from the *r* strand. Is this pattern of RNA synthesis reasonable? One can estimate that the molecular weight of the 12S RNA is approximately 250,000 (ref. 28), so that as a messenger RNA it could encode a protein of approximately 25,000–30,000 molecular weight. The size of the *N* gene protein is unknown, but

hybridize to its DNA. Fig. 8b shows the result of annealing RNA from the gradient of Fig. 8a to φ80iλ DNA: it is apparent that the 7S RNA is synthesized from wild type DNA and is not synthesized from λx₁₃ DNA. Hybridization of portions of the gradient to *l* strand DNA of λb₂ demonstrates that the 12S *l* strand RNA is synthesized from both templates. The x₁₃ mutation also reduces *r* strand specific RNA synthesis in an *in vitro* reaction without ρ-factor, so that, as in the case of λsex, the mutation cannot function by interacting with ρ-factor to terminate synthesis of the 7S RNA prematurely. It appears very likely that the 7S RNA is initiated at the site of mutation x₁₃, and that this site is the rightward promoter.

More Accurate RNA Synthesis with ρ-Factor

An example of the increased overall accuracy of *in vitro* RNA synthesis which the ρ-factor produces is provided by the mutant λc₁₇. The c₁₇ mutation creates a new promoter which allows constitutive synthesis of messenger RNA from the *x–O–P* region, even in the presence of phage repressor and the absence of the *N* gene product[1,26,27]. In an *in vitro* system using RNA polymerase and no ρ-factor it was found that approximately 60 per cent more RNA homologous to φ80iλ *r* strand DNA (which includes the *x–O–P* region) is synthesized from λc₁₇ DNA than from wild type DNA (ref. 1). It was argued that the effect of the c₁₇ mutation on *m*RNA synthesis *in vivo* is likely to be much greater than this, for the mutation has a profound effect on the behaviour of the phage *in vivo*. A comparison of *in vitro* RNA synthesis from wild type λ and λc₁₇ templates in the presence of ρ-factor is presented in Table 1. One measure of the *in vitro* effect of the c₁₇ mutation is the ratio of φ80iλ *r* strand specific RNA synthesis from λc₁₇ and λ wild type templates under conditions which produce identical *l* strand specific RNA synthesis; with ρ-factor this ratio is 3 to 4, whereas without ρ-factor it was approximately 1·6 (ref. 1). The apparent excess RNA specific to φ80iλ *r* DNA synthesized from λc₁₇ DNA is increased five-fold by ρ-factor. Because RNA homologous to φ80iλ *r* strand DNA includes the immunity specific 7S RNA which is synthesized from both wild type and λc₁₇, the actual excess synthesized from the *x–O–P* region of λc₁₇ DNA may be closer to ten times that from the wild type template.

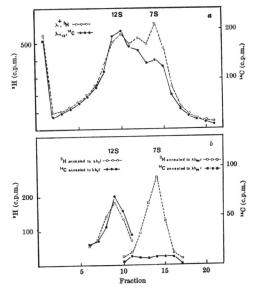

Fig. 8. Comparison of RNA product from λx₁₃ and wild type λ templates. Conditions of synthesis were as in Fig. 2, except that 190 μl. reaction mixtures contained 0·05 mM UTP (unlabelled), 2·4 μg DNA template, 2·5 μg RNA polymerase, and 1·5 μg ρ-factor. The mutant λx₁₃ was isolated by H. Eisen. λx₁₃ phage were prepared by mitomycin induction of a tandem double lysogen of λiᵢⁿᵗⁱⁿt and λc₁₅₅⁺x₁₃ constructed by M. Gottesman; the phage burst contains about 90 per cent λx₁₃ and 10 per cent λiⁱⁿᵗ (unpublished observation and personal communication from M. Gottesman). The λx₁₃ were separated from the λiⁱⁿᵗ by banding the phage to equilibrium in a CsCl density gradient. RNA synthesized from wild type λ DNA was labelled with ³H-ATP (0·15 mM at 0·9 mCi/μmole), and RNA synthesized from λx₁₃ template was labelled with ¹⁴C-ATP (0·15 mM at 110 μCi/μmole). The reaction mixtures were incubated, combined and centrifuged on a sucrose gradient as in Fig. 7, except that the centrifugation time was 5 h. Twenty-one fractions were collected. *a*, 50 μl. portions of each fraction were precipitated with TCA, filtered and counted as before. ○ - - - ○, Radioactivity (c.p.m.) ³H; ● — ●, in ¹⁴C. *b*, 50 μl. portions were annealed to *r* strand DNA of φ80iλ or to *l* strand DNA of λb₁, as described in the legend to Fig. 5. ○ - - - ○, Radioactivity (c.p.m.) in ³H annealed to λb₁ *l* DNA; ● — ●, in ¹⁴C annealed to λb₁ *l* DNA; □ - - - □, in ³H annealed to φ80iλ *r* DNA; ■ — ■, in ¹⁴C annealed to φ80iλ *r* DNA.

the size ascribed to the *m*RNA made *in vivo* for the *N* gene region is 225,000 (Kourilsky, P., Marcaud, L., Portier, M. M., aud Gros, F., manuscript in preparation). It remains to be determined if the species made *in vivo* and *in vitro* are indeed identical. The 7*S* RNA is estimated to have a molecular weight in the vicinity of 100,000, which might encode a protein of approximately 10,000 molecular weight. This protein may correspond to the turn-off function (*tof*) which limits expression of genes to the left of *N* after early stages of infection, and, like the 7*S in vitro* RNA, is both encoded by the immunity region and under the control of promoter p_R (ref. 30). Either the 7*S* RNA or *tof*, or both, may correspond to *CRN*, a gene which prevents repressor synthesis during lytic growth and is also synthesized from promoter p_R in the immunity region (H. Eisen, manuscript in preparation). An RNA species with the properties of the 7*S* RNA recently has been detected *in vivo* (S. Heinemann, in preparation), and it has been shown that most *in vivo* r strand transcription in the absence of the *N* function is restricted to the immunity region (S. Kumar and W. Szybalski, personal communication).

When ρ-factor is not present during *in vitro* synthesis, the 12*S* and 7*S* RNA species are not found and most of the RNA product is larger. Because the ρ-factor was shown to affect propagation rather than initiation of RNA chains, I infer that in its absence RNA synthesis is not terminated at these points, but instead proceeds into the areas indicated in Fig. 9 by the dotted lines. This model explains both the change in size of the RNA product and the decrease in net RNA synthesis which the ρ-factor produces.

The existence of the ρ-factor suggests a possible mechanism of action of the λ *N* protein, which also might be relevant to other systems such as the arabinose operon[31] which are under positive control. The *N* function greatly stimulates transcription in the regions to the left and right of the termination points in Fig. 9. Because these regions are directly adjacent to the termination points, the *N* protein might function simply by preventing the termination process and allowing RNA synthesis to proceed beyond the termination signals. For example, if the ρ-factor functions by binding to DNA at the termination point, the *N* protein might bind to the same site more tightly and thereby deny ρ-factor access to it. (The *N* function would have to be specific to the phage DNA and not to the ρ-factor or another host function, for different phages which grow in the same host are known to have incompatible *N* functions—H. Eisen, personal communication.) This "conditional termination" hypothesis requires that RNA synthesized under stimulation by *N* consists of molecules initiated at promoters p_L and p_R, and there is strong evidence that this is true (D. Luzatti, in preparation; J. Pero, in preparation; and refs. 16, 32). A reasonable explanation for the slight synthesis of messenger for genes *O* and *P* which occurs in the absence of *N* function is that termination on the right is not completely efficient, but instead occasionally allows an RNA polymerase molecule to pass the termination points. The conditional termination hypothesis makes several predictions: (1) Messenger RNA molecules for genes beyond the termination points should include the sequences of the 12*S* and 7*S* RNA species at their 5′ ends. (2) There should exist mutations which allow λ to grow in the absence of *N* function by destroying termination signals against which the *N* protein acts. *N*-independent mutants of λ exist[33–35], but their mechanism is unknown. (3) If the 7*S* RNA corresponds to the *tof* function, then the *tof* function should not be under control of the *N* gene product.

Why is the fraction of RNA specific to the region affected by the c_{17} mutation greater in the presence of ρ-factor? The c_{17} mutation is located to the right of the immunity region: it can be crossed into phage 434hy (ref. 26). Because the 7*S* RNA is synthesized within the im-

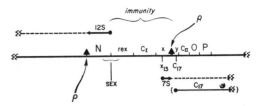
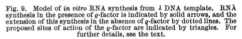

Fig. 9. Model of *in vitro* RNA synthesis from λ DNA template. RNA synthesis in the presence of ρ-factor is indicated by solid arrows, and the extension of this synthesis in the absence of ρ-factor by dotted lines. The proposed sites of action of the ρ-factor are indicated by triangles. For further details, see the text.

munity region, the factor must terminate RNA synthesis at a point within the immunity region or very close to its right boundary. Thus it is likely that the c_{17} mutation is located outside the site at which the ρ-factor acts, so that even in the presence of ρ-factor, RNA polymerase molecules which initiate at the c_{17} promoter can proceed through the region of the *O* and *P* genes. Transcription initiated at p_R does not continue beyond the termination site in the presence of ρ-factor. Whereas in the absence of factor the c_{17} promoter simply adds to transcription initiated at p_R and continuing through the *O*–*P* region, in the presence of factor it is essentially the only source of transcription of this region.

It seems safe to infer that the ρ-factor is responsible for termination events in the synthesis of early messenger RNA of λ in infected cells, and that it has the same function in bacterial RNA synthesis. It might, of course, be only one component of the complete termination system of *E. coli*: there could exist in addition several factors like ρ which recognize distinct classes of termination signals, or possibly several distinct termination processes. I have observed that at least one termination event occurs with λ DNA template in the absence of ρ-factor to produce an RNA species of approximately 22*S*, although its physiological significance is unknown. There is little information about the molecular mechanism of termination by the ρ-factor. It might bind to DNA and act when it is met by an RNA polymerase molecule progressing along the template; alternatively, it might form a complex with RNA polymerase and function when the complex reaches a termination point. A third possibility is that the configuration of RNA polymerase changes at the termination site, allowing ρ-factor to interact with it and remove it from the DNA. Whatever the mechanism, there must exist specific DNA sequences which function as termination signals in the presence of the ρ-factor.

Purification and Properties of ρ-Factor

All steps are performed at 0°–4° C. Frozen *E. coli* cells (50 g) are ground with 135 g alumina; to the paste is added 140 ml. of a buffer containing 0·05 M Tris (pH 7·4), 0·20 M KCl, 0·01 M MgCl₂, 0·0001 M dithiothreitol, 0·0001 M ethylenediaminetetraacetate, and 5 per cent glycerol. Electrophoretically pure deoxyribonuclease (600 μg; Worthington) is added and the mixture is allowed to stand for 15 min. Alumina and débris are removed by low speed centrifugation and ribosomes are removed by centrifugation for 2 h at 100,000g. To the supernatant is added 32 g ammonium sulphate per 100 ml., and precipitated proteins are dissolved in 0·05 M potassium phosphate buffer (pH 7·5), containing 0·0001 M dithiothreitol and 5 per cent glycerol. (The molar ratio of K₂HPO₄ to KH₂PO₄ in the buffer is 5·2.) The proteins are dialysed against this buffer, loaded on a 100 ml. phospho-cellulose column in the same buffer, and eluted with a gradient from 0·05 M to 0·50 M phosphate. Portions (5 μl.) of column fractions are assayed in a reaction mixture similar to that described in the legend to Fig. 1. Fractions around the peak of depression at approximately 0·16 M phosphate are pooled, precipitated with ammonium sulphate at 60 per cent saturation, and the precipitated protein is dissolved in 0·02 M potassium phosphate buffer (*p*H 7·5), containing 0·0001 M dithiothreitol and 5 per cent glycerol. The protein is dialysed into this buffer and loaded on a 20 ml. DEAE cellulose column equilibrated in the same buffer. Protein is eluted with a salt gradient from 0·02 M to 0·20 M phosphate, and 3 µl. of each fraction is assayed as before. Fractions around the peak of depression eluted at approximately 0·10 M phosphate are pooled, concentrated by ammonium sulphate precipitation as before, and dissolved in 400 µl. of a buffer containing 0·05 M Tris (*p*H 7·4), 0·10 M KCl, 0·0001 M dithiothreitol, 0·0001 M ethylenediaminetetraacetate, and 5 per cent glycerol. After a brief dialysis against this buffer, the sample is layered on to two 5 ml. glycerol gradients (10–30 per cent, by volume) in the same buffer and centrifuged approximately 12 h at 65,000 r.p.m. in the Spinco *SW* 65 rotor. Twenty-two fractions of each gradient are collected; the $A_{260\ nm}$ value of each fraction is determined and 2 µl. portions of fractions are assayed as before.

Fig. 10 illustrates the sedimentation profile obtained from the glycerol gradient step. The depression activity clearly coincides with a peak in $A_{280\ nm}$ which sediments at 8–10S and is well separated from the bulk of contaminating material. The assay is not linear over most of the concentration range in the peak, so that the activity appears to trail well beyond the absorbance peak. Because the factor cannot be assayed at the early steps of purification, and the assay has not been quantitated adequately for any step, the recovery of ρ-factor cannot be calculated; I suspect that it is small.

The ρ-factor purified to this extent appears to be a pure protein. Its $A_{260\ nm}$ to $A_{280\ nm}$ ratio is 1·5–2·0, which is characteristic of a protein and implies that very little nucleic acid is present. Heating the factor for 5 min at 50° C destroys the depression activity. It adheres to both DEAE-cellulose and phospho-cellulose at low ionic strength, so that it contains regions of both positive and negative charge. The sedimentation constant of 8–10S suggests a molecular weight in the vicinity of 200,000, assuming an average configuration. When fractions across the 8–10S peak of Fig. 10 are examined by polyacrylamide gel electrophoresis in SDS, one predominant band is found; its maximum intensity coincides with the absorbance peak. A gel resolution of purified ρ-factor is shown in Fig. 11; its purity is estimated to be greater than 95 per cent. The molecular weight of the polypeptide chain is approximately 50,000 measured on polyacrylamide gels in SDS by the method of Shapiro *et al.*[36], using the subunits of RNA polymerase as markers[11]. These molecular weights suggest that the 8–10S protein may be a tetramer.

I thank Professor Walter Gilbert, who suggested and guided these experiments; Professor J. D. Watson, for support and advice; Professor Klaus Weber, for helpful advice; Dr Max Gottesman, for providing phage strains and for suggesting that λ$_{sex}$ might be a promoter mutant; Drs R. Burgess, M. Ptashne, M. Schwartz and A. Travers, and R. Hendrix, N. Hopkins, R. Kamen and J. Pero for helpful conversations; Mrs Ann Soderquist for technical assistance; and Mrs C. Roberts for figure drawings. I was supported by a predoctoral fellowship from the US National Science Foundation and by grants to the laboratory from the US National Science Foundation and the US National Institutes of Health.

Received November 11, 1969.

Fig. 11. Polyacrylamide gel electrophoresis of purified ρ-factor. The electrophoresis and staining procedure were as described in ref. 11; the gel contained 5 per cent acrylamide and 0·1 per cent SDS.

[1] Roberts, J. W., *Nature*, **223**, 480 (1969).
[2] Milanesi, G., Brody, E. N., and Geiduschek, E. P., *Nature*, **221**, 1015 (1969).
[3] Naono, S., and Gros, F., *Cold Spring Harbor Symp. Quant. Biol.*, **31**, 361 (1966).
[4] Cohen, S. N., Maitra, U., and Hurwitz, J., *J. Mol. Biol.*, **26**, 19 (1967).
[5] Summers, W. C., and Siegel, R. B., *Nature*, **223**, 1111 (1969).
[6] Travers, A. A., and Burgess, R. R., *Nature*, **222**, 537 (1969).
[7] Bautz, E. K. F., Bautz, F. A., and Dunn, J. J., *Nature*, **223**, 1022 (1969).
[8] Richardson, J. P., *J. Mol. Biol.*, **21**, 115 (1966).
[9] Geiduschek, E. P., and Haselkorn, R., *Ann. Rev. Biochem.*, **38**, 647 (1969).
[10] Bremer, H., and Konrad, M. W., *Proc. US Nat. Acad. Sci.*, **51**, 801 (1964).
[11] Burgess, R. R., Travers, A. A., Dunn, J. J., and Bautz, E. K. F., *Nature*, **221**, 43 (1969).
[12] Glynn, I. M., and Chappell, J. B., *Biochem. J.*, **90**, 147 (1964).
[13] Radding, C., and Echols, H., *Proc. US Nat. Acad. Sci.*, **60**, 707 (1968).
[14] Kumar, S., Bøvre, K., Guha, A., Hradecna, Z., Maher, Sr. V. M., and Szybalski, W., *Nature*, **221**, 823 (1969).
[15] Eisen, H., Fuerst, C., Siminovitch, L., Thomas, R., Lambert, L., Pereira da Silva, L., and Jacob, F., *Virology*, **30**, 224 (1966).
[16] Taylor, K., Hradecna, Z., and Szybalski, W., *Proc. US Nat. Acad. Sci.*, **57**, 1618 (1967).
[17] Pereira da Silva, L. H., and Jacob, F., *Virology*, **33**, 618 (1967).
[18] Ogawa, T., and Tomizawa, J., *J. Mol. Biol.*, **38**, 217 (1968).
[19] Westmoreland, B. C., Szybalski, W., and Ris, H., *Science*, **163**, 1343 (1969).
[20] Ptashne, M., and Hopkins, N., *Proc. US Nat. Acad. Sci.*, **60**, 1282 (1968).
[21] Bolle, A., Epstein, R. H., Salser, W., and Geiduschek, E. P., *J. Mol. Biol.*, **31**, 325 (1968).
[22] Signer, E. R., Manly, K. F., and Brunstetter, M., *Virology*, **39**, 137 (1969).
[23] Szybalski, W., Bøvre, K., Fiandt, M., Guha, A., Hradecna, Z., Kumar, S., Lozeron, H. A., Maher, Sr. V. M., Nijkamp, H. J. J., Summers, W. C., and Taylor, K., *J. Cell Physiol.*, **74**, suppl. 1, 33 (1969).
[24] Szpirer, J., Thomas, R., and Radding, C. M., *Virology*, **37**, 585 (1969).
[25] Lozeron, H. A., and Szybalski, W., *Virology*, **39**, 373 (1969).
[26] Pereira da Silva, L. H., and Jacob, F., *Ann. Inst. Pasteur*, **115**, 145 (1968).
[27] Packman, S., and Sly, W. S., *Virology*, **34**, 778 (1968).
[28] Spirin, A. S., *Biokhimiya*, **26**, 511 (1961).
[29] Kourilsky, P., Marcaud, L., Sheldrick, P., Luzzati, D., and Gros, F., *Proc. US Nat. Acad. Sci.*, **61**, 1013 (1968).
[30] Pero, J., *Virology* (in the press).
[31] Sheppard, D., and Englesberg, E., *Cold Spring Harbor Symp. Quant. Biol.*, **31**, 345 (1966).
[32] Schwartz, M., *Virology* (in the press).
[33] Hopkins, N., *Virology* (in the press).
[34] Butler, B., and Echols, H., *Virology* (in the press).
[35] Court, D., and Sato, K., *Virology*, **39**, 348 (1969).
[36] Shapiro, A., Viñuela, E., and Maizel, J. V., *Biochem. Biophys. Res. Commun.*, **30**, 240 (1967).

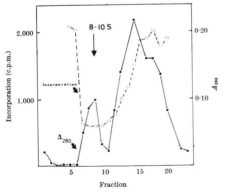

Fig. 10. Glycerol gradient centrifugation of ρ-factor. Details are given in the text. The direction of centrifugation is from right to left. ○ - - - ○, Radioactivity (c.p.m.) in ³H; ●—●, $A_{260\ nm}$.

14

Reprinted from *Nature*, **221**(5183), 838–841 (1969)

Two Compounds implicated in the Function of the RC Gene of *Escherichia coli*

by

MICHAEL CASHEL*
JONATHAN GALLANT
Department of Genetics,
University of Washington,
Seattle, Washington

Autoradiography has revealed in a stringent strain of *E. coli* two compounds that seem to be involved in the inhibition of the synthesis of RNA.

IN bacterial cells a regulatory mechanism termed "stringent control" reduces the rate of accumulation of RNA whenever protein synthesis is limited by the absence of any required amino-acid[1-3] or the failure to activate an amino-acid[4]. At least one gene involved in stringent control, the *RC* gene of *E. coli*, has been identified as the site of mutations to the "relaxed" phenotype in which RNA accumulation is nearly independent of amino-acid availability[5,6]. Although the function of the *RC* gene product is not known, the existence of isogenic pairs of strains carrying wild type (stringent) and mutant (relaxed) alleles of the *RC* gene makes it possible to isolate the *RC* function experimentally.

Results of early studies suggested that stringent control operated directly on the transcription reaction itself[5,7,8]. But subsequent studies of transcription *in vitro* (Nierlich and Gros, personal communication and refs. 9 and 10) or in decryptified whole cells[11] disclose little if any effect of amino-acid starvation, uncharged transfer RNA or mutation in the *RC* gene on RNA polymerase activity.

An alternative approach was suggested by the discovery that *RC* function results in inhibition of the biosynthesis of a number of phosphorylated metabolites, including ribonucleoside triphosphates and glycolytic intermediates[11-14]. The significance of these observations lies in the fact that direct inhibition of RNA polymerase activity does not provoke any such effects[13,14]. This finding indicates that the role of the *RC* function in controlling the synthesis of phosphorylated metabolites

cannot be explained as a secondary consequence of control at the RNA polymerase level. We therefore proposed that stringent control operated directly on some step in energy metabolism, and controlled the rate of RNA synthesis indirectly through substrate limitation. We postulated that a reaction normally involved in protein synthesis "idles" when any species of transfer RNA is not charged with its cognate amino-acid, and that this idling reaction, carried out by the *RC* gene product, generates an inhibitor of phosphorylation[13]. We report here the discovery of two compounds which may be products of the postulated idling reaction.

Amino-acid starvation affects the biosynthesis of many phosphorylated metabolites. Fig. 1 shows the results of an early experiment in our exploration of the metabolic consequences of stringent control. A stringent strain of *E. coli*, growing exponentially in medium fortified with an amino-acid mixture (0·2 per cent casein hydrolysate), was filtered and aliquots were cultivated in fresh medium in the presence and absence of the amino-acid mixture. Both aliquots were labelled for 20 min with $^{32}PO_4$, and the acid-soluble metabolites were submitted to thin-layer chromatography as described in the legend. Fig. 1 shows autoradiograms of these chromatograms.

In the presence of amino-acids, at least eighteen well resolved spots corresponding to various phosphate compounds can be discerned. Four of these are the ribonucleoside triphosphates, while many of the rest have not been identified. Comparison of the two autoradiograms reveals that each compound is much more heavily labelled in the culture supplemented with amino-acids. Our

* Present address: Laboratory of Molecular Biology, National Institute of Neurological Diseases and Stroke, Bethesda, Maryland.

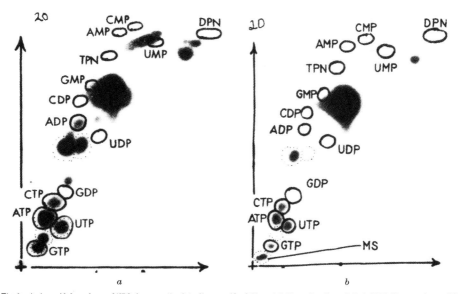

a   *b*

Fig. 1. ³²PO₄ incorporation into diverse acid soluble metabolites. A culture of strain B333 (thy⁻ pro⁻ leu⁻ met⁻), growing exponentially at 37° C in the presence of 20 μg/ml. of thymine and 0·2 per cent casein hydrolysate (Difco 'casamino'-acids) was filtered, washed, and resuspended in fresh medium containing 10⁻⁴ M phosphate and 10 μCi/ml. of ³²PO₄. One aliquot was labelled in the absence of amino-acids, the other was again supplemented with 0·2 per cent casein hydrolysate. After aeration at 37° C for 20 min, formic acid extracts were prepared as described previously[13] and chromatographed by stepwise elution in lithium chloride in the first dimension, and stepwise elution in sodium formate in the second dimension as described in Fig. 88 by Randerath[20]. Autoradiography was carried out as earlier described[13]. After 20 min there were 25 μmoles/μl. of TCA-precipitable phosphate in the culture supplemented with amino-acids (*a*), and 2 mμmoles/ml. in the culture of starved amino-acids (*b*).

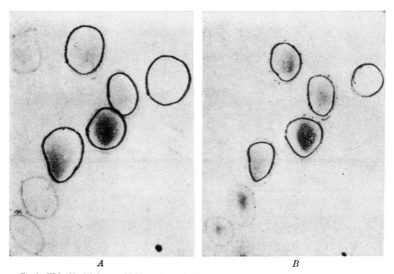

A   *B*

Fig. 2. ³²PO₄ (22 μCi/ml.) was added to a culture of CP78 (leu⁻ his⁻ art⁻ threo⁻ B1⁻ RC⁺ᵗ) growing exponentially in *tris* glucose medium containing 2 × 10⁻³ M phosphate, 40 μg/ml. of leucine, 20 μg/ml. of histidine, 100 μg/ml. of arginine, 200 μg/ml. of threonine and 1 μg/ml. of B1. After 91 min valine was added to a final concentration of 400 μg/ml. At various times before and after the addition of valine, formic acid extracts were prepared and chromatographed by the method[15], as modified by us[13]. *A*, Autoradiogram of a sample taken 2 min before the addition of valine; *B*, autoradiogram of a sample taken 3 min afterward. The dark lines show the positions of unlabelled marker nucleotides, chromatographed with the sample and located by ultraviolet absorption. The lighter lines (which were made on the film in pencil of a different colour) show the regions of the chromatogram cut out and counted (Fig. 4). The positions of the nucleotides on the chromatogram are:

113

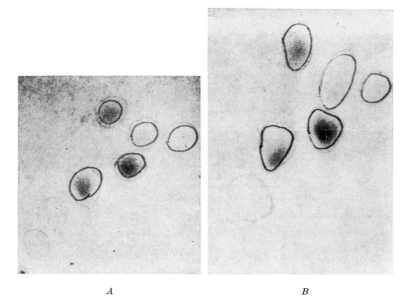

A   B

Fig. 3. ³²PO₄ (13·3 μCi/ml.) was added to a culture of CP79 (isogenic with CP78 except RCʳᵉˡ) growing exponentially in medium of the same composition as that of Fig. 2.   After 97 min, 400 μg/ml. of valine was added.
Formic acid extracts were prepared and chromatographed as in Fig. 2. *A*, Autoradiogram of a sample taken 2 min
before the addition of valine; *B*, autoradiogram of a sample taken 3 min later.

concern here, however, is with a nineteenth spot which is
detectably labelled only in the culture starved of amino-
acids. This material, which we call the magic spot, has
a low mobility in both dimensions and is to be found
between GTP, the most sluggish of the triphosphates, and
the origin. What interests us about this compound is
its unique pattern of synthesis: it is formed only in
conditions where the synthesis of other phosphorylated
metabolites is markedly inhibited by the operation of
stringent control. This is precisely what we would expect
of the postulated phosphorylation inhibitor.

Starvation for only a single amino-acid also elicits the
production of magic spot, which is resolved into two components in a solvent system developed for the separation
of nucleoside triphosphates[15]. Fig. 2 shows two such autoradiograms. A stringent strain of *E. coli* was cultivated
for a generation and a half in the presence of ³²PO₄, to
pre-label all nucleotide pools, and then valine, which
produces endogenous isoleucine deficiency, was added.
Samples were removed for chromatography 2 min before
the addition of valine (Fig. 2*A*) and 3 min later (Fig. 2*B*).
The autoradiogram of the latter sample reveals the appearance of two compounds, MS I and MS II, in the magic
spot region between GTP and the origin.

To determine the relationship between the synthesis
of the MS compounds and the function of the *RC* gene
product, we performed the same experiment with CP79,
a derivative of CP78 which differs from the parent strain
only in the presence of a mutation in the *RC* gene. The
autoradiograms of Fig. 3 show that valine inhibition does
not elicit MS synthesis in the relaxed mutant. Fig. 4
compares the effect of valine inhibition on the synthesis
of RNA and of the MS compounds in CP78 and CP79.
Formation of the MS compounds and a cessation of RNA
synthesis occur promptly in the stringent strain, while the
relaxed mutant shows neither response. A more detailed
survey, involving starvation for several different amino-
acids and comparison of three different stringent-
relaxed pairs of strains, has confirmed that MS synthesis

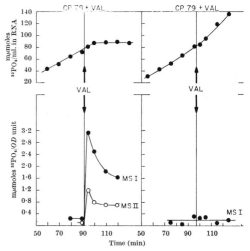

Fig. 4. Synthesis of RNA and MS production, CP78 and CP79. The
regions of the chromatograms corresponding to MS I and MS II in all
samples taken during the experiments of Figs. 2 and 3 were cut out and
counted in 'Liquifluor' in a Packard 'Tri-Carb' liquid scintillation
counter. The data are reported as mμmoles of phosphate per unit
optical density at 720 mμ (1 optical density units = 10⁹ cells). In
samples taken before the addition of valine in CP78, or at any time in
CP79, no darkening of the film was visible in the positions of MS I and
MS II. The appropriate regions of the chromatograms were nonetheless
cut out (only MS I in the case of CP79) and counted, but these low levels
of radioactivity were probably due to trailing of other metabolites rather
than basal levels of the MS compounds.   For measurement of phosphate
in RNA, duplicate 20λ samples were taken from the labelled cultures and
added to 1·0 mL of 10 per cent TCA, and 0·5 N NaOH, respectively. Total
TCA-precipitable, alcohol-insoluble ³²P (RNA + DNA) was determined in
the former sample by filtering and washing with 10 per cent TCA and
95 per cent ethanol. The quantity of DNA ³²P, determined in the
second sample by the method of Roodyn and Mandel³¹, was then subtracted from this value.

is invariably associated with the stringent response (M. C.'s unpublished results).

Two explanations can be suggested for this correlation between MS production and stringency. On the one hand, the synthesis of the MS compounds may be a direct consequence of *RC* function; for example, they may be products of the enzyme activity of the *RC* gene product. Alternatively, because the *RC* system blocks RNA synthesis in response to amino-acid starvation, the production of the MS compounds might be a consequence of inhibition of RNA synthesis, and thus only an indirect reflexion of *RC* function. If this were the case, we would expect direct inhibition of RNA synthesis to elicit the production of the MS compounds. But direct inhibition of RNA synthesis, by uracil starvation, has no such effect (Table 1). The failure of MS I to appear during uracil starvation is therefore evidence that its formation is not simply a response to blocked RNA synthesis, but rather a specific consequence of the function of the *RC* gene product. (We have detected no labelling of MS I by radioactive uridine, which indicates that it is not made from pyrimidine precursors.) The chemical composition and role in metabolism of the MS compounds should therefore be an important clue to the mechanism of stringent control.

Table 1. PRODUCTION OF MS I DURING URACIL STARVATION AND HISTIDINE STARVATION

Culture	mμmoles $^{32}PO_4/OD$ in		
	TCA precipitate	UTP	MS I
Starved of uracil	0·23	< 0·1	< 0·1
Uracil added back	24·0	5·8	0·2
Starved of histidine	1·1	—	2·8

A culture of *Escherichia coli* BD1, which is stringent and requires thymine, uracil and histidine, was cultivated in *tris*-glucose medium containing 2×10^{-4} M phosphate, 20 μg/ml. of thymine, 20 μg/ml. of histidine, and 4 μg/ml. of uracil. Exhaustion of this limiting supply of uracil stopped growth at an optical density of 720 mμ of 0·65, and the optical density then remained constant. After 30 min of uracil starvation, $^{32}PO_4$ was added to two aliquots of the culture, one of which received in addition 20 μg/ml. of uracil. After 20 min, 20 μl. samples were added to 1·0 ml. of 10 per cent TCA to measure TCA-precipitable, alcohol-insoluble phosphate (principally RNA), and 100 μl. samples were acidified with formate and chromatographed as in Figs. 2, 3 and 4. Autoradiograms showed barely visible labelling of MS I, which was cut and counted, and no labelling at all of MS II, which was not counted.

For comparison, the effect of histidine starvation on MS I production in the same strain was determined. A culture growing exponentially in medium containing sufficient histidine (20 μg/ml.) was filtered and resuspended in fresh medium lacking histidine. After 20 min of labelling with $^{32}PO_4$, samples were removed for measurement of TCA-precipitable material and MS I, as described. Data are expressed as mμmoles of phosphorus per unit optical density at 720 mμ at the moment of addition of $^{32}PO_4$.

Our original hypothesis, which led us to find the MS compounds, would identify them as products of the idling reaction. Which step in protein synthesis could fulfil this role? The work of Neidhardt and his collaborators[4] revealed that the *RC* gene product does not respond to the absence of free amino-acids, but rather to the absence of activated amino-acids. Thus the idling reaction must be a stage in protein synthesis which utilizes the product of amino-acid activation, that is, aminoacyl-transfer RNA. This step is the transfer of the aminoacyl residue from transfer RNA to growing peptide chain, with the concomitant hydrolysis of GTP[16-19]. The transfer factors which carry out this step interact directly with aminoacyl transfer RNA molecules, and could therefore be affected by the absence of one or more charged species (or the presence of uncharged ones).

The transfer step is therefore a likely candidate for the role of idling reaction. Furthermore, careful pool studies have shown that the GTP pool shrinks abruptly at the onset of the stringent response (unpublished work of J. G. and Harada). This suggests that the idling reaction consumes GTP, just as the transfer step does.

Accordingly, we anticipated that the products of the idling reaction would prove to be derivatives of GTP. Purification and preliminary characterization of MS I (to be described later by M. C. and Kalbacher) suggest that it is guanosine-5′-tetraphosphate. It is reasonable therefore to suppose that MS I is made by the *RC* gene product in an idling reaction of the sort postulated.

This work was supported by a grant from the National Institutes of Health, US Public Health Service. M. C. is a US Public Health Service trainee on leave from the Laboratory of Molecular Biology, National Institute of Neurological Diseases and Stroke, US National Institutes of Health. We thank Mrs Barry Harada for technical assistance.

Note added in proof. More recent evidence reveals that MS I is a periodate insensitive, alkali labile tetraphosphate of guanosine rather than guanosine-5′-tetraphosphate.

Received November 27, 1968.

[1] Gale, E. F., and Folkes, J. P., *Biochem. J.*, **53**, 493 (1953).

[2] Pardee, A., and Prestidge, L., *J. Bact.*, **71**, 677 (1956).

[3] Gros, F., and Gros, F., *Exp. Cell Res.*, **14**, 104 (1958).

[4] Neidhardt, F. C., *Bact. Rev.*, **30**, 701 (1966).

[5] Stent, G. S., and Brenner, S., *Proc. US Nat. Acad. Sci.*, **47**, 2005 (1961).

[6] Alföldi, L., Stent, G., and Clowes, R., *J. Mol. Biol.*, **5**, 348 (1962).

[7] Kurland, C. G., and Maaløe, O., *J. Mol. Biol.*, **4**, 193 (1962).

[8] Tissières, A., Bourgeois, S., and Gros, F., *J. Mol. Biol.*, **7**, 100 (1963).

[9] Olenick, J., and Hahn, F., *Biochim. Biophys. Acta*, **129**, 429 (1966).

[10] Bremer, H., Yegian, C., and Konrad, M., *J. Mol. Biol.*, **16**, 94 (1966).

[11] Gallant, J., and Cashel, M., *J. Mol. Biol.*, **25**, 545 (1967).

[12] Edlin, G., and Neuhard, J., *J. Mol. Biol.*, **24**, 225 (1967).

[13] Cashel, M., and Gallant, J., *J. Mol. Biol.*, **34**, 317 (1968).

[14] Irr, J., and Gallant, J., *J. Biol. Chem.* (in the press).

[15] Neuhard, J., Randerath, E., and Randerath, K., *Anal. Biochem.*, **13**, 211 (1965).

[16] Keller, E. B., and Zamecnik, P. C., *J. Biol. Chem.*, **221**, 45 (1965).

[17] Allende, J. E., Munro, R., and Lipmann, F., *Proc. US Nat. Acad. Sci.*, **51**, 1211 (1964).

[18] Conway, T. W., and Lipmann, F., *Proc. US Nat. Acad. Sci.*, **52**, 1462 (1964).

[19] Arlinghaus, R., Shaeffer, J., and Schweet, R., *Proc. US Nat. Acad. Sci. Wash.*, **51**, 1291 (1964).

[20] Randerath, K., *Thin Layer Chromatography*, 230 (Academic Press, New York, 1966).

[21] Roodyn, D., and Mandel, G., *Biochim. Biophys. Acta*, **41**, 80 (1960).

Part III
TRANSLATION: THE BACTERIAL RIBOSOME

Editor's Comments
on Papers 15 Through 22

The remainder of this volume is devoted to the components involved in the process by which mRNA is translated on the ribosome to yield the polypeptide gene products in their appropriate amino acid sequences. Part III is concerned with the structure and assembly of the bacterial ribosome itself. The mechanism of peptide bond formation, the genetic code, and the structure and function of tRNA are considered later.

The discovery of the bacterial ribosome dates to electron micrographic observations of Luria et al. (1943) on bacteriophage-infected *E. coli*. After cell lysis, phage were superimposed upon a background of small, uniform-sized particles that appeared to exist freely throughout the cytoplasm. In an exhautive chemical and ultracentrifugal analysis of extracts from several bacterial species, Schachman et al. (1952) demonstrated that these particles were composed of ribonucleoprotein and formed discrete boundaries that undoubtedly represented ribosomal subunits. The isolation of bacterial ribosomes and a precise determination of their sedimentation coefficients was reported by Tissières et al. (Paper 15). They found that in the presence of 10^{-2} M magnesium, *E. coli* ribosomes have a sedimentation coefficient of $70S$; in 10^{-3} M magnesium, the ribosomes dissociate into $50S$ and $30S$ subunits. Each subunit is composed of approximately two-thirds RNA and one-third protein. The $50S$ subunit has a particle weight of 1.8 $\times$ 10^6 daltons and contains $23S$ and $5S$ rRNA molecules and about 35 discrete proteins. The particle weight of the $30S$ subunit is 0.85 $\times$ 10^6 daltons; it consists of a single $16S$ rRNA and about 20 proteins (for reviews that include detailed discussions of ribosome structure, see Nomura, 1970; Maden, 1971; and Kurland, 1972).

It was shown by Tissières et al. (1960) that the $70S$ ribosome is the functional unit involved in protein synthesis. Subsequently, Guthrie and Nomura (1968) demonstrated that, at the onset of protein synthesis, the $30S$ ribosomal subunit combines with mRNA, the fully charged initiating tRNA species, and the soluble protein initiation factors to form the $30S$ initiation complex. This complex is then joined by the $50S$ subunit to form the $70S$ initiation complex, and polypeptide chain elongation commences. Several $70S$ ribosomes attach to a single mRNA molecule in the form of a polyribosome. After elongation and release of the polypeptide chain, $70S$ ribosomes are thought to dissociate from the mRNA as $30S$ and $50S$ subunits (Kaempfer, 1972). Reassociation of the freed subunits is apparently prevented by one of the initiation factors (see Part IV). The free subunits are thereby conserved for the initiation of new rounds of translation.

With regard to the RNA components of the *E. coli* ribosome, Kurland (1960) demonstrated that the $30S$ ribosomal subunit contains a single $16S$ rRNA molecule; a $23S$ rRNA was found as a component of the $50S$ subunit. It was later shown by Rosset and Monier (Paper 16) and Rosset et al. (1964) that the $50S$ subunit also contains a $5S$ rRNA with a molecular weight of 4 $\times$ 10^4 daltons. Intact $16S$ and $23S$ rRNA species were characterized by Stanley and Bock (Paper 17). They reported molecular weights of 1.1×10^6 and 0.55×10^6 daltons for the 23*S* and 16*S* components, respectively.* Small but significant differences were found in their base compositions, and it was concluded that the two molecular species are separate noninterconvertible RNA classes. The complete base sequence of the 5*S* rRNA has been elucidated (Brownlee et al., Paper 18), and the 16*S* rRNA has been partially sequenced (Fellner et al., 1972). The sequences of these species and that of 23*S* rRNA are apparently distinct (Nomura, 1970). Each rRNA component has an essential structural role: they are necessary for the correct assembly of functionally active ribosomal subunits through specific interactions with ribosomal proteins (see below). Possible direct functional roles for rRNA have not been established, but modifications in the methylation of crucial bases in both 16*S* and 23*S* rRNA are apparently responsible for ribosomal resistance to the antibiotics kasugamycin and lincomycin, respectively (Helser et al., 1971; Lai et al., 1973). In addition, Bowman et al. (1971) have demonstrated that inactivation of 30*S* subunits by colicin E3, a specific antibacterial peptide, is due to excision of a portion of 16*S* rRNA.

In contrast to studies on rRNA, the isolation and characterization of ribosomal proteins has, because of their large number (Waller, 1964), proved a more formidable task. It is further complicated by the possibility that an individual protein component might represent a nonribosomal contaminant or that it has arisen by *in vitro* modification of another protein. Several lines of evidence have suggested that the majority of the 21 proteins found associated with the 30*S* subunit are of ribosomal origin. This evidence includes the isolation of the proteins from rigorously purified subunits that have retained full functional activity (Hardy et al., 1969), establishment of functional and assembly roles for individual proteins in the reconstitution of functionally active 30*S* subunits from their isolated molecular components (see below), mutational alteration in proteins affecting ribosome function (Birge et al., 1969; Bollen et al., 1969; Ozaki et al., 1969), formation of site-specific complexes with rRNA (Kurland, 1972), and immunological evidence (Traut et al., 1969). Although it has not been possible to fulfill all these criteria with any of the proteins isolated from the 50*S* subunit, it is believed to contain at least 28 and possibly as many as 34 discrete proteins (Kurland, 1972). None of the ribosomal proteins is common to both the 30*S* and 50*S* subunits (Traut et al., 1969).

With the exhaustive characterization of the ribosomal proteins, it has become possible to estimate the number of copies of each protein

---

*Nearly identical apparent molecular weight values were reported by Midgley (1965) using an independent technique.

present per subunit. Thus, by summing the molecular weights of 19 different 30S associated proteins, an aggregate mass of 410,000 daltons was obtained; however, the 30S subunit contains only about 260,000 daltons of protein (Kurland et al., 1969). This suggested that the 30S subunit contains less than one copy of some ribosomal proteins. This was elegantly demonstrated by the stoichiometry measurements of Voynow and Kurland (Paper 19). At least eight of the 30S-associated proteins were present in amounts much less than one copy per ribosome. No protein was present in amounts that exceeded one copy per ribosome. These findings indicate that 30S subunits are structurally and functionally heterogeneous. The results of Kurland et al. (1969) suggest that some of the proteins exchange from one 30S subunit to another, causing functional heterogeneity within the ribosomal population. In contrast to the 30S, the 50S subunit appears to contain one copy of the majority of its proteins (Traut et al., 1969; Kurland, 1972).

In physical studies of *E. coli* ribosomes, Meselson et al. (1964) discovered that about one third of the proteins were split off during equilibrium buoyant density centrifugation in cesium chloride. This resulted in the formation of 23S and 40S core particles from 30S and 50S subunits, respectively. The core particles contained rRNA and the remaining proteins. Subsequently, it was shown by Staehelin and Meselson (1966) and Hosokawa et al. (1966) that under appropriate conditions the protein-synthesizing activity of the core structures could be restored in the presence of the homologous split proteins. The *in vitro* reconstitution of active particles from their partially resolved molecular components was thereby achieved. Such reconstitution systems have facilitated studies on the overall mechanism of ribosome assembly and on the structural and functional roles of individual ribosomal components.

In a later study, Traub and Nomura (Paper 20) dissociated 23S core particles into their RNA and protein components. When a mixture of 16S rRNA and the core proteins was incubated together with the split proteins, functionally active 30S particles were reconstituted. The total reconstitution reaction exhibited an absolute specificity for 16S rRNA (see below) and protein derived from the 30S particle. The core and split protein fractions could be replaced by the total ribosomal proteins obtained directly from the 30S subunit. These studies demonstrated that all the information necessary for the complete assembly of the 30S particle is contained within its molecular components.

With the availability of a system for the total reconstitution of the 30S subunit, it has become possible to study the assembly roles of the various protein components as well as their function in several aspects of the translation process. In the approach originated by Nomura and his colleagues (1969), the proteins are purified, and a single protein component is omitted during the reconstitution reaction. In this manner, at least five proteins (S4, S7, S8, S9, S16 and/or S17)* were shown to be strongly required for both assembly and polyphenylalanine synthesis directed by the synthetic mRNA polyuridylic acid (poly U). Single omission of these proteins caused the formation of incompletely reconstituted particles of diminished activity that sedimented at 20–25S. Functional and structural roles were also identified for another six proteins (S3, S5, S10, S11, S14, S19); in their absence, incomplete particles sedimenting at 26–30S were formed. Finally, most of the other proteins, although devoid of major assembly functions, were shown to be needed for full functional activity. On the basis of such studies, an assembly map for the 30S ribosomal proteins has been constructed (see Nomura, Paper 21). Recently, Held et al. (1973) have reported the complete resolution of all the 30S components and the reconstitution of highly active particles from the pure protein components and 16S rRNA. The above studies on the assembly of the 30S ribosomal subunit are reviewed and updated by Nomura (Paper 21). A revision of the assembly map found in Paper 21 has subsequently appeared (Held et al., 1974).

Kurland (1972) has provided an assessment of the single-omission method for establishing individual ribosomal protein functions. He has pointed out that such results are difficult to interpret in a functional sense because the role of an individual component by itself may not be well defined and separate. Thus, the omission of a single protein may simultaneously lead to a loss of several ribosomal functions. Nevertheless, the results of these studies have demonstrated highly cooperative functional interactions in the structural requirements for ribosome self-assembly.

Other approaches are now available for the elucidation of the structure–function roles of ribosomal proteins. These include the stimulation of ribosomal functions by adding extra copies of those proteins in which the ribosome is deficient. Accordingly, Randall-Hazelbauer and Kurland (1972) have found that proteins S2, S3, and S14 stimulate aminoacyl-tRNA binding and may form part of the acceptor (A) site of the ribosome. Unfortunately, this method is apparently restricted to those ribosomal proteins occurring in quantities of much less than one copy per 30S particle. Additional methods that have proved useful include genetic and chemical approaches, especially when employed in conjunction with reconstitution. The genetic method involves determining the structural basis for a mutation that alters a specific ribosomal function. Using this technique together

---

*Recent experiments reported by Nomura (Paper 21) suggest that S9, S16, and S17 are only required for assembly.

with the single-omission method, it has been found that S12 is altered by a mutation from streptomycin sensitivity to streptomycin resistance (Ozaki et al., 1969); an alteration in S12 was also responsible for the streptomycin dependence of another mutant strain (Birge and Kurland, 1969). Similarly, S5 was identified as the component altered in spectinomycin resistance (Bollen et al., 1969). The reconstitution technique has also been employed for the identification of protein components altered by chemical methods. For example, a role for S11 and S21 in the binding of mRNA and tRNA to 30S particles has been suggested from such a chemical modification approach (Craven et al., 1969). Specific antibodies have also been introduced as a reagent for elucidating ribosomal protein function (Kischa et al., 1971).

The complete reconstitution system has also permitted studies on the kinetics of 30S subunit assembly (see Nomura, Paper 21) and the specificity of the 16S rRNA requirement for ribosome assembly. Thus, in the studies of Traub and Nomura (Paper 20) it was shown that the 16S rRNA of *E. coli* could not be replaced by a "16S RNA" derived from the 23S species or the rRNA components from the smaller subunits of yeast and rat-liver cytoplasmic ribosomes. It was subsequently reported by Nomura et al. (1968) that the 16S rRNA of other bacterial species could replace that of *E. coli* in reconstitution of functionally active hybrid 30S particles; with the 16S rRNA of *E. coli* and proteins from the other species, active hybrid particles were also obtained. Since large portions of the base sequences of these 16S rRNA molecules differed, it was concluded that the specific interactions of ribosomal proteins with rRNA during assembly require only small conserved regions of base sequence homology or tertiary structure.

Unlike studies on the assembly of 30S ribosomal subunits, no reproducible procedure has been reported for the complete reconstitution of active 50S particles in *E. coli*. Successful reconstitution of active 50S subunits from dissociated components, however, has been achieved with the thermophilic bacterium *Bacillus stearothermophilus* in which the elevated temperatures necessary for complete assembly do not result in inactivation of functional constituents (Nomura and Erdmann, 1970). Using this system, it was demonstrated that 5S rRNA is essential for reconstitution of functionally active particles. In addition, *E. coli* 5S rRNA replaced that of *B. stearothermophilus*. Other studies with the 50S ribosomal reconstitution system are discussed by Nomura (Paper 21).

With regard to studies on the biosynthesis of the bacterial ribosome, several experimental approaches have yielded much valuable information. These include following the flow of radioactive precursors into mature ribosomes and studies with mutants defective in ribosome assembly. Using the former method, Mangiarotti et al. (1968) demonstrated that *in vivo* ribosome assembly occurs by a stepwise process in *E. coli*. Pulse labeling revealed that the rRNA molecules rapidly associate with ribosomal proteins to form ribonucleoprotein precursor particles. Two such intermediate particles with sedimentation coefficients of 32S and 43S were identified as precursors of the 50S subunit, whereas a precursor sedimenting at 26S was detected for the 30S subunit. An additional precursor sedimenting at 22S has been proposed for the 30S particle by Osawa et al. (1969). Although Monier et al. (1969) demonstrated that the 5S rRNA present in the 43S precursor had a primary structure identical to its counterpart from mature 50S subunits, Mangiarotti et al. (1968) reported that the other RNA species in the precursor particles were submethylated. Osawa et al. (1969) have also reported that ribosomal precursor structures are deficient in ribosomal proteins; appropriate conformational changes may then permit further protein binding and final conversion of the accumulated particles into mature ribosomes. This would be analogous to the stepwise assembly process demonstrated *in vitro* (see Nomura, Paper 21).

Conditional-lethal mutants have proved particularly useful in other *in vivo* studies on ribosome formation. Since *in vitro* reconstitution of ribosomes exhibits a marked dependence upon elevated temperatures, mutations affecting the assembly process would be manifested at lower temperatures. Accordingly, cold-sensitive mutants defective in ribosome assembly have been isolated from *E. coli* (Guthrie et al., 1969) and *Salmonella typhimurium* (Tai et al., 1969). At 20°C, various ribosomal subunit assembly defective (*sad*) mutants from *E. coli* accumulated a 21S precursor of the 30S subunit as well as 32S and 43S particles that function as precursors of the 50S subunit. Two of these mutants were defective in both 30S and 50S subunit assembly and accumulated 21S and 32S particles. Both of these assembly defects appear to result from a single mutation, which thereby suggests a coupling in 30S and 50S subunit biosynthesis. In Paper 21, a striking similarity in the protein composition of the 21S particle accumulated in *sad* mutants and that of the 21S reconstitution intermediate particle is seen. Thus, with regard to the assembly of ribosomal proteins, the *in vitro* reconstitution of the 30S particle may mimic the process within the cell. However, a major difference between the 21S particle of the cold-sensitive mutants and the reconstitution intermediate is that the former contains a precursor species of 16S rRNA, whereas the mature species is utilized for the latter. Recently, Wireman and Sypherd (1974) have reconstituted 30S ribosomal subunits using an unmethylated precursor of 16S rRNA and proteins derived from mature ribosomes, but these particles were inactive in polyphenylalanine synthesis, which suggests that during the biosynthesis of active 30S subunits a cooperative and sequential maturation of RNA precursor species occurs.

Major details of the overall mechanism involved in the biosynthesis of rRNA in *E. coli* have been elucidated only recently. It had been known that each of the rRNA molecular species arises directly from precursors slightly larger than their mature counterparts (reviewed by Pace, 1973). However, recent studies by Nikolaev et al. (Paper 22) and Dunn and Studier (1973) suggest that the immediate precursors of 16S and 23S rRNA are cleaved from a single large RNA molecule ("30S pre-rRNA") by ribonuclease III, a specific endonuclease. The existence of a single tandem transcript containing the rRNA precursors had been suggested by the clustering of rRNA genes (see Pace, 1973) and *in vitro* transcription studies (Pettijohn et al., 1970). The cleavage of this single large precursor species is analogous to the mechanism established for the synthesis of rRNA in eukaryotic organisms. However, the 30S pre-rRNA in normal *E. coli* is apparently cleaved during its formation, since pre-rRNA could only be demonstrated in a mutant of *E. coli* deficient in ribonuclease III activity. Whether 5S rRNA also arises from the pre-rRNA transcript remains to be established.

Although an extensive discussion on the genetics of the ribosomal RNA and protein components is not within the scope of this volume, a brief discussion on this aspect of the ribosome is presented in Paper 21. In addition, an extensive section on rRNA genes may be found in the recent review of Pace (1973).

REFERENCES

Birge, E. A., and C. G. Kurland (1969). Altered ribosomal protein in streptomycin-dependent *Escherichia coli. Science, 166,* 1282–1286.

——, G. R. Craven, S. J. S. Hardy, C. G. Kurland, and P. Voynow (1969). Structure determinant of a ribosomal protein: K locus. *Science, 164,* 1285–1286.

Bollen, A., J. Davies, M. Ozaki, and S. Mizushima (1969). Ribosomal protein conferring sensitivity to the antibiotic spectinomycin in *Escherichia coli. Science, 165,* 85–86.

Bowman, C. M., J. E. Dahlberg, T. Ikemura, J. Konisky, and M. Nomura (1971). Specific inactivation of 16S ribosomal RNA induced by colicin E3 *in vivo. Proc. Natl. Acad. Sci. U.S.A., 68,* 964–968.

Craven, G. R., R. Gavin, and T. Fanning (1969). The transfer RNA binding site of the 30S ribosome and the site of tetracycline inhibition. *Cold Spring Harbor Symp. Quant. Biol., 34,* 129–137.

Dunn, J. J., and F. W. Studier (1973). T7 early RNAs are cut from large precursor RNAs *in vivo* by ribonuclease III. *Proc. Natl. Acad. Sci. U.S.A., 70,* 3296–3300.

Fellner, P., C. Ehresmann, P. Stiegler, and J.-P. Ebel (1972). Partial nucleotide sequence of 16S ribosomal RNA from *Escherichia coli. Nature New Biol., 239,* 1–5.

Guthrie, C., and M. Nomura (1968). Initiation of protein synthesis: a critical test of the 30S subunit model. *Nature (London), 219,* 232–235.

——, H. Nashimoto, and M. Nomura (1969). Studies on the assembly of ribosomes *in vivo. Cold Spring Harbor Symp. Quant. Biol., 34,* 69–75.

Hardy, S. J. S., C. G. Kurland, P. Voynow, and G. Mora (1969). The ribosomal proteins of *Escherichia coli:* I. Purification of the 30S ribosomal proteins. *Biochemistry, 8,* 2897-2905.

Held, W. A., S. Mizushima, and M. Nomura (1973). Reconstitution of *Escherichia coli* 30S ribosomal subunits from purified molecular components. *J. Biol. Chem., 248,* 5720-5730.

———, B. Ballou, S. Mizushima, and M. Nomura (1974). Assembly mapping of 30S ribosomal proteins from *Escherichia coli.* Further studies. *J. Biol. Chem., 249,* 3103-3111.

Helser, T. L., J. E. Davies, and J. E. Dahlberg (1971). Mechanism of kasugamycin resistance in *Escherichia coli. Nature (London), 235,* 6-9.

Hosokawa, K., R. K. Fujimura, and M. Nomura (1966). Reconstitution of functionally active ribosomes from inactive subparticles and proteins. *Proc. Natl. Acad. Sci. U.S.A., 55,* 198-204.

Kaempfer, R. (1972). Initiation factor IF-3: a specific inhibitor of ribosomal subunit association. *J. Mol Biol., 71,* 583-598.

Kischa, K., W. Möller, and G. Stöffler (1971). Reconstitution of a GTPase activity by a 50S ribosomal protein from *E. coli. Nature New Biol., 233,* 62-63.

Kurland, C. G. (1960). Molecular characterization of ribonucleic acid from *Escherichia coli* ribosomes: I. Isolation and molecular weights. *J. Mol. Biol., 2,* 83-91.

——— (1972). Structure and function of the bacterial ribosome. *Ann. Rev. Biochem., 41,* 377-408.

———, P. Voynow, S. J. S. Hardy, L. Randall, and L. Lutter (1969). Physical and functional heterogeneity of *E. coli* ribosomes. *Cold Spring Harbor Symp. Quant. Biol., 34,* 17-24.

Lai, C.-J., B. Weisblum, S. R. Fahnestock, and M. Nomura (1973). Alteration of 23S ribosomal RNA and erythromycin-induced resistance to lincomycin and spiromycin in *Staphylococcus aureus. J. Mol. Biol., 74,* 67-72.

Luria, S. E., M. Delbrück, and T. F. Anderson (1943). Electron microscope studies of bacterial viruses. *J. Bacteriol., 46,* 57-77.

Maden, B. E. H. (1971). The structure and formation of ribosomes in animal cells. *Progr. Biophys. Mol. Biol., 22,* 129-177.

Mangiarotti, G., D. Apirion, D. Schlessinger, and L. Silengo (1968). Biosynthetic precursors of 30S and 50S ribosomal particles in *Escherichia coli. Biochemistry, 7,* 456-472.

Meselson, M., M. Nomura, S. Brenner, C. Davern, and D. Schlessinger (1964). Conservation of ribosomes during bacterial growth. *J. Mol. Biol., 9,* 696-711.

Midgley, J. E. M. (1965). The estimation of polynucleotide chain length by a chemical method. *Biochim. Biophys. Acta, 108,* 340-347.

Monier, R., J. Feunteun, B. Forget, B. Jordan, M. Reynier, and F. Varricchio (1969). 5S RNA and the assembly of bacterial ribosomes. *Cold Spring Harbor Symp. Quant. Biol., 34,* 139-148.

Nomura, M. (1970). Bacterial ribosome. *Bacteriol. Rev., 34,* 228-277.

———, and V. A. Erdmann (1970). Reconstitution of 50S ribosomal subunits from dissociated molecular components. *Nature (London), 228,* 744-748.

———, P. Traub, and H. Bechmann (1968). Hybrid 30S ribosomal particles reconstituted from components of different bacterial origins. *Nature (London), 219,* 793-799.

———, S. Mizushima, M. Ozaki, P. Traub, and C. V. Lowry (1969). Structure and function of ribosomes and their molecular components. *Cold Spring Harbor Symp. Quant. Biol., 34,* 49-61.

Osawa, S., E. Otaka, T. Itoh, and T. Fukui (1969). Biosynthesis of 50S ribosomal subunit in *Escherichia coli. J. Mol. Biol., 40,* 321-351.

Ozaki, M., S. Mizushima, and M. Nomura (1969). Identification and functional characterization of the protein controlled by the streptomycin-resistant locus in *E. coli. Nature (London), 222,* 333-339.

Pace, N. R. (1973). Structure and synthesis of the ribosomal ribonucleic acid of prokaryotes. *Bacteriol. Rev., 37,* 562-603.

Pettijohn, D. E., K. Clarkson, C. R. Kossman, and O. G. Stonington (1970). Synthesis of ribosomal RNA on a protein–DNA complex isolated from bacteria: a comparison of ribosomal RNA synthesis *in vitro* and *in vivo. J. Mol. Biol., 52,* 281-300.

Randall-Hazelbauer, L. L., and C. G. Kurland (1972). Identification of three 30S proteins contributing to the ribosomal A site. *Mol. Gen. Genet., 115,* 234-242.

Rosset, R., and R. Monier, and J. Julien (1964). Les ribosomes d'*Escherichia coli:* I. Mise en évidence d'un RNA ribosomique de faible poids moléculaire. *Bull. Soc. Chim. Biol., 46,* 87-109.

Schachman, H. K., A. B. Pardee, and R. Y. Stanier (1952). Studies on the macromolecular organization of bacterial cells. *Arch. Biochem. Biophys., 38,* 245-260.

Staehelin, R. T., and M. Meselson (1966). *In vitro* recovery of ribosomes and of synthetic activity from synthetically inactive ribosomal subunits. *J. Mol. Biol., 15,* 245-249.

Tai, P., D. P. Kessler, and J. M. Ingraham (1969). Cold-sensitive mutations in *Salmonella typhimurium* which affect ribosome synthesis. *J. Bacteriol., 97,* 1298-1304.

Tissières, A., D. Schlessinger, and F. Gros (1960). Amino acid incorporation into proteins by *Escherichia coli* ribosomes. *Proc. Natl. Acad. Sci. U.S.A., 46,* 1450-1463.

Traut, R. R., H. Delius, C. Ahmad-Zadeh, T. A. Bickle, P. Pearson, and A. Tissières (1969). Ribosomal proteins of *E. coli:* stoichiometry and implications for ribosome structure. *Cold Spring Harbor Symp. Quant. Biol., 34,* 25-38.

Waller, J. P. (1964). Fractionation of ribosomal protein from *Escherichia coli. J. Mol. Biol., 10,* 319-336.

Wireman, J. W., and P. Sypherd (1974). *In vitro* assembly of 30S ribosomal particles from precursor 16S RNA of *Escherichia coli. Nature (London), 247,* 552-554.

15

Reprinted from *J. Mol. Biol.*, 1(3), 221–233 (1959)

Ribonucleoprotein Particles from *Escherichia coli*

A. Tissières, J. D. Watson, D. Schlessinger and B. R. Hollingworth

Biological Laboratories, Harvard University, Cambridge, Mass., U.S.A.

(*Received 14 May 1959*)

In exponential cultures, 25% of the dry weight of *E. coli* is accounted for by RNA. Of this about 80 to 90% is present in ribonucleoprotein particles and 10 to 20% in the "soluble" or "non-sedimentable" fraction. Magnesium stabilizes the particles, and on varying its concentration, four kinds of components are observed, with sedimentation constants of 30, 50, 70 and 100 S. 70 S is formed of one 30 S and one 50 S; 100 S, of two 70 S. Each type of particle has been isolated. They all contain about 63% RNA and 37% protein, and have the same density. The molecular weights of the 30 S, 50 S and 70 S particles are about 0.7×10^6, 1.8×10^6, and 2.6×10^6 respectively.

1. Introduction

Bacteria release on lysis a large number of small particles of uniform size which in electron micrographs seem to constitute the bulk of the bacterial cytoplasm (Luria, Delbrück & Anderson, 1943). These particles contain most of the ribonucleic acid (RNA) of the cells; they form sharp boundaries in the analytical centrifuge, have a molecular weight of the order of 10^6, and are found in all the bacterial species examined (Schachman, Pardee & Stanier, 1952). They resemble in chemical composition and size the ribonucleoprotein particles which have been studied extensively in animal tissues (Petermann & Hamilton, 1952, 1957; Palade, 1955; Palade & Siekewitz, 1956a and b), in plants (Tso, Bonner & Vinograd, 1956, 1958), and in yeast (Chao & Schachman, 1956; Chao, 1957), and which are believed to be an important site of protein synthesis (Littlefield, Keller, Gross & Zamecnik, 1955; Littlefield & Keller, 1956, 1957; Schweet, Lamfrom & Allen, 1958).

Bacteria in the exponential phase of growth are one of the richest sources of ribonucleoprotein particles. In *Escherichia coli* dividing every 30 min, about 30% of the dry weight of the organism is accounted for by these particles (see below). While they are often firmly attached to the reticulum in animal tissues (Palade, 1955), they are free in bacterial extracts. *E. coli* is therefore convenient for the preparation of ribonucleoprotein particles in amounts sufficient for physical and chemical studies.

Magnesium stabilizes ribonucleoprotein particles from *E. coli* (Bolton, Hoyer & Ritter, 1958) from yeast (Chao, 1957) and from plants (Tso *et al.*, 1956, 1958). In the experiments reported here, in suitable concentrations of magnesium, four kinds of ribonucleoprotein particles were observed, with sedimentation coefficients of 30 S, 50 S, 70 S and 100 S. The isolation and some properties of each component will be described, and their molecular weights, based on sedimentation, diffusion, viscosity, and partial specific volume measurements, will be presented and discussed. A preliminary account of some of this work has appeared elsewhere (Tissières & Watson, 1958).

2. Methods

E. coli strain B was grown in enriched broth at 37°C under forced aeration. The broth had the following composition: 1% bactotryptone (Difco), 1% NaCl, 0.5% yeast extract

(Difco), 0·1% glucose, in distilled water. 1 ml. 1 N-NaOH was added to 1 l. of medium to adjust the pH to about 7·0. The cells were harvested in the exponential phase of growth and washed twice in the refrigerated centrifuge with the buffer used to make the extract. They could be kept frozen. Unless otherwise mentioned, the cell-free extract was made by grinding the well packed cells by hand in a cold mortar for 2 to 3 min with 3 parts (wt/wt) of alumina powder (Norton levigated alumina from Norton Abrasives, Worcester 6, Mass.) according to McIlwain (1948), and extracting with 3 vol of buffer. All manipulations were carried out at 0° to 4°C. Deoxyribonuclease (1 μg/ml.) obtained from Worthington Biochemical Corp., Freehold, New Jersey, was added to depolymerize the viscous bacterial deoxyribonucleic acid. The mixture of alumina, broken cells and buffer was centrifuged at 6000 g for 15 min, giving a sediment composed of two layers: a lower layer of alumina and some unbroken cells and a brown upper layer of large *cell debris*. The supernatant after this centrifugation, which is referred to as *crude extract*, contained the bulk of the ribonucleic acid of the cells, mostly in the form of ribonucleoprotein particles. The crude extract was usually centrifuged once more at 6000 g for 15 min to sediment any remaining alumina and cell debris. From this the ribonucleoprotein particles were then isolated by high speed centrifugation as described below.

A Model L Spinco ultracentrifuge was used for the preparation of the various particle fractions. The centrifugal forces given here represent g average, as calculated for the middle of the centrifuge tube. Analytical ultracentrifugation was done with a model E Spinco centrifuge, with either schlieren or ultra-violet optics. The films taken with the latter were analyzed for their optical density with a Spinco Analytrol densitometer. The centrifuge cell had a light path of 12 mm. The sedimentation constants were corrected to 20°C, water and zero concentration. They are expressed in Svedberg units,

$$S \; (= S°_{20w} \times 10^{-13} \; cm/sec).$$

Electrophoresis and diffusion measurements were performed with a Spinco model H apparatus, in an 11 ml. Tiselius cell at 1°C.

Viscosity was measured with Ostwald-Fenske type viscosimeters, solvent running time 350 to 370 sec, at 24·97 ± 0·02°C.

RNA was estimated by the orcinol method as described by Dische (1953), with correction, when necessary, for deoxyribonucleic acid (Schneider, 1945) by the diphenylamine reaction (Dische, 1955).

Protein was estimated by the biuret test (Gornall, Bardawill & David, 1949).

Dry weight measurements were done on aliquots dried at 110°C to constant weight. The dry weight of the buffer, in which the ribonucleoprotein particles were suspended, was measured in the same way and subtracted from the weight of the particle preparations.

Partial specific volumes were measured in a 25 ml. pycnometer.

The relative amount of each ribonucleoprotein component present in a solution was measured on plates from ultracentrifuge runs by determining (a) the area covered by each peak on the schlieren diagrams and (b) the optical density given by each component on pictures taken with the ultra-violet optics.

Crystalline pancreatic ribonuclease was obtained from Worthington Biochemical Corp.

0·01 M-phosphate buffer pH 7·0 was prepared by mixing 1 vol of 0·01 M-KH$_2$PO$_4$ with 3 vol of 0·01 M-Na$_2$HPO$_4$. The pH was then adjusted to 7·0 by addition of 0·01 M-KH$_2$PO$_4$ or Na$_2$HPO$_4$.

Tris(hydroxymethyl)aminomethane was Sigma 7–9 biochemical buffer from Sigma Chemical Company. The pH of tris solutions was adjusted by addition of 0·1 N-HCl. Unless otherwise mentioned, 0·01 M-tris, pH 7·4, was used.

The magnesium salt, added to stabilize the particles, was magnesium acetate.

3. Experimental

(a) *Amount of RNA in E. coli and in fractions from cell free extracts*

RNA and protein were estimated on *E. coli* cells and on the following three fractions: (a) *cell debris*, consisting of pieces of cell walls and cell membranes which sedimented as a brown layer on top of the alumina during the first centrifugation at 6000 g for 15 min. To obtain this fraction free of crude extract and of intact cells, which might well modify considerably the RNA content, it was washed five times in the centrifuge with the buffer used to make the extract. Each time the upper layer of the sediment was resuspended in buffer in order to eliminate the intact cells, unbroken during the grinding with alumina, which would, in view of their mass, centrifuge first to the bottom of the tube. This fraction was examined under the microscope and was found free of intact cells; (b) *crude extract*, or supernatant after centrifugation at 6000 *g* for 15 min; (c) *supernatant* obtained by carefully removing 2 ml. on top of the 11 ml. centrifuge tube, after centrifuging the crude extract at 100,000 *g* for 120 min.

TABLE 1

Amounts of RNA and protein in intact cells and fractions from cell free extract. The cells were ground with 3 parts of alumina and extracted with 0·001 M-*tris pH* 7·4 *and* 0·01 M-*magnesium acetate.*

(Similar results are obtained when the cells are extracted with 0·01 M-phosphate buffer pH 7·0 and 0·001 M-magnesium acetate. The *cell debris* is formed mostly of pieces from cell walls and cell membranes (see text). The *crude extract* is the fraction obtained after removing the alumina, the intact cells and the large cell debris by 15 min centrifugation at 6000 *g*. The crude extract, centrifuged for 120 min at 100,000 *g*, yields the *supernatant* and a small pellet formed essentially of ribonucleoprotein particles.)

		Protein	RNA	RNA/Protein
Intact cells (% of dry weight)		62·5	24·1	0·386
% of amount in intact cells	cell debris	15·0	0·9	0·06
	crude extract	85·0	97·0	1·17
% of amount in crude extract	} supernatant	45·0	17·0	0·378
mg/ml.	crude extract	5·2	3·44	0·66
	supernatant	2·34	0·585	0·25
Difference (mg sedimented in 120 min centrifugation at 100,000 *g* from 1 ml. crude extract)		2·86	2·855	0·99

The results given in Table 1 show that: (a) 24% of the dry weight of the cell is accounted for by RNA; (b) the bulk of the RNA of the cell is present in the crude extract; (c) the debris of cell wall and cell membrane, after washing with buffer, contain less than 1% of the amount of RNA present in the cells; (d) 17% of the RNA in the crude extract is in the form of " non-sedimentable RNA", and is found in the supernatant after centrifugation at 100,000 *g* for 120 min (this value ranged in various experiments from 15 to about 25%); (e) about equal amounts of RNA and protein are found in the unwashed fraction sedimented from the crude extract in a field of 100,000 g. All the RNA and most of the protein in this fraction belong to ribonucleoprotein particles, which are formed, as shown below, of about 63% of RNA and 37% of protein. A few per cent of the protein particles sedimenting with this fraction probably consist of fragments derived from the cell membrane.

(b) Sedimentation diagrams of crude extracts

The schlieren diagram of a crude extract made in 0·01 M-Mg^{++} and 0·01 M-tris buffer pH 7·4 showed two ribonucleoprotein peaks with sedimentation constants of 70 S and 100 S. The latter was predominant. With 0·001 M-Mg^{++}, the 70 S peak was the major one and in addition there were small amounts of 100 S, 50 S and 30 S peaks. With still lower magnesium concentrations (0·0001 to 0·0002 M), the ribonucleoprotein particles were present as 30 S and 50 S peaks. In the crude extracts the bulk of the soluble proteins of the cells appeared as a peak with a coefficient of about 5 S.

When the crude extract was made with 0·001 M-Mg^{++} and 0·01 M-phosphate buffer pH 7·0, the ribonucleoprotein particles were present as 30 S and 50 S components. This can be partly explained by the fact that some of the magnesium is bound to phosphate.

Thus four kinds of ribonucleoprotein particles could be observed in bacterial extracts, with sedimentation coefficients of 30 S, 50 S, 70 S and 100 S. The largest particles were predominant in presence of the highest concentration of magnesium. In low magnesium, only the 30 S and 50 S peaks were visible; these two components always appeared together and in the same ratio.

Schachman et al. (1952) found that extracts made by breaking the cells in different ways showed essentially similar sedimentation patterns. This was confirmed here: extracts made by sonic vibration or by lysing the protoplasts formed following the addition of penicillin to the culture medium according to Lederberg (1956) gave identical schlieren diagrams.

(c) Isolation of the 70 S component

The crude extract was made with 0·01 M-Mg^{++} and tris buffer pH 7·4. Under these conditions, the schlieren diagram showed a main 100 S component and a small amount of a 70 S peak. The 30 S and 50 S peaks were not visible. This extract was centrifuged at 78,000 g for 180 min. The supernatant was poured off and the yellow brown gelatinous pellet was resuspended in 0·005 M-Mg^{++} and 0·01 M-tris buffer by means of a teflon homogenizer used directly in the nitrocellulose centrifuge tube. The suspension was first centrifuged at 8000 g for 15 min in order to sediment and discard some aggregated material; then it was centrifuged at 78,000 g for 180 min. Washing the particles by resuspending them in buffer, followed by low and high speed centrifugation, as described above, was repeated twice. Finally the pellet was resuspended in tris buffer containing 0·001 M-Mg^{++} and the suspension was centrifuged at 8000 g for 15 min. The sediment was discarded. In the supernatant there was a major 70 S component, with in addition 10 to 15% of a 100 S peak and smaller amounts of 30 S and 50 S peaks. In one preparation 1 g (dry weight) of particles was obtained from 70 g (wet weight) of bacteria. The yield varied somewhat from preparation to preparation.

The 100 S particle is obtained by increasing the magnesium concentration of a 70 S preparation to 0·005–0·01 M. Sedimentation and viscosity measurements suggest that it is formed of two 70 *S* particles. This view is supported by electron micrographs of 100 *S* (Dr. C. Hall, personal communication) showing large numbers of two particle aggregates.

All the 70 *S* and 100 *S* pellets were yellow-brown, although the intensity of this color was reduced by repeated centrifugation. When a 70 *S* pellet was examined as such in a 3 mm thick cell, with the Cary spectrophotometer, in the presence and in the absence of a reducing agent, the typical absorption spectrum of all the cytochrome pigments found in *E. coli* (Keilin & Harpley, 1941) was observed. It is likely that the ribonucleoprotein particles are contaminated at this stage by nonparticle protein, to the extent of about 5%. This view is consistent with the percentages of RNA and protein reported in Table 2.

(d) *Conversion of 70 S particles into 30 S and 50 S components*

The observations with the crude extracts described above suggested that the 70 *S* and 100 *S* particles were formed by the aggregation of 30 *S* and 50 *S* particles. Experiments with isolated 70 *S* particles proved this conjecture and showed that as the magnesium concentration is decreased 70 *S* breaks down to 30 *S* and 50 *S*. Two procedures were used to effect this transition.

(i) A 70 *S* preparation was centrifuged at 78,000 *g* for 180 min and the sedimented particles were resuspended in sufficient 0·00025 M-Mg^{++} and tris buffer to give a concentration of ribonucleoprotein of about 5 mg/ml. This was then dialyzed overnight in the cold with stirring against 0·00025 M-Mg^{++} and tris buffer. In some cases this treatment led to a complete conversion of 70 *S* particles into 30 *S* and 50 *S* components. However, in many instances a small amount of the 70 *S* peak was still visible on the schlieren diagram. The amount of this remaining 70 *S* component can be further reduced by lowering the magnesium concentration or by exposure to 45°C for several hours.

(ii) The pellet of particles after high speed centrifugation was resuspended in 0·001 M-Mg^{++} and 0·01 M-phosphate buffer pH 7·0. This procedure gave a 100% conversion of 70 *S* into the two smaller units, and was therefore adopted in most cases for the preparation of 30 *S* and 50 *S*. The conversion could also be obtained by dialysis against the phosphate-magnesium mixture.

(e) *Relative amounts of 30 S and 50 S arising from 70 S*

70 *S* was split to 30 *S* and 50 *S* and the resulting solutions were examined in the analytical ultra-centrifuge. For two different preparations, the areas of the 30 *S* and 50 *S* peaks on the schlieren plates were in the ratio of about 1 : 2. Applying the Johnston & Ogston (1946) correction would somewhat lower this ratio. Since, as shown below, the RNA and protein composition of 30 *S* and 50 *S* is the same, the optical density for each component on films taken with the ultra-violet optics is also a measure of their relative proportions. In four such experiments the ratio of 30 *S* to 50 *S* varies from 1 : 2 to 1 : 2·5. The errors in these measurements were possibly as great as 20%.

The molecular weights of the 30 *S* and 50 *S*, calculated from sedimentation and diffusion rates (see Table 3), are in the ratio of 1 to 2·5. Furthermore the molecular weight of the 70 *S* is approximately the sum of those of the 30 *S* and 50 *S*. These data show clearly that the 70 *S* particle is formed of one 30 *S* and one 50 *S*.

(f) *Isolation of the* 30 *S component*

The mixture of 30 *S* and 50 *S*, in concentrations of 5 to 10 mg/ml., was centrifuged at 100,000 *g* for 360 min. 7/10ths of the supernatant fluid was carefully removed without stirring and discarded while the remaining fluid was shaken gently by hand for about 5 sec in order to resuspend the upper layer of the pellet. This fluid was poured off and saved. It consisted of 30 *S* particles with 10 to 30 % of 50 *S*. With one or two more cycles of centrifugation at 100,000 *g* for 360 min and gentle resuspension of the upper layer of the pellet, the ratio of 30 *S* to 50 *S* could be brought to about 20 to 1, as judged by the area covered by each peak under the schlieren curve. The 30 *S* preparation could be sedimented by centrifugation at 100,000 *g* for 360 min. The pellet thus obtained was colorless. The buffer was either tris with 0·00025 M-Mg^{++} or phosphate with 0·001 M-Mg^{++} depending on whether procedure (a) or (b) was used to make 30 *S* and 50 *S*.

(g) *Isolation of the* 50 *S component*

The pellet after the first centrifugation at 100,000 *g* for 360 min was resuspended in buffer and magnesium mixture (either 0·01 M-tris pH 7·4 and 0·00025 M-Mg^{++}, or 0·01 M-phosphate pH 7·0 and 0·001 M-Mg^{++}). It was then centrifuged at 15,000 *g* for 15 min, the sediment was discarded and the supernatant was centrifuged at 100,000 *g* for 90 min. This last step was repeated twice. Further purification could be achieved by lowering the pH of the preparation to 5·0 with 0·1 N-CH$_3$.COOH, centrifuging at 8000 *g* for 15 min and finally bringing the pH of the supernatant to 7·0 with 0·1 N-NH$_4$OH. This treatment removed some cell debris derived probably from the membrane and the cell wall (see below) and also small amounts of 30 *S* which may still be present at this stage. After this last step the preparation showed only one peak on the sedimentation diagram and upon high speed centrifugation formed a colorless pellet. In addition it moved as a single symmetrical peak on electrophoresis at pH 7·6 or 4·7.

(h) *Formation of* 70 *S particles from purified* 30 *S and* 50 *S*

30 *S* and 50 *S* particles (in tris buffer and 0·00025 M-Mg^{++}) were mixed together in equal number amounts and magnesium added to a final concentration of 0·001 M. No aggregation occurred. On raising magnesium to 0·005 M, 70 *S* particles appeared and at 0·01 M the majority of particles were in the 70 *S* form. Subsequent lowering of the magnesium level to 0·001 M reveals a majority of 70 *S* particles. Thus a higher magnesium concentration is required to form 70 *S* particles than to effectively stabilize them. Our data are summarized by the equation

$$\text{increasing Mg}^{++} \longrightarrow$$
$$2\,(30\,S) + 2(50\,S) \rightleftharpoons 2(70\,S) \rightleftharpoons (100\,S) \qquad (1)$$

When a mixture of peaks is observed, they are usually very sharp. This would not be true if the various particle types were in rapid equilibrium, as the peaks would blend into each other. This implies that the transitions need not occur immediately, but that an activation energy is required when 70 *S* splits into 30 *S* and 50 *S*. This is supported by the observation that incubation of 70 *S* particles in low magnesium (0·00025 M) does not lead to immediate breakdown and that the transition is hastened by incubation at 45°C. The same reason may explain why the 70 *S* particle is not formed at 0·001 M-Mg^{++} even though once formed it initially seems stable and can be isolated at this magnesium level. In fact, prolonged incubation (several days to a week) of 70 S particles at 45°C leads to their breakdown to 30 S and 50 S particles. It is thus probable that the 70 S particles are completely stable only at magnesium concentrations greater than 0·005 M where in fact they begin to further aggregate to 100 S particles.

(i) *Chemical composition*

The results of RNA and protein estimations, together with dry weights and extinctions at 260 mμ on a dry weight basis, are given in Table 2. The dry weights corresponded, within the limits of error, to the sum of the weights of RNA and protein found by chemical estimation. The errors in these estimations were probably 3 to 4%. Thus it appears that the particles are formed mostly, if not exclusively, of RNA and protein. It is likely that the ratio of RNA to protein is the same in the three kinds of particles, 30 S, 50 S and 70 S, and that the small variations shown in Table 2 are due to varying amounts of impurities. This view is supported by the finding (see below) that the densities of the three kinds of particles are very similar. The values for the extinction at 260 mμ suggest that the purest preparations are those of 30 S and of 50 S, the latter after treatment at pH 5·0.

TABLE 2

Dry weight, chemical composition and specific extinction coefficients at 260 mμ of 30 S, 50 S and 70 S ribonucleoprotein particle preparations.

Particle fraction	Dry weight mg/ml.	RNA mg/ml.	Protein mg/ml.	RNA + Protein mg/ml.	% RNA	% Protein	Specific extinction coefficient at 260 m$\mu = E_{1\,cm}^{1\%}$
30 S	3·72	2·38	1·50	3·88	62·0	38·0	167·0
30 S	1·58	0·95	0·60	1·55	61·3	38·7	162·0
50 S	2·26	1·39	0·83	2·22	62·5	37·5	—
50 S	1·06	0·63	0·37	1·00	63·4	36·6	182·0
50 S	2·82	1·80	1·00	2·80	64·0	36·0	—
50 S†	6·96	3·94	2·85	6·79	58·0	42·0	149·0
70 S	39·00	23·50	17·60	41·10	57·3	42·7	147·0
70 S	—	—	—	—	—	—	145·0
70 S	13·60	8·80	5·70	14·50	60·7	39·3	157·0

† 50 S: this preparation had not been treated at pH 5·0.

(j) *Absorption spectrum*

The absorption curves of the 30 S, 50 S and 70 S in the ultra-violet region were nearly identical. They all showed an absorption peak with a maximum at 259 mμ. The RNA in 30 S, 50 S and 70 S was found to react with formaldehyde, as described by Fraenkel-Conrat (1954) for several plant viruses: the absorption in the 260 mμ region increased by about 20% and the maximum was shifted about 3 mμ towards the longer wavelengths. The spectrum of a 50 S preparation and the effect of formaldehyde are shown in Fig. 1.

(k) *Relative densities*

A particle preparation containing 30 S and 50 S components was added to salt solutions to give an optical density of 1·0 at 260 mμ. The resulting samples contained

0·01 M-phosphate pH 7·0 and 0·001 M-Mg^{++} in 0·0, 0·5, 1·5, 3·5, and 5·0 molal CsCl. The sedimentation rates observed with ultra-violet optics are shown in Fig. 2. Corrected for the viscosity and density of the medium, they correspond to 30 S and 50 S. In all cases, the relative sedimentation constants of the two components were the same, indicating that they had similar if not identical partial specific volumes. Further experiments were performed in varying concentrations of sucrose and NaCl with identical results. In one experiment involving 70 S particles in NaCl solution, the sedimentation run was done before complete 70 S breakdown (see below) and the sedimentation rate of 70 S was reduced at the same rate as those of 30 S and 50 S.

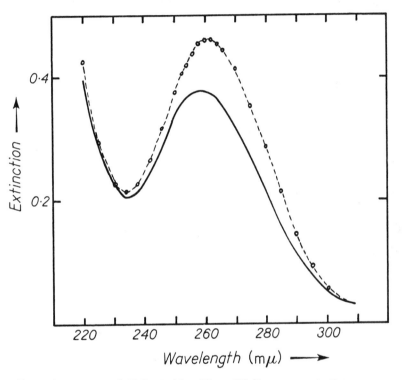

FIG. 1. Absorption spectra of 50 S particles. The solid line represents the spectrum of 50 S (20·6 μg/ml.) in 0·01 M-phosphate and 0·001 M-Mg^{++}, before and after 70 hr at 45°C. The dashed line shows the spectrum after similar incubation in 2% HCHO.

(1) Stability

(i) *Effect of magnesium.* The four kinds of particles were usually stable in the presence of magnesium for over a month in the cold in 10 μg/ml. streptomycin, although occasionally after a few weeks the particles were found to break down. Buffer was not necessary; preparations remained intact for several days in distilled water and magnesium.

Magnesium is required not only for the existence of the 70 S and 100 S, but also for the integrity of the 30 S and 50 S subunits. If 30 S or 50 S is dialyzed against buffer containing no magnesium, or against distilled water, or if 0·01 M-ethylene-diamine tetra-acetate (versene) is added, breakdown of the particles occurs.

It has been mentioned that 0·003 to 0·01 M-Mg^{++} causes aggregations to the 100 S component. The particles are stable up to 0·02 M-Mg^{++}, above which breakdown was observed.

(ii) *Effect of freezing.* The particles could be kept frozen for several months without any noticeable change of their properties.

(iii) *Effect of pH.* The 50 *S* was stable for 24 hr at pH 4·7 and 9·5. The 30 *S*, 70 *S* and 100 *S* broke down under these conditions.

If the pH of a suspension of 30 *S*, 50 *S* and 70 *S* (1·5 mg/ml.) was lowered to 4·4 with 0·01 M-ammonium acetate or 0·1 N-CH₃.COOH, and the resulting mixture dialyzed for 15 hr against 100 vol of 0·001 M-Mg⁺⁺ and 0·01 M-ammonium acetate pH 4·4, a completely diffuse peak, suggesting random aggregation, was seen in the

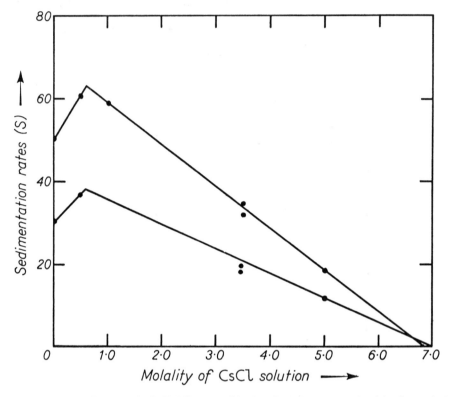

FIG. 2. Sedimentation rates in CsCl. Mixtures of 30 *S* and 50 *S* were examined in the analytical ultracentrifuge with ultra-violet optics, immediately after dilution in different concentrations of CsCl. The ratio of sedimentation rates was always 50 : 30. The graphs extrapolate to a particle density in CsCl of 1·73 ± 2% for both components. The initial rise in sedimentation rate is probably attributable to initial replacement of Na⁺ in the particles by Cs⁺.

analytical ultracentrifuge. If the same sample was then dialyzed for 15 hr against 0·01 M-tris, pH 7·4, a 50 *S* peak was again visible, with traces of 30 *S* and 70 *S*. If the pH was lowered further to 4·2, a white precipitate was observed. The precipitate did not revert to ribonucleoprotein components on raising the pH.

(iv) *Effect of ribonuclease.* 1 to 5 μg/ml. of pancreatic ribonuclease, at pH 7·6, had no effect on most particle preparations, even after 4 to 6 hr at room temperature, as judged by the appearance of the peak in the analytical ultracentrifuge. In some cases, however, the peak heights were reduced. The reason for these discrepancies is not clear.

(v) *Effect of NaCl and CsCl.* 30 *S*, 50 *S*, or 70 *S*, or mixtures of them, in 0·001 M-Mg⁺⁺ solutions, were allowed to stand at 4°C in different NaCl or CsCl concentrations. After

15 hr, samples were examined in the analytical centrifuge with the ultra-violet optics. 30 S and 50 S were stable below about 0·15 and above 2·0 molal NaCl and CsCl. They broke down in the intermediate range and consequently no sedimentation rate greater than 5 S was observed. 70 S was stable in low salt concentrations up to about 0·15 molal. In higher concentrations it split to 30 S and 50 S.

Elson (1958) has shown that a latent ribonuclease is present in ribonucleoprotein particle preparations from *E. coli*. The unusual stability of the particles in salt reported here may be partly explained by the presence of this enzyme, assuming that above 0·15 molal NaCl or CsCl the particle structure is altered in such a way that the endogenous ribonuclease can degrade the particulate RNA. The stability of the particles above 2·0 molal NaCl or CsCl might be due to inhibition of ribonuclease by high salt concentrations, or else simply to salt effect on the particles.

(m) *Molecular weights*

Values shown in Table 3 are based on the following data:

(i) *Sedimentation constants*. The coefficients for 30 S, 50 S and 70 S were measured both with the ultra-violet optics at concentrations of about 60 μg/ml., and with the schlieren optics at concentrations of 0·5 to 5·0 mg/ml., in 0·01 M-tris and the suitable magnesium concentrations. The results are shown in Fig. 3. The salt concentration had no appreciable effect between 0·01 and 0·05 M-phosphate and 0·001 and 0·05 M-tris.

(ii) *Diffusion coefficients*. Free diffusion was allowed to proceed for 2 to 3 days. Measurements were made on two different preparations of 30 S (both at 0·11%, in 0·01 M-tris), three of 50 S (at 0·06%, in 0·01 M-phosphate; 0·07%, in 0·05 M-phosphate; and 0·20%, in 0·01 M-phosphate), and one of 70 S (at 0·10% and 0·20%, in 0·01 M-tris).

TABLE 3

Physical constants and molecular weights. Percentages of error are given as the average deviation from the mean. The molecular weights are consistent with equation (1).

Compo-nents	Precise S $(S^0_{20,w} \times 10^{13}$ cm/sec)	D $(D^0_{20,w} \times 10^7$ cm²/sec)	$\bar{V}$	ηint. cm/dl.	Molecular weights from S & D (Svedberg & Pedersen, 1940)	from S & η (Scheraga & Mandelkern, 1953)
30 S	30·6 ± 1	2·95 ± 4 %	0·64	0·080 ± 8 %	0·7 × 10⁶ ± 8 %	1·0 × 10⁶ ± 16 %
50 S	50·0 ± 1	1·91 ± 1 %	0·64	0·054 ± 8 %	1·8 × 10⁶ ± 4 %	1·8 × 10⁶ ± 16 %
70 S	69·1 ± 1	1·83 ± 2 %	0·64	0·061 ± 10 %	2·6 × 10⁶ ± 6 %	3·1 × 10⁶ ± 18 %
100 S	100·0 ± 2		0·64	0·071 ± 12 %		5·9 × 10⁶ ± 20 %

Because of the impossibility of forming infinitely sharp initial boundaries with proteins, the calculated values of D decreased with time. Plots of D against $1/t$, which were linear, when extrapolated to infinite time, gave corrected values of D. Zero-time corrections were of the order of 20 to 50 min. The values of D were then corrected to 20°C in water. For 30 S, they were 2·81 and 3·10 × 10⁻⁷ cm²/sec; for 50 S, 1·88, 1·90, and 1·95 × 10⁻⁷; and for 70 S, 1·82 and 1·84 × 10⁻⁷. Averaged values are shown in Table 3.

(iii) *Partial specific volume*. The values obtained were 0·64 and 0·66 for two 50 *S* preparations, and 0·63 for one 70 *S*, at concentrations varying from 0·5 to 4%. Since the density and chemical composition of the three kinds of particles is probably the same (see above), an average value of 0·64 was used in all calculations.

(iv) *Viscosity*. The preparations used contained at least 85% of one ribonucleo-protein component, as measured on plates taken with the schlieren optics. For various 30 *S*, 50 *S* and 70 *S* preparations, and for one of 100 *S*, series of viscosity measurements were extrapolated to an average intrinsic viscosity (Table 3). Concentrations were calculated assuming that 67 μg/ml. corresponded to an optical density of 1·0 for 50 *S*, 70 *S* and 100 *S*, and 61 μg/ml. for 30 *S*. These values were deduced from the specific extinction coefficients at 260 mμ given in Table 2.

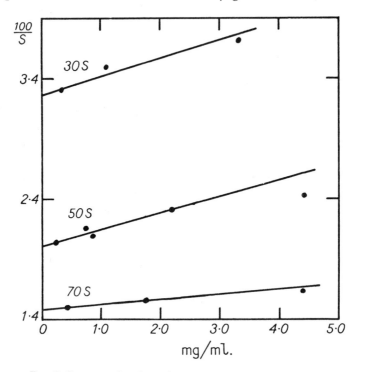

FIG. 3. Concentration dependence of sedimentation constants.

The molecular weight of the 50 *S* particles calculated from sedimentation, diffusion and partial specific volume data agrees well with that derived from sedimentation, viscosity and partial specific volume. As the value involving the diffusion coefficient is independent of the shape of the particle, while that using intrinsic viscosity assumes a spherical shape, this result, in agreement with the electron micrographs (Hall & Slayter, 1959), suggests a spherical shape for the 50 *S* particle.

In contrast the molecular weight of the 30 *S* particles obtained by using intrinsic viscosity seems high when compared to that based on diffusion values, and may reflect the less symmetrical form shown in electron micrographs.

4. Discussion

The construction of ribonucleoprotein particles from two unequal units which join in presence of a suitable magnesium concentration seems to be a general phenomenon.

With both pea seedlings (Tso *et al.*, 1956) and yeast (Chao, 1957) 80 *S* particles break down reversibly to 60 *S* and 40 *S*, and we have seen that in *E. coli* the 70 *S* component is made of one 30 *S* and one 50 *S*, the molecular weights of which are in the ratio of about 1 : 2·5. Viscosity measurements suggest greater asymmetry of 30 *S* than of 50 *S* and 70 *S*, and the electron micrographs (Hall & Slayter, 1959) show that 30 *S* is definitely asymmetric, while 50 *S* and 70 *S* appear nearly spherical. A simple model for the structure of 70 *S* would thus be a capped sphere or acorn shape.

The particles from *E. coli* seem to differ from those of other sources in at least three ways: (a) the molecular weight is lower (2·8 × 10^6 for 70 *S*, while values varying between 4 and 4·5 × 10^6 have been reported for animal (Dintzis, Borsook & Vinograd, 1958), plant (Tso *et al.*, 1956), and yeast particles (Chao & Schachman, 1956)); (b) the percentage of RNA (60 to 65%) is higher than that reported for particles from other sources (Tso *et al.*, 1956; Chao & Schachman, 1956; Petermann & Hamilton, 1957; Dintzis *et al.*, 1958) with the exception of particles from rabbit appendix (Takata & Osawa, 1957); (c) the magnesium requirement for the integrity of the *E. coli* particles appears to be greater.

If we take the weight of one *E. coli* cell to be 10^{-12} gm, with 25% RNA in exponential cultures, there is $2·5 \times 10^{-13}$ gm of RNA per cell. One 70 *S* particle contains $2·74 \times 10^{-18}$ gm RNA. Thus there are about 90,000 particles in one bacterium. In the stationary phase, or upon incubation in phosphate buffer, this number decreases by a factor of 4 to 5 (Mendelsohn & Tissières, 1959), or to about 20,000 particles per cell.

We wish to thank Dr. J. T. Edsall for his interest and encouragement of this work, and Dr. D. M. Skinner and Miss C. Laumont for their assistance in some of these experiments. This work has been supported by grants from the U.S. National Science Foundation and an institutional grant from the American Cancer Society.

REFERENCES

Bolton, E. T., Hoyer, B. H. & Ritter, D. B. (1958). *Microsomal Particles and Protein Synthesis*, p. 18. New York: Pergamon Press.
Chac, F. C. (1957). *Arch. Biochem. Biophys.* **70**, 426.
Chao, F. C. & Schachman, H. K. (1956). *Arch. Biochem. Biophys.* **61**, 220.
Dintzis, H. M., Borsook, H. & Vinograd, T. (1958). *Microsomal Particles and Protein Synthesis*, p. 95. New York: Pergamon Press.
Dische, Z. (1953). *J. Biol. Chem.* **204**, 983.
Dische, Z. (1955). *The Nucleic Acids*, Vol. I, p. 287. New York: Academic Press.
Elson, D. (1958). *Biochem. biophys. Acta*, **27**, 216.
Fraenkel-Conrat, H. (1954). *Biochem. biophys. Acta*, **15**, 307.
Gornall, A. G., Bardawill, C. T. & David, M. M. (1949). *J. Biol. Chem.* **177**, 751.
Hall, C. E. & Slayter, H. S. (1959). *J. Mol. Biol.* 1, in press.
Johnston, J. P. & Ogston, A. G. (1946). *Trans. Faraday Soc.* **42**, 789.
Keilin, D. & Harpley, C. H. (1941). *Biochem. J.* **35**, 688.
Lederberg, J. (1956). *Proc. Nat. Acad. Sci., Wash.* **42**, 574.
Littlefield, J. W. & Keller, E. B. (1956). *Fed. Proc.* **15**, 302.
Littlefield, J. W. & Keller, E. B. (1957). *J. Biol. Chem.* **224**, 13.
Littlefield, J. W., Keller, E. B., Gross, J. & Zamecnik, P. C. (1955). *J. Biol. Chem.* **217**, 111.
Luria, S. E., Delbruck, M. & Anderson, T. F. (1943). *J. Bact.* **46**, 57.
McIlwain, H. (1948). *J. Gen. Microbiol.* **2**, 288.
Mendelsohn, J. & Tissières, A. (1959). *Biochem. biophys. Acta*, in press.
Palade, G. (1955). *J. Biophys. Biochem. Cytol.* **1**, 59.
Palade, G. & Siekewitz, P. (1956a). *J. Biophys. Biochem. Cytol.* **2**, 171.
Palade, G. & Siekewitz, P. (1956b). *J. Biophys. Biochem. Cytol.* **2**, 671.

Petermann, M. L. & Hamilton, M. G. (1952). *Cancer Research*, **12**, 373.

Petermann, M. L. & Hamilton, M. G. (1957). *J. Biol. Chem.* **224**, 723.

Schachman, H. K., Pardee, A. B. & Stanier, R. Y. (1952). *Arch. Biochem. Biophys.* **38**, 245.

Scheraga, H. A. & Mandelkern, L. (1953). *J. Amer. Chem. Soc.* **75**, 179.

Schneider, W. C. (1945). *J. Biol. Chem.* **161**, 293.

Svedberg & Pedersen, K. O. (1940). *The Ultracentrifuge*, p. 5. Oxford: Clarendon Press.

Schweet, R., Lamfrom, H. & Allen, E. (1958). *Proc. Nat. Acad. Sci., Wash.* **44**, 1029.

Takata, K. & Osawa, S. (1957). *Biochem. biophys. Acta*, **24**, 207.

Tissières, A. & Watson, J. D. (1958). *Nature*, **182**, 778.

Tso, P. O. P., Bonner, J. & Vinograd, J. (1956). *J. Biophys. Biochem. Cytol.* **2**, 451.

Tso, P. O. P., Bonner, J. & Vinograd, J. (1958). *Biochem. biophys. Acta*, **30**, 570.

16

ON THE PRESENCE OF
LOW-MOLECULAR-WEIGHT RIBONUCLEIC ACID
IN THE RIBOSOMES OF *ESCHERICHA COLI*

R. Rosset and R. Monier

*Institute of Biological Chemistry Faculty of
Sciences, Marseille*

*This article was translated by Dr. K. S. Neuhauser, Rutgers
University, from "A propos de la présence d'acide ribonucléique
de faible poids moléculaire dans les ribosomes* d'Escherichia coli,"
Biochim. Biophys. Acta, **68**, 653–656 (1963)

The presence of low-molecular-weight RNA in purified ribosomes of diverse origins has been demonstrated by several authors[1-4] and, in particular, by Elson,[5] who established that this RNA, in *Escherichia coli,* possesses certain of the properties of transfer RNA.[6] The existence of transfer RNA bound to the ribosomes is moreover, one predictable consequence of the role that it plays in protein biosynthesis.[7] A detailed study of the RNA extracted by deproteinization with phenol from the ribosomes of *E. coli* ML308, washed in a medium containing 0.01 M Mg^{2+}, showed us nevertheless that the low-molecular-weight RNA that one always finds associated with the ribosomes under these conditions is in reality heterogeneous. About one half is made up of a product with clearly distinct properties from those of the transfer RNAs, as the results briefly reported here show.

The low-molecular-weight RNA extracted from the ribosome can be differentiated from the transfer RNAs isolated from the cytoplasm by chromatography on a column of kieselguhr impregnated with methylated serum albumin.[8] An important fraction of the light ribosomal RNA is eluted from these columns as a narrow peak, preceded by products with the same characteristics as cytoplasmic RNA (Figure 1A and B). The products in this peak labeled with ^{32}P have been recovered after chromatography and submitted to various tests, the results of which are summarized below.

1. The sedimentation coefficients of this RNA in a sucrose gradient[12] are slightly greater than those of cytoplasmic transfer RNAs, probably in the neighborhood of 5S. To distinguish this RNA from the transfer RNAs with a sedimentation coefficient of 4S, we are designating it 5S RNA.

2. The nucleotide composition of 5S RNA presents certain analogies with that of total transfer RNA (Table 1). It is very different from ribosomal 16S and 23S RNA and messenger RNA (DNA-like RNA of Midgley and McCarthy).[9] It is consequently improbable that 5S RNA can be considered a nonspecific degradation product of one of these other RNAs.

3. The composition of 5S RNA is also significantly different from the composition of the transfer RNA fraction eluted in the same region of the chromatograms

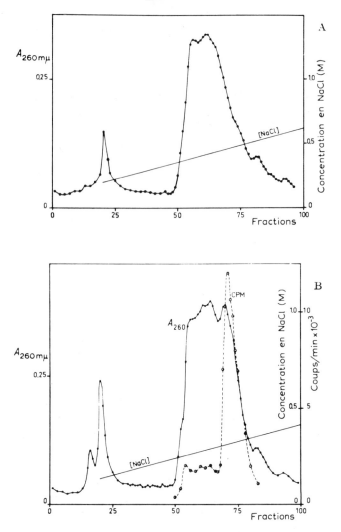

Figure 1 Chromatography on columns of methylated serum albumin.[8] Only that region of each chromatogram in which light RNAs are eluted has been presented. A. RNA extracted from the 105,000*g* supernatant of *E. coli* ML308 (11). B. RNA extracted from ribosomes of *E. coli* ML308 labeled with ^{32}P for over 10 generations (3 mg), added to 1.5 mg of RNA extracted from the 105,000*g* supernatant of unlabeled *E. coli* ML308. Radioactivity was measured on an aliquot of each fraction after precipitation of the RNA by 5 percent trichloracetic acid and millipore filtration.

in the course of chromatography such as that represented in Figure 1A (fraction IV of Table 1). In particular, alkaline hydrolysis of 5S RNA isolated by our technique liberates only low amounts of 2′,3′-pseudouridylic acid and guanosine (2′,3′),5′-diphosphate. On the contrary, another nucleoside diphosphate appears in the course of this hydrolysis. Certain of its properties suggest that it is uridine (2′,3′), 5′-diphosphate, but the quantities isolated as yet do not permit definitive identification.

Table 1 Nucleotide composition of different fractions of RNA from *E. Coli* ML308

The composition of the RNA was determined after alkaline hydrolysis of the ^{32}P-labeled material by measuring the radioactivity associated with the different nucleotides separated by chromatography on Dowex 1-X2, formate.[10] (Results expressed relative to 2', 3'-adenosine phosphate = 10.0.)

Nucleotide	Total[a]	Transfer RNA					Ribosomal RNA		
		I[b]	*II*[b]	*III*[b]	*IV*[b]	*V*[b]	*5S*[a]	*16S*[a]	*23S*[a]
C-2',3'-P	15.4	15.5	16.5	14.2	13.6	13.8	15.1	9.0	8.7
G-2',3'-P	16.7	16.3	17.8	16.7	15.2	16.6	17.4	13.2	12.9
U-2',3'-P[c]	10.2	8.8	9.5	10.0	8.6	8.8	8.5	9.0	9.0
Pseudouridine-2',3'-P	1.3	—	—	—	—	—	0.15	—	—
G-(2',3')5'-PP	0.61	—	—	—	—	—	0.08	—	—
Unidentified nucleoside diphosphate	0.09	—	—	—	—	—	0.47	—	—
Purines/pyrimidines	1.04	1.08	1.07	1.10	1.13	1.18	1.16	1.29	1.29

[a] RNA purified by chromatography on columns of methylated serum albumin. Transfer RNA is prepared from the 105,000*g* supernatant by the method of Tissieres.[11]
[b] I, II, III, IV, V correspond to different regions of the transfer RNA peak on columns of methylated serum albumin, fraction I being the first fraction eluted.
[c] In the case of transfer RNA, the values given correspond in reality to the sum of (2',3')-uridylic acid and (2',3')-thymine ribonucleotide.

These quantities would be compatible with a molecular weight greater than 30,000 daltons corresponding to a length of around 90–100 nucleotides. The analysis of total transfer RNA by our techniques leads to a determination of a molecular weight on the order of 26,000 daltons, corresponding to 77 nucleotides. These results are close to those of sedimentation coefficient values.

4. The kinetics of liberation of nucleotides in the course of degradation with snake venom phosphodiesterase is not in accord with the existence of a sequence . . . pCpCpA at the 3'-OH terminus of the 5S RNA.

5. The specific acceptor activity of light ribosomal RNAs with regard to amino acid activation, measured under normal conditions in the presence of a hydrolysate of algal ^{14}C protein, is around 50 percent of that of the cytoplasmic transfer RNAs.

This set of facts permits us to conclude that 5S RNA cannot be made up of a fraction of the transfer RNAs which are concentrated specifically in the ribosomes. It is, therefore, an RNA that probably does not possess the function of transfer RNA. The quantities isolated from the ribosomes are sufficient to account for around one molecule of this RNA per 70S ribosome. The experiments to elucidate the exact conditions of attachment of this RNA to the ribosomes and its physiological significance are now in progress.

This work has been supported by the Delegation for Scientific and Technical Research (61-FR-113), the Commissariat of Atomic Energy, the U.S. Public Health Service (Grant A-4642), and the Rockefeller Foundation.

References

1. T. Hultin and A. Von der Decken, *Exptl. Cell Res., 16,* 444 (1959).
2. L. Bosch, H. Bloemendal, and L. Sluyser, *Biochim. Biophys. Acta, 34,* 272 (1959).
3. M. Takanami, *Biochim. Biophys. Acta, 55,* 132 (1962).
4. K. C. Smith, *Biochemistry, 1, 866* (1962).
5. D. Elson, *Biochim. Biophys. Acta, 53,* 232 (1961).
6. D. Elson, *Biochim. Biophys. Acta, 61,* 460 (1962).
7. M. B. Hoagland and L. T. Comly, *Proc. Natl. Acad. Sci. U.S.A., 46,* 1554 (1960).
8. J. D. Mandell and A. D. Hershey, *Anal. Biochem., 1,* 66 (1960).
9. J. E. Midgley and B. J. McCarthy, *Biochim. Biophys. Acta, 61,* 696 (1962).
10. S. Osawa, K. Takata, and Y. Hotta, *Biochim. Biophys. Acta, 28,* 271 (1958).
11. A. Tissières, *J. Mol. Biol., 1,* 365 (1959).
12. K. McQuillen, R. B. Roberts, and R. J. Britten, *Proc. Natl. Acad. Sci. U.S.A., 45,* 1437 (1959).

17

Copyright © 1965 by the American Chemical Society

Reprinted from *Biochemistry,* 4(7), 1302–1311 (1965)

Isolation and Physical Properties of the Ribosomal Ribonucleic Acid of *Escherichia coli**

Wendell M. Stanley, Jr.,† and Robert M. Bock

ABSTRACT: Ribosomal ribonucleic acid preparations of exceptional stability were obtained from the ribosomes of *Escherichia coli* by a method of isolation employing phenol, sodium dodecyl sulfate, and a purified hectorite, Macaloid. The sedimentation and viscosity properties of the total ribosomal RNA and of the separated components were extensively investigated and molecular weight determinations were made by sedimentation viscosity, sedimentation equilibrium, and viscosity kinetics at elevated temperatures. Molecular weights of 1.07×10^6 and 0.55×10^6 g/mole were found for the 23 and 16 S components, respectively. The influence of RNA aggregation upon hydrodynamic parameters was evaluated and several methods (organic solvents, reaction with formaldehyde, low ionic strength, and heat) are suggested for the detection of aggregates within RNA preparations. Chromatography of undegraded ribosomal RNA upon DEAE-cellulose was found to be complicated by the formation at equilibrium of a nonelutable complex between the ion exchanger and the polynucleotide. Base composition analyses were performed upon the separated ribosomal RNA components, and slight but significant differences were found between the 23 and 16 S molecules. It is concluded that the ribosomal RNA of *E. coli* is composed of two classes of polyribonucleotide chains, each class being covalently continuous and thus not containing polynucleotide subunits.

Although several studies have appeared in the literature concerning the isolation and physical properties of *E. coli* ribosomal RNA (see, for example, Hall and Doty, 1958, 1959; Littauer and Eisenberg, 1959; Kurland, 1960), none of the RNA preparations investigated was of sufficient stability to warrant carrying out physical measurements which required prolonged periods of observation. Utilizing the negatively charged purified hectorite, Macaloid,[1] we have been able routinely to obtain RNA preparations which have exhibited a freedom from nuclease activity such that their stability has been limited only by the known chemical stability of the phosphodiester bond in polyribonucleotides (Eigner *et al.*, 1961; Ginoza, 1958). It thus became possible to undertake an extensive investigation into the physicochemical properties of such RNA preparations. Many of the results and techniques to be presented will be of value during the characterization of other polyribonucleotides, both natural and synthetic.

During the past few years a number of authors have suggested that ribosomal RNA is formed of "subunits" held together by secondary interactions (Takanami, 1958; Hall and Doty, 1959; Osawa, 1960; Chao, 1961; Aronson and McCarthy, 1961; Otaka *et al.*, 1961). The results reported in this paper are not consistent with this concept. We have found, instead, that the two molecules of ribosomal RNA, the so-called 23 and 16 S RNA, are in fact continuous phosphodiester chains. In addition, we have been able to produce particles which hydrodynamically resemble intact ribosomal RNA chains but which have been subjected under controlled conditions to a limited number of hydrolytic scissions. These modified RNA particles are quite stable under conditions favoring strong secondary interactions; however, they dissociate, revealing fragments of reduced molecular weight, when placed in environments which have often been employed to demonstrate the existence of ribosomal RNA "subunits."

Materials and Methods

Growth and Lysis of Bacteria. Escherichia coli strain W3101, obtained from Dr. J. Lederberg, Stanford University, was grown under forced aeration at 37° in the following medium: Difco Bacto-tryptone, 10 g/liter; dextrose, 5.6 g/liter; Na_2HPO_4, 6 g/liter; KH_2PO_4, 3 g/liter; NaCl, 5 g/liter. Cultures in late log phase were chilled to 0° by the addition of chipped

---

* From the Department of Biochemistry, University of Wisconsin, Madison. *Received November 30, 1964; revised March 8, 1965.* Taken from the dissertation submitted by W. M. Stanley, Jr., in partial fulfillment of the requirements for the Ph.D. degree at the University of Wisconsin, Madison, Wis. For descriptions of experimental techniques in detail greater than that set forth here, the reader is referred to the dissertation "Physical Studies on the Ribosomes and Ribosomal Ribonucleic Acid of *Escherichia coli*," Dissertation Abstracts No. 64-611.

† Present address: Department of Biochemistry, New York University, New York, N.Y. 10016.

[1] Macaloid is a trade name for a purified hectorite (sodium magnesium lithofluorosilicate) obtained from the American Tansul Co., Baroid Division, National Lead Co., 2404 Southwest Freeway, P.O. Box 1675, Houston, Texas 77001. Stock suspensions in dilute buffer were made at 90 to 100° in a high shear mixer.

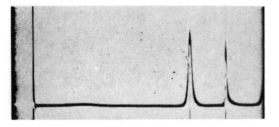

FIGURE 1: A photograph of the schlieren pattern of unfractionated ribosomal RNA in 0.1 M KCl, 0.05 M Tris, pH 7. Sedimentation is from left to right. The $s_{20,w}$ of the two main components are 14.5 and 17.4 S. The RNA concentration is 6.7 mg/ml.

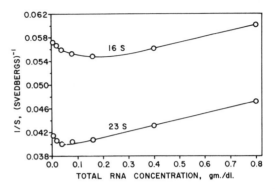

FIGURE 2: The dependence of the sedimentation coefficients of unfractionated ribosomal RNA upon the RNA concentration at 20° in 0.1 M KCl, 0.05 M Tris, pH 7. The RNA concentration is expressed as the sum of the two components. The intercepts, $s_{20,w}^0$, are 17.5 and 24.1 S. The significance of the negative slopes at the lower concentrations is not clear.

ice (from distilled water) and harvested by continuous flow centrifugation. The bacterial paste was frozen on solid carbon dioxide and stored at −20°. No differences were found in either the ribosomes or ribosomal RNA prepared from frozen versus fresh bacterial cells.

Thawed bacterial paste was washed twice in ten volumes of 0.01 M Tris, 0.01 M MgCl₂, pH 7, by centrifugation at 4°. To each kg of washed cells was added 600 ml of the above buffer and 960 ml of 120-μ diameter glass beads. The cells were ruptured by circulating the suspension through an Eppenbach colloid mill (Garver and Epstein, 1959) operated at 10,000 rpm with a gap setting of 0.081 cm. The lysate was maintained between 0 and 8° and complete cell breakage was obtained within 20 minutes.

Isolation of the Ribosomes. All operations were carried out between 0 and 4°. The lysate, freed of glass beads by decantation, was clarified by three successive centrifugations of 20 minutes each at 20,000 rpm in the No. 30 rotor of the Spinco Model L ultracentrifuge. The ribosomes were then sedimented at 30,000 rpm for 4 hours in the No. 30 rotor. Following resuspension in 0.01 M Tris, 0.01 M MgCl₂, pH 7, and a 20-minute centrifugation at 20,000 rpm in the 40 rotor (sediment discarded), the ribosomes were repelleted at 40,000 rpm for 3 hours in the No. 40 rotor. These final pellets were resuspended at a nucleoprotein concentration of from 15 to 20 mg/ml.

Isolation of the Ribosomal RNA. All operations were carried out between 0 and 4° unless otherwise indicated. The combination of phenol and SDS[2] employed by Hall and Doty (1959), with the modifications noted below, was used for the separation of the RNA from the protein. The nucleoprotein solution was made 0.5% (w/v) in SDS by the addition of 10% SDS. The flask was quickly warmed to 20°, held for 5 minutes, and chilled to 0° by swirling in an ice bucket. The solution was transferred to a separatory funnel containing an equal volume of buffer-equilibrated phenol, and a Macaloid suspension was added to give a final Macaloid concentration of 0.1% (w/v). The mixture was

shaken vigorously for 5 minutes and then the phases were separated by centrifugation at 4980 rpm for 20 minutes in the K-6 swinging-bucket rotor of the Spinco Model K centrifuge stopped without braking. The upper (aqueous) phase was removed with a large syringe equipped with a 6-cm square-tipped 12-gauge needle. Extreme care was used to avoid disturbing the interfacial region, and the lower 10% of the aqueous layer was discarded along with the phenol phase. The recovered aqueous phase was returned to a clean separatory funnel, and phenol and Macaloid were added as described above. The extraction was repeated twice more, each time avoiding the interfacial region and each time discarding the 10% of the aqueous phase nearest the phenol.

Dissolved phenol was removed from the final aqueous solution by three extractions with equal volumes of peroxide-free ethyl ether and the residual ether was removed from the RNA solution by bubbling nitrogen gas through the liquid until the odor of ether was no longer discernible. The RNA solution was then dialyzed for 36 hours against a 50-fold volume of 0.1 M KCl, 0.05 M Tris, pH 7, changed every 6 hours. This solution was either stored frozen at −20° or lyophilized directly and stored as RNA plus salt powder at −20°. No differences in physical properties were found between fresh RNA, frozen RNA, or lyophilized RNA solutions.

Handling of RNA Solutions. Provided that the procedures given above, from the lysis of the bacteria through the final isolation of the RNA, were carried out in one continuous operation, RNA preparations of high molecular weight were routinely obtained which contained no detectable ribonuclease activity. It was essential for the stability of the RNA during subsequent manipulations, however, that all equipment coming into contact with the RNA solutions was both clean and sterile. Routinely, this involved cleaning glassware

---

[2] Abbreviations used in this work: SDS, sodium dodecyl sulfate; DMSO, dimethyl sulfoxide.

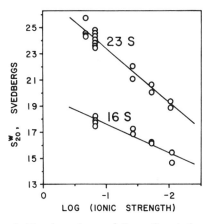

FIGURE 3: The dependence of the sedimentation coefficients of unfractionated ribosomal RNA upon the ionic strength at 20°. The RNA concentration was 40 μg/ml for all analyses.

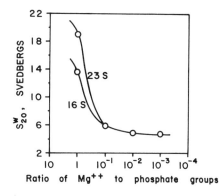

FIGURE 4: The influence of MgCl₂ upon the sedimentation coefficients at 20° of the unfractionated ribosomal RNA in 4×10^{-4} M KCl, 2×10^{-4} M Tris, pH 7.2. The RNA in this buffer was passed through a column of Dowex A-1 chelating resin in the K⁺ form prior to the addition of the requisite magnesium. The RNA concentration was 40 μg/ml for all analyses.

with an acid-chromate cleaning solution, rinsing with distilled water, and drying under vacuum. Plastic materials, such as Kel-F centerpieces for the analytical cells of the Spinco Model E ultracentrifuge, were cleaned at 80° for 3 to 4 minutes in a 0.1% SDS-0.1% Na₂EDTA solution, rinsed with distilled water, and dried under vacuum. Dialysis tubing was boiled for 15 minutes in the SDS-EDTA solution, boiled in distilled water, and rinsed with distilled water.

Sedimentation Measurements. VELOCITY. The Spinco Model E ultracentrifuge was equipped with both phase-plate schlieren and ultraviolet absorption optics. Cell centerpieces were of Kel-F. Sedimentation coefficients measured in either aqueous or organic solvents were corrected to water at 20° assuming a partial specific volume of 0.53 ml/g.

EQUILIBRIUM. Runs were of 18-hours duration in 30-mm epoxy double-sector cells in the 30-mm AN-E rotor operated at 5° in the Spinco Model E ultracentrifuge. Solution column heights were 1 mm (Williams *et al.*, 1958). A speed of 2095 rpm was employed except for an initial period of 1350 seconds at 6166 rpm (Hexner *et al.*, 1961). Schlieren photographs of the refractive index gradients were obtained on Eastman Kodak Kodaline CTC pan plates.

ZONE VELOCITY. Exponential gradients of 5 to 20% sucrose in 0.1 M KCl, 0.05 M Tris, 0.001% SDS, pH 6.7, were formed at 4°. The character of the gradient was $c = 23 - 18e^{-v/10}$, where c is the concentration, in per cent, of the sucrose being delivered at a volume equal to v ml. For preparative runs, 6 mg of RNA in 0.6 ml of 0.1 M KCl, 0.05 M Tris, pH 7, was layered over each gradient tube, and the tubes were centrifuged in the SW-25 rotor of the Spinco Model L ultracentrifuge at 4° for 20 to 24 hours at 25,000 rpm. The rotor was stabilized by hand during acceleration and deceleration below 5000 rpm and was allowed to stop at the end of the run without braking. The contents of the tubes were

analyzed by siphoning the gradients through a 1-mm flow cell mounted in a Beckman DB spectrophotometer operated at 260 mμ. The output of the spectrophotometer was recorded automatically as a function of time. Separated RNA components were recovered from their respective fractions by the addition of one volume of 2-propanol and chilling to −20°.

Viscosity Measurements. Viscosity measurements were performed in a Cannon Ubbelohde semimicro dilution viscometer requiring a minimum of 1 ml of sample and having an outflow time for distilled water at 20° of 317 seconds. Temperature control at 20° was maintained to within ±0.01° in an oil bath regulated by a Sargent S-82055 Model SV Thermonitor. Temperature control at temperatures other than 20° was regulated in a water bath controlled by a LaPine Tempunit to within ±0.1°. Samples were withdrawn from the viscometer following measurements at 20°, and both the concentration and sedimentation coefficients were determined.

Spectrophotometric Measurements. Concentrations were routinely determined by recording the optical density at 20° in a Model 11 or a Model 15 Cary recording spectrophotometer. From the base composition of the RNA and a hyperchromic shift of 40.1 ± 0.2% following alkaline hydrolysis and return to pH 7, the extinction of the potassium salt of 1 mg of RNA/ml of 0.1 M KCl, 0.05 M Tris, pH 7, was calculated to be 22.3 ± 0.2 at 260 mμ. Ultraviolet absorption measurements at temperatures other than 20°, *i.e.*, melting curves, were determined in a Beckman DB spectrophotometer. The temperature of the solution being analyzed was monitored by a LaPine Type 402 thermister probe connected to a Yellow Spring Model 43-TC Telethermometer.

Base Ratio Determinations. The base compositions

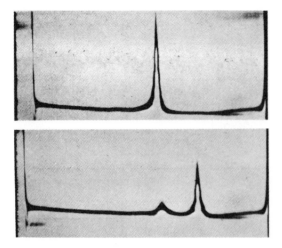

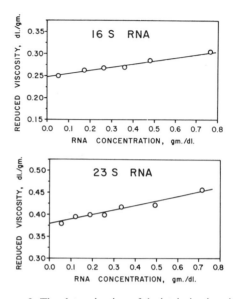

FIGURE 6: The determination of the intrinsic viscosities at 20° in 0.1 M KCl, 0.05 M Tris, *p*H 7, of the separated components of ribosomal RNA. The intrinsic viscosity of the 16 S component was found to be 0.248 dl/g and that of the 23 S component 0.379 dl/g.

FIGURE 5: Photographs of the schlieren patterns of the components of ribosomal RNA in 0.1 M KCl, 0.05 M Tris, *p*H 7, following one cycle through sucrose gradient zone velocity centrifugation. Sedimentation is from left to right. The upper pattern is of the 16 S fraction. The RNA concentration is 2 mg/ml and the $s_{20,w}$ 14.8 S. The lower pattern is of the 23 S fraction. The RNA concentration is 2 mg/ml and the $s_{20,w}$ equal to 16.6 and 20.7 S. The 23 S fraction was recycled through a second sucrose density gradient centrifugation before use in subsequent experiments.

of RNA samples were determined after either acid or alkaline hydrolysis. In the case of acid hydrolysis (1 hour, 100°, 1 M HCl) the bases and nucleotides were separated by paper chromatography (HCl-2-propanol-water (Wyatt, 1951)), eluted, and quantitated spectrophotometrically. Following alkaline hydrolysis (18 hours, 37°, 0.3 M KOH) RNA samples were neutralized to between *p*H 4 and 5, diluted 20-fold, and applied to a 225- × 0.9-cm column of Biorad AG 1-X8 resin (−400 mesh, 3.2 meq/g). The nucleotides were eluted at 45° with a 2-liter linear gradient of from 0 to 0.6 M ammonium formate, *p*H 4.00, at 25°.

TABLE I: Effect of $MgCl_2$ upon the Sedimentation Coefficients of Ribosomal RNA.[a]

$MgCl_2$ Concentration (M)	$s_{20,w}$	
	16 S Component	23 S Component
0.00	18.2	24.2
0.01	21.2	28.2

[a] Sedimentation runs were performed in 0.1 M KCl, 0.05 M Tris, *p*H 7, plus the indicated concentration of $MgCl_2$. The RNA had been pretreated with Dowex A-1 chelating resin and was analyzed at a concentration of 40 μg/ml.

Results

Sedimentation Properties. The schlieren pattern of the unfractionated ribosomal RNA is depicted in Figure 1. This distribution is characteristic of *E. coli* ribosomal RNA and has been widely quoted in the literature (Littauer and Eisenberg, 1959; Hall and Doty, 1959; Aronson and McCarthy, 1961). The dependence of the sedimentation coefficients upon the total RNA concentration is illustrated in Figure 2 and the effect of ionic strength in Figure 3. The effect of the divalent magnesium cation at low and high ionic strengths is indicated by the data in Figure 4 and Table I, respectively.

The sedimentation coefficients of the two main ribosomal RNA components isolated by zone velocity sedimentation through sucrose gradients were found to agree with those observed in unfractionated material when measured at low concentrations (40 μg/ml). The schlieren patterns of the fractionated material following one cycle through sucrose gradients are illustrated in Figure 5. The 16 S component was employed for further experiments at this stage, but the 23 S material was rerun on a second series of sucrose gradients. Material of greater than 95% purity (as judged by velocity sedimentation) was thus obtained and was employed for subsequent experiments.

Viscosity Properties. The intrinsic viscosity of unfractionated ribosomal RNA at 20° in 0.1 M KCl,

TABLE II: Molecular Weights and Radii of Gyration of the Components of Ribosomal RNA.

| Parameter | Method of Determination | RNA Component | |
		16 S	23 S
Molecular weight, g/mole	S and η	0.55×10^6	1.07×10^6
	Sedimentation equilibrium	—	1.07×10^6
	η kinetics	0.6–1.1×10^6	
Radius of gyration, A	Sedimentation	189	275
	Viscosity	161	233

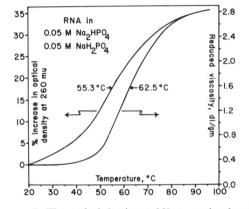

FIGURE 7: The optical density at 260 mμ and reduced viscosity of unfractionated ribosomal RNA as a function of temperature. The temperatures of the midpoints of the two transitions are indicated on the graph. The data for the optical density were obtained at temperature intervals of 2.5° and were corrected for the thermal expansion of the solution. The RNA concentration was 40 μg/ml. The data for the reduced viscosity were obtained at temperature intervals of 5° and were extrapolated to zero time at each temperature by plots of the logarithm of the reduced viscosity versus time. The RNA concentration was 1.9 mg/ml. The experimental points are not indicated since the scatter of each set of data was less than the width of the curves.

0.05 M Tris, pH 7.3, was found to vary between 0.33 and 0.42 dl/g depending upon the history of the sample. Freshly prepared RNA gave the lower value, while heated and cooled material gave high values. The intrinsic viscosities of the separated components of ribosomal RNA were also determined and are represented in Figure 6. It is to be noted that, in all measurements of the intrinsic viscosities, the value of the Huggins constant k in the equation

$$\eta_{\mathrm{sp}}/c = [\eta] + k[\eta]^2 c$$

fell between the values of 0.6 and 1.1.

In accord with the results of Spirin and co-workers

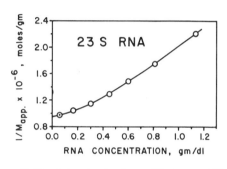

FIGURE 8: The determination by equilibrium centrifugation of the apparent molecular weight of the 23 S component of ribosomal RNA as a function of concentration. The double point at the lowest concentration results from duplicate analyses, one in the presence of a supporting column of silicone oil, and the other in the absence of oil. Following the equilibrium runs, all RNA samples were checked for possible degradation by recovery, dilution, and velocity sedimentation. All samples were found to be monodisperse with sedimentation coefficients of 24 ± 0.5 S. The extrapolation of $1/M_{\mathrm{app}}$ to zero concentration yields a molecular weight for the 23 S component of 1.07×10^6 g/mole.

(Spirin, 1962; Bogdanova et al., 1962; Shakulov et al., 1962), the reduced viscosity of ribosomal RNA was found to be a function of temperature. This variation, along with the variation of optical density, is illustrated in Figure 7.

Molecular Weight Determinations. The values of the sedimentation coefficients ($s_{20,w}^0$) and the intrinsic viscosities of the separated ribosomal RNA components were combined in the Scheraga-Mandelkern equation (Scheraga and Mandelkern, 1953) to yield their hydrodynamic molecular weights. β was assumed to be 2.16×10^6 and v equal to 0.53 ml/g. The intrinsic viscosities and the sedimentation coefficients were also utilized to calculate the radii of gyration of the two components. These results are summarized in Table II.

The weight average covalent molecular weight of the unfractionated ribosomal RNA was also estimated through a kinetic analysis of viscometric data obtained

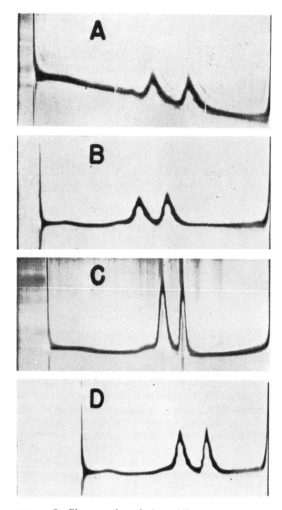

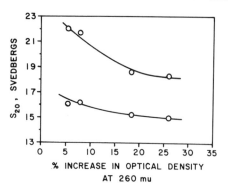

FIGURE 10: The effect of reaction with formaldehyde upon the sedimentation coefficients of unfractionated ribosomal RNA. RNA at a concentration of 0.4 mg/ml was allowed to react at 20° for various times with 1% formaldehyde between *p*H 9.6 and 10.7 in 0.005 M KCl, 0.0025 M potassium phosphate. The reactions were stopped by making them 30 μg/ml in RNA, 1% in formaldehyde, 0.1 M in KCl, 0.05 M in potassium phosphate, *p*H 6.9. The per cent increases in optical density at 260 mμ and the sedimentation coefficients were measured in this solvent. The increases in the optical densities were found to be stable for at least several hours. Complete reaction of the RNA with formaldehyde corresponds to a per cent increase in the optical density at 260 mμ of 32%.

FIGURE 9: Photographs of the schlieren patterns of unfractionated ribosomal RNA in formamide and dimethyl sulfoxide and after return to aqueous medium. Sedimentation is from left to right. (A) Ribosomal RNA in 0.2 M LiCl in formamide; RNA concentration is 2 mg/ml and $s_{20,w}$ 16.5 and 21.1 S. (B) The sample above returned to 0.1 M KCl, 0.0005 M MgCl$_2$, 0.05 M Tris, *p*H 7; RNA concentration is 2 mg/ml and the $s_{20,w}$ equal to 15.9 and 20.3 S. (C) Ribosomal RNA in 0.2 M LiCl in dimethyl sulfoxide; RNA concentration is 3.6 mg/ml and $s_{20,w}$ 13.6 and 16.4 S. (D) The sample above returned to 0.1 M KCl, 0.0005 M MgCl$_2$, 0.05 M Tris, *p*H 7; RNA concentration is 3.4 mg/ml and $s_{20,w}$ 15.4 and 19.3 S.

upon the RNA at 80°. Utilizing the equations for the initial stages of the random degradation of a high molecular weight polymer (see, for example, Tanford, 1961) and making the following assumptions: (a) RNA at 80° in 0.05 M Na$_2$HPO$_4$, 0.05 M NaH$_2$PO$_4$ is a flexible linear chain, (b) the ratio of the reduced viscosities measured at any two times is equal to the ratio of

the intrinsic viscosities at the same two times, and (c) that a similar relationship is true for the viscosity average and the weight average molecular weights, the following equation is obtained:

$$-\log \eta_t = -\log \eta_0 + (a) \log \left(1 + \frac{M_0 k t}{2R_w}\right)$$

where η_t is the reduced viscosity at time t, η_0 is the reduced viscosity at time zero, a is a parameter whose value lies between 0.5 and 0.8 for flexible linear chains, M_0 is the initial weight average molecular weight of the polymer undergoing degradation, k is the rate constant for the hydrolyzable bond in the polymer, t is the time of reaction, and R_w is the residue weight of the polymer. Taking 362 g/mole as the value of R_w, 8.6×10^{-8} sec^{-1} as the value of k (Eigner *et al.*, 1961), and 0.5 and 0.8 as the limits of a for a flexible linear chain, it may be calculated that the limits of M_0, the initial average molecular weight of the unfractionated ribosomal RNA, lie between 0.6×10^6 and 1.1×10^6 g/mole. From quantitative zone velocity sedimentation analyses and velocity sedimentation analyses at low concentrations, the mass ratio of 16 S to 23 S RNA in the ribosomal RNA is between 0.67 and 1.0. Taking this ratio to be one, a weight average molecular weight of 0.83×10^6 g/mole may be calculated for the unfractionated ribosomal RNA. The excellent agreement between this calculation and the results of the viscosity kinetics provides strong evidence that the hydrodynamic and covalent molecular weights of ribosomal RNA are very similar and probably identical.

The weight average molecular weight of the larger of the two main ribosomal RNA components was also measured by the technique of equilibrium sedimentation. This method yielded a value of 1.07×10^6 g/mole (Figure 8), also in agreement with the value derived from sedimentation viscosity measurements.

Aggregation. Once RNA preparations have been obtained free from nuclease activity, aggregation becomes the principal limitation on the accuracy of physical measurements. Both the sedimentation coefficients and the reduced viscosities increase in moderately concentrated RNA solutions following brief exposure to elevated temperatures. This behavior is outlined in Table III, and is undoubtedly due to the formation of intermolecular aggregates during the reestablishment of secondary interactions. It may also be seen that the presence of these aggregates prevents the viscosity at 25° from being a valid measure of the covalent molecular weight of the RNA. Thus Table III indicates that,

TABLE III: Effect of Thermal Treatment upon the Viscosity and Sedimentation Coefficients of Ribosomal RNA.

Minutes at 80°	$M_t/M_0{}^a$	Relative Reduced Viscosity at 25° after Heating[b]	$s_{20,w}{}^c$ 16 S	23 S
0	1.000	1.000	16.2	20.7
1	0.985	1.216	17.1	21.1
2	0.975	1.213	—	—
14	0.864	1.161	—	—
31	0.725	1.100	—	—
50	0.647	1.090	—	—
78	0.514	1.029	—	—
108	0.423	1.000	—	—
154	0.307	0.920	—	—

[a] The reduced viscosity, measured at 80° at the beginning and at the end of the incubation at 80°, was used to estimate M_t/M_0 from the relationship $M = K[\eta]^{1/a}$ with the assumption that $a = 0.5$. [b] The RNA concentration for the viscosity measurements was 3 mg/ml and the solvent was 0.1 M KCl, 0.05 M Tris, pH 7. [c] The RNA concentration for the sedimentation measurements was 9.3 mg/ml and the solvent was 0.1 M KCl, 0.05 M Tris, pH 7.

for a given reduced viscosity, the covalent molecular weight of an RNA preparation may actually vary over at least a factor of two.

The possibility that undetected aggregation may

exist in ribosomal RNA as it is usually isolated was demonstrated by introducing a few scissions into the once continuous phosphodiester chains by either thermal or enzymatic treatment under conditions favoring stable secondary interactions. These modified RNA particles are no longer stable (see Table IV) toward treatments which dissociate RNA aggregates

TABLE IV: Production and Demonstration of Limited Scissions in Ribosomal RNA.[a]

Experiment	Treatment of RNA	$s_{20,w}$
1A[b]	None	20.7, 16.2
	+ 1 minute at 80°	21.1, 17.1
1B	After 368 hours at 25°	20.9, 16.6
	+ 1 minute at 80°	15.4[d]
2A[c]	None	24.8, 18.2
	+ Passage through formamide	25.8, 19.5
2B	After ribonuclease	25.5, 18.2
	+ Passage through formamide	15.0[d]

[a] Scissions were introduced into the ribosomal RNA either by extended incubation at 25° (experiment 1) or by limited digestion with pancreatic ribonuclease (experiment 2). The demonstration of scissions was either by thermal treatment (experiment 1) or by solution in, and recovery from, formamide (experiment 2). [b] The RNA in experiment 1 was treated and analyzed in 0.1 M KCl, 0.05 M Tris, pH 7, at a concentration of 9.3 mg/ml. [c] In experiment 2, the ribonuclease treatment was for 15 minutes at 20° in 0.1 M KCl, 0.01 M $MgCl_2$, 0.05 M Tris, pH 7.3, at an RNA concentration of 1.12 mg/ml and a pancreatic ribonuclease (Pentex 5 × crystallized) concentration of 2.8×10^{-3} μg/ml. Following the digestion, the RNA was reisolated by the SDS–phenol–Macaloid procedure described under Methods. Passage through formamide consisted of precipitation from aqueous medium, solution in formamide for 30 minutes at 25°, precipitation (see legend, Table VI), and solution in 0.1 M KCl, 0.05 M Tris, pH 7.3. Sedimentation analyses were carried out in this solvent at a RNA concentration of 40 μg/ml. [d] Broad distribution of sedimentation coefficients.

and duplicated, in many respects, the ribosomal RNA preparations which earlier workers have believed to have been composed of "subunits" (Hall and Doty, 1958, 1959; Takanami, 1958; Osawa, 1960; Brown *et al.*, 1960; Chao, 1961; Aronson and McCarthy, 1961; Otaka *et al.*, 1961).

Several relatively mild procedures were adapted to reveal aggregation when it existed and to provide additional tests of the concept that ribosomal RNA is

TABLE V: Effect of Extended Dialysis upon Ribosomal RNA.[a]

RNA Treatment	$s_{20,w}$
After dialysis without EDTA	24.7, 17.8
+ 5 minutes at 78°	25.0, 19.2
After dialysis with EDTA	23.9, 17.8
+ 5 minutes at 78°	23.7, 18.2

[a] The RNA samples were dialyzed against 200 volumes of 1.25×10^{-2} M KCl, 6.25×10^{-3} M Tris, *p*H 7.2, with and without added 1×10^{-4} M EDTA. Dialyses were continued at 4° for 2 weeks with the dialysis buffers renewed once every 24 hours. Samples were made 0.1 M in KCl and 40 μg/ml in RNA before the sedimentation analyses.

(see Figure 7); further supportive data are given in Table VII. Taken together, these data provide very strong evidence that the two main components of *E. coli* ribosomal RNA are each continuously covalently linked chains.

Interaction with DEAE-Cellulose. Attempts to chromatograph ribosomal RNA upon DEAE-cellulose (Brown Co., 1 meq/g) were unsuccessful. The data presented in Figures 11 and 12 show that this failure was due to the fact that ribosomal RNA only slowly equilibrated with the ion exchanger and that, when equilibrated, formed a nonelutable complex. Additional studies showed that the RNA was not eluted by 1 M KCl at any temperature up to 80°, nor at 25° by 3 M solutions of the most competitive anions.

Base Compositions. The base compositions of the unfractionated ribosomal RNA and of the two separated

TABLE VI: Sedimentation Coefficients of Ribosomal RNA in, and Recovered from, Formamide and Dimethyl Sulfoxide.[a]

RNA Recovered from	Sedimentation Solvent	RNA (mg/ml)	$s_{20,w}$
Aqueous solution	$HCONH_2$ + 0.2 M LiCl	2.0	21.1, 16.5
$HCONH_2$ + 0.2 M LiCl	0.1 M KCl, 0.05 M Tris, *p*H 7	2.0	20.3, 15.9
$HCONH_2$	0.1 M KCl, 0.05 M Tris, *p*H 7	0.04	22.9, 16.8
Aqueous solution	DMSO + 0.2 M LiCl	3.6	16.4, 13.6
DMSO + 0.2 M LiCl	0.1 M KCl, 0.05 M Tris, *p*H 7	3.4	19.3, 15.4
DMSO	0.1 M KCl, 0.05 M Tris, *p*H 7	5.0	15.6, 12.9

[a] RNA samples were dissolved in the organic solvents following precipitation from aqueous solution with ethyl alcohol and washing with absolute ethyl alcohol followed by ethyl ether. Residual ethyl ether was removed under vacuum. Samples were recovered from organic solution by the addition of one volume of ethyl alcohol and two volumes of ethyl ether. The precipitates were washed three times with ethyl alcohol and the residual alcohol was removed under vacuum. The precipitates were then dissolved in the aqueous buffer system. All precipitations of the RNA were at 0°. All sedimentation coefficients have been corrected to water at 20° as the solvent assuming a partial specific volume of 0.53 ml/g.

composed of two classes of continuous chains of molecular weight 1.07×10^6 and 0.55×10^6 g/mole. Long term dialysis, with and without EDTA, was shown to be without effect on the RNA (Table V). RNA when dissolved in either formamide or dimethyl sulfoxide in the presence of a supporting electrolyte exhibited in the ultracentrifuge the usual two components; these were recovered unchanged when returned to an aqueous medium (Table VI and Figure 9). RNA recovered from either of these two organic solvents in the absence of salt also showed the normal distribution of components. Progressive reaction of the bases of the RNA chains with formaldehyde was found only slightly to diminish the sedimentation coefficients, suggesting that the RNA molecules were becoming more extended but not releasing "subunits" of lower molecuar weight (Figure 10). Finally, the effect of heat has already been discussed under the section on viscosity properties

components are summarized in Table VIII. The differences between the 16 and 23 S components are greater than experimental error and are considered significant. It is to be noted, however, that these differences are only significant when the components to be compared are derived from ribosomal RNA preparations obtained from a single bacterial harvest.

Discussion

The results presented in this paper lead to the following concept of *E. coli* ribosomal RNA. The two main components, the 16 and the 23 S RNA's, are separate, noninterconvertible classes and are formed of continuously covalently linked residues. In solution, the chains appear to be coiled or folded particles whose configurations are dependent upon secondary interactions which at moderate temperatures and ionic

TABLE VII: Thermal Expansion of Polyribonucleotides.

Polynucleotide[a]	Buffer[b]	RNA (mg/ml)	Reduced Viscosity (dl/g) Measured at 25°	Measured at 80°	Ratio 80°/25°
Ribosomal RNA	Phosphate	1.88	0.331	2.65	8.0
Ribosomal RNA	Phosphate	3.03	0.345	2.79	8.1
Ribosomal RNA	Phosphate	3.22	0.348	2.82	8.1
Ribosomal RNA	Phosphate	3.55	0.343	2.88	8.4
Ribosomal RNA	Tris	2.97	0.338	2.40	7.1
Ribosomal RNA	Tris	3.07	0.347	2.42	7.0
s-RNA	Phosphate	8.93	0.090	0.215	2.4
Poly-U	Phosphate	3.73	0.959	0.969	1.0

[a] Polynucleotides: ribosomal RNA, *E. coli* ribosomal RNA; s-RNA, yeast transfer RNA, Lot 2451-D, General Biochemicals; poly-U, polyuridylic acid, Lot 4729, Miles Laboratories ($s_{20,w}$ = 8 S). [b] Buffers: phosphate = 0.05 M NaH$_2$PO$_4$, 0.05 M Na$_2$HPO$_4$; Tris = 0.1 M KCl, 0.05 M Tris, *p*H 6.7 at 25°.

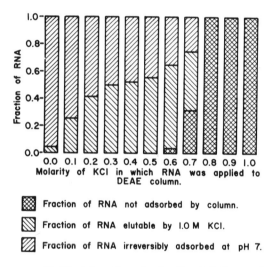

Fraction of RNA not adsorbed by column.

Fraction of RNA elutable by 1.0 M KCl.

Fraction of RNA irreversibly adsorbed at pH 7.

FIGURE 11: The interaction of unfractionated ribosomal RNA with DEAE cellulose at 20° as a function of salt concentration. Eleven DEAE-cellulose (0.9 meq/g) columns (0.7 × 4 cm) were prepared and equilibrated with 0.0, 0.1, 0.2, 0.3, . . . M KCl, 0.001 M Tris, *p*H 6.6. Eleven samples of ribosomal RNA (1 mg each) in 0.0, 0.1, 0.2, 0.3, . . . M KCl, 0.001 M Tris, *p*H 6.6, were introduced into the respective columns. Each column was washed with five bed volumes of the buffer in which the RNA had been applied to the column. The RNA emerging with this wash was labeled "Fraction of RNA not adsorbed by column." Each column was then washed with five bed volumes of 1 M KCl, 0.001 M Tris, *p*H 6.6. The RNA emerging with this wash was labeled "Fraction of RNA elutable by 1.0 M KCl." The RNA not accounted for by the sum of the first two fractions was labeled "Fraction of RNA irreversibly adsorbed at *p*H 7."

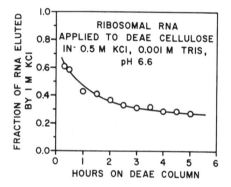

FIGURE 12: The interaction of unfractionated ribosomal RNA with DEAE-cellulose at 20° as a function of time. Eleven columns (0.7 × 4 cm) of DEAE-cellulose (0.9 meq/g) were equilibrated with 0.5 M KCl, 0.001 M Tris, *p*H 6.6. Eleven samples of ribosomal RNA (1 mg each) in 0.5 M KCl, 0.001 M Tris, *p*H 6.6, were introduced into the columns. Each of the columns was washed with five bed volumes of 0.5 M KCl, 0.001 M Tris, *p*H 6.6; no RNA was recovered in these washes. At successive time intervals one of the columns was washed with five bed volumes of 1.0 M KCl, 0.001 M Tris, *p*H 6.6. The fraction of RNA recovered in these washes was labeled "Fraction of RNA eluted by 1 M KCl."

strengths are only slowly equilibrated with their environment. This is indicated by the slowness of the equilibration with DEAE-cellulose and by the fact that the reduced viscosity, measured at 25°, continues to rise for many hours following an exposure to elevated temperatures. The configurations of the two components

TABLE VIII: Base Compositions of Ribosomal RNA.[a]

RNA Preparation[b]	RNA	Method of Hydrolysis	No. of Analyses	Base Composition as Per Cent of Total Bases			
				A	G	U	C
1	16 + 23 S	OH⁻	3	25.1	32.4	21.1	21.4
1	16 S	OH⁻	2	24.2	32.1	21.3	22.3
1	23 S	OH⁻	2	25.5	32.5	21.0	21.0
1	16 + 23 S	H⁺	2	25.1	32.3	21.4	21.2
2	16 + 23 S	H⁺	2	25.5	31.2	21.1	21.3
3	16 + 23 S	OH⁻	2	24.0	33.6	21.3	21.1

[a] The bases and nucleotides were separated, identified, and quantitated as described under Methods. [b] Ribosomal RNA preparations 1, 2, and 3 were obtained from independently grown and harvested cultures of *E. coli*.

when freshly isolated appear to be more compact than might be expected for a purely random coil. This is indicated by the rather high sedimentation coefficients and low intrinsic viscosities and also by the high value of the Huggins constant (0.6 to 1.1 as opposed to a value of approximately 0.35 predicted by a random coil model (Huggins, 1942)). It is quite possible that the configurations observed in solutions of freshly prepared RNA reflect the residual structure of the RNA as it existed in the ribosomes.

Several of the physical characteristics of ribosomal RNA which are measured by hydrodynamic parameters and which have been presented here were found to agree with values previously reported in the literature. This is true for those properties not sensitive to the covalent integrity of the RNA particles. Many of the other results, however, were dependent upon the absence of nuclease contamination and upon the high covalent molecular weight of these *E. coli* ribosomal RNA preparations. It is hoped that some of the techniques which have been reported here will prove useful in studies of conformation and structural integrity of other high molecular weight polyribonucleotides, both natural and synthetic.

References

Aronson, A. I., and McCarthy, B. J. (1961), *Biophys. J. 1*, 215.

Bogdanova, E. S., Gavrilova, L. P., Dvorkin, G. A., Kisselev, N. A., and Spirin, A. S. (1962), *Biokhimiya 27*, 387.

Brown, R. A., Ellem, K. A. O., and Colter, J. C. (1960), *Nature 187*, 509.

Chao, F. (1961), *Biochim. Biophys. Acta 53*, 64.

Eigner, J., Boedtker, H., and Michaels, G. (1961), *Biochim. Biophys. Acta 51*, 165.

Garver, J. C., and Epstein, R. L. (1959), *Appl. Microbiol. 7*, 318.

Ginoza, W. (1958), *Nature 181*, 958.

Hall, B. D., and Doty, P. (1958), Microsomal Particles and Protein Synthesis, New York, Pergamon, p. 27.

Hall, B. D., and Doty, P. (1959), *J. Mol. Biol. 1*, 111.

Hexner, P. E., Radford, L. E., and Beams, J. W. (1961), *Proc. Natl. Acad. Sci. U.S. 47*, 1848.

Huggins, M. L. (1942), *J. Am. Chem. Soc. 64*, 2716.

Kurland, C. G. (1960), *J. Mol. Biol. 2*, 83.

Littauer, U. Z., and Eisenberg, H. (1959), *Biochim. Biophys. Acta 32*, 320.

Osawa, S. (1960), *Biochim. Biophys. Acta 43*, 110.

Otaka, E., Oota, Y., and Osawa, S. (1961), *Nature 191*, 598.

Scheraga, H. A., and Mandelkern, L. (1953), *J. Am. Chem. Soc. 75*, 179.

Shakulov, R. S., Ajtkhozhin, M. A., and Spirin, A. S. (1962), *Biokhimiya 27*, 744.

Spirin, A. S. (1962), Acides Ribonucleiques et Polyphosphates, Structure, Synthèse et Fonctions, Colloques Internationaux du Centre National de la Récherche Scientifique No. 106, Paris, Centre National de la Récherche Scientifique, p. 73.

Takanami, M. (1958), *Biochim. Biophys. Acta 29*, 430.

Tanford, C. (1961), Physical Chemistry of Macromolecules, New York, Wiley.

Williams, J. W., Van Holde, K. E., Baldwin, R. L., and Fugita, H. (1958), *Chem. Rev. 58*, 715.

Wyatt, G. R. (1951), *Biochem. J. 48*, 584.

18

Reprinted from *Nature*, **215**, 735–736 (Aug. 1967)

NUCLEOTIDE SEQUENCE OF 5S-RIBOSOMAL RNA FROM *ESCHERICHA COLI*

G. G. Brownlee, F. Sanger, B. G. Barrell

MRC Laboratory of Molecular Biology, Cambridge

In 1964 Rosset, Monier and Julien[1] described and characterized a low molecular weight ribonucleic acid (5S RNA) that is present in the ribosomes of *Escherichia coli* in addition to the two larger components, the 16S and 23S RNA. It contains 120 nucleotide residues and in contrast to transfer RNA contains no "minor" bases. Using a two-dimensional fractionation procedure for the separation of nucleotides labelled with phosphorus-32 (ref. 2), we have determined the sequences of all the oligonucleotides obtained by complete digestion of ³²P-labelled 5S RNA with ribonuclease T1 and ribonuclease A (pancreatic ribonuclease)[3]. In order to arrange these nucleotides in the unique sequence of the 5S RNA, we

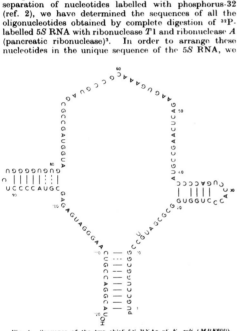

Fig. 1. Sequence of the two chief 5S RNAs of *E. coli* (*MRE*600) showing regions involved in base-pairing. A solid line indicates a "standard" (G—C or A—U) pair and a dashed line a G—U pair, for example, between residues 81 and 95.

have now studied the products of various partial ribonuclease digestions. This has involved the development of a number of new techniques which will be described in detail later. From the large number of partial digestion products obtained, we deduced the unique sequence shown in Figs. 1 and 2.

Fig. 1 is drawn to show the residues which are believed to be involved in base-pairing. These base-paired regions were identified as four sequences which were particularly resistant to digestion by ribonucleases. The longest double-stranded region is believed to be formed by base-pairing between the two ends of the molecule, and there are also two smaller "loops". Base-pairing between the two ends of the molecule is also found in transfer RNA, but otherwise there is less base-pairing in 5S RNA than in transfer RNA.

It will be noted that two residues are shown as occupying position 13. It appears that there are two 5S RNAs, presumably controlled by separate genes, one having a G in position 13 and the other a U. This was found in one strain of *E. coli* (*MRE*600, obtained from Dr H. E. Wade of the MRE Experimental Station, Porton, Wiltshire), while in another strain (*CA*265, obtained from Dr S. Brenner of this laboratory) a difference has been found in another position. It is probable that there are also other minor heterogeneities and therefore Fig. 1 illustrates the structure only of the two principal components of 5S RNA in *E. coli*, *MRE*600.

There are two sequences of ten and eight residues, respectively, that are repeated twice in the molecule. In Fig. 2 the structure is written so that the common sequences are aligned. There is considerable homology between the two parts of the chain, indicated by the boxed regions. This observation suggests that the 5S RNA may have evolved from a smaller RNA by a duplication of a part of the DNA sequence within the gene. There also appears to be some homology between the two ends of the molecule as shown by the underlining in Fig. 2. This could be explained by a separate duplication.

Previous work on RNA sequences has been confined to transfer RNAs and a number of complete sequences have been reported[4-7]. The 5S RNA is 120 residues long compared with 75–85 residues in the transfer RNAs, and the absence of "minor" bases makes interpretation somewhat more difficult. This work shows, however, that it is possible, using the small-scale techniques which we have developed, to determine the nucleotide sequence of an RNA labelled with phosphorus-32.

[1] Rosset, R., Monier, R., and Julien, J., *Bull. Soc. Chim. Biol.*, **46**, 87 (1964).
[2] Sanger, F., Brownlee, G. G., and Barrell, B. G., *J. Mol. Biol.*, **13**, 373 (1965).
[3] Brownlee, G. G., and Sanger, F., *J. Mol. Biol.*, **23**, 337 (1967).
[4] Holley, R. W., Apgar, J., Everett, G. A., Madison, J. T., Marquisee, M., Merrill, S. H., Penswick, J. R., and Zamir, A., *Science*, **147**, 1462 (1965).
[5] Zachau, H. G., Dutting, D., and Feldmann, H., *Hoppe Seyler's Z. Physiol. Chem.*, **347**, 212 (1966).
[6] Madison, J. T., Everett, G. A., and Kung, H., *Science*, **153**, 531 (1966).
[7] RajBhandary, U. L., Chang, S. H., Stuart, A., Faulkner, R. D., Hoskinson, R. M., and Khorana. H. G., *Proc. US Nat. Acad. Sci., Wash.*, **57**, 751 (1967).

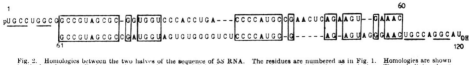
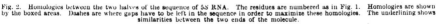

Fig. 2. Homologies between the two halves of the sequence of 5S RNA. The residues are numbered as in Fig. 1. Homologies are shown by the boxed areas. Dashes are where gaps have to be left in the sequence in order to maximize these homologies. The underlining shows similarities between the two ends of the molecule.

19

Reprinted from *Biochemistry,* **10**(3), 517–524 (1971)

Stoichiometry of the 30S Ribosomal Proteins of *Escherichia coli**

P. Voynow and C. G. Kurland†

ABSTRACT: The average number of copies per ribosome of each of eighteen proteins from the 30S ribosomal subunit of *Escherichia coli* has been measured by two independent procedures. The stoichiometric data for the sum of another pair of 30S proteins has also been obtained. The data indicate that there are at least two stoichiometric classes of proteins in the 30S subunit. One class of 12 proteins is present in amounts close to one copy per ribosome; these are called unit proteins. Another class of eight proteins are present in amounts significantly less than one copy per ribosome; these are called fractional proteins. No protein of the 30S subunit seems to be present in amounts corresponding to more than 1 copy/ribosome. We conclude that the 30S ribosomal subunits are structurally heterogeneous and that the proteins are arranged dissymetrically in this ribosomal subunit. The implications of these structural features of the ribosome are discussed in relation to the mechanism of protein synthesis.

A detailed analysis of the structure and function of the bacterial ribosome has become possible now that several laboratories have begun to purify and characterize the ribosomal proteins of *Escherichia coli* (Kaltschmidt *et al.*, 1967; Fogel and Sypherd, 1968; Moore *et al.*, 1968; Hardy *et al.*, 1969). Chemical and physical characterizations of isolated components indicate that there are at least 19 different proteins in the 30S ribosomal subunits (Craven *et al.*, 1969; H. G. Wittmann, personal communication) and roughly 30 different proteins in the 50S ribosomal subunits (G. Mora *et al.*, in preparation 1970; H. G. Wittman, personal communication).

The *in vitro* reconstitution procedure of Traub and Nomura (1968) has been applied to purified ribosomal components and has yielded valuable information about the 30S ribosomal subunits. Thus, a preliminary functional classification of the 30S ribosomal proteins has been obtained by Nomura *et al.* (1969) and several genetically determined alterations of individual ribosomal proteins have been identified (Birge *et al.*, 1969; Birge and Kurland, 1969; Bollen *et al.*, 1969; Ozaki *et al.*, 1969; Sypherd, 1969). As such data accumulate it becomes progressively more important to be able to describe the arrangement of the proteins within the functional ribosome. In particular, we wish to know whether or not all of the ribosomes from a single source have the same structure and whether any of the proteins are represented in amounts different from 1 copy/particle.

Stoichiometric data for the ribosomal proteins were first reported by Moore *et al.* (1968), who concluded that the 30S subunit contains 1 copy of each of 13 proteins. A similar conclusion has been drawn by Sypherd *et al.* (1969). However, our identification of at least 19 different 30S proteins with an aggregate mass of 410,000 daltons is inconsistent with the conclusion that there is 1 copy of each of these proteins in every ribosome, because the 30S subunits contain only about 260,000 daltons of protein (Hardy *et al.*, 1969; Craven *et al.*, 1969). The major discrepancies between our data and that of Moore *et al.* (1968) were the molecular weights that were assigned to six of the proteins purified by Moore *et al.* (1968). However, the latter data have since been corrected and now there is agreement between both laboratories that the 30S subunits are probably heterogeneous (Kurland *et al.*, 1969; Traut *et al.*, 1969).

We now wish to know how many different kinds of 30S particles there are in *E. coli* and how their protein complements differ. Unfortunately, it is not yet possible to obtain such data directly. However, we have measured the average number of copies of each protein per 30S particle and we find both with crude as well as salt-washed ribosomes that there are at least two stoichiometric classes of proteins. One group of proteins is present in amounts corresponding to 1 copy/ribosome and a second group of 30S proteins is present in amounts much less than 1 copy/ribosome. The present data strongly support the conclusion that the purified ribosomes are structurally heterogeneous. This in turn implies that the ribosomes may be functionally heterogeneous. Furthermore, no protein of the 30S subunit seems to be represented more than once per ribosome. Therefore, we conclude that the bacterial ribosome is both structurally and functionally dissymetric.

Methods

Preparation of Radioactive Proteins. Uniformly labeled [^{14}C]protein was obtained from cells grown in a medium consisting of M9 salts (60 mg/l.), adenine (60 mg/l.), uracil (5 g/l.), glucose, a mixture of amino acids that consisted of 2.1×10^{-4} M alanine, 0.88×10^{-4} M arginine, 1.6×10^{-4} M aspartic acid, 0.32×10^{-4} M cysteine, 1.7×10^{-4} M glutamic acid, 1.3×10^{-4} M glycine, 0.16×10^{-4} M histidine, 0.72×10^{-4} M isoleucine, 1.3×10^{-4} M leucine, 1.1×10^{-4} M lysine, 2.2×10^{-4} M methionine, 0.56×10^{-4} M phenylalanine, 0.72×10^{-4} M proline, 0.96×10^{-4} M serine, 0.80×10^{-4} M threonine, 0.16×10^{-4} M tryptophan, 0.32×10^{-4} M tyrosine, and 0.88×10^{-4} M valine. This mixture is proportional to the amino acid composition of total *E. coli* protein as described by Roberts *et al.* (1955), except that the methionine concentration was quadrupled. The medium also contained about 1 mCi/l. of ^{14}C-reconstituted protein hydrolysate (Schwarz BioResearch). A specific activity for each 30S protein was calculated from its amino acid composition and the specific activities of the amino acids in the medium; it was assumed

* From the Department of Zoology, University of Wisconsin, Madison, Wisconsin. *Received July 20, 1970.* Supported by U. S. Public Health Research Research Grants GM 13832 and GM 12411, as well as by funds supplied by the Wisconsin Alumni Research Foundation.

† To whom to address correspondence.

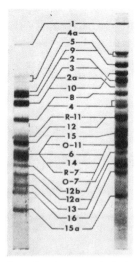

FIGURE 1: The electrophoretic pattern obtained with 30S protein migrating on soft (left) and hard (right) polyacrylamide gels. The proteins migrate toward the cathode at the bottom of the photograph. The composition of the gels and the method of electrophoresis are described in the text. The prefixes "O" and "R" correspond to the oxidized and reduced forms, respectively, of the indicated proteins. The "O" forms do not appear in reduced samples, and therefore are not visible in the gels depicted in Figure 2.

that three generations of growth is sufficient to equilibrate the amino acids in the precursor pools with those in the medium (McCarthy and Britten, 1962; McCarthy et al., 1962). These calculations indicate that all of the 30S proteins had the same specific activity within ±5%.

^{3}H labeling was achieved by growing bacteria in M9 salts containing 100 mg/l. of Casamino Acids (Difco), 60 mg/l. of adenine, 60 mg/l. of uracil, 5 g/l. of glucose, and about 5 mCi/l. of [^{3}H]lysine (Schwarz BioResearch). The bacteria were grown at 37° with rapid stirring and forced aeration. Logarithmic growth was allowed to proceed to a concentration of 1 g wet weight of cells/l. The cells were harvested by centrifugation and the pellets were frozen at −70°.

Purified ribosomes were prepared as described by Kurland (1966), except for a few minor changes. The frozen bacterial paste (weighing 1–2 g) was thawed in 10 ml of TSM (0.01 M Tris–0.003 M succinic acid–0.01 M MgCl$_2$, pH 8.0) containing 6 μg/ml of DNase. The bacterial suspension was disrupted in a French pressure cell at about 8000 psi. From this point on the procedure was scaled down to a working volume of 10 ml. MgCl$_2$ was used throughout instead of MgSO$_4$ and the puromycin incubation of the crude extract was omitted.

30S subunits were prepared by dialyzing ribosomes overnight against gradient buffer (0.01 M Tris–0.05 M KCl–0.3 mM MgCl$_2$–6 mM 2-mercaptoethanol, pH 7.6) layering them on a gradient of 5–30% sucrose in the same buffer and centrifuging for 11 hr at 25,000 rpm in a Spinco SW 25.2 rotor. The bottom of the centrifuge tube was punctured and 1.5–2-ml fractions were collected dropwise. A 5-μl aliquot of each sample was taken for counting; 10 ml of BBS-3 cocktail (50 ml of Beckman BioSolve BBS-3 and 5 g of 2,5-diphenyloxazole/l. of toluene) was added, and the vials were shaken and counted in a Beckman LS-250 liquid scintillation counter. The peak of 30S particles was pooled and dialyzed overnight

against TSM. Usually 1–2% of the starting radioactivity was recovered in this 30S fraction. The 30S subunits were stored at −70°.

For some experiments crude ribosomes were required. Frozen cells were thawed, suspended, lysed in a French pressure cell, and subjected to low-speed centrifugation exactly as in the procedure for purified ribosomes. This crude extract was dialyzed into gradient buffer and subjected to zone centrifugation on a sucrose gradient as described above to obtain crude subunits.

Proteins were extracted with acetic acid as described in Hardy et al. (1969). However the procedure was scaled down to accomodate a 20-mg sample. Nonradioactive 30S protein was added as carrier to the labeled protein to achieve this sample size. A cellulose phosphate column 1 × 25–30 cm was eluted with a gradient of 0.0–0.5 M NaCl in 400–600 ml of standard urea buffer, pH 5.8 (6 M urea, 50 mM NaH$_2$PO$_4$, 12 mM methylamine, and 0.8 mM β-mercaptoethanol), at a flow rate of 4 ml/hr. Fractions (1 ml) were collected and 50-μl aliquots were counted with 10 ml of BBS-3 cocktail in the scintillation counter.

When the purpose of the chromatography was the purification of [^{3}H]proteins, the starting material contained 5–10 × 10^7 cpm of ^{3}H 30S protein. For isotope dilution experiments the sample contained about 10^5 cpm of the pure [^{3}H]protein and about 10^6 cpm of ^{14}C total 30S protein.

Electrophoresis. As in Hardy et al. (1969) discontinuous electrophoresis in polyacrylamide gels at pH 4.5 was used to check purity of chromatographic fractions. However, since radiochemical purity was desired, a simple examination of the staining pattern was not sufficient. For this reason all gels containing test proteins were sliced and counted (as described below) to make sure that a single band of stain corresponded to a single band of radioactivity.

Gels for stoichiometry measurements were run as in Hardy et al. (1969) except for some minor changes. The gels were usually 10 × 0.6 cm, but were occasionally longer (up to 20 cm). Two different methylene–bisacrylamide concentrations were used depending on the protein being studied; 0.15% (w/v) was used for "soft" gels, and 0.75% (w/v) for hard gels. "Hard" and "soft" gel patterns for 30S protein are shown in Figure 1. The gels were fixed, stained, and destained as in Hardy et al. (1969).

Between 20,000 and 100,000 cpm of ^{14}C total 30S protein were applied to the gels used for "recovery" or "dilution" experiments. In addition, 1000–3000 cpm of a pure [^{3}H]-protein was included for dilution experiments. Since the radioactive proteins had a high specific activity, the quantities applied to the gel were insufficient to yield well-stained bands. For this reason 0.1–0.2 A_{230} of nonradioactive 30S protein was added to each sample.

Gel Slicing and Counting. Gels were fractionated by placing them in a glass tube (8-mm i.d. for soft gels and 7-mm i.d. for hard gels) and then slicing off thin sections with a razor blade as the gel was allowed to emerge from one end of the tube. The tube was clamped along one leg of an aluminum angle iron. The gap between the end of the tube and the other leg of the angle iron determines the thickness of the gel slices, since the leg stops the gel from sliding any farther out of the tube. The apparatus is basically a miniature salami slicer. The gels were kept wet during slicing by periodically pouring 7% acetic acid into the slicing tube. Such lubrication facilitates the sliding of the gel down the tube as well as even slicing.

Slices were placed in scintillation vials and the slice number

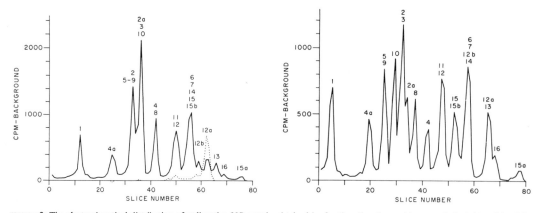

FIGURE 2: The electrophoretical distribution of radioactive 30S protein obtained by fractionating the proteins on soft (a, left) and hard (b, right) polyacrylamide gels. The gels were sliced and counted as described in the text. The solid line represents ^{14}C-labeled protein and the dashed line represents ^{3}H-labeled protein.

of each stained band was recorded. To each vial was added 0.15 ml of water and then 1 ml of NCS (Nuclear-Chicago Corp.) solubilizer. The vials were capped and shaken at 37° for at least 8 hr. This treatment causes the gel slices to swell and appears to extract the protein from the gels. Scintillation fluid (10 ml) consisting of 5 g of 2,5-diphenyloxazole (Packard Institute Corp.)/l. of toluene was added to each vial and they were shaken and counted in a Beckman LS-250 scintillation counter using the AQC (automatic quench correction) system.

Dilution Experiments. After a gel containing doubly labeled proteins is sliced and counted the ^{3}H and ^{14}C content of each slice is calculated and plotted (see Figure 2) and the ^{3}H/^{14}C ratio is computed for each fraction in the ^{3}H peak. Next the total ^{14}C counts in the gel (C_T) is summed, and so is the total of ^{3}H counts in the protein of interest (H_T); ^{3}H contaminants are not included.

The mass fraction, f_i, of the protein of interest is given by $f_i = (H_T/C_T)/(^{3}H/^{14}C)_{peak}$. The product of the mass fraction of a particular protein and the total protein content (P) of the particle gives the mass of that protein per particle. When this value is divided by its molecular weight (M_i), the average number of copies (N_i) of that protein per particle is obtained: $N_i = f_iP/M_i$.

Recovery Experiments. Stoichiometry measurements using the recovery method were made on both singly (^{14}C) and doubly (^{3}H and ^{14}C) labeled gels (in which case the presence of the ^{3}H was ignored). Whenever a peak in a ^{14}C gel profile was sufficiently well resolved from its neighbors, an estimate was made of the ^{14}C in that peak (C_i). Then C_i/C_T was taken as an estimate of the mass fraction, and the number of copies per particle was calculated as described above.

Dilution Experiments with Columns. The isotope dilution experiments performed on phosphocellulose columns are completely analogous to those on acrylamide gels. First the purified [^{3}H]protein and ^{14}C total protein components of the starting material are counted with the NCS system; *i.e.*, 150-μl aliquots are incubated with 1 ml of NCS at 37° for about 1 hr with occasional agitation and then counted in 10 ml of 5 g of 2,5-diphenyloxazole/l. of toluene. The ratio ^{3}H/^{14}C of these counts corresponds to H_T/C_T in the gel experiment.

Since counting with the NCS system is very time consuming

and expensive, the preliminary analysis of the column was done using the BBS-3 counting system even though the system is much less reliable. After the column had been eluted and counted in BBS-3, the ^{3}H and ^{14}C content of each fraction was calculated as it was in the gel experiment, and the ^{3}H counts were plotted.

The [^{3}H]protein peak was checked for purity by gel electrophoresis, slicing and counting. If the protein was significantly contaminated (by either [^{3}H]- or [^{14}C]proteins that were not the protein of interest), then purification by rechromatography was carried out as in Hardy *et al.* (1969). The ^{3}H and ^{14}C contents of the rechromatography fractions were calculated and plotted as described above, and once again the purity of the [^{3}H]protein peak was checked by electrophoresis. Once the [^{3}H]protein was obtained pure, the ^{3}H/^{14}C ratio was determined using the NCS system and, as in the gel experiment, the average number of copies per particle was determined.

Concentration Measurements. Protein concentrations were measured by the method of Lowry *et al.* (1951) using egg-white lysozyme (Worthington Biochemical Corp.) as a standard. A micro-Kjeldahl (Hiller *et al.*, 1948) nitrogen assay was run on total ribosomal protein, and this was compared to the Lowry assay. It was found that the Lowry reaction (with lysozyme standard) gave 93% of the Kjeldahl value for ribosomal protein concentration.

Ribosome concentrations were determined by measuring the A_{260} of ribosomes in TSM. A conversion factor of 44 μg of RNA/A_{260} unit was used (Kurland, 1960).

Results

Recovery Measurements. Our first measurements of the stoichiometry of the 30S proteins were made using the recovery method (see Methods section) because it is a quick and simple procedure. Figure 2 shows typical profiles of ^{14}C-labeled 30S protein that has been electrophoresed on hard and soft gels. Some of the proteins are sufficiently well resolved on one or the other kind of gel to permit an accurate estimate of their mass fractions by the recovery method. Thus, proteins 30S-1, 4a, 4, 5 + 9, 12 (in nonreduced material), 15a, and possibly 8 are resolved well enough for our purposes. How-

TABLE I: Stoichiometric Data for 18 Proteins of the 30S Ribosomal Subunit.[a]

| Protein | Mol Wt | Purified Ribosomes | | Crude Ribosomes | | Class |
		Dilution	Recov	Dilution	Recov	
1	65,000	0.14[1]	0.29[5]		0.95[5]	F
2[b]	18,000		0.80[2]			M
2a	17,600	0.90[2]		0.99[1]		U
3	24,000	0.80[1]				M
4	16,000	0.79[2]	0.70[6]	0.92[1]	0.91[3]	M
4a	30,000	0.55[2]	0.47[10]	1.10[1]	0.83[5]	F
5 + 9[c]	32,000	0.71[2]	0.77[6]	0.91[1]	0.84[2]	M
6	13,500	0.89[1]				U
7	10,700	0.83[2]	0.90[1]			U
8	21,500	0.89[2]	0.73[4]		0.91[3]	U
10	26,700	0.89[2]	0.87[6]	1.10[1]	1.08[2]	U
11	18,300	0.40[1]				F
12	21,000	1.06[1]	0.83[2]		1.00[2]	U
12a	14,600	0.73[2]	0.73[4]	0.74[1]	0.80[2]	M
12b	15,600	0.52[1]	0.46[3]	0.61[1]	0.57[2]	F
13	15,000	0.60[2]	0.56[4]		0.48[2]	F
15a	13,000	0.34[2]	0.31[5]	0.36[2]	0.38[5]	F
16	14,000	0.61[2]	0.48[3]	0.41[1]	0.50[2]	F

[a] The stoichiometric data presented as the average number of copies per 30S subunit were obtained by both the recovery and isotope dilution methods. The data for purified and crude ribosomes are included. The number of independent measurements is indicated in each entry by a superscript. All of the data was obtained on polyacrylamide gels, except for 30S-3 and 6; these values were obtained on cellulose phosphate columns. [b] The data for protein 30S-2 were obtained indirectly as described in the text. [c] The values for 30S-5 + 9 are the sums of two components which are considered modified forms of a single protein (Craven et al., 1969). The methods and calculations are described in detail in the text. U, M, and F designate unit, marginal, and fractional proteins as defined in the text.

ever, proteins 30S-7 (in nonreduced material), 10, 12a, 12b, 13, and 16 are not as well resolved from their neighbors. Consequently, it is necessary to estimate the small amounts of overlapping material that contaminate these components from neighboring ones (see Figure 2). This was done by extrapolating the shoulders of the neighboring peaks. The eight remaining proteins migrate on the gels in overlapping groups of two or more proteins; therefore, individual mass fraction estimates for these components cannot be obtained by the present technique.

Extensive recovery measurements were made on as many of the 30S proteins as possible; these data are summarized in Table I. Proteins 30S-7, 10, and 12 are present in amounts between 0.8 and 1.0 copy per 30S particle. Such deviations from perfect 1:1 stoichiometry are within our estimates of error ($\pm25\%$). We call such components unit proteins.

Proteins 30S-1, 4a, 12b, 13, 15a, and 16 are present in amounts corresponding to 0.5 copy/ribosome or less. This deviation from 1:1 stoichiometry is well outside of our error. We call such components fractional proteins.

Proteins 30S-4, 5 + 9, 8, and 12a appear to be present in intermediate amounts: between 0.7 and 0.8 copy per ribosome. The present stoichiometric estimates have an uncertainty of $\pm25\%$, because our mass fraction measurements are reproducible within $\pm10\%$ and there is an additional uncertainty of $\pm15\%$ in the molecular weight estimates. Therefore, a protein which appears to be present in three out of four ribosomes could be a unit protein or a fractional protein which is present on half of the ribosomes. Such proteins cannot be classified and we call these marginal proteins.

Since we rarely recover 100% of the radioactivity applied to the polyacrylamide gels and as much as 40% of the radioactivity can on occasion be lost, it is necessary to be certain that the data are not distorted by the selective loss of one or more proteins during electrophoresis. One indication that this is not a problem is the observation that the mass fractions of different proteins are the same for gels in which 60% of the radioactivity is recovered and for gels in which 90% of the radioactivity is recovered.

The loss of radioactivity seems to be due to a lower efficiency of counting when the proteins are embedded in gels. This is illustrated by control experiments in which the recovery of radioactivity was measured after adding labeled protein to gels before polymerization and measuring the effect of the gel on the recovery of radioactivity without performing electrophoresis. The relative recovery of radioactivity from such gel samples varies considerably from one gel batch to another. It makes no difference whether or not the gels have been fixed prior to counting the radioactivity. Although such control experiments suggest that the failure to recover all of the radioactivity that is applied to the gels is due to the lower counting efficiency for radioactive proteins embedded in polyacrylamide compared to simple liquid samples, it is possible that other artifacts distort the data.

There is one potential source of error in the stoichiometry measurements that is common to both the recovery method and the dilution method to be described below. In both cases the mass fraction of material in the gel is measured. If some of the material never enters the gel or is lost from the gel during fixation and staining it will be ignored in the measurement. This possibility cannot be ruled out by calculating the counts recovered from the gels because the efficiency of counting gel slices is lower than for liquid samples. However, in all experiments the spacer gel was sliced and counted and was found to contain much less than 1% of the total counts. In a control experiment the anode end of the gel tube was capped with dialysis tubing to trap any protein migrating toward the anode. After electrophoresis there were no detectable counts in this anode trap. We must conclude then that at least 99% of the starting sample runs into the gel. If the remaining <1% is all one protein, then its stoichiometry measurement might be distorted, but the overall stoichiometry pattern could not be effected.

Similarly, aggregation of some proteins would selectively remove material from components that are nominally identified on the gels as single components. We have previously identified such artifacts and we can control their appearance (Hardy et al., 1969). Such aggregates would appear in well-defined positions in the gels but they were not observed in these experiments. Therefore, we do not believe that such an artifact distorts the data.

In order to be more confident that the stoichiometric data are correct, we have employed a second independent procedure that also has the advantage of permitting measurements with proteins that migrate close together on polyacrylamide gels.

Isotope Dilution Experiments. The great advantage of the isotope dilution procedure (see Methods) is that even a small aliquot of the repurified, doubly labeled protein will suffice for an accurate measure of the isotope ratios of the protein. Therefore, when two proteins migrate close to one another on polyacrylamide gels, a small region of gel in which they do not overlap is sufficient for an unambiguous estimation of the mass fraction of the pure protein.

The dilution measurements for 17 of the 30S proteins are given in Table I. (Since 30S-3 and 30S-6 are not pure on acrylamide gels, isotope dilution experiments were performed with cellulose phosphate chromatography for these proteins.) The proteins have been grouped as before into three classes; unit proteins present in amounts greater than 0.8 copy/particle; marginal proteins present in amounts between 0.7 and 0.8 copy per particle; and fractional proteins present in amounts less than 0.6 copy/particle. Again we have the result that some proteins are present in amounts which are considerably less than 1 copy/particle. The measurements were made on two completely independent protein preparations from two different batches of radioactive cells. The reproducibility appears to be within ±10%.

The isotope dilution method requires the simultaneous measurement of ^3H and ^{14}C label. Therefore, this procedure is susceptible to errors due to selective self-absorption of one isotope as well as differential quenching effects. However, control experiments with doubly labeled samples embedded in gel suggest that the Automatic Quench Correction system is adequate to compensate for these effects. Furthermore, the recovery method is free of these difficulties. Since the agreement between the two sets of data in Table I is good, it is unlikely that serious errors were incurred through the double-label measurements.

Table I does not provide stoichiometric data for all the 30S proteins. Since 30S-15 and 30S-15b are not resolved on acrylamide gels and are difficult to purify on phosphocellulose, only the mass fraction of their sum was measured. There are only 22,500 daltons of 30S-15 + 30S-15b, and since they both have a molecular weight of about 19,000, it is clear that at least one of them is a fractional protein. If one of them is in fact a unit protein, it might be tempting to assume that the unit is 30S-15, since that is the protein responsible for the streptomycin phenotype of the 30S subunit (Ozaki *et al.*, 1969; Birge and Kurland, 1969). Protein 30S-15b seems to be recovered in very small amounts from the phosphocellulose columns; therefore, we suspect that it is a fractional protein.

Unfortunately, there are no direct stoichiometric data for 30S-2 and 30S-14. Neither of these proteins is obtained as pure components on polyacrylamide gels, and their mass fractions must be measured by isotope dilution on phosphocellulose columns. In both cases attempts to purify the final doubly labeled component failed for trivial technical reasons. However, it is possible to obtain an indirect estimate for the mass fraction of 30S-2. This protein has the same electrophoretic mobility as 30S-5 + 9 on soft gels. Recovery measurements indicate that 30S-5 + 9 + 30S-2 represents 40,000 daltons/30S particle. Since 30S-5 + 9 contributes *ca.* 25,000 daltons, 30S-2 must have *ca.* 15,000 daltons or 0.83 copy/30S particle. Also, recovery measurements on hard gels show that there are 49,000 daltons of 30S-2 + 30S-3 + 30S-2a. The contribution of 30S-3 + 30S-2a is 35,000 daltons, therefore 30S-2 must account for 14,000 daltons or 0.78 copy/30S particle. Thus, we have classified 30S-2 as a marginal protein.

The isotope dilution data for 30S-8 indicate that it is a unit protein; in contrast, the recovery data for this protein

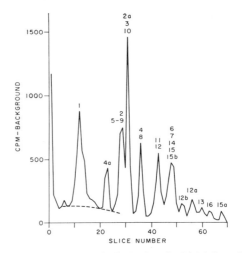

FIGURE 3: The electrophoretic distribution of ^{14}C-labeled protein obtained by fractionating the proteins of crude 30S ribosomal subunits on soft polyacrylamide gels. The dashed line indicates the relatively acidic, contaminating protein not present in 30S protein extracted from purified 30S particles (see Figure 2a).

suggest it is a marginal protein. Since the isotope dilution procedure provides a more accurate measurement of the mass fraction, we are inclined to classify 30S-8 as a unit protein.

Crude Ribosomes. Assuming that the stoichiometry measurements are reliable, it is still possible that the existence of the fractional class of proteins is an artifact resulting from the ribosome purification procedure. Perhaps the fractional proteins are really unit proteins that have been partially removed from the 30S particle during the extensive treatment with $(NH_4)_2SO_4$. To investigate this possibility it is necessary to do stoichiometry determinations on ribosomes that have not experienced any salt washings.

Stoichiometric measurements of proteins from crude (unwashed) ribosomes are presented in Table I. The tabulation includes data obtained by both the recovery and dilution methods. It is evident that crude ribosomes have the same pattern of fractional and unit proteins as purified ribosomes do, with the exception of proteins 30S-1 and 4a.

Hardy *et al.* (1969) have shown that crude ribosomes contain extra acidic components that migrate in the region of 30S-1 and 4a. It is, therefore, entirely possible that the extra material present in crude ribosomes is bound, supernatant protein having the same electrophoretic mobility as 30S-1 and 4a. A typical soft gel profile of ^{14}C-labeled protein from crude 30S particles is shown in Figure 3. If this profile is compared to that of protein from purified particles (Figure 2), the presence of extra material in the upper one-third of the gel is apparent. One can attempt to correct for the presence of this putative contamination of 30S-4a by subtracting the extrapolated base line indicated by the dashed line in Figure 3. This correction yields a value of 0.45 copy of 30S-4a/particle, which is similar to the value from purified particles. This observation is consistent with the hypothesis that the stoichiometry of 30S-4a is unchanged by salt washing, however, the question is by no means settled. The situation for 30S-1 is even more uncertain, since it has been observed that the amount of 30S-1 varies considerably from one preparation to another (Hardy *et al.*, 1969). Once again, it is difficult to decide whether we are dealing with the behavior of the protein itself or of a contaminant that has the same electrophoretic mobility. Therefore, we are obliged to leave a final decision on the status of 30S-1 and 4a until more data are available.

Proteins 30S-4, 5 + 9, and 12a were classified as marginal proteins on the basis of the data obtained with purified particles. However, when crude ribosomes are studied, these same proteins tend to have mass fractions that are consistent with their classification as unit proteins. This could be a consequence of contaminants which migrate together with these proteins on gels. However, it seems more likely that these proteins were partially stripped from the ribosome during purification with salt. Therefore, we tentatively suggest that the marginal proteins may in fact be unit proteins.

Discussion

We have previously reported that the 30S ribosomal subunit contains a minimum of 19 proteins with an aggregate mass of 410,000 daltons (Hardy et al., 1969; Craven et al., 1969). At that time the status of 30S-7 and of 30S-15b was uncertain because they were recovered in very small amounts; consequently, these two proteins were omitted from the catalog of 30S proteins. However, the present data indicate that 30S-7 is a unit protein; therefore, we now wish to include it in the catalog. Protein 30S-15b has been purified several times since our earlier report; therefore, we also wish to include this protein in the catalog. The addition of these 2 proteins brings the complement of 30S proteins to 21 with an aggregate mass of 440,000 daltons.

The 30S subunits purified by our procedure contain approximately 260,000 daltons of protein. Since the sum of the molecular weights of the 30S proteins is almost twice this amount, we concluded that every 30S particle could not contain 1 copy of each nominal 30S protein (Hardy et al., 1969; Craven et al., 1969). Therefore, it is not surprising that the stoichiometric measurements establish the existence of a class of fractional proteins. On the contrary, this is a necessary consequence of the manner in which the molecular weight data enter into the stoichiometry calculations. The purpose of the present experiments is to reveal which proteins, if any, are *not* fractionals. It must be understood that the present results do not, in themselves, constitute confirmation of our earlier molecular weight measurements. They merely provide an initial insight into the pattern of the heterogeneity.

The validity of the stoichiometric classifications depends, in part, on the accuracy of the molecular weight estimates for the individual proteins. Our confidence in the accuracy of these estimates is based on the reproducibility of the measurements, the agreement between chemical and physical molecular weight estimates (Craven et al., 1969) as well as the agreement of our data with the recent measurements of Traut et al. (1969). In order for the putative class of fractional proteins to be a consequence of errors in the molecular weight measurements, the requisite errors would have to be of the order of 200–300%. This is quite a bit larger than the estimates of ±15% for the uncertainty in the molecular weight measurements of individual proteins, and we feel that ±15% is a conservative estimate of error.

Another potential source of error in the present calculations is the value used for the average mass of protein per 30S subunit. We obtain an average mass of 260,000 daltons of protein/ purified 30S subunit (Hardy et al., 1969). This value is based on the molecular weight of 16S RNA and the chemical composition of the particles. Ribosomes which are obtained without exposure to high concentrations of salt contain much more protein, but this extra material is mostly contaminating supernatant protein (Hardy and Kurland, 1966; Kurland, 1966; Hardy et al., 1969).

Ribosomes obtained by a washing procedure which is not as extreme as ours also seem to contain somewhat more protein: approximately 350,000 daltons of protein/30S particle, which is based on the mass of these particles (Hill et al., 1969) and the molecular weight of 16S RNA (Kurland, 1960; Midgley, 1965; Stanley and Bock, 1965). Furthermore, the protein/RNA ratios of the 30S and 50S subunits obtained from these ribosomes are different (Hill et al., 1969). However, the 30S and 50S subunits prepared by our procedure are characterized by indistinguishable protein/RNA ratios. Thus, the differences in the amounts of protein per 30S subunit reported by different laboratories seem to be real. These differences are probably not due to errors in the chemical estimates but are more likely to be due to the different ways that the ribosomes were prepared.

The previous considerations strengthen our confidence in the conclusion that purified 30S particles are structurally heterogeneous. However, it is still possible that this heterogeneity is an artifact introduced during the purification of the ribosomes. For example, several proteins that we classify as marginal ones in purified ribosomes (30S-4, 5 + 9, and 12a) appear to have mass fractions in crude ribosomes which qualify them as unit proteins. Similarly, two fractional proteins (30S-1 and 4a) have much larger apparent mass fractions in crude ribosomes than in purified ones. Thus, the selective loss of some proteins during the purification of ribosomes might account for the apparent heterogeneity of the 30S subunits. However, the presence of contaminating supernatant proteins on the crude ribosomes (see Figure 3) could account for the larger apparent mass fractions of 30S-1 and 4a in crude ribosomes compared to purified ones. Furthermore, the mass fractions of four other fractional proteins (30S-12b, 13, 15a, and 16) are the same in crude and purified ribosomes. These observations suggest that the heterogeneity is real.

If the difference between the aggregate mass of all the 30S proteins (440,000 daltons) and the average mass of protein per purified 30S particle (260,000 daltons) were due to the partial degradation of the ribosomes during isolation, roughly half as much 30S protein as is present in the recovered ribosomes should be found in the supernatant fractions or adsorbed to the bacterial wall membrane fraction. We have searched for free ribosomal proteins; our failure to detect such material suggested that it must be present in amounts less than 10% of the mass of ribosomal protein in the bacteria. A more sensitive immunochemical assay for free ribosomal protein has detected such material in amounts corresponding to approximately 1% of the mass of ribosomal protein in the bacteria (G. Stöffler, personal communication). Therefore, it seems unlikely that the heterogeneity that we have observed is a consequence of a partial degradation of ribosomes during their isolation from the bacteria.

Our confidence in this conclusion is considerably strengthened by the data of Nomura et al. (1969). They have studied the contribution of most of the 30S proteins to the activity and structural integrity of the 30S particle by examining reconstituted particles missing each of the proteins in turn. They have found that the proteins fall into four classes: (1) those that are required for the formation of a particle with a sedimentation coefficient close to 30S; (2) those that are not required for particle formation but are necessary for *in vitro* activity; (3) those that stimulate *in vitro* activity; and (4) those that have no appreciable effect with respect to the activities assayed *in vitro*. Table II summarizes the findings of Nomura *et al.* (1969) and correlates them with the stoichiometric classification of the proteins. The correlation of unit proteins with those required for ribosome assembly is striking, particularly since the two classifications were obtained completely independently. Since none of the proteins which are required for the assembly of the 30S particle have been identified as fractional proteins, the absence of certain proteins from some ribosomes presents no problem for the assembly of these ribosomes. However, we must assume that proteins which are present on the average once per ribosome are indeed present once in every ribosome rather than twice in half of the ribosomes, etc. This assumption is attractive because the alternative seems so difficult to reconcile with the observation that no 30S particle can be reconstituted in the absence of any single unit protein that has been implicated in the assembly process (Nomura *et al.*, 1969).

If there were one copy for each unit protein as well as each suspected unit protein (marginal protein, see Table II) per 30S particle, this would correspond to roughly 230,000 daltons of protein. Since there are approximately 260,000 daltons of protein/30S particle, this leaves room for one or more fractional proteins for each ribosome. Now, the problem is to determine the functional significance of the fractional proteins.

We have considered two extreme views of the structural heterogeneity of the 30S subunits (Kurland *et al.*, 1969). One of these mutually compatible models, the static model, employs the fractional proteins to functionally differentiate classes of ribosomes. For example, a preference for initiating protein synthesis at the terminal cistron of a polycistronic messenger as opposed to a preference to initiate protein synthesis at more distal cistrons would be one of many conceivable functional specializations of the ribosome. The protein complement of the different classes of ribosomes is imagined to be fixed. According to this model the omission of a class-specific fractional protein from an *in vitro* reconstruction mixture should result in the production of a mixture of ribosomes that are only partially active. Only the classes of ribosomes that normally do not contain the omitted protein would be expected to be functional. Fractional proteins 30S-4a, 15a, and 16 might qualify as potential class-specific proteins. These three proteins are dispensable but are needed for the recovery of maximum activity from the reconstituted ribosomes (see Table II). It has been suggested that such proteins are required for the function of subclasses of ribosomes (Kurland, 1970).

The alternative that we have considered is the steady-state hypothesis. Here we postulate a functional cycle in which the complement of fractional proteins on a ribosome changes as a given ribosome proceeds through the different operational modes. One example of such a cycle would entail different sets of fractional proteins associated with the ribosomes during chain initiation, propagation, termination, and a rest mode. Some of the proteins which are involved in such a cycle might be required for protein synthesis by all ribosomes. Thus, when a ribosome cannot proceed through one of the early stages of protein synthesis it might jam. Fractional proteins 30S-11, 12b, and 13 superficially qualify as such functionally required proteins (see Table II).

We have observed *in vitro* an exchange of free ribosomal proteins with proteins of the intact ribosome (Kurland *et al.*, 1969). This exchange of exogenous and endogenous proteins is associated with an enhanced synthetic activity for the ribo-

TABLE II: Classification of 30S Ribosomal Proteins.[a]

30S Protein	Nomura's Code	Functional Class	Stoichiometric Class
2a	P4b	Required for assembly	U
6	P9*	Required for assembly	U
7	P9*	Required for assembly	U
8	P5	Required for assembly	U
10	P4a	Required for assembly	U
12	P8	Required for assembly	U
4	P6	Required for function	(U)
5 + 9	P3	Required for function	(U)
11	P7	Required for function	F
12b	P11	Required for function	F
13	P13	Required for function	F
15	P10	Required for function	(U)
3	P4	Dispensable	(U)
4a	P2	Dispensable	(F)
12a	P12	Dispensable	(U)
14	P10a	Dispensable	
15a	P15	Dispensable	F
16	P14	Dispensable	F
1	P1	Not known	(F)
2	P3a	Not known	(U)
15b		Not known	(F)

[a] The stoichiometric and functional classifications of the 30S proteins are based on the data in Table I as well as that of Nomura *et al.* (1969). Here, U designates a unit protein, F designates a fractional protein. (U) and (F) designate the tentative stoichiometric classification of proteins that are ambiguous for one reason or another (see text).

somes. Although such a reaction does not provide unambiguous support for the steady-state model, it does suggest that the structure of the ribosome may be a dynamic one. Therefore, the steady-state model seems worthy of further study.

Twelve of the twenty-one 30S proteins have been tentatively identified as unit proteins, the remaining 30S proteins may all be fractional proteins. There is no evidence either in the present experiments or those of Sypherd *et al.* (1969) and Traut *et al.* (1969), which suggests that any 30S protein is represented more than once per ribosome. Furthermore, the *in vitro* assembly data of Nomura *et al.* (1969) suggest that each unit protein that is required for assembly of ribosomes is present in every ribosome. Therefore, we conclude that there is at most one copy of any given 30S protein per ribosome. This would mean that the proteins of the 30S subunit are arranged dissymetrically in the functional ribosome. Consequently, models of protein synthesis which require symetrical arrangements of proteins in the 30S subunit are probably not valid.

References

Birge, E. A., Craven, G. R., Hardy, S. J. S., Kurland, C. G., and Voynow, P. (1969), *Science 164*, 1285.
Birge, E. A., and Kurland, C. G. (1969), *Science 166*, 1282.
Bollen, A., Davies, J., Ozaki, M., and Mizushima, S. (1969), *Science 165*, 85.

Craven, G. R., Voynow, P., Hardy, S. J. S., and Kurland, C. G. (1969), *Biochemistry 8*, 2906.

Fogel, S., and Sypherd, P. S. (1968), *Proc. Nat. Acad. Sci. U. S. 59*, 1329.

Hardy, S. J. S., Kurland, C. G., Voynow, P., and Mora, G. (1969), *Biochemistry 8*, 2897.

Hill, W. E., Rossetti, G. P., and Van Holde, K. E. (1969), *J. Mol. Biol. 44*, 263.

Hiller, A., Plazin, J., and Slyke, D. D. (1948), *J. Biol. Chem. 176*, 1401.

Kaltschmidt, E., Dzionara, M., Donner, D., and Wittmann, H. G. (1967), *Mol. Gen. Genet. 100*, 364.

Kurland, C. G. (1960), *J. Mol. Biol. 2*, 83.

Kurland, C. G. (1966), *J. Mol. Biol. 18*, 90.

Kurland, C. G. (1970), *Science* (in press).

Kurland, C. G., Voynow, P., Hardy, S. J. S., Randall, L., and Lutter, L. (969), *Cold Spring Harbor Symp. Quant. Biol. 34*, 17.

Lowry, O. H., Rosebrough, N. J., Farr, A. J., and Randal, R. J. (1951), *J. Biol. Chem. 193*, 265.

McCarthy, B. J., and Britten, R. J. (1962), *Biophys. J. 2*, 57.

McCarthy, B. J., Britten, R. J., and Roberts, R. B. (1962), *Biophys. J. 2*, 57.

Midgley, J. E. M. (1965), *Biophys. Biochim. Acta 108*, 340.

Moore, P. P., Traut, R. R., Noller, H., Pearson, P., and Delius, H. (1968), *J. Mol. Biol. 31*, 441.

Nomura, M., Mizushima, S., Ozaki, M., Traut, P., and Lowry, C. V. (1969), *Cold Spring Harbor Symp. Quant. Biol. 34*, 49.

Ozaki, M., Mizushima, S., and Nomura, M. (1969), *Nature (London) 222*, 333.

Roberts, R. B., Abelson, P. H., Cowie, D. B., Bolton, H. T., and Britten, R. J. (1955), *Carnegie Inst. Washington Pap.*, 607.

Stanley, W. M., and Bock, R. M. (1965), *Biochemistry 4*, 1302.

Sypherd, P. S. (1969), *J. Bacteriol. 99*, 379.

Sypherd, P. S., O'Neil, D. M., and Taylor, M. M. (1969), *Cold Spring Harbor Symp. Quant. Biol. 34*, 77.

Traub, P., and Nomura, M. (1968), *Proc. Nat. Acad. Sci. U. S. 59*, 777.

Traut, R. R., Delius, H., Ahmed-Zadeh, C., Bickle, T. A., Pearson, P., and Tissieres, A. (1969), *Cold Spring Harbor Symp. Quant. Biol. 34*, 25.

20

Reprinted from *Proc. Natl. Acad. Sci.,* 59(3), 777–784 (1968)

STRUCTURE AND FUNCTION OF E. COLI RIBOSOMES, V. RECONSTITUTION OF FUNCTIONALLY ACTIVE 30S RIBOSOMAL PARTICLES FROM RNA AND PROTEINS*

By P. Traub and M. Nomura

LABORATORY OF GENETICS, UNIVERSITY OF WISCONSIN, MADISON

Communicated by James F. Crow, January 16, 1968

Ribosomes are structurally complex cell organelles consisting of ribosomal RNA and a number of different protein molecules. The complexity of these particles raises two major questions: What is the relationship between structure and function of the ribosome, and what is the mechanism of assembly of the ribosome? It is obvious that the study of these two questions would be greatly aided if a method could be found for assembling active ribosomes from their dissociated macromolecular constituents.

Partial dissociation of ribosomes into inactive smaller ribonucleoprotein particles ("core" particles) and proteins ("split proteins") and the reconstitution of functionally active ribosomal particles from these components were reported previously.[1-3] This system has been used in this laboratory to study the two problems mentioned above. Split proteins were fractionated and various artificial ribosome derivatives deficient in specific protein components were prepared.[4-6] Analysis of the activity of these particles has revealed the functional significance of each of the purified ribosomal split proteins. Some proteins have been shown to be essential for ribosomal functions, others to be dispensable but to have stimulatory activity in *in vitro* polypeptide synthesis. It has also been shown that assembly of ribosomes from the split proteins and the core particles is spontaneous, supporting the hypothesis of spontaneous self-assembly of ribosomes from subribosomal components.[4, 7]

In extending this type of approach further, we have now been able to dissociate 23S core particles (derived from 30S ribosomal particles) into free ribosomal RNA (16S RNA) and proteins ("core proteins" from 30S, designated as CP30), and to reconstitute functionally active "30S" particles from free 16S RNA, CP30, and split proteins from 30S (SP30). It has been shown that neither yeast 16S ribosomal RNA nor rat liver 18S ribosomal RNA can replace *coli* 16S RNA in the reconstitution of "30S" particles. Degraded *coli* 16S RNA or degraded *coli* 23S ribosomal RNA are also ineffective. It has also been shown that core proteins from 50S ribosomal particles (CP50) cannot replace CP30 proteins. Thus, the specificity of the ribosomal RNA and the core proteins has been demonstrated.

The efficient reconstitution of functional 30S ribosomal particles in the present system indicates that the entire information for the correct assembly of the ribosomal particles is contained in the structure of their molecular components, and not in some other nonribosomal factors.

Materials and Methods.—The following buffer-salt mixtures were used: TMA I (10^- M Tris-HCl, pH 7.8, 10^{-2} M MgCl$_2$, 3×10^{-2} M NH$_4$Cl, 6×10^{-3} M mercaptoethanol):

TMA II (same as TMA I, but concentration of $MgCl_2$ is 3×10^{-4} M); PMK I (5×10^{-3} M H_3PO_4 neutralized to pH 7.4 with KOH, 1.7×10^{-2} M $MgCl_2$, 3×10^{-3} M $CaCl_2$ 1 M KCl, 6×10^{-3} M mercaptoethanol), PMK II and III (same as PMK I, but concentration of KCl is 0.5 M and 0.25 M, respectively), and SSC (0.15 M NaCl, 0.015 M Na citrate, pH 7.0).

Ribosomes were prepared from *E. coli* strain Q13. The methods of preparation of ribosomes and of polyacrylamide gel electrophoresis of proteins have been described previously.[5] Both 23*S* core particles and SP30 proteins were prepared by centrifugation of 30*S* particles in CsCl as described before.[1, 5, 8] The SP30 proteins were recovered from the top of the gradient, and 23*S* core particles from the band near the middle of the gradient.

16*S* ribosomal RNA was prepared as follows: 23*S* core particles recovered from CsCl gradients were first dialyzed against TMA I and then PMK II. The dialyzed 23*S* particles were shaken at 5°C for 10 min with an equal volume of phenol saturated with PMK II. The water phase containing RNA was taken and the phenol treatment was repeated 5 additional times. The RNA was then dialyzed against PMK III, with frequent changes of the outer solution, until no more phenol was detected in the outer solution. In some experiments, RNA was precipitated with alcohol and then dialyzed against PMK III. The same results were obtained with such RNA preparations. 23*S* ribosomal RNA was prepared in a similar way from 40*S* core particles.

CP30 proteins were prepared as follows: 23*S* core particles (in CsCl solution) were mixed with an equal volume of 8 M urea–4 M LiCl solution and allowed to stand in an ice bath for 24–48 hr. Precipitated RNA was removed by centrifugation (18,000 rpm for 10 min), and the supernatant was dialyzed against PMK II. SP30 proteins recovered from the top of CsCl gradient were dialyzed first against a buffer (6 M urea, 2 M LiCl, 10^{-2} M Tris-HCl, pH 7.5, 6×10^{-3} M mercaptoethanol) for about 8 hr and then against PMK I. Both CP50 and total ribosomal proteins from 30*S* were prepared from 40*S* core particles and 30*S* particles, respectively, in a way similar to that for CP30, and were dialyzed against PMK I.

In typical experiments, reconstitution of "30*S*" particles from 16*S* RNA and proteins was done as follows: 16*S* RNA (75 OD[260] units) was diluted to 15 ml with PMK III and heated to 37°C. Two ml of CP30 (79 OD[260] equivalents in PMK II) was added dropwise to RNA solution while the RNA solution was constantly stirred. The mixing took about 6 min and the resultant mixture was kept at 37°C for an additional 4-min period. The mixture was then left in the cold to cool slowly to 4°C. About 1 hr later, 0.6 ml of SP30 (90 OD[260] equivalents in PMK I) was added to the mixture with stirring. The final mixture was dialyzed against PMK I for 9 hr in the cold, and then against TMA I for 8 hr. The preparation was centrifuged at 30,000 rpm for 12 hr in a Spinco no. 30 rotor and the pellets were suspended in TMA I. Insoluble materials were removed by centrifugation (18,000 rpm, 5 min). The resultant supernatant was analyzed for OD[260] to calculate recovery of added RNA, and its concentration was suitably adjusted (usually 50 OD[260] units/ml). The control "native" 30*S* particles were also dialyzed against PMK I, and then against TMA I in the same way.

Results.—(1) *Physical reconstitution of "30S" particles from 16S ribosomal RNA and ribosomal proteins:* RNA (16*S* RNA) and two protein fractions (CP30 and SP30) were prepared from 30*S* particles as described in *Materials and Methods*. The SP30 was previously shown to be free from RNA. The amount of residual RNA contained in the CP30 preparation was analyzed by the orcinol reaction. The RNA content was less than 0.5 per cent of an equivalent amount of 23*S* core particles. The amount of residual protein in the 16*S* RNA preparation was also analyzed using the Lowry method. It was less than 2 per cent of that in an equivalent amount of 23*S* core particles, or less than 0.6 per cent of the amount of RNA. The sedimentation pattern of the 16*S* RNA preparation was analyzed on a sucrose gradient together with a reference C^{14}-RNA preparation, and the result is shown in Figure 1*a*.

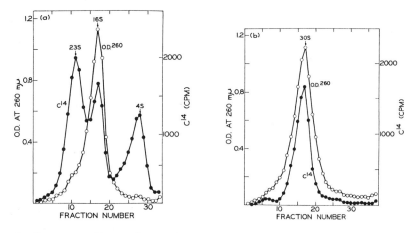

FIG. 1.—Sucrose gradient sedimentation pattern of 16S RNA used for the reconstitution (a) and particles recovered from the reconstitution mixture (b).

In (a), 16S RNA preparation (about 12 OD[260] units) was mixed with a small amount of reference C^{14}-labeled RNA preparation (containing about 1 OD[260] unit of a mixture of 23S, 16S, and 4S RNA) and dialyzed overnight against SSC containing EDTA (0.01 M). The sample was then put on a sucrose gradient (5–20% in SSC containing 0.01 M EDTA) and centrifuged at 39,000 rpm for 5 hr in the cold.

In (b), the reconstituted "30S" particles (about 15 OD[260] units) were mixed with a small amount of standard C^{14}-labeled 30S particles (containing about 0.5 OD[260] unit) and dialyzed overnight against TMA II buffer. The sample was then put on a sucrose gradient (5–20% in TMA II buffer) and centrifuged at 39,000 rpm for 3 hr in the cold. Fractions were collected from the bottom of the tubes, suitably diluted, and both OD[260] and acid-precipitable radioactive RNA were analyzed.

The 16S RNA was first mixed with CP30 proteins at 37°C and then with SP30 proteins in the cold as described in the *Materials and Methods*. After dialysis, the particles were sedimented by high-speed centrifugation and the pellets were resuspended in the TMA I buffer. Any insoluble materials were removed by low-speed centrifugation. The resultant preparation contained 70–95 per cent of the amount of the original RNA added. An aliquot of this preparation was mixed with reference C^{14}-30S particles and examined by zone centrifugation on a sucrose gradient. As shown in Figure 1b, the original free 16S RNA was not observed, and the RNA now sedimented in a single peak, which coincided with the peak of the reference C^{14}-30S particles. Although the spreading of the reconstituted "30S" peak is slightly greater than that of the reference peak, it is clear that the addition of both CP30 and SP30 proteins to 16S RNA converted most of the RNA to particles sedimenting at 30S.

The protein composition of the reconstituted "30S" particles was analyzed by polyacrylamide gel electrophoresis. Figure 2 shows that the reconstituted particles do contain all of the 30S proteins and that the relative amounts of these proteins are approximately the same as those found in native 30S particles.

(2) *Biological activity of the reconstituted "30S" particles:* The "30S" particles reconstituted from 16S RNA, CP30, and SP30 proteins were analyzed for their activity in polypeptide synthesis. As shown in Table 1, the particles showed activity nearly identical to that of the native 30S particles in poly U-

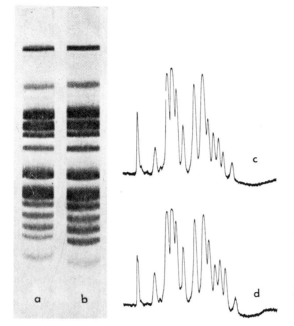

FIG. 2.—Polyacrylamide gel electrophoresis of the proteins of the native 30S particles (a) and the reconstituted "30S" particles (b). The tracings of negative photographs of these gels are shown in (c) (the native 30S particles) and (d) (reconstituted "30S" particles).

directed polyphenylalanine synthesis, phage f2 RNA-directed polypeptide synthesis, and poly U-directed Phe-tRNA binding. Omission of CP30 proteins from the reconstitution mixture resulted in preparations which are completely inactive in these reactions (Table 1, expt. 1). Omission of SP30 proteins from the reconstitution mixture resulted in preparations which are completely inactive in both poly U-directed polyphenylalanine synthesis and poly U-directed Phe-tRNA binding, but are weakly active in the assay of f2 RNA-directed polypeptide synthesis. This weak activity can be explained by the presence of small amounts of SP30 proteins in the "initiation factor" preparations which were used in the assay of f2 RNA-directed polypeptide synthesis.[9]

(3) *Examination of conditions necessary for effective reconstitution:* Some factors which might affect efficiency of reconstitution of "30S" particles were examined. It was first discovered that the temperature at which CP30 proteins were mixed with 16S RNA is important. When CP30 proteins were mixed with 16S RNA at about 0°C instead of 37°C and, about 15 minutes later (2 hr later in some experiments), SP30 proteins were added to the mixture, the activity of the final preparation was not high (Table 1, expt. 3). Recovery of added RNA in the final soluble particle preparation was also low (40–70% compared to 70–95% in the reconstitution with the mixing at 37°C). A considerable fraction of the added RNA was lost as insoluble aggregates in this case. It is probable that warming of the mixture of RNA and CP30 proteins at 37°C accelerates the rate of rearrangement of inactive nonspecific aggregate structures into a functionally active structure which is presumably the thermodynamically most stable one, and thus results in highly efficient reconstitution.

It was originally thought that additions of proteins to RNA should be done in the order: CP30 then SP30. However, reversing the order, that is, adding

TABLE 1. *Functional capacities of particles reconstituted from 16S RNA, 30S split proteins (SP30), and 30S core proteins (CP30).*

Expt. no.	Conditions for Reconstitution Components used	Temp.* (°C)	Poly U-Directed Phe Incorporation (cpm)	(%)	f2 RNA-Directed Val Incorporation (cpm)	(%)	Poly U-Directed Phe-tRNA Binding (cpm)	(%)
1	16S, CP30, SP30	37	13,168	113	1740	97	719	82
	16S, CP30	37	63	0.6	290	16	<5	<0.6
	16S, SP30	37	0 (−58)	<0.3	7	0.4	<5	<0.6
	(Native 30S)	—	(11,697)	(100)	(1,802)	(100)	(877)	(100)
2	16S, CP30, SP30	37	10,526	98				
	16S, total 30S proteins	37	10,158	94				
	(Native 30S)	—	(10,777)	(100)				
3	16S, CP30, SP30	37	12,665	107				
	16S, CP30, SP30	0	2106	18				
	(Native 30S)	—	(11,864)	(100)				

Equivalent amounts of 16S RNA, CP30 proteins, and SP30 proteins were mixed as indicated in the table, and reconstituted particles were prepared as described in the *Materials and Methods* section. Where indicated, "total 30S ribosomal proteins" obtained directly from 30S particles were used instead of CP30 plus SP30. Particles recovered from the reconstitution mixture were resuspended in TMA I buffer and aliquots containing 1 OD²⁶⁰ unit were analyzed for their activities in poly U-directed C¹⁴-phenylalanine incorporation, and f2 RNA-directed C¹⁴-valine incorporation (both assayed in the presence of 2 OD²⁶⁰ units of native 50S particles), and for poly U-directed C¹⁴-Phe-tRNA binding (assayed in the absence of 50S particles). C¹⁴-phenylalanine (10 μc/μmole), C¹⁴-valine (50 μc/μmole), and C¹⁴-Phe-tRNA (177 mμc of phenylalanine with the sp. act. of 366 μc/μmole attached to 1 mg of *E. coli* B tRNA mixture; obtained from New England Nuclear Corp.) were used for these assays. The methods for these assays have been described previously.[1, 5, 9] The values given in the table are corrected for the background incorporation values in the absence of "30S particles" (222 cpm for poly U system and 125 cpm for f2 RNA system in expt. 1, 237 cpm in expt. 2, and 272 cpm in expt. 3) or the background binding in the absence of poly U (Phe-tRNA binding in expt. 1).

* Temperature at which CP30 proteins were mixed with 16S RNA solution.

SP30 first and then CP30, did not cause any significant decrease in efficiency of the reconstitution as long as the temperature at the time of mixing of proteins was kept at 37°C. In fact, "total 30S proteins" prepared directly from 30S particles, without fractionation into SP30 and CP30, could be used in reconstitution of active 30S particles from free 16S RNA (Table 1, expt. 2).

(4) *Specificity of the reconstitution:* The reconstitution of "30S" particles from 16S RNA requires the presence of both CP30 proteins and SP30 proteins. It was shown previously that functionally active "30S" particles can be reconstituted from 23S core particles and SP30 proteins, and that split proteins from 50S (SP50) cannot replace SP30 proteins.[1, 2] In the present reconstitution system, it has been shown that core proteins from 50S (CP50) cannot replace CP30 proteins (Table 2, expt. 1). Thus CP30 proteins are functionally different from CP50 proteins.

Specificity of RNA was also examined (Table 2, expt. 2). Degradation of 16S RNA either by heating or by brief alkali treatment into smaller pieces (average sedimentation coefficient 6S) completely abolished the reconstitution activity of 16S RNA.

Next, "16S RNA" was prepared from *E. coli* 23S ribosomal RNA. It has been known that 23S RNA is converted to RNA with a sedimentation coefficient 16S under appropriate conditions, and a suggestion has been made that 23S

TABLE 2. *Specificity of the reconstitution of "30S" ribosomal particles.*

Expt. no.	Components Used for Reconstitution RNA	Proteins	Per cent recovery	Sedimentation behavior	"Particles" Recovered Poly U-Directed Phe Incorporation Activity* (cpm)	(%)
1	16S	CP30, SP30	92		10,526	98
	16S	CP50, SP30	26		0 (−41)	<0.3
	(native 30S)		—		(10,777)	(100)
2	16S	SP30, CP30	96	30S (cf. Fig. 1)	9484	92
	"Heated 16S" (ca. 6S)†	SP30, CP30	17	ca. 10S	4	<0.3
	"Degraded 23S" (ca. 16S)‡	SP30, CP30	12	ca. 16S (10–30S, heterogeneous)	0 (−33)	<0.3
	Yeast 16S§	SP30, CP30	16	ca. 17S (heterogeneous, see Fig. 3)	0 (−50)	<0.3
	Rat liver 18S§	SP30, CP30	<1	—		
	(native 30S)				(10,335)	(100)

In expt. 1, 50S core proteins (CP50) were used instead of CP30 and compared with the control reconstitution. The amount of CP50 used was the same as that of CP30. In expt. 2, various ribosomal RNA preparations were substituted for 16S *E. coli* RNA. The standard reconstitution method was used, and the particles were recovered as described in the *Materials and Methods*. The particle preparation was analyzed for RNA content by measuring OD^{260}. Per cent recovery of the particles was calculated from these values and from the amount (50 OD^{260} units in this experiment) of RNA originally used in the reconstitution mixture. Aliquots (1 OD^{260} unit) were analyzed for their poly U-directed phenylalanine incorporation activity. As indicated in the table, no particles containing RNA were recovered when rat liver 18S was used; and hence, the activity assay was not performed. Another aliquot of various particle preparations was mixed with a standard C^{14}-labeled 30S particle preparation and centrifuged on a sucrose gradient. The sedimentation pattern of the particles was compared with that of standard 30S particles and a rough S value of the particles was calculated.

* The values given in the table are corrected for the background incorporation values obtained in the absence of "30S particles" (237 cpm in expt. 1 and 261 cpm in expt. 2).

† *E. coli* 16S RNA was heated at 85°C for 5 min. Sedimentation analysis on a sucrose gradient showed that the 16S RNA was degraded and sedimented at about 6S. It is possible that the original 16S RNA, although it was fully active in the reconstitution, had breaks in the chain, but that at low temperature secondary forces held chain fragments together in the 16S form.

‡ *E. coli* 23S RNA was degraded to "16S" RNA according to the procedure described by Midgley.[10] The 23S RNA solution (in SSC) was mixed with an equal volume of 0.1 M glycine buffer (pH 10) and incubated for 2.5 hr at 37°C. Time of incubation was determined by preliminary experiments and was just sufficient to cause a complete conversion of 23S RNA into more slowly sedimenting (average about 16S) RNA. After the incubation, the reaction mixture was adjusted to pH 7.4 with acetic acid and the RNA was recovered by alcohol precipitation, resuspended in TMA II, and dialyzed against PMK III.

§ Yeast 16S RNA was prepared as follows: frozen yeast cells (*Saccharomyces lactis*) were a gift from Dr. R. M. Bock. Cells were ground with alumina and extracts were prepared in TMA I. Ribosomes were sedimented by centrifugation at 50,000 rpm for 1.5 hr and resuspended in TMA II. Bentonite (0.1%) and duponol (0.5%) were added and the mixture was incubated for 5 min at 37°C, and then cooled in ice. Proteins were removed by two successive treatments with phenol saturated with SSC. RNA in the water phase was precipitated with alcohol, dissolved in SSC containing EDTA (10^{-2} M). RNA was subjected to sucrose gradient centrifugation (5–20% sucrose in SSC + EDTA (10^{-2} M), 25,000 rpm for 16 hr in a Spinco SW25.1 rotor). Fractions containing 16S RNA were pooled and dialyzed against SSC. RNA was then precipitated with alcohol, resuspended in TMA II, and finally dialyzed against PMK III. Rat liver 18S RNA was prepared in the same way from rat liver polysomes. The rat liver polysomes were a gift from Dr. H. C. Pitot.

RNA is perhaps a dimer of 16S RNA molecules.[10] 23S RNA was treated with glycine buffer (0.05 M, pH 10) as described by Midgley.[10] Upon analysis by sucrose gradient sedimentation, the resultant preparation showed a heterogeneous sedimentation pattern, but the peak sedimented at about 16S. This "16S RNA" preparation was used in the reconstitution, replacing the 16S RNA prepared from 30S particles. It was completely inactive.

Smaller ribosomal RNA (16S RNA)[11] from yeast was prepared and substituted for *coli* 16S RNA in the reconstitution mixture. As shown in Table 2 (expt. 2), only a small fraction (16% in the experiment of Table 2) of the added RNA was recovered in the soluble particle fraction and no activity was detected with the recovered particles. Sucrose gradient sedimentation analysis of the recovered particles revealed that the particles are heterogeneous with a peak at about 17S. Formation of particles with a sedimentation behavior similar to 30S ribosomal particles was very small, if any (Fig. 3).

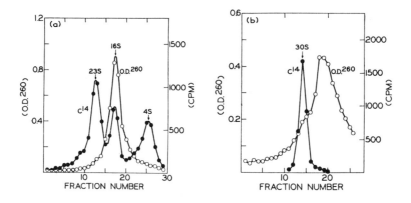

Fig. 3.—Sucrose gradient sedimentation of yeast 16S RNA (*a*) and "particles" recovered from the reconstitution mixture containing yeast 16S RNA, *coli* SP30, and *coli* CP30 (*b*). Both yeast 16S RNA and the "particles" recovered were those used for expt. 2 in Table 2. As references small amounts of a C14-labeled *E. coli* RNA preparation and C14-labeled 30S particles were used in (*a*) and (*b*), respectively. Experimental procedures were similar to those described in Fig. 1.

Similarly, 18S ribosomal RNA prepared from rat liver ribosomes could not substitute for *E. coli* 16S RNA in this reconstitution system (Table 2, expt. 2). Only insoluble aggregates were produced and no soluble particles resembling 30S particles were recovered.

Discussion and Conclusions.—The results presented in this paper conclusively demonstrate the reconstitution of functionally active 30S particles from free 16S ribosomal RNA and mixture of ribosomal proteins.

The present reconstitution system provides a system in which the structural and functional significance of ribosomal RNA can be studied. The experimental results described in this paper have already shown that 16S RNA must be specific and intact. Although similar in their sedimentation properties, yeast 16S ribosomal RNA or "16S" RNA from the 23S *coli* ribosomal RNA cannot replace the *coli* 16S RNA in the reconstitution of "30S" particles. Even inactive particles having similar size (about 30S) are not produced in a soluble form with these heterologous ribosomal RNA's. Further systematic studies using several related but heterologous bacterial ribosomal RNA's as well as *E. coli* 16S RNA modified in specific ways would give more information about the structural and functional role of ribosomal RNA in the ribosome.

The function of core proteins in the ribosome could not be studied previously. The present reconstitution system has enabled us to study their function. The results described in the present paper show that CP30 proteins are functionally different from CP50 proteins. It is now possible to separate the various CP30 proteins and to study their functional roles individually, as we have done with respect to proteins contained in the SP30 fraction.[6] Thus, eventual identification of all functionally essential 30S ribosomal proteins is technically feasible.

One use of the present reconstitution system is the identification of altered components in mutationally or physiologically altered ribosomes. It has been shown previously that ribosomes isolated from a streptomycin (Sm)-resistant mutant of *E. coli* are more resistant to Sm than those from the Sm-sensitive parental strain,[12, 13] and that the sensitivity is associated with the 23S core part of the 30S particles.[14, 15] Using the present reconstitution system, we have now shown that the alteration induced by Sm-resistant mutation resides in the CP30 proteins and not in the 16S ribosomal RNA.[16]

The present reconstitution of 30S particles from free RNA and proteins is highly efficient and takes place in the absence of any nonribosomal components. Thus we conclude that the information for the assembly of the 30S ribosomal particles is contained in the structure of their molecular components, that is, 16S RNA and ribosomal proteins.

Summary.—The reconstitution of functionally active 30S particles has been achieved from 16S ribosomal RNA and mixture of ribosomal proteins obtained from 30S particles. The component specificity for the reconstitution has been demonstrated.

We thank Drs. C. G. Kurland and M. Susman for their reading of the manuscript, Miss H. Bechmann and Mrs. U. Traub for their excellent technical assistance. This investigation was supported (in part) by U.S. Public Health Service research grant no. GM-15422 from the National Center for Urban and Industrial Health, and National Science Foundation grant GB-3947.

* Paper no. 1184 of the Laboratory of Genetics, University of Wisconsin. Paper no. IV in this series has been published (see ref. 6).

[1] Hosokawa, K., R. Fujimura, and M. Nomura, these Proceedings, **55**, 198 (1966).

[2] Staehelin, T., and M. Meselson, *J. Mol. Biol.*, **16**, 245 (1966).

[3] Spirin, A. S., and N. V. Belitsina, *J. Mol. Biol.*, **15**, 282 (1966).

[4] Nomura, M., and P. Traub, in *Organizational Biosynthesis*, ed. H. J. Vogel, J. O. Lampen, and V. Bryson (New York: Academic Press, 1966).

[5] Traub, P., and M. Nomura, manuscript submitted to *J. Mol. Biol.*

[6] Traub, P., K. Hosokawa, G. R. Craven, and M. Nomura, these Proceedings, **58**, 2430 (1967).

[7] Nomura, M., and P. Traub, manuscript submitted to *J. Mol. Biol.*

[8] Meselson, M., M. Nomura, S. Brenner, C. Davern, and D. Schlessinger, *J. Mol. Biol.*, **9**, 696 (1964).

[9] Traub, P., D. Söll, and M. Nomura, manuscript submitted to *J. Mol. Biol.*

[10] Midgley, J. E. M., *Biochim. Biophys. Acta*, **108**, 348 (1965).

[11] Bruening, G., and R. M. Bock, *Biochim. Biophys. Acta*, **149**, 377 (1967).

[12] Flaks, J. G., E. C. Cox, M. L. Witting, and J. R. White, *Biochem. Biophys. Res. Commun.*, **7**, 390 (1962).

[13] Speyer, J. F., P. Lengyel, and C. Basilio, these Proceedings, **48**, 684 (1962).

[14] Traub, P., K. Hosokawa, and M. Nomura, *J. Mol. Biol.*, **19**, 211 (1966).

[15] Staehelin, T., and M. Meselson, *J. Mol. Biol.*, **19**, 207 (1966).

[16] Traub, P., and M. Nomura, manuscript in preparation.

21

Reprinted from *Science,* **179,** 864–873 (Mar. 1973)

Assembly of Bacterial Ribosomes

In vitro reconstitution systems facilitate study
of ribosome structure, function, and assembly.

Masayasu Nomura

The structure of ribosomes is complex. There are two ribosomal subunits, named according to their sedimentation coefficients. In *Escherichia coli,* the smaller, $30S$ subunit consists of one $16S$ RNA molecule and about 20 protein molecules. The larger, $50S$ subunit consists of one $23S$ RNA molecule, one $5S$ RNA molecule, and about 30 to 35 protein molecules (*1–3*). The development in recent years of systems permitting the reconstitution of ribosomes

in vitro has advanced considerably our knowledge of the structure, function, and assembly of these organelles. Because the reconstitution of $30S$ subunits was achieved first and because our knowledge of this subunit is much more extensive than our knowledge of the $50S$ subunit, I center my discussion on the $30S$ subunits and refer only briefly to the $50S$ subunits; other aspects of ribosome research have been discussed more extensively elsewhere (*1–4*).

Assay of Ribosomal Functions

The activity of ribosomes and reconstituted ribosomes is routinely assayed by means of cell-free polypeptide synthesizing systems that are directed by natural messenger RNA or by synthetic mRNA such as polyuridylic acid. In addition, several partial reactions are used to assess the activity of ribosomes and reconstituted particles. These reactions include the binding of formyl-methionyl-transfer RNA directed by natural mRNA or the initiation codon AUG (5), the binding of other aminoacyl-tRNA's directed by synthetic mRNA, the peptidyl transferase reaction, the GTP binding that is dependent on G factor, and the UAA binding that is dependent on R factor. Details of these partial reactions are described in several reviews on the mechanism of protein synthesis (3, 6).

Partial Reconstitution of Ribosomes

The first step in the analysis of ribosome structure was the development of a system for reconstituting ribosomes that had been partially disassembled. The studies that led to this development were conducted by Staehelin and Meselson (7) and by workers at my laboratory (8) and were based on the observation that about 30 to 40 percent of the proteins (split proteins) in ribosomes are split off during density-gradient centrifugation of ribosomes in 5M cesium chloride. The functionally inactive nucleoprotein particles (core particles) that remain can then be mixed with the split proteins, at which time the splitting process is reversed and active ribosomes are again produced. This partial reconstitution reaction is rapid and relatively insensitive to experimental conditions; the reaction is complete within a few minutes at 37°C (9). In these respects, the partial reconstitution reaction is very different from total reconstitution that will be discussed later.

Of the seven proteins split from 30S subunits (see Table 1, column 1), five [S3(P3), S5(P4), S9(P8), S10(P6), and S14(P11)] were purified and characterized (10–12). That three of these proteins [S3(P3), S10(P6), and S14-(P11)] are essential for certain ribosome functions (10) has been shown by reconstituting particles from 23S core particles mixed with fractionated split proteins, the split protein to be tested being omitted from the reconstitution mixture. Such experiments are called single component omission experiments. Omission of any one of the three essential proteins results in nearly complete abolishment of polypeptide synthesis directed by synthetic as well as by natural mRNA, and of the binding of tRNA directed by mRNA. The presence of S5(P4) is required for full activity in

The author is the Conrad A. Elvehjem professor of life sciences at the Institute for Enzyme Research and the departments of biochemistry and genetics, University of Wisconsin, Madison 53706.

protein synthesis. I discuss the effect of omitting S9(P8) later in this article. Because the two remaining proteins, S1(P1) and S2(P2), had not been purified, we were not able to examine them individually.

The protein S14(P11) is a fractional protein, as defined by Kurland *et al.* (13); one 30S particle contains only 0.5 molecule of this protein (14) (Table 1). The biological significance of this heterogeneity of isolated ribosome populations is still unknown. The partial reconstitution experiments (10) strongly suggest, however, that all functionally active particles must contain S14(P11). Therefore, the heterogeneity of the ribosome populations with respect to this protein, for example, probably means that the population contains both active and inactive particles, the inactive particles being devoid of S14-(P11). Stimulation of the activity of isolated ribosomes upon addition of extra ribosomal proteins (13, 15) is consistent with this interpretation.

Total Reconstitution of 30S Subunits

The development of a partial reconstitution system was only a first step toward the functional analysis of ribosomal components. Comprehensive analysis of ribosomal structure and function would require reconstitution of the organelles from free RNA and individual proteins. In late 1967 the 30S subunit from E. coli was reconstituted from 16S RNA and a mixture of unseparated 30S ribosomal proteins (16). These experiments established the concept that the information for the correct assembly of ribosomal particles is contained in the structure of their molecular components, and not in nonribosomal factors.

Subsequently, 30S ribosomal proteins were separated and purified by several groups of investigators (17, 18) and reconstitution from mixtures of 16S RNA and 21 purified proteins was demonstrated (18–20). We can now reconstitute 30S particles which have activity

Table 1. Proteins from the 30S subunits of *Escherichia coli* ribosomes [for nomenclature, see (11)]. Proteins involved in the rate-limiting temperature-dependent step are compared with those found in the isolated reconstitution intermediate (RI) particles and with the 21S particles that are assembled in vivo. Unit (0.8 to 1.0 copy per ribosome), marginal (0.7 to 0.8 copy per ribosome), and fractional (0.5 or less copy per ribosome) proteins refer to the stoichiometric designations of Voynow and Kurland (14). Data on S15(P10b), S12(P10), and S13(P10a) are not available. Proteins that split off during centrifugation in 5M CsCl are indicated by SP (split proteins); those that remain associated with RNA in the 23S core particles are indicated by C (10). The extent to which each protein is required for the first step in the two-step reconstitution process (see text) is shown in the fourth column by ++, strongly required; +, moderately required; ±, weakly required; −, not required (20). The presence of each protein in the isolated RI particles (the 21S particles in vitro) is indicated in the fifth column by ++, present; +, present in reduced amounts; −, absent or almost absent (20, 43). Both S11(P7) and S9(P8) are probably present, but it is difficult to differentiate between these two proteins by electrophoresis on polyacrylamide gel. The protein composition of 21S particles accumulated by some cold-sensitive mutants (44) is shown in the sixth column: ++, present in amounts comparable to those in 30S subunits; +, present in reduced amounts; ±, found only in some preparations; −, not detected. For RI*, see page 868.

Protein	Stoichiometry	CsCl treatment	Required for RI* particle formation	Present in isolated RI particles	Present in 21S particles in vivo
S4(P4a)	Unit	C	++	++	+
S8(P4b)	Unit	C	++	++	++
S7(P5)	Unit	C	++	++	+
S16(P9a)	Unit	C	++	++	+
S19(P13)	Fractional	C	++	+	+
S17(P9b)	Unit	C	+	++	+
S15(P10b)		C	+	++	++
S5(P4)	Marginal	SP	±	−	−
S11(P7)	Fractional	C	±	(++)	−
S9(P8)	Unit	SP	±	(++)	±
S12(P10)		C	±	−	−
S18(P12)	Marginal	C	±	++	±
S20(P14)	Fractional	C	−	++	+
S13(P10a)		C	−	++	+
S6(P3b,c)	Marginal	C	−	++	+
S1(P1)	Fractional	SP	−	−	−
S2(P2)	Fractional	SP	−	−	−
S3(P3)	Marginal	SP	−	−	−
S10(P6)	Marginal	SP	−	−	−
S14(P11)	Fractional	SP	−	−	−
S21(P15)	Fractional	C	−	−	−

as high as the original 30S subunits in the several ribosomal functions tested. Thus, we believe that in the list of 30S ribosomal proteins shown in Table 1 we have not omitted any significant macromolecular component that has an important function in polypeptide synthesis. However, this does not necessarily mean that all proteins listed in Table 1 are genuine ribosomal proteins, especially since several of them are present in amounts less than a single molecule per particle in isolated ribosomal preparations [fractional proteins (13, 14); see Table 1]. Our reconstitution experiments have shown that almost all of the proteins listed in Table 1 are, in fact, required for full activity of ribosomes in various functional assays (10, 18, 20). The main exception is S1(P1). Omission of S1(P1) from the reconstitution mixture does not cause a reduction in any of the ribosomal functions tested (18, 20). This protein usually fails to become incorporated into the reconstituted ribosome under the conditions of the reconstitution reaction, which include the use of buffers having high ionic strength (21). Isolated 30S ribosomal particles contain only about 0.1 to 0.3 copy of this protein per 30S particle (14). Thus, S1(P1) might not be a true ribosomal protein, although the presence of S1(P1) in assay mixtures has been shown to stimulate certain ribosomal functions (22). The accurate definition of ribosomal proteins is therefore difficult (1, 2).

Single Component Omission Experiments

Four possible roles can be considered for the components of the ribosome. (i) A given component may be essential for the ribosome assembly but not required for any function; once assembly has occurred the component may be removed without loss of function. (ii) A given component may be required indirectly for some ribosomal function because its presence maintains an active center in a proper configuration in the ribosome structure. (iii) A given component may be a part of the active center, playing a direct role in a given ribosomal function. (iv) A given component may be required both for the assembly of the ribosome and for ribosomal functions.

To assign a particular role to any one component by single component omission experiments has proved difficult; omission of a single protein usually affects several functions while a given function can be shown to require the presence of each of several proteins. Nevertheless, single component omission experiments have provided us with some useful information. For example, the omission of protein S12(P10) results in the production of particles which show a pronounced reduction in translational error frequency under several conditions, whereas the omission of protein S11(P7) results in particles which show a pronounced increase in translational error frequency (18). Although the actual mechanism responsible for translational errors is not known, each of these two proteins must have some unique role in this mechanism since none of the other proteins which could be tested affect the quency of translation errors (18, 23, 24).

We now believe that some proteins are required only for efficient assembly [role (i) above], but not for the efficient functioning of the assembled particles. In our earlier studies several proteins were found to be important for the assembly of 30S subunits. Omission of any one of these proteins produced particles sedimenting at 20S to 25S rather than at 30S under the standard reconstitution conditions employed (incubation at 40°C for 20 minutes) (18). As expected, these particles showed greatly reduced activity in all the functions tested. On this basis, proteins S4(P4a), S8(P4b), S7(P5), S9(P8), and a mixture (called P9 at that time) of S16(P9a) and S17(P9b), were called "assembly proteins." However, the question of whether these proteins are required only for assembly or both for assembly and for ribosomal function was not answered.

Recent experiments strongly suggest that both S16(P9a) and S17(P9b), and also possibly S9(P8), have roles in the assembly reaction, but do not participate directly in ribosomal functions (20). After conducting a series of single component omission experiments we compared the results obtained from the partial reconstitution system with the results obtained from the total reconstitution system. With the partial reconstitution system the omission of S9(P8) did not have much effect on the functional ability of the reconstituted particles (10), whereas omission of the same protein in the total reconstitution system produced particles sedimenting at about 25S with very

weak activity in several ribosomal functions (18). This suggests that S9(P8) is important in providing a "core structure" during the assembly reaction, but has no direct functional role. Apparently, the 23S core particles obtained by CsCl treatment retain the core structure, although they have lost S9(P8). Hence, S9(P8) may be dispensable in the partial reconstitution system.

Reconstitution in the absence of S16(P9a), or in the absence of both S16(P9a) and S17(P9b), resulted in particles which varied considerably depending on the duration of incubation. Systematic studies showed that in the absence of S16(P9a) and S17(P9b) the rate of reconstitution is very slow compared to the complete system containing these proteins. Nonetheless, 30S-like particles, apparently with full activity, are eventually produced. Thus, the proteins appear to be required for assembly but not for function. In early experiments, in which the time of incubation for the reconstitution was short, an inactive intermediate 21S particle was the dominant product. Thus, it is possible that other proteins also play a role only in the assembly process, their absence from the finished ribosome structure not seriously affecting function.

These findings may help to explain the evolution of the present ribosomal structure with so many protein subunits. Of course, it is still possible that the postulated assembly-specific proteins, such as S16(P9a) and S17(P9b), have some unknown functions in the finished ribosome structure which cannot be detected in vitro by assays we have used.

Other Functional Analyses of Ribosomal Components

There are several other approaches to the functional analysis of ribosomal components. In one approach the ribosomes are altered by chemical means and the resultant functional alterations are correlated with the chemical alterations of the components. In another approach, mutants with altered ribosome functions are isolated and the components responsible for the altered phenotypes are identified. The first ribosomal protein to be identified in this way was S12(P10), which was altered by the mutation of a streptomycin-sensitive strain of *E. coli* to a streptomycin-resistant strain (25). Both approaches are now being used in extensive studies of ribosomal functions in many laboratories (26), the reconstitution technique being used to identify the components altered by chemical means or by mutation (1).

In an approach similar to those above, the reconstitution technique is used to prepare particles containing a single protein (or RNA) component with a specific modification. With such ribosomal particles, the role of the modified component (and the modified portions of that component) in the ribosome function or assembly can be analyzed. Components modified either in vivo or in vitro by chemical or enzymatic means can be used for such experiments.

In another approach the technique known as affinity labeling, which is used to study the structure of active centers in many enzymes (27), has been used in studies of ribosome function by Cantor and his co-workers (28). These workers used *n*-bromoacetylphenylalanyl-tRNA, an analog of peptidyl-tRNA, and identified two proteins L2 and L27, in *E. coli* 50*S* subunits as the proteins situated at or near the site which interacts with peptidyl-tRNA (the P-site).

Functional Role of Ribosomal RNA

Reconstitution experiments have established that intact 16*S* RNA is essential for the assembly of 30*S* subunits (16). For example, 18*S* RNA from the smaller subunits of yeast and rat liver cytoplasmic ribosomes cannot replace *E. coli* 16*S* RNA in the reconstitution. An RNA prepared from bacterial 23*S* RNA with the size similar to the 16*S* RNA is also ineffective as a replacement (16). Yet, some 16*S* ribosomal RNA from distantly related bacterial species, such as *Azotobacter vinelandii* or *Bacillus stearothermophilus* can replace *E. coli* 16*S* RNA and form functionally active hybrid 30*S* particles with *E. coli* 30*S* proteins (29). Although 16*S* RNA's from these three different bacterial species have some common base sequences, large portions are different. Thus, the requirement for a specific base sequence in ribosomal RNA is not absolute. Perhaps only some portions of the 16*S* RNA are directly involved in specific interaction with ribosomal proteins, these regions having structures (primary or tertiary) which are identical or are very similar among

bacterial species (29). These sites on 16*S* RNA are being studied by several workers.

The reconstitution system for the 50*S* subunit from *B. stearothermophilus* has also been investigated. For reconstitution to occur in this system, 5*S* RNA as well as 23*S* RNA is essential (30). The 5*S* RNA's from *E. coli* and *Pseudomonas fluorescens* are both as active as *B. stearothermophilus* 5*S* RNA (31, 32). Although the primary sequence of *B. stearothermophilus* 5*S* RNA is not known, the complete primary sequences of both *E. coli* and *P. fluorescens* 5*S* RNA's are known (33). The sequences of these two 5*S* RNA's are alike over only about two thirds of the molecules. In this case, therefore, about one third of the base sequence can be altered in suitable ways without any detectable loss of function. It is conceivable that many more bases could be altered without affecting function. *Bacillus stearothermophilus* 23*S* RNA can also be replaced by *Staphylococcus aureus* 23*S* RNA is this system (41).

Studies of 5*S* RNA modified by chemical means have also shown that bases at several positions in the 5*S* RNA molecule can be modified without loss of function, as analyzed by reconstitution of the 50*S* subunit from *B. stearothermophilus* (31). Because the elimination of the 3'-end nucleoside of 5*S* RNA or the chemical modification of the ribose moiety of this terminal nucleoside does not affect the activity of 5*S* RNA (31), it has been suggested that models of protein synthesis (34) that invoke peptidyl-5*S* RNA as an intermediate are invalid.

By contrast, chemical modification (for example, by nitrous acid or monoperphthalic acid) of 16*S* RNA indicates that only a few base alterations (per molecule) are sufficient to destroy completely the reconstitution activity of the molecule (29, 35). Thus a major portion of the base sequences (exposed to these reagents) of the RNA appears to be important. However, in these experiments the inactivation appeared to be caused by the formation of subunits with abnormal configurations.

That ribosomal RNA has a direct functional role as well as structural role has not been shown by any chemical modification experiments. Yet, several other types of experiments suggest that ribosomal RNA may also have a direct role. First, 30*S* subunits isolated from colicin E3–treated *E. coli* are inactive in polypeptide synthesis in vitro (36).

The inactive 30*S* subunits are indistinguishable from the active subunits in their physical behavior and retain all the ribosomal proteins. Reconstitution experiments show that all proteins are functionally intact, and the alteration responsible for inactivation resides in 16*S* RNA (37). The alteration appears to involve a single nucleolytic event at a site about 50 nucleotides from the 3'-end of the 16*S* RNA (37, 38).

The second indication of a functional role for RNA comes from the analysis of kasugamycin-resistant mutants of *E. coli* (39). The ribosomes from these mutants are resistant to kasugamycin during polypeptide synthesis in vitro. The alteration resides in the 30*S* subunit (40). By chemical analysis and reconstitution studies, it was shown that the absence of a methyl group in the normal sequence, $m_2^6Am_2^6$ ACCUG (5), of 16*S* RNA is responsible for the mutant ribosome behavior (39). The importance of methylation of 23*S* RNA with respect to lincomycin resistance in *S. aureus* has also been shown by the reconstitution technique (41).

With the availability of techniques for determining RNA sequences, as well as the reconstitution assay technique, it is somewhat surprising that relatively few investigators have attempted to determine the possible direct roles of ribosomal RNA in protein synthesis.

Mechanism of Assembly in vitro

Using individual purified 30*S* ribosomal proteins, we have studied the assembly of 30*S* subunits. Under the conditions of reconstitution, only seven of the proteins become bound to the RNA. Certain other proteins become bound only after some of the first seven proteins are bound (19, 21). [Other investigators, however, have presented results suggesting that only five out of these seven proteins are specific initial binding proteins (42).] The remaining proteins require the presence of proteins in both of the above groups in order to become bound. In this way, we have analyzed the sequence of addition of proteins to the 16*S* RNA molecule and have constructed the assembly map shown in Fig. 1 (19, 21).

The arrows connecting proteins to proteins in Fig. 1 represent only the associations that we detected during the experiments that were conducted to construct the map. Many other protein-protein interactions which stabilize ribosomal structure probably occur. In addition, it is conceivable that proteins other than the initial binding proteins also interact with 16S RNA in the finished ribosome structure.

It is probable that the "sequence" described in the assembly map corresponds at least approximately to the temporal sequence of assembly. Other evidence for this assembly sequence comes from the early kinetic studies of the assembly of 30S subunits which showed that there is a rate-limiting unimolecular reaction that requires a high activation energy (43). At lower temperatures, a subset of the 30S ribosomal proteins (called "RI proteins") interacts with 16S RNA, and intermediate particles (reconstitution intermediate, or "RI particles") accumulate. The isolated "RI particles" sediment at 21S in a buffer having a low concentration of magnesium ions; they are deficient in several proteins (called "S proteins" because they remain in the supernatant after 21S particles sediment), and have no functional activity. From this and other experiments done on the isolated "RI particles," the following reaction scheme was proposed (43):

$$16S\ RNA \xrightarrow[\text{proteins}]{+\ RI} RI\ particles \xrightarrow{Heat}$$

$$RI^*\ particles \xrightarrow{+\ S\ proteins} 30S\ ribosomal\ subunits$$

In this scheme, the step, RI particle → RI* particle, represents the rate-limiting unimolecular reaction, and is thought to involve a large conformational change in the structure of the intermediate, the RI particle. The 21S RI particles discussed here should not be confused with various particles (20S to 25S) which accumulate when reconstitution occurs in the absence of one of the "assembly proteins" (see above). However, some 20S to 25S particles may have accumulated because of the effect on the same rate-determining step and they may be similar to the 21S RI particles.

We have attempted to determine the protein composition of the RI particle required for this presumed conformational change, that is, the RI proteins. First we determined the composition of the isolated RI particles (Table 1) (20, 44, 45). While some of the required proteins may have been lost during isolation, it is also possible that additional proteins become bound but are not required for the conformational change. Therefore, to determine which proteins must be present during the heating step in order to get the highest

degree of reconstitution, we used a two-step reconstitution method (20). The 21 purified proteins were divided into two groups in various ways. One group of proteins was incubated with 16S RNA at high temperatures and, after cooling, the remaining proteins were added and incubated at low temperatures for a short time. In this way, we identified the proteins required during heating for efficient reconstitution to occur (Table 1) and observed some differences between the proteins identified in this way and the proteins found in isolated RI particles. First, S20(P14), S13(P10a), and S6(P3b,c), which are found in isolated RI particles, are unnecessary for the formation of RI* particles. These proteins can be added either before or after the presumed rate-determining conformational change has taken place, without any effect on the rate of reconstitution. Second, some proteins such as S12(P10), S19(P13), and S5(P4) are required for efficient RI* formation and yet are present only in small amounts or are not detectable in isolated RI particles. These proteins might constitute a part of the true RI particle that is simply lost during isolation, or they might aid in stabilizing the particles after the conformational change. Despite these observed differences, however, it should be emphasized that most of the proteins required for the formation of RI* particles are among those found in the isolated RI particles. Thus, the real intermediate particles that undergo the rate-determining conformational change must be similar to the isolated RI particles, with perhaps some additional proteins [such as S12(P10), S19(P13), and S5(P4)]. Proteins above the dotted line in the assembly map (Fig. 1) indicate the proteins found in the isolated RI particles and the additional three proteins mentioned above. Thus the incorporation of other proteins (under the dotted line) into the reconstituted 30S particles does appear to depend on the conformational change discussed above.

Several proteins have only a moderate or weak influence on the formation of RI* particles (Table 1). These proteins can be added after the heating step, with only a small decrease in the efficiency of reconstitution. It is probable that there are several different routes for assembly of 30S ribosomal subunits. Differences in free energy of activation among these several alternative routes may be rather small. Supporting this general conclusion is the striking observation that 30S-like particles with

nearly full activity are slowly produced even in the absence of S16(P9a) and S17(P9b), both of which play major roles in the assembly process. It has not been determined whether such flexibility in the assembly process exists in vivo.

The tertiary structure of free 16S RNA and free 23S RNA is quite different from that within the ribosomes (1, 46). Thus, the RNA must undergo conformational changes during the assembly reaction. Some data suggest that the major change in ribosomal RNA (rRNA) takes place during the rate-limiting step (20); further investigations of the conformational changes in RNA are in progress.

As already mentioned, certain specific proteins are the first to bind to 16S rRNA (21, 42). Once properly formed, the RNA-protein complexes do not dissociate during subsequent assembly steps (21). Investigators in several laboratories are attempting to identify the sites on the 16S rRNA which bind these proteins (47, 48). For example, Zimmerman et al. have obtained several RNA fragments which interact with one or several of the initial binding proteins (48). From the chemical analysis of these RNA fragments, the approximate position of the binding sites for several initial binding proteins has been located on the 16S rRNA (48). It is noteworthy that five of the initial binding proteins [S4(P4a), S8(P4b), S13-(P10a), S15(P10b), S20(P14)] bind to the 5'-terminal 900 nucleotides of the RNA and only one, S7(P5), attaches to the 3'-terminal third of the molecule. Such studies are certain to produce useful information both on the structure of ribosomes and on the mechanisms by which the proteins recognize specific RNA structures.

Three-Dimensional Structure

Although x-ray or electron microscopic methods may eventually prove useful in the complete elucidation of the three-dimensional structure of ribosomes, these techniques have provided little useful information so far. This is due both to the lack of techniques for making ribosome crystals suitable for x-ray methods and to the asymmetry and complexity of ribosome structure. Most investigators now study three-dimensional structure by chemical methods, such as the use of bifunctional cross-linking reagents that reveal relationships among neighboring protein components. The relationships among proteins that become bound, as revealed in the assembly map of 30*S* subunits (Fig. 1) may also reflect topological relationships among ribosomal proteins in the ribosomal structure (*21*). Thus, results obtained by chemical and other methods can be compared with the relationships shown in the assembly map. For example, Craven and his co-workers examined the protein components of subparticles produced by mild ribonuclease digestion of *E. coli* 30*S* subunits (*49*) and found that those proteins that cluster together in such subparticles are, in general, the proteins which show interrelationships in the assembly map. Similar results were obtained by Cox and his co-workers (*50*). The striking correlation between the assembly map and the results obtained by chemical studies supports the suggestion that the assembly map reflects the topological relationships of ribosomal proteins.

In most chemical studies of ribosome structure it is assumed (i) that the ribosomal preparation contains a homogeneous population of active ribosomes and (ii) that the ribosome structure remains intact during the chemical treatment used. In fact, only a fraction (at most 50 percent) of the ribosomal particles in most ribosome preparations is active and several proteins [fractional proteins (*13, 14*)] are actually missing from some (presumably inactive) ribosomal particles. Moreover. assumption

(ii) is usually difficult to prove. Therefore, the information obtained by chemical methods on the three-dimensional arrangement of proteins within the ribosome structure should still be treated with caution.

Cross-linking reagents are currently employed by several investigators for the study of ribosome structure. For example, using tetranitromethane, Shih and Craven showed cross-linkage of S18-(P12), S11(P7), and S21(P15) which are interrelated on the assembly map (*51*). Traut and his co-workers, as well as several other groups, used the bifunctional reagent, bis-methyl suberimidate, and obtained results that do not correlate well with the assembly map (*52*). Some of the relationships shown among neighboring proteins in these studies may include protein-protein interactions that have not been revealed during the construction of the assembly map.

Some investigators have been studying the reactivity of 30*S* ribosomal proteins toward other chemical reagents, enzymes, or antibodies in the hope that relatively exposed proteins might be distinguished from relatively unexposed proteins within the ribosome structure (*53*). These studies have given somewhat conflicting results so far. Obviously, the results may vary depending on the kind of reagents used. In addition, as noted above, some of the reagents used may themselves have caused changes in the ribosome structure.

Several other techniques are being considered for use in the study of ribosome configuration. One is the use of fluorescent probes as molecular measures of distance [singlet-singlet energy transfer; see (*54*)] between proteins in the ribosomes. In preliminary experiments Cantor and his co-workers (in collaboration with us) have found that several 30*S* ribosomal proteins covalently labeled with suitable fluorescent dyes can be incorporated into functionally active reconstituted 30*S* subunits. Thus it appears to be possible, at least with some protein combinations, to measure the distance between two different ribosomal proteins (or between a protein and a particular position on 16*S* rRNA, for example, 3'-end) within functionally active 30*S* subunits.

Similarly, distances between proteins can be measured by neutron diffraction. Because the neutron-scattering properties of hydrogen are different from all other normally encountered atoms including deuterium, it is theoretically possible to measure the distance between two hydrogen-rich proteins in reconstituted ribosomes that otherwise contain heavily deuterated molecular components (*55*). If the distances between enough pairs of proteins can be measured by these techniques, it should be possible to determine the unique arrangement of proteins in the ribosome. In addition, the use of functionally active proteins labeled with fluorescent

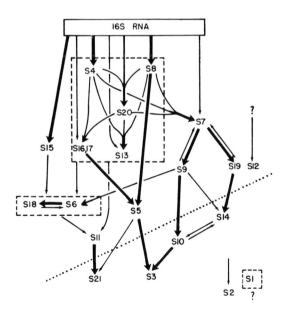

Fig. 1. The assembly map of 30*S* ribosomal proteins. The assembly map is based on the work of Mizushima and Nomura (*21*), slight modifications having been made as a result of subsequent work (*19, 42, 44*). The nomenclature of Wittmann *et al.* (*12*) is used instead of the previous nomenclature used in my laboratory. Arrows between proteins indicate the facilitatory effect on binding of one protein on another; a thick arrow indicates a major facilitatory effect. The map may be used to indicate the following relationships. The thick arrow from 16*S* RNA to S4 indicates that S4(P4a) binds directly to 16*S* RNA in the absence of other proteins. The thin arrow from 16*S* RNA to S7 indicates that S7(P5) binds weakly to 16*S* RNA in the absence of other ribosomal proteins. Thin arrows pointing toward S7 from S4(P4a), S8(P4b), S20(P14), S9(P8), and S19(P13) indicate that the latter proteins all help the binding of S7(P5) to RNA. The thick arrow from S7 to S19 indicates that in the absence of S7(P5), S19(P13) fails to bind the 16*S* rRNA containing complexes even in the presence of all other proteins under the standard reconstitution conditions. The arrow to S11(P7) from large box with dashed outline indicates that S11(P7) binding depends on some of the proteins enclosed in the box; it is not known exactly which proteins. The binding of S2(P2) and S12(P10) takes place at a later stage in the assembly sequence, but the exact position of these proteins in the map is not known (see *21*). S1(P1) does not bind under the conditions used. S16 and S17 were previously studied as a mixture, S17(P9b), but not S16(P9a), binds directly to 16*S* RNA (*47, 88*). The specificity of this binding has not been proved, and some investigators have not observed the direct binding of S13 to 16*S* RNA (*47*). Proteins above the dotted line are those either required for the formation of RI* particles or found in the isolated 21*S* RI particles (see text).

dyes should allow sensitive detection of conformational changes in ribosomal structure during assembly and during the various steps in protein synthesis. The resolution provided by this technique should far exceed that provided by any technique now in use. It is thus evident that many different approaches can soon be expected to yield much information on the three-dimensional structure of ribosomes.

Total Reconstitution of 50S Ribosomal Subunits

Although we encountered great difficulty in reconstituting *E. coli* 50S subunits in vitro, we recently succeeded in reconstituting 50S subunits obtained from the thermophilic organism, *B. stearothermophilus* (*32*).

In view of the results obtained with 30S reconstitution, it seemed reasonable to suppose that the difficulty with 50S reconstitution might reflect the greater complexity of the assembly reaction, as well as higher kinetic energy barriers which might be overcome only by longer incubation at even higher temperatures. The ribosomal components of *E. coli* 50S subunits or partially assembled intermediate particles might be too unstable to tolerate higher incubation temperatures. We found that reconstitution of 50S subunits in the *B. stearothermophilus* system, although possible, proceeds much more slowly than reconstitution of 30S subunits even at the optimum temperature (60°C) (*32*).

Initially, we used RNA and protein fractions obtained by dissociation of 50S subunits with 2M lithium chloride in 4M urea. The RNA fraction was shown to contain one protein still tightly bound to 23S rRNA. This protein, called L3, was not present in the 50S protein fraction obtained by this method. The protein L3 can be removed from 23S rRNA by treatment with acetic acid (66 percent) or with magnesium acetate at *p*H 2. We have subsequently demonstrated that functional 50S subunits can be reconstituted from a mixture of protein-free RNA with the 50S protein fraction plus the extracted protein L3 (*56*). The presence of L3 is essential for the reconstitution.

Studies on the effects of certain mutations on the assembly in vivo of *E. coli* ribosomes suggested that 30S subunits (or their precursors or components) play some crucial role in the assembly of 50S subunits (*57*). In earlier studies, the addition of 30S ribosomal subunits to the 50S reconstitution system occasionally stimulated the reconstitution of 50S subunits in vitro (*32*) but this stimulatory effect was not reproducible. Furthermore, when 50S and 30S subunits are reconstituted simultaneously from RNA and protein components derived from 70S ribosomes, the activity of the reconstituted 50S subunits is not significantly different from the activity obtained when the 50S subunits are reconstituted according to the standard procedure (*58*). Thus, the reconstitution of 50S subunits in vitro under the conditions described does not require the presence of 30S subunits (or their components). The role of 30S subunits in the assembly of 50S subunits in vivo remains unknown.

Studies of the 50S ribosome assembly system have shown that the presence of 5S RNA is required for reconstitution and have provided some information on the function of 5S RNA (*30, 31*). We have also demonstrated the role of methylation of 23S rRNA in the resistance of *S. aureus* to lincomycin (and to erythromycin and other related antibiotics) using a "hybrid reconstitution" system. By using dimethyladenine-containing 23S rRNA from resistant *S. aureus* strains, and other components (proteins and 5S RNA) from *B. stearothermophilus*, we obtained hybrid 50S subunits which were resistant to lincomycin; by replacing this 23S rRNA with 23S rRNA (lacking dimethyladenine) from a sensitive strain, we obtained particles that were sensitive to the drug (*41*).

To analyze the functions of 50S proteins, partial reconstitution systems obtained from *E. coli* have also been used. In earlier studies several 50S proteins were removed by high concentrations of CsCl as in the case of 30S subunits mentioned above, and functional analyses were done on these proteins (*59*). In pursuing this kind of approach, Nakamoto and his co-workers recently found that ethanol treatment removes a protein (or proteins) from 50S subunits and that this protein is essential for ribosomal functions involving both G and T factors, but is not required for the nonenzymatic binding of aminoacyl-tRNA or for the peptidyl transferase reaction (*60*). Similar results with what are presumably the same 50S proteins were obtained by other workers (*61*).

Several experiments have been done on RNA-protein interactions among components derived from the 50S subunits of *E. coli* under conditions which are defined to be optimum for 30S ribosomal reconstitution. For example, under such conditions 8 out of 34 50S proteins became bound to 23S RNA (*62*). Binding of 5S RNA to 23S RNA required the presence of only a few 50S proteins (*63*). However, because we lack a system for reconstitution of active *E. coli* 50S subunits, we cannot be sure whether the interactions studied are really pertinent to the structure of biologically active 50S ribosomal subunits.

There is a recent report that functionally active *E. coli* 50S subunits can be reconstituted from a mixture of 5S RNA, 23S RNA, and 50S proteins (*64*). We have not yet been able to reproduce these results and therefore cannot ascertain the importance of such a system in studies of *E. coli* 50S subunits.

Other Possible Ribosomal Functions

Since ribosomes comprise as much as a quarter of the total mass (dry weight) of a bacterial cell, and since they have a complex structure consisting of as many as 50 to 60 macromolecules, it is not unreasonable to suppose that ribosomes are engaged not only in protein synthesis, but also in other, unknown functions in vivo. Such multifunctional properties could be advantageous for coordination of several biosynthetic reactions. We should perhaps be willing to entertain this possibility.

An enzyme that participates in phospholipid biosynthesis in *E. coli*, phosphatidyl serine synthetase, is known to be associated with ribosomes and cannot be removed even after washing with buffers of high ionic strength (*65*). The significance of this association in phospholipid metabolism is unknown, but the possibility that it plays a role in coordinating lipid synthesis with protein synthesis has been suggested (*65*).

It has also been suggested (*66*) that ribosomes are involved in the regulation of rRNA synthesis by their direct participation in the metabolism of guanosine tetra- and pentaphosphates. This hypothesis is supported by the recent experiments of Haseltine et al. (*67*) who have shown that guanosine tetra- (and penta-) phosphates can be synthesized in vitro on the ribosome using guanosine diphosphate (and guanosine triphosphate) and ATP as substrates. Further studies on this subject would undoubtedly clarify the role of ribosomes in the regulation of rRNA synthesis.

Assembly of Ribosomes in vivo

The reconstitution of bacterial ribosomes in vitro has suggested the ordered sequence by which the 30S subunit is assembled from 16S RNA and its component proteins. It is likely that there is a similar ordered sequence of events in the assembly of 50S subunits in vitro (*19, 58*). That the sequential assembly of ribosomes also takes place in vivo is suggested by kinetic studies in which intermediate particles produced during ribosome synthesis can be detected in vivo by isotope labeling for limited periods (*68*). The assembly sequence has also been confirmed by the isolation of mutants which accumulate ribonucleoprotein precursors because of defects in assembly (*44, 57, 69*). In several instances, such ribonucleoprotein precursors were isolated and their protein compositions were analyzed (*44, 45, 70*). In the case of 30S precursors, protein compositions were compared with the map constructed for ribosome assembly in vitro. The 21S particles accumulated by some cold-sensitive mutants have protein compositions very similar to that of 21S RI particles isolated from reconstitution mixtures in vitro (*44*) (Table 1). On this basis, it has been concluded that the order of addition of proteins during 30S assembly in vivo is similar, if not identical, to the order of addition of proteins during the reconstitution of 30S subunits in vitro (*44*).

Although studies of reconstitution in vitro have provided much useful information on the assembly process of ribosomal particles, it is evident that the assembly processes in vivo and in vitro are not identical. One clear difference is the use of precursor 16S RNA and perhaps unmodified proteins in vivo. The RNA in the ribonucleoprotein precursor particles is slightly larger than mature 16S RNA, as determined by electrophoretic mobility in polyacrylamide gels (*71*) and by studies of the nucleotide sequence (*72*). The RNA is also submethylated or not methylated at all (*72*). During assembly of 30S subunits in vivo, some specific cleavages at both ends of the precursor 16S RNA must take place in addition to the methylation. Investigators are now attempting to identify and characterize these reactions in vitro (*73*). One reaction already identified is the formation of dimethyladenine. As mentioned before, kasugamycin-resistant mutants of *E. coli* lack methyl groups that are present in the dimethyladenine residues in the 16S RNA of wild-type kasugamycin-sensitive strains (*39*). An RNA methylase which has been isolated from the kasugamycin-sensitive strains is responsible for this methylation reaction and is absent from resistant mutants (*39*).

Some ribosomal proteins are known to have NH_2-terminal amino acids in which the NH_2 group of the amino acid is blocked (for example, *N*-acetylserine) or rare amino acids such as *N*-methyllysine (*74*). Thus some modifications of ribosomal proteins must occur after translation. During the head assembly of phage T4, some specific protein cleavages take place (*75*). Whether similar specific protein cleavages take place during ribosome assembly has not been studied.

Regulation of Ribosome Biosynthesis

The analysis of mutants that show defects in ribosome assembly (*69*) will probably provide much information on the detailed sequence of biochemical reactions and on the factors, ribosomal or nonribosomal, which control ribosome assembly. Useful information concerning intermediate particles and the possible "coupling" of 30S and 50S assembly, has already been obtained from such studies (*44, 57, 69, 76*). However, many problems still exist and have been discussed in detail in a recent review on the genetics of ribosomes (*4*). I will describe some of these problems very briefly.

Although the synthesis of rRNA has been studied extensively (*77*) the exact mechanism that regulates the synthesis is still unknown. Physiological studies (*78*) as well as electron microscopic studies (*79*) strongly suggest that the genes for 16S, 23S, and 5S rRNA's are in one single operon and are transcribed as a unit. This probably ensures the coordinated synthesis of 5S, 16S, and 23S rRNA. Such coordination would be necessary to make equal numbers of 30S and 50S ribosomal subunits in a growing cell. A simple and ingenious theory involving a protein factor known as ψ and guanosine tetra- (and penta-) phosphate has also been proposed (*80*) to explain the regulation of rRNA gene expression. Although the reported specific stimulation of rRNA gene transcription by the ψ factor is still controversial (*81*), there is a strong correlation between guanosine tetra- (and penta-) phosphate formation and the suppression of rRNA synthesis in vivo (*82*).

A formal genetic analysis of rRNA genes has never been made. Not only the genetic structure of a single rRNA gene operon and its regulation, but also the significance of the redundancy of such rRNA genes in a bacterial cell could well be studied by genetic approaches. This would be a challenging problem for bacterial geneticists.

Little information is available on the regulation of synthesis of ribosomal proteins (*4, 83*). The coordinated expression and the regulation of several nonribosomal bacterial genes have been studied by a variety of genetic approaches. One of the major obstacles to the use of such genetic approaches in studies of ribosomal protein genes is that many mutations of these genes are lethal, healthy ribosomes being essential for cell growth. To overcome this obstacle, studies have been initiated in our laboratory on ribosomal gene organization. For these studies we are using partial diploid *E. coli* strains which are heterozygous for the *str-spc* region on the chromosome. It is known that many genes coding for ribosomal proteins occur at the *str-spc* region (*4, 84*). Recent investigations of drug-resistant mutations induced by mu phage in such diploid strains have shown that a single mutational event caused by the insertion of mu phage affects the expression of more than one gene (*85*). It appears that many genes coding for ribosomal proteins which occur near the locus of *str* gene are transcribed as a single unit (ribosomal protein operon). Further studies of such diploid systems should provide useful information on the initiation site (promoter) of the ribosomal protein operon and the regulatory mechanisms controlling the expression of these genes. The transcriptional gene products, that is, the mRNA's for ribosomal proteins, have never been isolated. Even if we accept the existence of a large ribosomal protein operon, as deduced from the mu insertion experiments, this does not mean that all the proteins are synthesized on a single polycistronic mRNA. It is possible that the initial transcriptional product undergoes posttranscriptional cleavages and modifications.

Finally, there is the problem of how the regulatory processes for the expression of rRNA genes and ribosomal protein genes are coordinated. That such coordination occurs in vivo seems highly likely. Possible mechanisms for coordination have been discussed previously (*4, 83*) but there is no experimental proof to support any of these mechanisms exist.

Conclusions

I have not mentioned the remarkable progress made mainly by Fellner and his co-workers (86) in the elucidation of the primary structure of rRNA's and by Wittmann and his co-workers (87) in determining the structure of several ribosomal proteins. Such knowledge of primary structures is certainly the basis of complete understanding of the structure of the ribosome. With the current progress in technology, complete elucidation of the primary structure of all the ribosomal components is probably a matter of time. As indicated in this article, a rough approximation of the three-dimensional structure of ribosomes is likely to emerge soon. Although not mentioned in this article, studies of ribosomes from higher organisms are also progressing. We must, therefore, consider what further studies should be conducted and what kinds of questions we would like to solve.

Some groups of investigators aim to elucidate the complete three-dimensional structure of ribosomes and to find out how these complex cell organelles function; they hope to determine the conformational changes of many of the component molecules within the ribosome structure in response to external macromolecules and cofactors engaged in protein synthesis. Such knowledge will also be important in enabling us to understand the regulation of translation of genetic messages. Other groups of investigators aim to elucidate the complex series of events which originate in the transcription of the more than 60 genes and culminate in the formation of the specific structure of the organelle. Complete reproduction in vitro of all the assembly events that occur in vivo should not be difficult to achieve in principle. It should then become possible to study in vitro any factor regulating the biogenesis of the organelle. Although we do not know whether such studies would reveal any new fundamental principle that governs the complex circuits of interconnected macromolecular interactions, the achievement of such a complete in vitro system would represent a necessary step in the comprehensive understanding of biogenesis of organelles, and eventually, of the more complex behavior and genesis of cells (89).

References and Notes

1. M. Nomura, *Bacteriol. Rev.* **34**, 228 (1970).
2. C. G. Kurland, in *Protein Synthesis: A Series of Advances*, E. McConkey, Ed. (Dekker, New York, 1970), vol. 1, pp. 179–228; C. G. Kurland, *Annu. Rev. Biochem.* **41**, 377 (1972);

B. E. H. Maden, *Progr. Biophys. Mol. Biol.* **22**, 129 (1971).
3. Papers in *Cold Spring Harbor Symp. Quant. Biol.* **34** (1969).
4. J. Davies and M. Nomura, *Annu. Rev. Genet.* **6**, 203 (1972).
5. Notes on abbreviations: In the initiation codon AUG, A is adenine, U, uracil, and G, guanine; GTP is guanosine triphosphate; G factor is the elongation factor G and is related to the translocation function of the 50S subunit; UAA is a termination codon; R factor is the protein release factor, the R factor dependent UAA binding reaction being related to the chain termination function; m₂⁶Am₀⁶ ACCUG is a hexanucleotide containing N⁶-dimethyladenine, N⁶-dimethyladenine, cytosine, cytosine, uracil, and guanine; T factor is the elongation factor T and is involved in the binding of aminoacyl-transfer RNA to the ribosomes; ATP is adenosine triphosphate.
6. P. Lengyel and D. Söll, *Bacteriol. Rev.* **33**, 264 (1969); J. Lucus-Lenard and F. Lipmann, *Annu. Rev. Biochem.* **40**, 409 (1971).
7. T. Staehelin and M. Meselson, *J. Mol. Biol.* **16**, 245 (1966).
8. K. Hosokawa, R. Fujimura, M. Nomura, *Proc. Nat. Acad. Sci. U.S.A.* **55**, 198 (1966).
9. M. Nomura and P. Traub, *J. Mol. Biol.* **34**, 609 (1968).
10. P. Traub, K. Hosokawa, G. R. Craven, M. Nomura, *Proc. Nat. Acad. Sci. U.S.A.* **58**, 2430 (1967).
11. The nomenclature used here for the 30S proteins includes both that adopted by Wittmann *et al.* (12) and, in parentheses, the nomenclature used in my laboratory, because although the different nomenclatures have been correlated (12) there are slight uncertainties with some proteins. In the nomenclature by Wittmann *et al.* (12) proteins from the small (30S) subunits are named S1, S2, and so on, and those from the large (50S) subunits are named L1, L2, and so on.
12. H. G. Wittmann, G. Stöffler, I. Hindennach, C. G. Kurland, L. Randall-Hazelbauer, E. A. Birge, M. Nomura, E. Kaltschmidt, S. Mizushima, R. R. Traut, T. A. Bickle, *Mol. Gen. Genet.* **111**, 327 (1971).
13. C. G. Kurland, P. Voynow, S. J. S. Hardy, L. Randall, L. Lutter, *Cold Spring Harbor Symp. Quant. Biol.* **34**, 17 (1969).
14. P. Voynow and C. G. Kurland, *Biochemistry* **10**, 517 (1971).
15. L. L. Randall-Hazelbauer and C. G. Kurland, *Mol. Gen. Genet.* **115**, 234 (1972).
16. P. Traub and M. Nomura, *Proc. Nat. Acad. Sci. U.S.A.* **59**, 777 (1968).
17. S. J. S. Hardy, C. G. Kurland, P. Voynow, G. Mora, *Biochemistry* **8**, 2897 (1969); I. Hindennach, G. Stöffler, H. G. Wittmann, *Eur. J. Biochem.* **23**, 7 (1971); R. R. Traut, H. Delius, C. Ahmad-Zadeh, T. A. Bickle, P. Pearson, A. Tissieres, *Cold Spring Harbor Symp. Quant. Biol.* **34**, 24 (1969); P. S. Sypherd, D. M. O'Neill, M. M. Taylor, *ibid.*, p. 77.
18. M. Nomura, S. Mizushima, M. Ozaki, P. Traub, C. V. Lowry, *Cold Spring Harbor Symp. Quant. Biol.* **34**, 49 (1969).
19. M. Nomura, *Fed. Proc.* **31**, 18 (1972).
20. W. Held and M. Nomura, in preparation.
21. S. Mizushima and M. Nomura, *Nature* **226**, 1214 (1970).
22. J. Van Duin and C. G. Kurland, *Mol. Gen. Genet.* **109**, 169 (1970).
23. For further information on the role of ribosomes in translational fidelity, see L. Gorini, *Cold Spring Harbor Symp. Quant. Biol.* **34**, 101 (1969); *Nature New Biol.* **234**, 261 (1971).
24. Functional analysis of ribosome components has been discussed in detail (1).
25. M. Ozaki, S. Mizushima, M. Nomura, *Nature* **222**, 333 (1969).
26. Recent studies on the analysis of altered ribosomal components from several mutants are reviewed in (4). Examples of chemical modification studies are as follows: G. R. Craven, R. Gavin, T. Fanning, *Cold Spring Harbor Symp. Quant. Biol.* **55**, 457 (1971); J. A. Retsema and T. W. Conway, *J. Mol. Biol.* **60**, 169 (1971); H. F. Noller, C. Chang, G. Thomas, J. Aldridge, *ibid.* **61**, 669 (1971).
27. S. J. Singer, *Advan. Protein Chem.* **22**, 1 (1967).
28. M. Pellegrini, H. Oen, C. R. Cantor, *Proc. Nat. Acad. Sci. U.S.A.* **69**, 837 (1972); C. R. Cantor, personal communication.
29. M. Nomura, P. Traub, H. Beckmann, *Nature* **219**, 793 (1968).

30. V. Erdmann, S. Fahnestock, K. Higo, M. Nomura, *Proc. Nat. Acad. Sci. U.S.A.* **68**, 2932 (1971).
31. V. Erdmann, H. G. Doberer, M. Sprinzl, *Mol. Gen. Genet.* **114**, 89 (1972); S. Fahnestock and M. Nomura, *Proc. Nat. Acad. Sci. U.S.A.* **69**, 363 (1972); G. Bellemare, R. Monier, S. Fahnestock, M. Nomura, in preparation.
32. M. Nomura and V. Erdmann, *Nature* **228**, 744 (1970).
33. G. G. Brownlee, F. Sanger, B. G. Barrell, *ibid.* **215**, 735 (1967); D. Bernard and S. M. Weissman, *J. Biol. Chem.* **246**, 747 (1971).
34. I. D. Raacke, *Proc. Nat. Acad. Sci. U.S.A.* **68**, 2357 (1971).
35. W. Held and M. Nomura, unpublished data; see also (21).
36. J. Konisky and M. Nomura, *J. Mol. Biol.* **26**, 181 (1967).
37. C. M. Bowman, J. E. Dahlberg, T. Ikemura, J. Konisky, M. Nomura, *Proc. Nat. Acad. Sci. U.S.A.* **68**, 964 (1971).
38. B. W. Senior and I. B. Holland, *ibid.*, p. 959.
39. T. L. Helser, J. E. Davies, J. E. Dahlberg, *Nature* **233**, 12 (1971); *ibid.* **235**, 6 (1972).
40. P. F. Sparling, *Science* **167**, 56 (1968).
41. C. J. Lai, B. Weisblum, S. Fahnestock, M. Nomura, *J. Mol. Biol.*, in press.
42. H. W. Schaup, M. Green, C. G. Kurland, *Mol. Gen. Genet.* **109**, 193 (1970); *ibid.* **112**, 1 (1971); R. A. Garrett, K. H. Rak, L. Daya, G. Stöffler, *ibid.* **114**, 112 (1971).
43. P. Traub and M. Nomura, *J. Mol. Biol.* **40**, 391 (1969).
44. H. Nashimoto, W. Held, E. Kaltschmidt, M. Nomura, *ibid.* **62**, 121 (1971).
45. H. E. Homann and K. H. Nierhaus, *Eur. J. Biochem.* **20**, 249 (1971).
46. S. H. Miall and F. O. Walker, *Biochim. Biophys. Acta* **174**, 551 (1969).
47. H. W. Schaup and C. G. Kurland, *Mol. Gen. Genet.* **114**, 350 (1972); H. W. Schaup, M. Sogin, C. Woese, C. G. Kurland, *ibid.*, p. 1.
48. R. A. Zimmerman, A. Muto, P. Fellner, C. Ehresmann, C. Branlant, *Proc. Nat. Acad. Sci. U.S.A.* **69**, 1282 (1972).
49. P. Schendel, P. Maeba, G. R. Craven, *ibid.*, p. 544.
50. R. Brimacombe, J. M. Morgan, R. A. Cox, *Eur. J. Biochem.* **23**, 52 (1971).
51. C. T. Shih and G. R. Craven, in preparation. See also (49). In relation to this, it should be noted that Kurland and his co-workers observed cross-linking of S18(P12) and S21(P15) using *N,N'-p*-phenylenedimaleimide [C. G. Kurland, M. Green, H. W. Schaup, D. Donner, L. Lutter, E. A. Birge, *Fed. Eur. Biochem. Soc. Symp.* **23**, 75 (1972)]. Similar experiments were also done by Chang and Flaks [F. N. Chang and J. G. Flaks, *J. Mol. Biol.* **68**, 177 (1972)].
52. T. A. Bickle, J. W. B. Hershey, R. R. Traut, *Proc. Nat. Acad. Sci. U.S.A.* **69**, 1327 (1972).
53. G. R. Craven and V. Gupta, *ibid.* **67**, 1329 (1970); F. N. Chang and J. G. Flaks, *ibid.*, p. 1321; *J. Mol. Biol.* **61**, 387 (1971); P. Spinik-Elson and A. Breiman, *Biochim. Biophys. Acta* **254**, 457 (1971); R. R. Crichton and H. G. Wittmann, *Mol. Gen. Genet.* **114**, 95 (1971); L. Kahan and E. Kaltschmidt, *Biochemistry* **11**, 2691 (1972); K. Huang and C. R. Cantor, *J. Mol. Biol.* **67**, 265 (1972).
54. L. Streyer, *Science* **162**, 526 (1968).
55. D. M. Engelman and P. Moore, *Proc. Nat. Acad. Sci. U.S.A.* **69**, 1997 (1972).
56. S. Fahnestock, V. Erdmann, M. Nomura, *Biochemistry* **12**, 220 (1973).
57. H. Nashimoto and M. Nomura, *Proc. Nat. Acad. Sci. U.S.A.* **67**, 1440 (1970); G. Kreider and B. L. Brownstein, *J. Mol. Biol.* **61**, 135 (1971).
58. S. Fahnestock, W. Held, M. Nomura, in *Proceedings of the First John Innes Symposium on Generation of Subcellular Structures*, R. Markham, Ed. (Innes Institute, Norwich, England, in press).
59. P. Traub and M. Nomura, *J. Mol. Biol.* **34**, 575 (1968); T. Staehelin, D. Maglott, R. E. Monro, *Cold Spring Harbor Symp. Quant. Biol.* **34**, 39 (1969).
60. E. Hamel, M. Koka, T. Nakamoto, *J. Biol. Chem.* **347**, 805 (1972).
61. K. Kischa, W. Möller, G. Stöffler, *Nature New Biol.* **233**, 62 (1971); N. Brot, E. Yamasaki, B. Redfield, H. Weissbach, *Arch. Biochem. Biophys.* **148**, 148 (1972); H. Weissbach, B. Redfield, E. Yamazaki, R. C. Davis, Jr., S. Pestka, N. Brot, *ibid.* **149**, 110 (1972).
62. G. Stöffler, L. Daya, K. H. Rah, R. A. Garrett, *Mol. Gen. Genet.* **114**, 125 (1972).

63. P. N. Gray and R. Monier, *Fed. Eur. Bio-chem. Soc. Lett.* **18**, 145 (1971); P. N. Gray, R. A. Gerrett, G. Stöffler, R. Monier, *Eur. J. Biochem.* **28**, 412 (1972).
64. H. Maruta, T. Tsuchiya, D. Mizuno, *J. Mol. Biol.* **61**, 123 (1971).
65. C. R. Raetz and E. P. Kennedy, *J. Biol. Chem.* **247**, 2008 (1972).
66. B. Hall and J. Gallant, *Nature New Biol.* **237**, 131 (1972); H. A. DeBoer, G. Raue, M. Gruber, *Biochim. Biophys. Acta* **237**, 131 (1972); E. Lund and N. O. Kjeldgaard, *Eur. J. Biochem.* **28**, 316 (1972).
67. W. A. Haseltine, R. Block, W. Gilbert, K. Weber, *Nature* **238**, 381 (1972).
68. R. J. Britten and B. J. McCarthy, *Biophys. J.* **2**, 49 (1962); G. Mangiarotti, D. Apirion, D. Schlessinger, L. Silengo, *Biochemistry* **7**, 456 (1968); S. Osawa, *Annu. Rev. Biochem.* **37**, 109 (1968).
69. C. Guthrie, H. Nashimoto, M. Nomura, *Proc. Nat. Acad. Sci. U.S.A.* **63**, 384 (1969); P. Tai, D. P. Kessler, J. Ingraham, *J. Bacteriol.* **97**, 1298 (1969).
70. S. Osawa, E. Otaka, T. Itoh, T. Fukui, *J. Mol. Biol.* **40**, 321 (1969); L. J. Lewandowski and B. L. Brownstein, *ibid.* **41**, 277 (1969).
71. N. B. Hecht and C. R. Woese, *J. Bacteriol.* **95**, 986 (1968); M. Adesnik and C. Levinthal, *J. Mol. Biol.* **46**, 281 (1969); A. Dahlberg and A. C. Peacock, *ibid.* **55**, 61 (1971); H. Nashimoto and M. Nomura, *Proc. Nat. Acad. Sci. U.S.A.* **67**, 1440 (1970).
72. M. Sogin, B. Pace, N. R. Pace, C. R. Woese, *Nature New Biol.* **232**, 48 (1971); G. G. Brownlee and E. Cartwright, *ibid.*, p. 50; C. V. Lowry and J. E. Dahlberg, *ibid.*, p. 52; F. Hayes, D. Hayes, P. Fellner, C. Ehresmann, *ibid.*, p. 55.
73. A. Yuki. *J. Mol. Biol.* **56**, 435 (1971); *ibid.* **62**, 321 (1971); G. Corte, D. Schlessinger, D. Longo, P. Venkov, *ibid.* **60**, 325 (1971).
74. C. Terhorst, B. Wittmann-Liebold, W. Möller, *Eur. J. Biochem.* **25**, 13 (1972).
75. U. K. Laemmli, *Nature* **227**, 680 (1970); R. C. Dickson, S. L. Barnes, F. A. Eiserling, *J. Mol. Biol.* **53**, 461 (1970); J. Hosoda and R. Cone, *Proc. Nat. Acad. Sci. U.S.A.* **66**, 1275 (1970).
76. R. Rosset, C. Vola, J. Feunteun, R. Monier, *Fed. Eur. Biochem. Soc. Lett.* **18**, 127 (1971).
77. O. Maaløe and N. O. Kjeldgaard, *Control of Macromolecular Synthesis* (Benjamin, New York, 1966); G. Edlin and P. Broda, *Bacteriol. Rev.* **32**, 206 (1968); R. R. Burgess, *Annu. Rev. Biochem.* **40**, 711 (1971).
78. M. L. Pato and K. von Meyenberg, *Cold Spring Harbor Symp. Quant. Biol.* **35**, 497 (1970); W. F. Doolittle and N. R. Pace, *Nature* **228**, 125 (1970); H. Bremer and L. Berry, *Nature New Biol.* **234**, 81 (1971); C. R. Kossman, T. D. Stamato, D. E. Pettijohn, *Nature* **234**, 102 (1971).
79. O. L. Miller, Jr., and B. A. Hamkalo, *Int. Rev. Cytol.* **33**, 1 (1972).
80. A. Travers, R. Kamen, M. Cashel, *Cold Spring Harbor Symp. Quant. Biol.* **35**, 415 (1970); A. Travers, R. I. Kamen, R. F. Schleif, *Nature* **228**, 749 (1970); A. Travers, *Nature New Biol.* **229**, 69 (1971).
81. W. A. Haseltine, *Nature* **235**, 329 (1972); C. Hussey, J. Pero, R. G. Schorenstein, R. Losik, *Proc. Nat. Acad. Sci. U.S.A.* **69**, 407 (1972); D. E. Pettijohn, *Nature New Biol.* **235**, 204 (1972).
82. M. Cashel and J. Gallant, *Nature* **221**, 838 (1969); R. A. Lazzarini and M. Cashel, *J. Biol. Chem.* **246**, 4381 (1971); M. Cashel, *ibid.* **244**, 3133 (1969).
83. O. Maaløe, *Develop. Biol. Suppl.* **3**, 33 (1969).
84. P. S. Sypherd, D. M. O'Neil, M. M. Taylor, *Cold Spring Harbor Symp. Quant. Biol.* **34**, 77 (1969); S. Dekio, R. Takata, S. Osawa, *Mol. Gen. Genet.* **109**, 131 (1970); S. Osawa, in "Functional units in protein synthesis," *Proc. Fed Eur. Biol. Soc. 7th 1971, Varna, Bulgaria,* in press.
85. M. Nomura and F. Engbaek, *Proc. Nat. Acad. Sci. U.S.A.* **69**, 1526 (1972). Phage mu is known to cause mutation in many bacterial genes by the direct integration into the gene [see A. L. Taylor, *ibid.* **50**, 1043 (1963); W. Boram and J. Abelson, *J. Mol. Biol.* **62**, 171 (1971); A. J. Bukhari and D. Zipser, *Nature New Biol.* **236**, 240 (1972)]. The insertion of the mu prophage DNA into any bacterial operons should interfere with the transcription of the cistrons distal to the initiation site (the "promotor"), but not the cistrons proximal to the promotor. This principle has been confirmed experimentally [see E. Jordon, H. Saedler, P. Starlinger, *Mol. Gen. Genet.* **102**, 353 (1968); A. Touissaint, *ibid.* **106**, 89 (1969)] and was used in the experiment of Nomura and Engbaek to test if four drug resistant genes at the *str-spc* region are in the same operon.
86. P. Fellner, *Biochimie* **53**, 573 (1971); P. Fellner, C. Ehresmann, P. Stiegler, J. P. Ebel, *Nature*, in press.
87. G. Funatsu, E. Schiltz, H. G. Wittmann, *Mol. Gen. Genet.* **114**, 106 (1972); G. Funatsu, W. Puls, E. Schiltz, J. Reinbolt, H. G. Wittmann, *ibid.* **115**, 131 (1972); G. Funatsu, K. Nierhaus, B. Wittmann-Liebold, *J. Mol. Biol.* **64**, 201 (1972); G. Funatsu and H. G. Wittmann, *ibid.* **68**, 547 (1972).
88. B. Ballou, S. Mizushima, M. Nomura, unpublished data.
89. This manuscript was written in April 1972. Several papers which are pertinent to the discussion but were published thereafter may not be mentioned. Supported by grant GM-15422 from the National Institutes of Health and by grant GB-31086X from the National Science Foundation. This is Paper No. 1553 of the Laboratory of Genetics.

22

Reprinted from *Proc. Natl. Acad. Sci.*, 70(12), Pt. 1, 3361–3365 (1973)

Synthesis of a Large Precursor to Ribosomal RNA in a Mutant of *Escherichia coli*

(ribonuclease III/DNA–RNA hybridization)

N. NIKOLAEV*, L. SILENGO†, AND D. SCHLESSINGER

Department of Microbiology, Washington University School of Medicine, St. Louis, Missouri 63110

Communicated by Paul A. Marks, July 10, 1973

ABSTRACT   A mutant of *E. coli*, isolated by Kindler and Hofschneider as a strain defective in RNase III activity, forms a 30S precursor of ribosomal RNA ("30S pre-rRNA"). The half-life of the 30S pre-rRNA in growing cells at 30°, estimated by the rate of specific ³[H]uridine incorporation, is about 1 min. In rifampicin-treated cells, the RNA is metabolized to mature rRNA with a half-life of about 2 min.

The 30S pre-rRNA has been highly purified. DNA–RNA hybridization tests demonstrate that it contains both 16S and 23S rRNA sequences. Also, in cultures treated with rifampicin, the cleavage products of radioactive 30S pre-rRNA include 25S and 17.5S RNA species, destined to become 23S and 16S rRNA. Thus, each 30S chain probably contains one 16S and one 23S RNA sequence, as well as additional sequences. Two independent techniques indicate that the additional portions account for about 27% of the total length: (*1*) By comparison to the sedimentation rate and electrophoretic mobility of marker RNAs, the 30S pre-RNA has an apparent molecular weight of $2.3 \times 10^6 \pm 5\%$, or 28% more than the sum of 16S and 23S rRNA; (*2*) 27% of the 30S pre-rRNA is not competed away from hybridization by mature 16S and 23S rRNA.

Thus, bacteria appear to make a pre-rRNA similar in some respects to that observed in eukaryotes; though in normal *E. coli* cells, the pre-rRNA is ordinarily cleaved endonucleolytically during its formation.

AB105, isolated by Hofschneider and his collaborators (1), was originally characterized as a strain of *Escherichia coli* that grew poorly even in rich medium, and yielded extracts which showed little or no attack on the double-stranded replicative form of RNA from R17 phage. Such endonucleolytic attack is attributed to RNase III (2).

One suggested process in which endonuclease function might be required is the formation of 23S, 16S, and perhaps 5S RNA from longer transcripts, possibly even starting from a single initiation point (3, Pace, N. R., review in preparation). In growing bacteria, this process has been inferred only from indirect experiments that sometimes allowed conflicting interpretations (4, Pace, N. R., review in preparation). In AB105, the process is slowed and a precursor containing both 16S and 23S rRNA becomes directly observable.

MATERIAL AND METHODS

E. coli strain AB105, from Dr. P. Hofschneider, and its parent strain A19 (5) were grown at 30° with aeration, in minimal

---

* On leave from the Institute of Biochemistry, Bulgarian Academy of Sciences, Sofia, Bulgaria.
† Present address: Department of Biochemistry, 2nd Faculty of Medicine, University of Naples, Italy.

salts medium plus glucose, supplemented with 0.8% technical grade Difco casamino acids.

RNA was labeled either with [¹⁴C]uracil (52 Ci/mol) or [³H]uridine (30 Ci/mmol, Schwarz BioResearch), or both, as indicated in the text. The labeled RNA was extracted at 60° with phenol and 1% sodium dodecyl sulfate, in 0.2 M Tris·acetate buffer (pH 5.0) containing 5 mM ethylenediaminetetraacetate (6). After two to three deproteinizations and chloroform extraction, the RNA in the aqueous phase was precipitated with 2.5 volumes of ethanol at −20° for at least 12 hr. For electrophoretic analysis, the precipitated RNA was redissolved in electrophoresis buffer (0.036 M Tris·HCl (pH 7.6)–0.03 M sodium dihydrogen phosphate–1 mM ethylenediaminetetraacetate–0.2% sodium dodecyl sulfate. For sedimentation analysis, the RNA was dissolved in 0.1 M Tris·acetate (pH 7.5) containing 0.5% sodium dodecyl sulfate.

30S precursor to ribosomal RNA, pre-rRNA, was purified by two successive zonal sedimentation runs. 38-ml 10–30% sucrose gradients were used in the SW27 rotor of the L2-65B Spinco ultracentrifuge (see Fig. 4). The buffer in the gradient was 0.1 M Tris·acetate (pH 7.5) containing 0.5% sodium dodecyl sulfate, which permitted centrifugation at 4° without salt precipitation. Centrifugation in each run was for 23 hr at 25,000 rpm. 0.75-ml Fractions were collected by pumping from the bottom of the tubes. The fractions from the second gradient containing the 30S pre-rRNA (as in Results, Fig. 4) were again precipitated with ethanol.

For DNA–RNA hybridization analyses, the purified 30S pre-rRNA was resuspended at 2 μg/ml in 0.01 M Tris·HCl (pH 7.5)–0.01 M MgSO₄. The RNA was then treated with 10 μg/ml of DNase (30 min at 37°). 0.1 M NaCl was then added along with 10 μg/ml of Pronase [predigested for 30 min at 37° and tested against [³H]poly(U) for absence of RNase activity]. After 30 min of Pronase treatment at 37°, the solution was again deproteinized twice with phenol, and the RNA was precipitated with ethanol and resuspended in standard saline–citrate [0.15 M sodium chloride, 0.015 sodium citrate (pH 7.0)] (7) containing 5 mM ethylenediaminetetra-acetate. Once again the RNA was precipitated with ethanol and resuspended in standard saline–citrate for hybridization. The electrophoretic mobility of 30S pre-rRNA was retained throughout the purification procedure, with a final yield of about 50% of the initial material in each of two trials.

Unlabeled 16S and 23S rRNA for use as competitors in hybridization trials (Fig. 6) were prepared by phenol extrac-

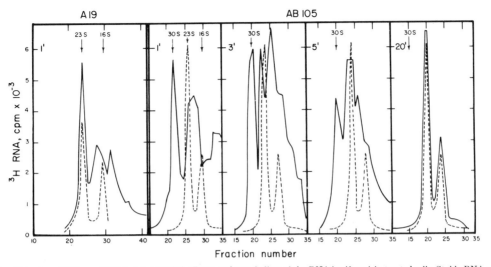

FIG. 1.   Detection of 30S pre-rRNA in pulse-labeled AB105, and metabolism of the RNA in rifampicin-treated cells. Stable RNA was pre-labeled in a 20-ml growing culture of A19 (*left panel*) or the derivative AB105 (*right panels*) for two generations with 0.05 μCi/ml of [^{14}C]uracil. At an optical density at 420 nm of 0.6, 5 μCi/ml of [^{3}H]uridine was added (see *Methods*). 1 min later, 5-ml portions were chilled on crushed ice containing 100 μg/ml of chloramphenicol. The rest of the AB105 culture received 300 μg/ml of rifampicin, and further portions were harvested at the indicated times (' = min). Cells were collected from each sample and RNA was extracted with phenol. The RNA was then analyzed in agarose-acrylamide gel electrophoresis (see *Materials and Methods*). (——), ^{3}H-pulse-labeled RNA; (– – –), ^{14}C-labeled marker RNA; ^{14}C cpm are plotted multiplied by 10.

tion of gradient-purified 30S and 50S ribosomes (5). Before use the RNA was treated with DNase and Pronase, and reextracted with phenol by the same protocol employed for 30S pre-rRNA (see above).

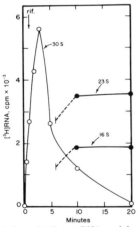

FIG. 2. Metabolism of 30S pre-rRNA, and formation of 16S and 23S rRNA, in a rifamicin-treated culture of AB105. In an experiment like that in Fig. 1, RNA samples were extracted from cells before or after rifampicin addition. The ^{14}C-marker label in a set of gel analyses was then normalized to constancy ($\pm 5\%$), and the relative amount of [^{3}H]RNA with the mobility of 30S, 23S, and 16S RNA was estimated. At early times, no figure is given for 23S and 16S RNA species, because of the apparent presence of immature rRNA precursors and mRNA with similar mobilities.

DNA–RNA hybridization was according to Kennell (7). *E. coli* K12 DNA was obtained from General Biochemicals, Chagrin Falls, Ohio.

Electrophoresis in 0.5 agarose–1.8% acrylamide gels was according to a modification of the method of Peacock and Dingman (8, Wimmer, E., manuscript in preparation).

DNase was from Worthington Biochemicals, pancreatic

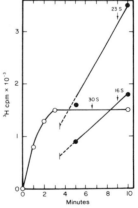

FIG. 3.   Saturation with [^{3}H]uridine of 30S pre-rRNA during continuous labeling of AB105. In an experiment like that of Fig. 1, but with no addition of rifampicin, RNA was extracted from serial samples and analyzed by gel electrophoresis. The ^{3}H content in 30S pre-rRNA (and, at late times, in mature 16S and 23S rRNA) was then added up and plotted, normalized to the constant content of ^{14}C-pre-labeled marker RNA.

RNase from Sigma Chemical Co. Rifampicin was the gift of Gruppo Lepetit, Milan.

Gel samples were extracted with undiluted commercial NH₄OH (0.5 ml per gel slice) before counting in Bray's solution. Other samples were precipitated with 5% CCl₃COOH, collected on Millipore filters, and counted in toluene–0.5% 2,5-diphenyloxazole–0.03% 1,4-bis[2-(5-phenyloxazol-yl)]-benzene scintillation fluid with the counter set appropriately for simultaneous measurements of ³H and ¹⁴C.

RESULTS

Fig. 1 shows acrylamide gel analyses of pulse-labeled RNA extracted from strain A19 (*left panel*) and the derivative strain AB105 (*other panels*). Total RNA in the cultures was first labeled uniformly with [¹⁴C]uracil; [³H]uracil was then added. The distribution for AB105 shows a sharp peak with a mobility corresponding to about 30 S, and a relative deficiency of label at the positions of marker ¹⁴C-16S and 23S rRNA.

When 30S RNA was pulse-labeled for 1 min in growing cells, and 300 μg/ml of rifampicin was then added to block further initiation of RNA chain synthesis, the fate of the RNA could be inferred from the subsequent distribution of radioactivity. Fig. 1 shows some analyses at increasing times after rifampicin addition. The regions of 16 S and 23 S at early times are obscured by mRNA species, but in Fig. 1 (and also in Fig. 4 below), the newly synthesized RNA appears to progress through the sequence 30 S, then 25 S and 17.5 S, before appearing as mature 23S and 16S RNA. Presenting these data as in Fig. 2 illustrates that during rifampicin treatment 30S RNA decays with a half-life of about 2 min.

These results suggested that the 30S RNA might be a precursor of 16 S and 23 S rRNA. The suggestion is further supported by labeling studies and DNA–RNA hybridization

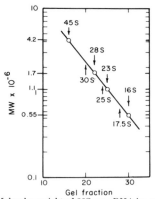

FIG. 5. Molecular weight of 30S pre-rRNA by comparison of electrophoretic mobility to that of known RNA markers. Purified ¹⁴C-labeled 45S rRNA and 28S rRNA extracted from cultured HeLa cells, and ¹⁴C-*E. coli* 16S and 23S rRNA, were added as known markers to acrylamide gels containing ³H-labeled 30S pre-rRNA. After electrophoresis and gel slicing, positions were plotted on a semi-logarithmic scale against gel fraction number (according to ref. 19). The molecular weights of the markers were used to construct the scale of the ordinate. Positions of the "25S" and "17.5S" RNA species (compare Fig. 4) are also indicated.

analysis. If labeling with [³H]uridine was continued for longer times, without rifampicin addition, analysis of extracted RNA showed a gradual saturation of the 30S RNA peak. The ³H content of 16S and 23S RNA is again obscured at early times, but there is a progressive increase in 16S and 23S rRNA thereafter. A number of the analyses are summarized in Fig. 3. From the kinetics of saturation, we estimate that the half-life of the 30S pre-rRNA in growing cells is about 1 min.

The very large size of 30S RNA has simplified its purification. The resolution from other RNA is as good in sucrose gradients as in gels. When the fractions comprising the 30S peak from a sucrose gradient were pooled, concentrated, and analyzed again, the RNA was essentially pure by centrifugal (Fig. 4) or electrophoretic analysis (data not shown). In these experiments, the initial RNA fraction contained mature 16S and 23S RNA labeled with [¹⁴C]uracil as a marker; the purified ³H-30S RNA contained no detectable ¹⁴C label, confirming its purity as >98%.

A finer estimate of the size of the 30S RNA was obtained from comparisons of its mobility and sedimentation to those of added *E. coli* or HeLa cell RNA markers. For example, Fig. 5 shows a comparison of mobilities in gels; on this basis, the large RNA from *E. coli* has a molecular weight of 2.3 × 10⁶ ± 5%, and will be called 30S pre-rRNA. A very similar estimate was derived from additional sedimentation analysis.

The availability of pure 30S pre-RNA made possible independent, definitive proof that it contains the sequences of mature rRNA. Fig. 6 shows the results of a sample DNA–RNA hybridization trial. Both 16S and 23S RNA competed with 30S pre-rRNA sequences in the trials. Though difficulties with cross-hybridization make it impossible to quantitate the relative amounts of 16S and 23S-specific sequences, the maximum competition is clear (Fig. 6, *bottom*

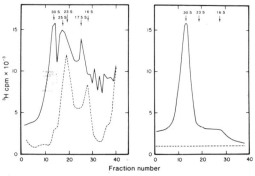

FIG. 4. Zonal sedimentation analysis and purification of 30S pre-rRNA in sucrose gradients. Cells were labeled as in Fig. 1, but with 50 μCi/ml instead of 5 μCi/ml of [³H]uridine. 2 min after rifampicin addition, the 10-ml culture was harvested and RNA was extracted and fractionated in sucrose gradients (for details, see *Methods*). The *left panel* shows the sedimentation profile in a sucrose gradient of the extracted RNA (compare the electrophoretic analysis of the 3-min sample in Fig. 1). The fractions from the shaded area were pooled, the RNA precipitated with ethanol, resuspended and reanalyzed. The 30S peak of the second gradient (*right panel*) was used as 30S pre-rRNA for the analyses of Fig. 6 below.

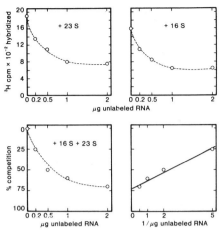

FIG. 6. Competition for DNA–RNA hybridization by 30S pre-rRNA and mature 16S and 23S rRNA. *Top panels* show the competition of labeled 30S pre-rRNA by increasing levels of unlabeled 23S and 16S rRNA. *Bottom left,* competition by total unlabeled ribosomal RNA, plotted as percent of the hybridization in absence of competitor. *Bottom right,* the data for competition by total rRNA, plotted as a function of the reciprocal of added competitor concentration. Hybridization reactions contained, in 0.5 ml of 6-times concentrated standard saline–citrate, 0.1 μg of labeled 30S pre-rRNA (3500 cpm) and a filter bearing 100 μg of alkali-denatured unlabeled *E. coli* DNA. After 20 hr of incubation at 66°, the filters were treated with 40 μg/ml of pancreatic RNase (pre-heated to inactive DNase) for 1 hr in 2-times concentrated standard saline–citrate (9), then washed with 50 ml of the same buffer, dried, and counted in toluene scintillation fluid.

panels): total unlabeled ribosomal RNA competed with 73% of the sequences in 30S pre-rRNA.

DISCUSSION

There seems little doubt that the 30S RNA species formed in strain AB105 is a pre-rRNA that is a primary transcription product. The sharpness of the peak suggests that the RNA is uniform in structure. From the sum of the results, each chain likely contains one 16S and one 23S rRNA molecule, as well as additional sequences. This transcript will probably be valuable for studies of sequences in rRNA and of steps in ribosome formation.

Estimates from zonal sedimentation and acrylamide gel analyses have agreed, so that the estimated molecular weight of 2.3×10^6, or about 7700 nucleotides, is probably not distorted by unusual conformational effects, and should be accurate to ±5%. The sum of one 16S and one 23S rRNA is 1.65×10^6 (9), which corresponds to 72% of the 30S pre-rRNA. Therefore every molecule of 30S pre-rRNA would contain, besides 16S and 23S RNA sequences, sequences adding up to 0.65×10^6 (28% of 30S pre-rRNA). These may include sequences associated with 5S RNA (Pace, N. R., review in preparation).

DNA–RNA hybridization analyses are consistent with these estimates. High levels of DNA were required for efficient hybridization of 30S pre-rRNA, supporting the idea of a relatively homogeneous fraction. At high efficiencies of

hybridization, 27% hybridized as if different from mature rRNA, once again equivalent to a molecular weight of about 0.6×10^6 per chain.

On the basis of these results and a more complete kinetic analysis of formation of 16S and 23S rRNA (unpublished), we anticipate several points: (*1*) From Figs. 2 and 3, we estimate that the half-life of 30S pre-RNA is at least 1 min, and this RNA species represents at least 0.5% of the total rRNA in cells doubling every 180 min. The kinetics indicate that all, or very nearly all, the 16S and 23S rRNA chains in strain AB105 start out as part of 30S pre-rRNA chains. (*2*) The peaks at 25S and 17.5S (Fig. 4) appear to be initial cleavage products of 30S pre-rRNA (see below), which give rise respectively to 23S and 16S RNA. (*3*) The results help to settle a puzzle about the origin of 23S RNA: During continuous labeling of cells with [³H]uracil, the rates of labeling of 16S and 23S rRNA became maximal with a very similar time course (10, 11). These results contrasted with the expectation of a synthetic time half as long for the 16S RNA. However, if the 23S RNA sequences are located near the 3′-terminus of a 30S pre-rRNA, then complete 30S chains will first be labeled in their 23S RNA portion, with a lag before 16S RNA sequences are appreciably labeled. A split in the 30S pre-RNA when RNA polymerase is part way through the 23S RNA sequence could then produce the observed rates of labeling of mature 16S and 23S rRNA.

A similar suggestion has been made by Pettijohn and Kossman, who have observed formation of long RNA chains *in vitro*, apparently by extension of those begun *in vivo*; the chains contain both 16S and 23S RNA sequences in tandem (manuscript in preparation). The largest precursor they report ("p30S") is rather smaller than the 30S pre-rRNA reported here (1.75×10^6 instead of 2.3×10^6); but their suggestions that RNA polymerase can read an entire rRNA transcription unit as such, with no internal initiations, and that the product is then cleaved, are demonstrated here *in vivo*.

Since strain AB105 shows a specific lesion in RNase III (unpublished results), it seemed possible that RNase III was directly involved in the processing of the 30S pre-rRNA. We have found that highly purified RNase III produces 25S and 17.5S RNA peaks when incubated with pure 30S pre-rRNA; the products comigrate in gels and sucrose gradients with the 25S and 17.5S RNAs seen *in vivo* (Fig. 4). RNase III has been considered completely specific for double-stranded RNA species; but in addition to our data, Dunn and Studier report a possible involvement of RNase III in the cleavage of RNA polymerase products formed on purified T7 DNA (12). The continued slow growth of strain AB105 might then be due to a low remaining level of the important RNase III, or to the weak action of another, "substituting" RNase that provides a slow rate of ribosome formation.

If one assumes that the 30S RNA in AB105 is truly a full transcript of a ribosomal RNA segment of the genome, then the apparent difference in rRNA formation in bacteria and eukaryotic cells becomes much less. The rRNA precursor in plants, for example (13), is almost the same as 30S pre-rRNA in its mobility in acrylamide gel electrophoresis. Snyder *et al.* (14) have already suggested that self-complementary loops may be involved in processing of rRNA precursors in eukaryotic cells. It remains to be seen whether RNase III or an analogous enzyme in higher cells can initiate processing of pre-rRNA in those cells as well.

It thus appears that in eukaryotes (15) and *E. coli* (16–18) the sequence of events in rRNA maturation is fundamentally similar, based on a combination of endonuclease and exonuclease action. Concerning mRNA, a comparable processing has been described in eukaryotes (19), thought to have developed during the evolution of nuclear organization. However, we have also indications from strain AB105 that even that process may occur in *E. coli*.

NOTE ADDED IN PROOF

In formamide gel electrophoresis [according to Staynov, D. Z., Pinder, I. C. & Gratzer, W. B. (1972) *Nature New Biol.* **235**, 108–110], the molecular weights of 17.5S, 25S, and 30S pre-rRNA are proportionally lower than in agarose-acrylamide gels (0.65, 1.2, and 2.1×10^6 compared to the standards in Fig. 5).

Some of our data on RNase III action on mRNA and 30S pre-rRNA appeared in *J. Biol. Chem.*, issue of Oct. 25, 1973. After acceptance of our manuscripts, we learned of overlapping independent work, in substantial agreement, by Dunn and Studier (this issue, p. 3298).

We are grateful to Dr. P. H. Hofschneider for providing us with strain AB105 and communicating unpublished results. Dr. S. Gotoh in our laboratory kindly prepared the purified 45S and 28S RNA species from HeLa cells used in the studies for Fig. 5. The work was supported by Grant GB 23052 from the National Science Foundation.

1. Keil, T. U. & Hofschneider, P. H. (1973) *Biochim. Biophys. Acta,* **312**, 297–310.
2. Robertson, H. D., Webster, R. D. & Zinder, N. D. (1968) *J. Biol. Chem.* **243**, 82–91.
3. Kossman, C., Stamato, T. & Pettijohn, D. (1971) *Nature New Biol.* **234**, 102–104.
4. Bleyman, M., Kondo, M., Hecht, N. & Woese, C. (1969) *J. Bacteriol.* **99**, 535–543.
5. Gesteland, R. F. (1966) *J. Mol. Biol.* **18**, 356–371.
6. Hadjiolov, A. A., Nikolaev, N. & Shulga, S. (1972) *Int. J. Biochem.* **3**, 509–517.
7. Kennell, D. & Kotoulas, A. (1968) *J. Mol. Biol.* **34**, 71–84.
8. Peacock, A. C. & Dingman, C. W. (1968) *Biochemistry* **7**, 668–674.
9. Kurland, C. G. (1960) *J. Mol. Biol.* **2**, 83–91.
10. Mangiarotti, G., Apirion, D., Schlessinger, D. & Silengo, L. (1968) *Biochemistry* **7**, 456–472.
11. Adesnik, M. & Levinthal, C. (1969) *J. Mol. Biol.* **46**, 281–303.
12. Dunn, J. J. & Studier, F. W. (1973) *Proc. Nat. Acad. Sci. USA* **70**, 1559–1563.
13. Grierson, D., Rogers, M. E., Sartirama, M. L. & Loening, N. E. (1970) *Cold Spring Harbor Symp. Quant. Biol.* **35**, 589–598.
14. Snyder, A. L., Kann, H. E. & Kohn, K. W. (1971) *J. Mol. Biol.* **58**, 555–565.
15. Maden, B. E. H. (1971) *Progr. Biophys. Mol. Biol.* **22**, 127–177.
16. Corte, G., Schlessinger, D., Longo, D. & Venkov, P. (1971) *J. Mol. Biol.* **60**, 325–338.
17. Venkov, P., Waltschewa, L. & Schlessinger, D. (1972) *FEBS Symp.* **23**, 379–394.
18. Yuki, A. (1971) *J. Mol. Biol.* **62**, 321–329.
19. Darnell, J. E., Philipson, L., Wall, R. & Adesnik, M. (1971) *Science* **174**, 507–510.

Part IV
TRANSLATION: MECHANISM OF PEPTIDE BOND FORMATION

Editor's Comments
on Papers 23 Through 30

In Part IV, details of the mechanism by which specific sequences of peptide bonds are formed in the polypeptide gene products are presented. The translation process has been divided into three stages: initiation, elongation of the polypeptide chain, and chain termination.

An early finding, which suggested catalytic requirements for polypeptide polymerization, was made in the pioneering studies of Zamecnik et al. (1958). Using a cell-free system from liver, they showed that the incorporation of amino acids from aminoacyl-tRNA into protein could occur only in the presence of a soluble fraction distinct from aminoacyl-tRNA synthetase. Since the advent of cell-free amino acid incorporation systems in bacteria (Lamborg and Zamecnik, 1960; Tissières et al., 1960; Nirenberg and Matthaei, 1961), the nature of the catalytic activities necessary for the formation of peptide bonds has been elucidated in eukaryotic and prokaryotic organisms (for reviews, see Lucas-Lenard and Lipmann, 1971; Haselkorn and Rothman-Denes, 1973). It is now possible to study the initiation, elongation, and termination processes as distinct partial reactions of the overall translation process (see Figure 1).

A requirement for three protein factors in the initiation of translation was demonstrated by Iwasaki et al. (Paper 23) and Revel et al. (1968a). These initiation factors (IF-1, IF-2, and IF-3) are associated with the ribosomes of *Escherichia coli* and are removed by washing with 0.5 to 1.0 M NH_4Cl solutions. All three initiation factors appear to act cooperatively to give maximal initiation (steps 1 and 2 of Figure 1). IF-3 contains two polypeptide species and is required for the binding of mRNA to the 30S ribosomal subunit (Paper 23; Revel et al., 1968b). IF-2, which also contains two classes of polypeptide chains, promotes the binding of the initiator tRNA (see below) and GTP to the 30S initiation complex at low magnesium concentrations (Salas et al., 1967). In addition, IF-2 functions as a ribosome-dependent GTPase (Thach et al., 1969). IF-1 also combines with the 30S initiation complex; it is released when the complex is joined by the 50S ribosomal subunit (Thach et al., 1969). This is accompanied by GTP hydrolysis and results in the formation of the 70S initiation complex. IF-1 is believed to function in the dissociation of IF-2 from the 70S initiation complex (GTP hydrolysis is also required), and thereby promotes cyclic reutilization of IF-2 for further initiator tRNA binding (Haselkorn and Rothman-Denes, 1973). After termination of the translation process and release of the ribosomal subunits, reversible interaction of the released IF-3 with the 30S subunit prevents its reassociation with the 50S subunits, which thereby maintains a pool for the formation of further initiation complexes (see Part III).

The signal for initiation of the translation process is provided by the codons for the initiator tRNA species. Both the AUG and GUG codons

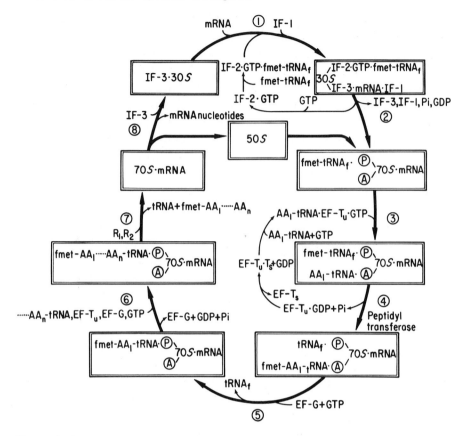

Figure 1 Scheme for the prokaryotic translation cycle. Steps 1 and 2 represent the formation of the 30S and 70S initiation complexes, respectively. A more detailed scheme for these steps is presented by Haselkorn and Rothman-Denes (1973). The binding of aminoacyl (AA)-tRNA to the ribosomal A site is depicted in step 3. The cyclic scheme for the reutilization of EF-T$_u$ is essentially that presented by Lucas-Lenard and Lipmann (1971). In step 4, the formation of the initial peptide bond is shown. This reaction is catalyzed by peptidyl transferase, an integral part of the 50S ribosomal subunit. Steps 5 to 8 have been adapted from a schematic summary presented by Pestka (1971). Step 5 represents the translocation of the growing peptide chain to the P site. In this process, the next codon in the mRNA is exposed. Step 6 shows subsequent stages of the translation cycle in which a repetition of steps 3 to 5 occurs. When a terminator codon is placed in the A site, termination and release of the completed polypeptide chain takes place (step 7). Step 8 represents dissociation of the 70S ribosome into 30S and 50S ribosomal subunits. The definition of other abbreviations and a further description of each step are presented in the text.

function as start signals (Clark and Marcker, 1966; Ghosh et al., 1967). Formylmethionyl tRNA (fmet-tRNA$_f$), which was discovered by Marcker and Sanger (1964), serves as the polypeptide chain initiator (Adams and Capecchi, Paper 24; Webster et al., 1966; Clark and Marcker, 1966). The formylation of the initiator is catalyzed by met-tRNA transformylase; N^{10}-formyltetrahydrofolate serves as the formyl donor (Paper 24; Marcker, 1965; Dickerman et al., 1967). A second tRNA for methionine (tRNA$_m$) has also been found (Clark and Marcker, 1966; Berg et al., Paper 33). Although it is aminoacylated by the same met-tRNA synthetase, it is unable to serve as a formyl acceptor and functions in the recognition of internal AUG codons and the positioning of methionine in the interior of growing polypeptide chains.

Even though the genetic message is read in the $5' \rightarrow 3'$ direction (Smith et al., 1966), none of the bacteriophage RNAs whose sequences have been determined starts with AUG or GUG (Roblin, 1968; De-Wachter and Fiers, 1969). In the phage Qβ RNA coding for the coat protein, AUG does not appear until bases 62–64 (Billeter et al., 1969). To test if the initiation machinery is directed to the mRNA start signal by the primary sequence of the bases preceding it, initiation regions of phage R17 and Qβ RNA have been sequenced and compared with the N-terminal sequence of their protein products (Steitz, Paper 25; Hindley and Staples, 1969). Each of the initiation regions contained AUG and not GUG as the initiation codon.* No extensive sequence homologies were detected around the initiation codons that might serve as recognition sites. These results suggest that primary structure plays no direct role in the specificity of initiation. Since the nucleotides at the beginning of the message for R17 and Qβ coat protein are capable of forming a looped structure, it was suggested that secondary structure governs initiation specificity. Accordingly, only initiating AUG (and presumably GUG) codons would be exposed in the loop with internal AUG codons sequestered within the interior of the messenger molecule. The internal codons become available for translation as the molecule unfolds during elongation of the polypeptide chain.

That such a model is not restricted to the RNA bacteriophages is suggested by the recent studies of Ricard and Salser (1974), who detected a high degree of base pairing in mRNA coding for the lysozyme of the DNA phage T4. It was suggested that the overall secondary structure of the mRNA also controls initiation at the ribosome attachment sites. However, a recent determination of the nucleotide sequence of the initiating region of the *E. coli* lactose operon mRNA did not reveal any obvious secondary structure in which the AUG initiation codon could be placed in a loop (Maizels, 1974). Similarly, no satisfactory secondary structure could be proposed for the initiating region of RNA synthesized *in vitro* from phage T7 (Arrand and Hindley, 1973). Thus, factors other than mRNA conformation may govern initiation specificity, and both the initiation factors and the ribosome itself may have roles in this regard.

---

*A GUG codon was subsequently found in the initiator region of phage MS2 RNA coding for the A protein (Volckaert and Fiers, 1973).

After formation of the 70*S* initiation complex, the polypeptide chain elongation cycle commences (steps 3 to 6 of Figure 1). As the peptide chain grows, it is apparently shunted between two ribosomal sites: the aminoacyl or acceptor (A) site and the donor or peptidyl (P) site (Watson, 1964). The elongation cycle consists of three steps: (1) codon-directed binding of aminoacyl-tRNA to the A site (step 3); (2) formation of a peptide bond between the free amino group of the aminoacyl-tRNA at the A site and the terminal carboxyl group of the fmet-tRNA$_f$ or peptidyl-tRNA in the P site (step 4); (3) release of the tRNA from the P site followed by translocation of peptidyl tRNA from A to P site, and movement of mRNA such that the next codon is available for interaction with aminoacyl-tRNA (step 5). Elongation requires three soluble protein elongation factors (EF-T$_u$, EF-T$_s$, and EF-G), GTP, and the enzyme peptidyl transferase, a component of the 50*S* ribosomal subunit.

A specific requirement for soluble factors in the elongation cycle was suggested by the early findings of Nathans and Lipmann (1961). They noted that washing *E. coli* ribosomes with solutions of high ionic strength resulted in a requirement for a supernatant fraction in the transfer of amino acids from charged tRNA to the growing end of the polypeptide chain. The required soluble fraction was resolved into two complementary elongation factors (EF-T and EF-G) by Nishizuka and Lipmann (1966). EF-T was further fractionated into its T$_s$ and T$_u$ components by Lucas-Lenard and Lipmann (1966); both of these components were required for poly-U-directed polyphenylalanine synthesis in the presence of EF-G. It was subsequently shown by Ravel (1967) that EF-T is involved in the catalysis of a GTP-dependent binding of aminoacyl-tRNA to the ribosome at appropriate magnesium concentrations. Cooper and Gordon (Paper 26) have presented evidence suggesting that EF-T$_u$ from *E. coli* functions with GTP and aminoacyl-tRNA in the direct formation of a ternary complex (see step 3 of Figure 1). The three components in their associated form serve as an intermediate in the binding of aminoacyl-tRNA to the ribosome–messenger complex. Upon transfer of aminoacyl-tRNA to the A site, the GTP is hydrolyzed to GDP and inorganic phosphate (Gordon, 1969; Lucas-Lenard et al., 1969; Shorey et al., 1969; Ono et al., 1969). The nonhydrolyzable GDP analogue, 5′ guanylyl (methylene diphosphonate) or GDPCP, substitutes for GTP in the aminoacyl-tRNA binding reaction; however, no peptide bond is formed and both EF-T$_u$ and GDPCP remain bound to the ribosome (Lucas-Lenard et al., 1969; Skoultchi et al., 1970). This suggests that GTP hydrolysis is necessary for removal of EF-T$_u$ from the ribosome, thus permitting peptide bond formation (Haselkorn and Rothman-Denes, 1973). Both EF-T$_u$ and GDP are removed from the ribosome as an EF-T$_u$-GDP complex (see Weissbach et al., Paper 27). The tightly bound GDP is removed through interaction of EF-T$_s$ with EF-T$_u$. The regenerated EF-T can then further interact with GTP and aminoacyl-tRNA to reform the ternary complex containing EF-T$_u$. A scheme for the cyclic reutilization of EF-T$_u$ is presented in Paper 27. Although an EF-T$_u$–GTP complex forms part of this scheme, the affinity of EF-T$_u$ for GTP is approximately 100 times less than the EF-T$_u$ for EF-T$_s$ or GDP (Miller and Weissbach, 1970; see also Paper 26).

Following the binding of aminoacyl-tRNA to the ribosomal A site, a peptide bond is formed between its α-amino group and the terminal carboxyl group provided by the initiator- or peptidyl-tRNA from the P site (step 4 of Figure 1). This reaction is catalyzed by peptidyl transferase, an integral part of the 50S ribosomal subunit (Maden et al., 1968). A model system has been developed for the peptidyl transferase reaction that permits study of the 50S subunit as a separate entity (Monro et al., 1968). Using this assay and the dissociation of the 50S subunit into a series of split protein and core particle fractions by cesium chloride buoyant density centrifugation, Staehelin et al. (Paper 28) have localized the peptidyl transferase activity in a protein fraction that requires the remaining core particle for activity. This finding suggests a cooperative interaction of 50S components in the overall reaction.

Immediately after the peptidyl transferase reaction, the peptidyl-tRNA product is localized in the ribosomal A site. No additional aminoacyl-tRNA can bind the ribosome unless the A site is vacated and the next codon is exposed. To permit formation of the next peptide bond, the tRNA released in the peptidyl transferase reaction is dissociated from the P site (Lucas-Lenard and Haenni, 1969) and peptidyl-tRNA is translocated to the empty site (Haenni and Lucas-Lenard, 1968) in EF-G- and GTP-dependent reactions (step 5 of Figure 1). Movement of the message to a new codon also occurs (Gupta et al., 1971). The role of EF-G in the translocation reaction was established in the studies of Erbe et al. (Paper 29) using synthetic messages of known sequence. The presence of EF-G was essential for translocation of dipeptidyl-tRNA and the availability of a third codon for tripeptide formation. In the absence of EF-G, only dipeptide formation occurred. During translocation, a molecule of GTP is hydrolyzed to GDP and inorganic phosphate in a ribosome-dependent reaction catalyzed by EF-G (Nishizuka and Lipmann, 1966; Modolell et al., 1973). That the hydrolysis occurs on the ribosome surface is suggested by the isolation of complexes between EF-G and ribosomes containing both GTP and GDP (see Kaziro et al., 1969). It has been suggested that GTP hydrolysis is necessary for removal of EF-G from the ribosome (Inoue-Yokosawa et al., 1974). Recent evidence is consistent with an interdependence of the EF-T$_u$- and EF-G-associated GTPase activities. Two protein components from the 50S subunit are essential for both of these GTPase activities (Sander et al., 1972), and that of IF-2 as well (Fakunding et al., 1973).* In addition, there are antibiotics that prevent the binding and GTPase reactions associated with both these elongation factors (Modolell et al., 1971; Weissbach et al., 1972). Also, the presence of one factor on the ribosome inhibits the activity of the other (Richter, 1972). As in the case of IF-2 and EF-T$_u$, EF-G is also cyclically reutilized during polypeptide chain formation (Haselkorn and Rothman-Denes, 1973).

We shall terminate this section with a brief discussion of the polypeptide chain termination process (step 7 of Figure 1). After addition of the carboxyl terminal amino acid to the nascent peptide chain and translocation of the completed peptidyl-tRNA to the P site, a terminator codon becomes localized in the A site. Release of the peptidyl moiety of the peptidyl-tRNA occurs after an interaction between the terminator codon (*nonsense* triplet) and soluble protein release factors (Scolnick et al., 1968). Based upon studies with bacteriophage mutants in which certain polypeptide chains are prematurely terminated, UAG (Stretton and Brenner, 1965), UAA (Brenner and Beckwith, 1965) and UGA (Brenner et al., 1967) have been identified as the terminator codons. Subsequent genetic alterations of the chain termination mutants resulted in the translation of *nonsense* triplets as an amino acid and completion of the prematurely terminated peptide (Brenner et al., 1965; Kaplan et al., 1965). The recognition of the *nonsense* codons is performed by suppressor tRNAs, which contain mutational alterations in the base sequences of their anticodons (see Part V). By using such suppressor mutants, it was possible to determine the amino acid replacements in the released proteins and thereby deduce the sequences of the *nonsense* codons. Verification of terminator codon assignments has come from sequence studies on the RNA of bacteriophage R17 by Nichols (1970); the tandem stop signal (UAA · UAG) was found. In contrast, Weiner and Weber (1973) have presented evidence for a single UGA termination signal for phage Qβ coat protein.

A soluble fraction that caused peptide release was isolated from *E. coli* by Capecchi (1967) using as message R17 RNA with a premature chain termination signal. The identity of the soluble release factors was established by Scolnick et al. (1968) using a simplified assay for termination that responded to the addition of *nonsense* triplets. Two release factors were isolated: R1 involved in the translation of UAA and UAG, and R2, which translates UAA and UGA. An

---

*Other evidence recently presented by Lockwood and Maitra (1974) suggests that the IF-2-specific GTPase site on the ribosome may be distinct from the proposed common or overlapping site for the GTPase associated with EF-G and EF-T$_u$.

additional protein factor that apparently stimulates the binding of terminator codons has also been isolated (Milman et al., 1969). These studies are detailed by Caskey et al. (Paper 30) and by Capecchi and Klein (1969). Two steps have been identified in the termination reaction: terminator-codon-dependent R factor binding to the ribosome and a peptidyl-tRNA cleavage reaction (Lucas-Lenard and Lipmann, 1971). The latter reaction is apparently catalyzed by peptidyl transferase, which acts as a hydrolase by transferring the cleaved peptide to a hydroxyl group from water instead of the free α-amino group of an aminoacyl-tRNA.

REFERENCES

Arrand, J. R., and J. Hindley (1973). Nucleotide sequence of a ribosome binding site on RNA synthesized *in vitro* from coliphage T7. *Nature New Biol., 244,* 10–13.

Billeter, M. A., J. E. Dahlberg, H. M. Goodman, J. Hindley, and C. Weissmann (1969). Sequence of the first 175 nucleotides from the 5′ terminus of Qβ RNA synthesized *in vitro. Nature (London), 224,* 1083–1086.

Brenner, S., and J. Beckwith (1965). Ochre mutants, a new class of suppressible nonsense mutants. *J. Mol. Biol., 13,* 629–637.

——, A. O. W. Stretton, and S. Kaplan (1965). Genetic code: the "nonsense" triplets for chain termination and their suppression. *Nature (London), 206,* 994–998.

——, L. Barnett, E. R. Katz, and F. H. C. Crick (1967). UGA: a third nonsense triplet in the genetic code. *Nature (London), 213,* 449–450.

Capecchi, M. R. (1967). Polypeptide chain termination *in vitro:* isolation of a release factor. *Proc. Natl. Acad. Sci. U.S.A., 58,* 1144–1151.

——, and H. A. Klein (1969). Characterization of three proteins involved in polypeptide chain termination. *Cold Spring Harbor Symp. Quant. Biol., 34,* 469–477.

Clark, B. F. C., and K. A. Marcker (1966). The role of *N*-formyl-methionyl-sRNA in protein biosynthesis. *J. Mol. Biol., 17,* 394–406.

DeWachter, R., and W. Fiers (1969). Sequences at the 5′ terminus of bacteriophage Qβ RNA. *Nature (London), 221,* 233–235.

Dickerman, H. W., E. Steers, B. G. Redfield, and H. Weissbach (1967). Methionyl soluble ribonucleic acid transformylase: I. Purification and partial characterization. *J. Biol. Chem., 242,* 1522–1525.

Fakunding, J. L., R. R. Traut, and J. W. B. Hershey (1973). Dependence of initiation factor IF-2 activity on proteins L7 and L12 from *Escherichia coli* 50S ribosomes. *J. Biol. Chem., 248,* 8555–8559.

Ghosh, H. P., D. Söll, and H. G. Khorana (1967). Studies on polynucleotides: LXVII. Initiation of protein synthesis *in vitro* as studied by using ribopolynucleotides with repeating nucleotide sequences as messenger. *J. Mol. Biol., 25,* 275–298.

Gordon, J. (1969). Hydrolysis of guanosine 5′-triphosphate associated with binding of aminoacyl transfer ribonucleic acid to ribosomes. *J. Biol. Chem., 244,* 5680–5686.

Gupta, S. L., J. Waterson, M. L. Sopori, S. M. Weissman, and P. Lengyel (1971).

Movement of the ribosome along the messenger ribonucleic acid during protein synthesis. *Biochemistry, 10,* 4410-4421.

Haenni, A.-L., and J. Lucas-Lenard (1968). Stepwise synthesis of a tri-peptide. *Proc. Natl. Acad. Sci. U.S.A., 61,* 1363-1369.

Haselkorn, R., and L. B. Rothman-Denes (1973). Protein synthesis. *Ann. Rev. Biochem., 42,* 397-438.

Hindley, J., and D. H. Staples (1969). Sequence of a ribosome binding site in bacteriophage Qβ-RNA. *Nature (London), 224,* 964-967.

Inoue-Yokosawa, N., C. Ishikawa, and Y. Kaziro (1974). The role of guanosine triphosphate in translocation reaction catalyzed by elongation factor G. *J. Biol. Chem., 249,* 4321-4323.

Kaplan, S., A. O. W. Stretton, and S. Brenner (1965). Amber suppressors: efficiency of chain propagation and suppressor specific amino acids. *J. Mol. Biol., 14,* 528-533.

Kaziro, Y., N. Inoue, Y. Kuriki, K. Mizumato, M. Tanaka, and M. Kawakita (1969). Purification and properties of factor G. *Cold Spring Harbor Symp. Quant. Biol., 34,* 385-393.

Lamborg, M. R., and P. C. Zamecnik (1960). Amino acid incorporation into protein by extracts of *E. coli. Biochim. Biophys. Acta, 42,* 206-211.

Lockwood, A. H., and U. Maitra (1974). Relation between the ribosomal sites involved in initiation and elongation of polypeptide chains. Evidence for two guanosine triphosphatase sites. *J. Biol. Chem., 249,* 346-352.

Lucas-Lenard, J., and A.-L. Haenni (1969). Release of tRNA during peptide chain elongation. *Proc. Natl. Acad. Sci. U.S.A., 63,* 93-97.

———, and F. Lipmann (1966). Separation of three microbial amino acid polymerization factors. *Proc. Natl. Acad. Sci. U.S.A., 55,* 1562-1566.

———, and F. Lipmann (1971). Protein biosynthesis. *Ann. Rev. Biochem., 40,* 409-448.

———, P. Tao., and A.-L. Haenni (1969). Further studies on bacterial polypeptide elongation. *Cold Spring Harbor Symp. Quant. Biol., 34,* 455-462.

Maden, B. E. H., R. R. Traut, and R. E. Monro (1968). Ribosome-catalyzed peptidyl transfer: the polyphenylalanine system. *J. Mol. Biol., 35,* 333-345.

Maizels, N. (1974). *E. coli* lactose operon ribosome binding site. *Nature (London), 249,* 647-649.

Marcker, K. (1965). The formation of *N*-formyl-methionyl-sRNA. *J. Mol. Biol., 14,* 63-70.

———, and F. Sanger (1964). *N*-formyl-methionyl-sRNA. *J. Mol. Biol., 8,* 835-840.

Miller, D. L., and H. Weissbach (1970). Studies on the purification and properties of factor T_u from *E. coli. Arch. Biochem. Biophys., 141,* 26-37.

Milman, G., J. Goldstein, E. Scolnick, and T. Caskey (1969). Peptide chain termination: III. Stimulation of *in vitro* termination. *Proc. Natl. Acad. Sci. U.S.A., 63,* 183-190.

Modolell, J., B. Cabrer, A. Parmeggiani, and D. Vázquez (1971). Inhibition by siomycin and thiostrepton of both aminoacyl-tRNA and factor G binding to ribosomes. *Proc. Natl. Acad. Sci. U.S.A., 68,* 1796-1800.

———, B. Cabrer, and D. Vázquez (1973). The stoichiometry of ribosomal translocation. *J. Biol. Chem., 248,* 8356-8360.

Monro, R. E., J. Cerná, and K. A. Marcker (1968). Ribosome-catalyzed peptidyl transfer: substrate specificity at the P site. *Proc. Natl. Acad. Sci. U.S.A., 61,* 1042-1049.

Nathans, D., and F. Lipmann (1961). Amino acid transfer from aminoacyl-ribonucleic acids to protein on ribosomes of *Escherichia coli. Proc. Natl. Acad. Sci. U.S.A., 47,* 497–504.

Nichols, J. L. (1970). Nucleotide sequence from the polypeptide chain termination region of the coat protein cistron in bacteriophage R17 RNA. *Nature (London), 225,* 147–151.

Nirenberg, M. W., and J. H. Matthaei (1961). The dependence of cell-free protein synthesis in *E. coli* upon naturally occurring or synthetic polyribonucleotides. *Proc. Natl. Acad. Sci. U.S.A., 47,* 1588–1602.

Nishizuka, Y., and F. Lipmann (1966). Comparison of guanosine triphosphate split and polypeptide synthesis with a purified *E. coli* system. *Proc. Natl. Acad. Sci. U.S.A., 55,* 212–219.

Ono, Y., A. Skoultchi, J. Waterson, and P. Lengyel (1969). Peptide chain elongation: GTP cleavage catalyzed by factors binding aminoacyl-transfer RNA to the ribosome. *Nature (London), 222,* 645–648.

Pestka, S. (1971). Inhibitors of ribosomes functions. *Ann. Rev. Microbiol., 25,* 487–562.

Ravel, J. M. (1967). Demonstration of a guanosine triphosphate-dependent enzymatic binding of aminoacyl-ribonucleic acid to *Escherichia coli* ribosomes. *Proc. Natl. Acad. Sci. U.S.A., 57,* 1811–1816.

Revel, M., J. C. Lelong, G. Brawerman, and F. Gros (1968a). Function of three protein factors and of ribosomal subunits in the initiation of protein synthesis in *E. coli. Nature (London), 219,* 1016–1021.

——, M. Herzberg, A. Becarevic, and F. Gros (1968b). Role of a protein factor in the functional binding of ribosomes to natural messenger RNA. *J. Mol. Biol., 33,* 231–249.

Ricard, B., and W. Salser (1974). Size and folding of the messenger for phage T4 lysozyme. *Nature (London), 248,* 314–317.

Richter, D. (1972). Inability of *E. coli* ribosomes to interact simultaneously with bacterial elongation factors EF T_u and EF G. *Biochem. Biophys. Res. Commun., 46,* 1850–1856.

Roblin, R. (1968). Nucleotides adjacent to the 5′-terminus of bacteriophage R17 RNA. *J. Mol. Biol., 36,* 125–135.

Salas, M., M. B. Hille, J. A. Last, A. J. Wahba, and S. Ochoa (1967). Translation of the genetic message: II. Effect of initiation factors on the binding of formylmethionyl-tRNA to ribosomes. *Proc. Natl. Acad. Sci. U.S.A., 57,* 387–394.

Sander, G., R. C. Marsh, and A. Parmeggiani (1972). Isolation and characterization of two acidic proteins from the 50S subunit required for GTPase activities of both EF-G and EF-T. *Biochem. Biophys. Res. Commun., 47,* 866–873.

Scolnick, E., R. Tompkins, T. Caskey, and M. W. Nirenberg (1968). Release factors differing in specificity for termination codons. *Proc. Natl. Acad. Sci. U.S.A., 61,* 768–774.

Shorey, R. L., J. M. Ravel, C. W. Garner, and W. Shive (1969). Formation and properties of an aminoacyl-transfer ribonucleic acid–guanosine triphosphate-protein complex: an intermediate in the binding of aminoacyl-transfer ribonucleic acid to ribosomes. *J. Biol. Chem., 244,* 4555–4564.

Skoultchi, A., Y. Ono, J. Waterson, and P. Lengyel (1970). Peptide chain elongation: indications for the binding of an amino acid polymerization factor, guanosine 5′-triphosphate–aminoacyl transfer ribonucleic acid complex to the messenger–ribosome complex. *Biochemistry, 9,* 508–514.

Smith, M. A., M. Salas, W. M. Stanley, Jr., A. J. Wahba, and S. Ochoa (1966). Direction of reading of the genetic message, II. *Proc. Natl. Acad. Sci. U.S.A.,* *55,* 141–147.

Stretton, A. O. W., and S. Brenner (1965). Molecular consequences of the amber mutation and its suppression. *J. Mol. Biol., 12,* 456–465.

Thach, R. E., J. W. B. Hershey, D. Kolakofsky, K. F. Dewey, and E. Remold-O'Donnell (1969). Purification and properties of initiation factors F_1 and F_2. *Cold Spring Harbor Symp. Quant. Biol., 34,* 277–284.

Tissières, A., D. Schlessinger, and F. Gros (1960). Amino acid incorporation into proteins by *Escherichia coli* ribosomes. *Proc. Natl. Acad. Sci. U.S.A., 46,* 1450–1463.

Volckaert, G., and W. Fiers (1973). Studies on the bacteriophage MS2. GUG as the initiation codon of the A-protein cistron. *FEBS Letter, 35,* 91–96.

Watson, J. D. (1964). The synthesis of proteins upon ribosomes. *Bull. Soc. Chim. Biol., 46,* 1399–1425.

Webster, R. E., D. L. Engelhardt, and N. D. Zinder (1966). *In vitro* protein synthesis: chain initiation. *Proc. Natl. Acad. Sci. U.S.A., 55,* 155–161.

Weiner, A. M., and K. Weber (1973). A single UGA codon functions as a natural termination signal in the coliphage Qβ coat protein cistron. *J. Mol. Biol., 80,* 837–855.

Weissbach, H., B. Redfield, E. Yamasaki, R. C. Davis, Jr., S. Pestka, and N. Brot (1972). Studies on the ribosomal sites involved in factors T_u- and G-dependent reactions. *Arch. Biochem. Biophys., 149,* 110–117.

Zamecnik, P. C., M. L. Stephenson, and K. I. Hecht (1958). Intermediate reactions in amino acid incorporation. *Proc. Natl. Acad. Sci. U.S.A., 44,* 73–78.

23

Reprinted from *Arch. Biochem. Biophys.*, **125**(2), 542–547 (1968)

Translation of the Genetic Message

VII. Role of Initiation Factors in Formation of the Chain Initiation Complex with *Escherichia coli* Ribosomes[1]

KENTARO IWASAKI, STEVEN SABOL,[2] ALBERT J. WAHBA, AND
SEVERO OCHOA

Department of Biochemistry, New York University School of Medicine, New York, New York 10016

Received January 15, 1968

Washing of *Escherichia coli* Q13 ribosomes with 1.0 M ammonium chloride yields, in addition to the known initiation factors F_1 and F_2, a third factor F_3. All three factors are required for translation of natural messenger RNA (mRNA), e.g., phage RNA, in the cell-free *E. coli* system. They function in formation of the polypeptide chain initiation complex. Incubation of purified ribosomes with ^{32}P-labeled Q_β phage RNA and methionine-^{14}C–labeled formylmethionyl-transfer RNA, at low magnesium ion concentrations, followed by zonal centrifugation analysis in sucrose gradients, showed that formation of the initiation complex involves at least two steps: (*1*) The F_3-dependent binding of mRNA to the ribosome, and (*2*) the F_2-dependent binding of formylmethionyl-transfer RNA to the mRNA-ribosome complex to form the initiation complex proper.

Previous work has shown the requirement of two factors (F_1 and F_2), isolated from 0.5 M ammonium chloride washes of *Escherichia coli* ribosomes, for translation of natural messenger RNA (mRNA), e.g., MS2 RNA, Q_β RNA, and of synthetic oligonucleotide messengers having an AUG codon at the 5′-terminus (1). Both factors were found to function in the binding of formylmethionyl~tRNA$_F$ (Fmet~tRNA$_F$)[3] to ribosomes, in the presence of the trinucleoside diphosphate, ApUpG (AUG), indicating

To Richard S. Schweet, pioneer in the study of protein biosynthesis—in memoriam.

[1] This work was aided by grants AM-01845, FR-05099, and GM-01234 from the National Institutes of Health, U.S. Public Health Service, and E. I. du Pont de Nemours and Company, Inc.

[2] M.D.-Ph.D. student. Supported by Medical Scientists Training Grant 5TO5 GM-01668 from the National Institutes of Health, U.S. Public Health Service.

[3] met-tRNA$_F$ refers to the methionyl-transfer RNA species whose methionine residue can be formylated.

their involvement in polypeptide chain initiation (2). This view is supported by recent work in other laboratories (3, 4). Revel *et al.* (4) found further that a third factor is needed for translation of natural mRNA and provided evidence for its participation in the binding of mRNA to ribosomes.

We found that washing of the ribosomes with 1.0 M ammonium chloride yields a third factor, F_3. In line with the findings of Revel *et al.* (4), a system containing ribosomes purified by DEAE-cellulose chromatography, after washing with 1.0 M ammonium chloride, requires addition of all three factors for maximal rates of polypeptide synthesis with phage RNA as messenger.

We have endeavored to throw further light on the role of the initiation factors in formation of the chain initiation complex (mRNA - ribosome - Fmet~tRNA$_F$). This complex is formed on the 30S ribosomal subunit (5, 6), followed by attachment of the 50S subunit to make a 70S couple (5, 7). Purified *E. coli* Q13 ribosomes (30 and 50S subunits) were incubated, at low Mg^{++} concentration (5 mM), with ^{32}P-labeled Q_β RNA and Fmet$\sim$tRNA$_F$ labeled with ^{14}C in the methionine residue, in the absence or presence of factors, and ribosomal binding was analyzed by zonal centrifugation in sucrose density gradients. The results show that, in the presence of the three factors, both mRNA and Fmet$\sim$tRNA$_F$ are bound to ribosomes. In the absence of F_2, Fmet $\sim$ tRNA$_F$ is not bound.

MATERIALS AND METHODS

Ribosomes and supernatant fraction. Escherichia coli Q13, obtained from Dr. W. Gilbert, Harvard University, was grown as described by Haruna and Spiegelman (8). All operations were carried out at 0–4° unless otherwise stated. Fifty grams of fresh cells were washed once with 0.2 M NH$_4$Cl, adjusted to pH 7.0 with NH$_4$OH, and once with a buffer (buffer A) containing 0.02 M Tris-HCl, pH 7.8, 0.01 M magnesium acetate, and 0.01 M 2-mercaptoethanol. The washed cells were ground with 100 gm of alumina (Alcoa, A301) and extracted with 80 ml of buffer A. Cell fragments and alumina were removed by centrifugation at 30,000g for 30 minutes in a Servall angle centrifuge. The supernatant was incubated with 3.0 μg of DNase (Worthington, electrophoretically purified) per milliliter for 20 minutes at 37° and centrifuged for 30 minutes at 30,000g (S30 extract). The S30 extract was centrifuged for 2.5 hours at 150,000g in the No. 50 rotor of the Spinco model L ultracentrifuge. The upper two-thirds of the resulting supernatant fraction (about 35 ml) was collected and dialyzed for 14–18 hours against buffer A. The dialyzed supernatant fraction was used for the amino acid incorporation experiments. The lower third of the S150 supernatant fraction was discarded; the ribosomal pellet was suspended in about 50 ml of a buffer (buffer B), containing 1.0 M NH$_4$Cl, 0.01 M Tris-HCl, pH 8.1, 0.01 M magnesium acetate, and stirred overnight. The suspension was centrifuged for 20 minutes at 30,000g and the sediment was discarded. The supernatant fraction was centrifuged for 2.5 hours at 150,000g. The resulting supernatant fluid was used for preparation of the initiation factors.

For further purification of the ribosomes, a dark brown layer on top of the ribosomal pellet was carefully scraped off and discarded. The pellet was then suspended in 50 ml of buffer B and stirred overnight. The suspension was then centrifuged for 20 minutes at 30,000g, and the supernatant fluid centrifuged once more for 2.5 hours at 150,000g. The pellet was suspended in a buffer containing 0.25 M NH$_4$Cl, 0.01 M Tris-HCl, pH 8.1, and 0.005 M magnesium acetate, to give a concentration of ribosomes of approximately 100 A_{260} units/ml. About 4000 A_{260} units of this ribosomal suspension were applied, at a rate of about 2 ml/minute, to a column (1.6 × 45 cm) of DEAE cellulose (Serva, DEAE-SH cellulose, lot B3999, 0.83 meq/gm) previously equilibrated with the above buffer. The column was washed with 500 ml of the same buffer; the ribosomes were subsequently eluted with buffer B and recovered by centrifugation for 2.5 hours at 150,000g. The resulting pellet was suspended in a buffer containing 0.5 M NH$_4$Cl, 0.01 M Tris-HCl, pH 8.1, 0.002 M magnesium acetate, to give a final concentration of about 1500 A_{260} units/ml, and the suspension (consisting of 30 and 50 S subunits) was stored at 0–4°. Under these conditions the ribosomes kept their activity for at least 2 weeks.

Initiation factors. The procedure for isolation and partial purification of the initiation factors (9) was slightly modified. The S150 supernatant fraction from the first washing of the ribosomes with buffer B was used for isolation of the factors. Batches corresponding to 200 gm (wet weight) of cells were mostly used for this purpose. The factors were precipitated with ammonium sulfate. The fraction between 0.3 and 0.7 ammonium sulfate saturation (about 200 mg of protein) was dissolved in a small amount of 0.01 M Tris-HCl, pH 7.6 (buffer C) and the solution was dialyzed overnight against the same buffer. The dialyzed solution was brought to 20 ml and applied (at a rate of about 0.4 ml/minute) to a column (1.1 × 36 cm) of DEAE-cellulose (Serva, DEAE-SH cellulose,

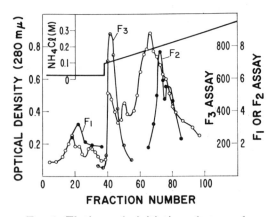

Fig. 1. Elution of initiation factors from DEAE-cellulose. F_1 or F_2 assay (AUG-dependent binding of Fmet $\sim$ tRNA$_F$ to ribosomes), $\mu\mu$moles of Fmet(^{14}C) $\sim$ tRNA$_F$ bound. F_3 assay (translation of MS2 RNA), $\mu\mu$moles of lysine-^{14}C incorporated into acid-insoluble material. $\bigcirc$, A_{280}; $\bullet$, F_1 or F_2 activity; $\odot$, F_3 activity.

lot B3999, 0.83 meq/gm) equilibrated with buffer C. The column was washed with buffer C, containing 0.025 M NH$_4$Cl, until A_{280} of the wash became nil. The wash contains F$_1$. A linear gradient of NH$_4$Cl (0.1–0.35 M) in buffer C was then started. A typical elution pattern is shown in Fig. 1. As previously reported (9), F$_2$ is eluted at about 0.2 M NH$_4$Cl; F$_3$ is eluted at about 0.1 M NH$_4$Cl.

F$_1$ and F$_2$ were assayed by their effect on the AUG-dependent binding of Fmet$\sim$tRNA$_F$ to purified ribosomes (6) at 3 mM Mg^{++}. This assay will be described in detail elsewhere. The assay for F$_3$ was based on the stimulation of lysine incorporation into acid-insoluble material, in the cell-free *E. coli* Q13 system, with the supernatant fluid and purified ribosomes described above and MS2 RNA or Q$_\beta$ RNA as messenger, in the presence of F$_1$ and F$_2$. Factors F$_1$ and F$_2$ were mostly used after concentration of the DEAE-cellulose eluates by precipitation with ammonium sulfate at 0.9 saturation followed by overnight dialysis against buffer C containing 0.01 M 2-mercaptoethanol. F$_3$ was used as the DEAE-cellulose eluate. Except for F$_1$, that had little or no contamination with F$_2$, each of the factors, in particular F$_3$, appeared to have varying degrees of contamination with the other two.

Formation of initiation complex. The samples contained, in a volume of 0.125 ml, Tris-HCl buffer, pH 7.8, 50 mM; NH$_4$Cl, 50 mM; magnesium acetate, 5 mM; 2-mercaptoethanol, 16 mM; GTP, 0.3 mM; *E. coli* W tRNA (when present), 125 μg; purified *E. coli* Q13 ribosomes, 8.5 A_{260} units; ^{32}P-Q$_\beta$ RNA, 1.0 A_{260} unit (6320 cpm); Fmet (^{14}C)$\sim$tRNA$_F$ (specific radioactivity, 200 μC/μmole), 38 μμmoles; factors (when present, in μg of protein), F$_1$, 34; F$_2$, 94; F$_3$, approximately 10. After incubation for 15 minutes at 37°, 0.1-ml aliquots were layered on 5 ml of a linear 5–20% sucrose gradient in a solution containing 50 mM Tris-HCl buffer, pH 7.8, 50 mM NH$_4$Cl, and 5 mM magnesium acetate. The samples were centrifuged for 100 minutes, at 39,000 rpm and 4°, in the Spinco SW39 rotor. Fractions were collected and analyzed for optical density at 260 mμ and for radioactivity. The ^{32}P and ^{14}C radioacivity was measured in a Packard Tri-Carb liquid scintillation spectrometer with Bray's solution (10) as scintillation fluid.

Miscellaneous. Amino acid incorporation into acid-insoluble material was measured as done previously (1). The samples contained, in a volume of 0.125 ml, Tris-HCl buffer, pH 7.8, 65 mM; NH$_4$Cl, 70 mM; magnesium acetate, 14 mM; 2-mercaptoethanol, 18 mM; ATP, 1.3 mM; GTP, 0.3 mM; phosphocreatine, 17.3 mM; creatine kinase, 3.1 μg; tRNA (*E. coli* W), 125 μg; MS2 RNA, 40 μg; purified ribosomes (*E. coli* Q13), 7–8 A_{260} units;

TABLE I

EFFECT OF FACTORS ON TRANSLATION OF MS2 RNA

The experimental conditions are given under MATERIALS AND METHODS. Incubation, 20 minutes at 37°. Dialyzed (NH$_4$)$_2$SO$_4$ fraction of 1.0 M NH$_4$Cl ribosomal wash (when used) 25 μg of protein. The amounts of F$_1$ and F$_2$ were 48 and 56 μg of protein, respectively, in all cases. The amounts of F$_3$ (in μg of protein) are shown in parentheses for each sample. Blanks without MS2 RNA, without factors, or without either, averaged 20 μμmoles. This value was subtracted throughout to give the net effect of factor additions.

Experiment	Factor additions	Lysine incorporation (μμmoles/sample)
1	(NH$_4$)$_2$SO$_4$ fraction	960
	F$_1$ + F$_2$	30
	F$_1$ + F$_2$ + F$_3$ (12.5)	965
	F$_3$ (12.5)	308
	F$_1$ + F$_3$ (12.5)	320
	F$_2$ + F$_3$ (12.5)	467
2	F$_1$ + F$_2$	25
	F$_1$ + F$_2$ + F$_3$ (12.5)	1062
	F$_3$ (12.5)	335
3	F$_1$ + F$_2$	28
	F$_1$ + F$_2$ + F$_3$ (2.5)	130
	F$_1$ + F$_2$ + F$_3$ (5)	339
	F$_1$ + F$_2$ + F$_3$ (12.5)	708

S150 supernatant fraction (*E. coli* Q13), with 0.3 mg of protein, lysine-^{14}C (specific radioactivity, 10 μC/μmole), 0.1 mM; and the remaining (unlabeled) 19 amino acids, each 0.1 mM. Incubation was for 20 minutes at 37°. *Escherichia coli* W tRNA was purchased from Schwarz BioResearch, Inc. Crude Fmet$\sim$tRNA$_F$ labeled with ^{14}C in the methionine residue, was prepared by acylation of *E. coli* W RNA with methionine-^{14}C (specific radioactivity 200 μC/μmole) in the presence of N^5-formyltetrahydrofolic acid (Leucovorin, American Cyanamid Company) and a *Lactobacillus arabinosus* supernatant fraction (2). Samples of MS2 RNA, Q$_\beta$ RNA, and ^{32}P-labeled Q$_\beta$ RNA were prepared as in previous work from this laboratory (11); some of these were provided by Dr. Charles Weissmann. The sedimentation coefficient of the bulk of the MS2 RNA was 27S, that of the Q$_\beta$ RNA, 30S. A highly purified sample of tRNA^{phe}: was kindly made available to us by Dr. G. D Novelli, Oak Ridge National Laboratory, and of tRNA$_F^{met}$ by Dr. B. F. C. Clark, Laboratory of Molecular Biology, University of Cambridge, England. Other materials and methods were as in earlier work (2, 6, 9).

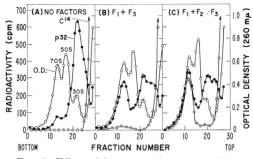

Fig. 2. Effect of factors on formation of chain initiation complex. Experimental procedure as described under MATERIALS AND METHODS. All samples contained tRNA. $\bigcirc$, A_{260}; $\bullet$, ^{32}P; $\odot$, ^{14}C.

RESULTS

Translation of phage RNA. As seen from Table I, with the present system maximal rates of translation of MS2 RNA required the presence of all three factors. This effect equaled that obtained with crude $(NH_4)_2SO_4$ fraction of the 1.0 M NH_4Cl ribosomal wash. The requirement for F_3 is particularly prominent as little translation was observed with only F_1 and F_2 and addition of F_3 increased the rate of translation about 30-fold. Despite contamination of the factors with one another, indicated by the substantial effects of addition of either F_3 alone, $F_1 + F_3$, or $F_2 + F_3$, the results strongly suggest that all three factors are required for translation of natural mRNA.

Formation of initiation complex. In order to clarify the role of the various factors in formation of the chain initiation complex we examined, at low Mg^{++} concentration, the binding of natural mRNA and Fmet$\sim$tRNA$_F$ to ribosomes, by incubation of ribosomes, ^{32}P-Q_β RNA, and ^{14}C-Fmet$\sim$tRNA$_F$, followed by zonal centrifugation analysis in sucrose gradients. Little or no binding occurred in the absence of factors. In the presence of factors, depending on the combination used, binding led to the transfer of either ^{32}P radioactivity, or both ^{32}P and ^{14}C radioactivity to the 70S ribosomal peak.

The sedimentation patterns of three samples with (a) no added factors, (b) with addition of F_1 and F_3, and (c) with addition of F_1, F_2, and F_3, are shown in Fig. 2. It is apparent that the mRNA-directed binding of Fmet$\sim$tRNA$_F$ to ribosomes does not

TABLE II

FORMATION OF CHAIN INITIATION COMPLEX

Experimental procedure as in Fig. 2. The data were plotted and the bound ^{32}P and ^{14}C radioactivity was calculated from the corresponding areas under the 70S ribosomal peak. Values in sample 1, in the absence of added factors (^{32}P, 400 cpm; ^{14}C, 50 cpm), were subtracted.

Sample	Factor additions	Radioactivity under 70S peak (cpm)	
		^{32}P	^{14}C
2	F_1	360	0
3	F_3	990	70
4	$F_1 + F_3$	1270	140
5	F_2	230	190
6	$F_1 + F_2$	540	550
7	$F_3 + F_2$	840	570
8	$F_1 + F_3 + F_2$	1550	1620

occur in the absence of F_2 for only the complete system [sample (c)] showed incorporation of ^{14}C label in the 70S ribosomal peak. On the other hand, F_2 is not required for binding of mRNA to the ribosomes. This is borne out by sample (b) showing good incorporation of ^{32}P, but not ^{14}C radioactivity, in the presence of F_1 and F_3.

When Fmet (^{14}C)$\sim$tRNA$_F$ is omitted, F_1 and F_3 still promote virtually the same incorporation of ^{32}P radioactivity in the 70S region. This shows that the presence of initiator aminoacyl$\sim$tRNA is not necessary for formation of the mRNA-ribosome complex. However, under these conditions addition of tRNA was required for incorporation of ^{32}P radioactivity in the 70S peak. Crude tRNA could be replaced by (uncharged) tRNA$_F^{met}$ but not by tRNA^{phe}, both highly purified. Since Q_β RNA peaks at 30S and, as we have observed, binding of ^{32}P-Q_β RNA to 30S ribosomes results in little or no shift of radioactivity to a higher S value,[4] it is possible that tRNA$_F^{met}$ promotes the addition of a 50S ribosomal subunit to the mRNA-30S ribosome complex, to form mRNA-70S ribosome, thus enabling detection of the complex. This hypothesis finds support in the observation (7) that tRNA promotes the formation of 70S couples from 30 and 50S ribosomes, but other explanations

[4] M. J. Miller, unpublished results.

are possible. Experiments now in progress on the binding of labeled oligonucleotide messengers to 30S ribosomes should throw some light on this matter.

The results of several experiments with various combinations of factors are summarized in Table II. The table gives the ^{32}P and ^{14}C radioactivity found in the area of of the 70S ribosomal peak. Samples 1, 4, and 8, correspond to samples (a), (b), and (c) of Fig. 2. The results of Fig. 2 showed that F_2 is required for the mRNA-directed binding of Fmet~tRNA$_F$ to ribosomes but not for formation of the mRNA-ribosome complex. Comparison of samples 3 (F_3) and 7 ($F_3 + F_2$) of Table II indicates that formation of the mRNA-ribosome complex is F_3-dependent. An additional requirement for F_1 in the over-all reaction is also apparent from the data in Table II (cf. samples 7 and 8). It would thus appear that the requirement of all three factors for translation of natural mRNA is largely a consequence of their participation in formation of the chain initiation complex.

DISCUSSION

The results presented in this paper clearly show that, in *E. coli* systems, formation of the chain intiation complex with natural mRNA involves at least two separable steps, (1) binding of mRNA to the ribosome and (2) attachment of Fmet~tRNA$_F$ to this complex to form the initiation complex proper. Step (1) requires F_3; step (2) requires F_2. The over-all reaction also requires F_1. However, due to contamination of the factors with one another it cannot at present be decided unequivocally whether F_1 functions in step (1) or in step (2). If, as must be assumed, the initiation steps are reversible

$$\text{Ribosome} + \text{mRNA} \rightleftharpoons \text{ribosome} - \text{mRNA} \quad (1)$$

$$\text{Ribosome} - \text{mRNA} + \text{Fmet~tRNA}_F \rightleftharpoons \text{initiation complex,} \quad (2)$$

the additive effect of F_1 and F_3, observed in the presence of F_2 both for ^{32}P and ^{14}C incorporation in the 70S ribosomal peak (Table II, cf. samples 6, 7, 8), could be due to stimulation by F_1 of either step (1) or step

(2). However, we are inclined to believe that F_1 is involved in step (1) because in the absence of F_2, when step (2) does not take place, F_1 and F_3 also have an additive effect on ribosomal binding of ^{32}P-Q$_\beta$ RNA, i.e., on step (1) (Table II, cf. samples 2, 3, 4).

Our results indicate that F_3 is involved, like factor C of Revel *et al.* (4), in the binding of mRNA to the ribosome. However, our results suggest further that F_1 also functions in this step. Further purification of the factors should clarify this point. One other discrepancy concerns the elution pattern of the factors from DEAE-cellulose for whereas F_3 is eluted at 0.1 M NH$_4$Cl, factor C is eluted at 0.22 M. This discrepancy may be more apparent than real and might be accounted for by differences in the early stages of preparation of the factors.

The molar ratio of Q$_\beta$ RNA to Fmet~tRNA$_F$ in the initiation complex can be calculated from the specific radioactivities of ^{32}P-Q$_\beta$ RNA (mol. wt., 1.1×10^6) and methionine-^{14}C in Fmet~tRNA$_F$. The net values for sample 8, Table II (1550 cpm ^{32}P, 1620 cpm ^{14}C) correspond to 9.5 $\mu\mu$moles of Q$_\beta$ RNA and 8.1 $\mu\mu$moles of methionine, i.e., a ratio of approximately 1. This result is of interest for it shows that although phage RNA is a polycistronic messenger it has, at least under the conditions of these experiments, only one site at which initiation can occur, presumably at or near the 5'-terminus.

ACKNOWLEDGMENTS

We are indebted to Dr. Maria Antonia Günther-Sillero for help with some of the factor preparations, to Miss Eva-Marie Webner for skillful technical assistance, and to Mr. Horace Lozina for growth of bacterial cells.

NOTE ADDED IN PROOF

Ribosomal binding of synthetic messengers with an initiator codon at the 5'-end (^{3}H-labeled AUGAAA...AAA) resembles that of Q$_\beta$ RNA in its requirement for tRNA but differs from it in that it occurs in the absence of initiation factors (F_1 and F_3). Thus, F_3 appears to be specifically required for the binding of natural messenger. The results also suggest that F_1 functions in the second step of formation of the initiation complex, i.e., the binding of F met ~ tRNA$_F$ to the ribosome-messenger complex.

REFERENCES

1. STANLEY, W. M., JR., SALAS, M., WAHBA, A. J., AND OCHOA, S., *Proc. Natl. Acad. Sci. U.S.* **56**, 290 (1966).
2. SALAS, M., HILLE, M. B., LAST, J. A., WAHBA, A. J., AND OCHOA, S., *Proc. Natl. Acad. Sci. U.S.* **57**, 387 (1967).
3. ANDERSON, J. S., DAHLBERG, J. E., BRETSCHER, M. S., REVEL, M., AND CLARK, B. F. C., *Nature* **216**, 1072 (1967).
4. REVEL, M., HERZBERG, M., BECAREVIC, A., AND GROS, F., *J. Mol. Biol.*, in press.
5. NOMURA, M., AND LOWRY, C. V., *Proc. Natl. Acad. Sci. U.S.* **58**, 946 (1967).
6. HILLE, M. B., MILLER, M. J., IWASAKI, K., AND WAHBA, A. J., *Proc. Natl. Acad. Sci. U.S.* **58**, 1652 (1967).
7. SCHLESSINGER, D., MANGIAROTTI, G., AND APIRION, D., *Proc. Natl. Acad. Sci. U.S.* **58**, 1782 (1967).
8. HARUNA, I., AND SPIEGELMAN, S., *Proc. Natl. Acad. Sci. U.S.* **54**, 579 (1965).
9. SALAS, M., MILLER, M. J., WAHBA, A. J., AND OCHOA, S., *Proc. Natl. Acad. Sci. U.S.* **57**, 1865 (1967).
10. BRAY, G. A., *Anal. Biochem.* **1**, 279 (1960).
11. WEISSMANN, C., AND FEIX, G., *Proc. Natl. Acad. Sci. U.S.* **55**, 1264 (1966).

24

Reprinted from *Proc. Natl. Acad. Sci.*, **55**, 147–155 (1966)

N-FORMYLMETHIONYL-sRNA AS THE INITIATOR OF PROTEIN SYNTHESIS

By Jerry M. Adams* and Mario R. Capecchi†

THE BIOLOGICAL LABORATORIES, HARVARD UNIVERSITY

Communicated by J. D. Watson, November 22, 1965

The work several years ago of J. P. Waller[1] revealed a bizarre fact about N-terminal groups of bacterial proteins. Instead of a random mixture, Waller found that the great majority of N-terminal groups were either methionine or alanine. This finding suggested that methionine and alanine constituted start signals for the initiation of polypeptide chains. Alternatively, the remaining amino acids were not detected because of acylation of their amino groups. The discovery of N-formylmethionyl-sRNA by Marcker and Sanger[2] provided a means for further study of the problem.

In this compound, the amino group of methionine is formylated, thereby prohibiting its use in polypeptide chain elongation and at the same time making it an attractive candidate for initiation of polypeptide chains. Knowing of Waller's observations, Marcker and Sanger also looked for N-formylalanyl-sRNA. None, however, was detected. If chain initiation required formyl amino acids, the terminal alanine end groups of *E. coli* proteins remained unexplained. There was, of course, the possibility that N-formylmethionine was unrelated to protein synthesis.

The direct way to test involvement of N-formylmethionyl-sRNA in chain initiation is to add this compound to *in vitro* extracts which are carrying out protein synthesis. Here we report experiments in which N-formylmethionyl-sRNA labeled in the formyl group was used with phage R17 RNA as the messenger RNA template in an *E. coli* extract. Labeled formyl groups were incorporated into at least two, if not all, of the several proteins coded by the R17 RNA. Formyl-methionyl-sRNA is thus likely to initiate the synthesis of all the R17 specific proteins.

This is a surprising result since alanine is known to be the N-terminal amino acid of coat protein isolated from intact virus particles.[3] Resolution of the problem came from experiments in which the coat protein made *in vitro* was digested by pronase to see which amino acid was N-formylated. All of the labeled formyl groups were found in N-formylmethionine. Further, these experiments demonstrated that the amino acid adjacent to N-formylmethionine was alanine. We thus suspect that several of the R17 specific proteins, and conceivably a majority of *E. coli* proteins, start with the sequence N-formylmethionylalanine. After synthesis, the terminal formylmethionine residues of certain classes of proteins are enzymatically removed, yielding proteins which have N-terminal alanine.

Materials and Methods.—(a) *H³-formyl-tetrahydrofolate:* H³-formate (specific activity either 31 or 2000 $\mu c/\mu$mole) was activated using a dialyzed $(NH_4)_2SO_4$ fraction of pigeon liver supernatant.[4] Activation was completely dependent on added tetrahydrofolate (THFA) and proceeded to at least 90% of completion. The reaction was stopped by adjusting the solution to 1% PCA, after which all the protein was removed by low-speed centrifugation. The acid converts the product N^{10}-formyl-THFA into N^5, N^{10}-methenyl-THFA.[5] The former is regenerated for use in the transformylation reaction by neutralizing a portion of the solution a few minutes before the reaction. The methenyl form is much more resistant to air oxidation than the N^{10}-formyl compound[5] and in our experience can be stored at $-20°C$ for weeks without significant oxidation.

(b) *Purification of formyl-THFA:* The donor was extensively purified on a column of Whatman cellulose powder.[6] A very large peak of material absorbing strongly at 280 mμ was eluted first, representing mainly THFA and its oxidation products. This peak was well resolved from the 360 mμ-absorbing material (methenyl-THFA). After two passages through this column, the donor had a UV spectrum in acid which agreed closely with the published spectrum of methenyl-THFA.[5]

(c) *Supernatant protein:* The supernatant protein used in the transformylation was prepared by grinding frozen *E. coli* B cells (Grain Processing Co.) with alumina and centrifuging for 5 hr at 78,000 g. The supernatant was then dialyzed for 24 hr against 0.01 M tris, pH 7.5, at 4°C and stored in small aliquots at $-20°C$.

(d) *Transformylation reaction:* The reaction for the transformylation of aminoacyl-sRNA contained 10 μmoles tris, pH 7.2, 1.0 μmole $MgCl_2$, 0.5 μmole ATP, 2.0 μg each amino acid, 0.5 mg sRNA (stripped), 2.0 μmoles of mercaptoethanol, 0.03 μmole of H³-formyl-THFA, and 15 μg of *E. coli* supernatant protein in a total volume of 0.10 ml. The reaction mixture was incubated for 10 min at 37°C and then precipitated with cold 5% TCA on Millipore filters, which were dried and counted in a liquid scintillation counter.

(e) *H³-formyl-sRNA:* sRNA was isolated from the reaction mixture by phenol extraction, precipitated several times with 66% ethanol, and dialyzed for 24 hr against 0.002 M potassium acetate, pH 5.0. The label in the purified sRNA became 96% acid-soluble on addition of RNase or dilute base. About 90% of the label became volatile after treating the product with 0.5 N HCl for 15 min at 100°C. Double-labeled sRNA was made similarly using C¹⁴-methionine and the other 19 C¹² amino acids.

(f) *Pronase digestion of R17 coat protein synthesized in vitro:* Fractions from a SW25 sucrose gradient corresponding to the R17 coat protein were pooled and digested with pancreatic RNase (20 μg/ml) in the presence of 0.02 M EDTA. The protein was precipitated and washed with 7% TCA. The TCA was removed by washing with ethanol-ether followed by two ether washes. The dried protein was resuspended in 0.05 M NH_4HCO_3, pH 7.9, and digested with pronase (0.5 mg/ml) for 15 hr at 37°C. After digestion, the pronase and salt were removed by filtration on a G25 Sephadex column. The samples were lyophilized and resuspended in a small volume of water

(20–30 μl). Aliquots of this material were spotted on Whatman no. 3 MM paper for electrophoretic analysis.

(g) *Electrophoresis and chromatography:* High-voltage electrophoresis was done on a cooled plate (10°C) at 28 v/cm for 3 hr. The electrophoresis buffer contained per liter: 25 ml of glacial acetic acid and 25 ml of pyridine (pH 4.7). Ascending chromatography was done with a pyridine, isobutanol, and H_2O (35:35:30) solvent at 20°C for 24 hr.

(h) *Chemicals:* d,1-Tetrahydrofolic acid (sealed under nitrogen) and *E. coli* sRNA were obtained from General Biochemicals; N-formylmethionine, N-formylalanine, Cyclo Chemical Corp.; methionylalanine, Mann Research; pronase, Calbiochem; H^3-formate, (2 c/mmole), Tracer Lab; C^{14}-alanine (123 mc/mmole), New England Nuclear; C^{14}-methionine (200 mc/mmole), Schwarz BioResearch.

Results.—Identification of an active formyl donor: Marcker and Sanger[2] first observed that methionyl-sRNA could be partially formylated by *E. coli* extracts. Formate itself was not the formyl donor since addition of labeled formate did not result in incorporation of radioactivity into aminoacyl-sRNA. This is not surprising, since most biological transformylations use N^{10}-formyl-tetrahydrofolic acid (N^{10}-formyl-THFA) as the immediate donor.[5] To test whether this compound might be responsible for the formylation of methionyl-sRNA, H^3-labeled N^{10}-formyl-THFA was prepared and incubated with an *E. coli* supernatant fraction supplemented with uncharged sRNA and the 20 amino acids (see *Materials and Methods*). Table 1 shows excellent transfer of the labeled formyl groups to material identified as aminoacyl-sRNA by its sensitivity to pancreatic RNase and by its sedimentation constant (i.e., 4S) on a sucrose gradient. The transformylation proceeded in a linear fashion until the reaction was about three quarters complete. The final level varied somewhat with different batches of sRNA but typically was about 0.6 mμmole of formate per mg of sRNA. The same sRNA preparations could accept about 1.0 mμmole of methionine per mg of sRNA.

If the purified donor is subjected to air oxidation for 1 hr, 85 per cent of its donor capacity is lost. Thus the donor cannot be N^5-formyl-THFA formed in small

TABLE 1

CONDITIONS FOR THE TRANSFER OF
FORMATE FROM FORMYL-THFA
TO AMINOACYL-sRNA

Conditions	Cpm incorporated
Complete system (see *Materials and Methods*)	1202
	1213
No supernatant	31
	38
Boiled supernatant	55
	41
Incubation at 0° for 5 min	448
	468
+ RNase (10 μg/ml)	55
	47
Incubation of the complete system, then hydrolysis at pH 9.5 for 30 min at 37°	27

The transformylation reaction was done as described in *Materials and Methods*, except that in this case the supernatant protein was 3 mg/ml and each sample was extracted with phenol prior to precipitation with cold TCA. The H^3-formyl-THFA used had a specific activity of 31 μc/μmole.

TABLE 2

DEMONSTRATION THAT FORMATE IS TRANSFERRED
ONLY TO METHIONYL-sRNA

Addition of amino acids	Cpm incorporated	% of incorporation with all amino acids added
No added amino acids	1013	14
	909	12
Each amino acid separately except methionine	754 to 1110	11 to 15
All amino acids except methionine	950	13
	932	13
Methionine only	7300	99
	8200	111
All amino acids	8100	110
	6640	90

The transformylation reaction was done as described in *Materials and Methods*, except that amino acids were added only as indicated in the table. The amount of sRNA was 0.25 mg per tube, and the specific activity of the formate 500 μc/μmole.

amounts on neutralization of methenyl-THFA[5] (see *Materials and Methods*) because the N[5]-formyl derivatives are stable to air oxidation.   Therefore, the donor of our system is most likely N[10]-formyl-THFA or possibly methenyl-THFA.   Conceivably the natural donor may be one of the poly-γ-glutamyl derivatives of N[10]-formyl-THFA.[5]   That N[10]-formyl-THFA functions as a donor has been found independently by Marcker.[7]

Evidence that the H[3]-formyl groups are attached to methionyl-sRNA: Transfer of H[3]-formyl groups from formyl-THFA to aminoacyl-sRNA is strongly dependent upon the presence of methionine in the transformylation reaction mixture.   Table 2 shows that the final level of incorporation in the absence of any amino acid supplementation or in the presence of all the amino acids except for methionine is only 13 per cent of that found when methionine is present.   Particularly important is the lack of stimulation by alanine.   The much lower level of incorporation in the absence of added methionine most likely reflects traces of methionine not removed by dialysis.

Additional evidence that the product of the transformylation reaction is N-formylmethionyl-sRNA comes from treatment of H[3]-labeled aminoacyl-sRNA with mild alkali.   This releases N-formyl methionine as shown by coelectrophoresis of this material with a formyl S[35]-methionine standard, prepared by formylation of the amino acid.[8]   As a control the same H[3]-labeled material was also run with formyl-C[14]-alanine.   The formyl-alanine ran sufficiently ahead of the H[3] peak to allow us to conclude that at most 2 per cent of the H[3]-formyl groups could be in formylalanine.

Stability of H[3]-formyl-C[14]-methionyl-sRNA: Further evidence that the reaction product is N-formylmethionyl-sRNA comes from experiments using sRNA which has been incubated with C[14]-methionine as well as the H[3]-formyl donor.   Figure 1 indicates the rate of release of the two labels in 0.1 M tris, pH 8.6, from sRNA. The methionine is clearly present in two forms, which we interpret as methionyl-sRNA and formylmethionyl-sRNA.   Since there is considerable variation in the rate constants for the hydrolysis of the various aminoacyl-sRNA's,[9] the fact that the H[3]-formate-labeled sRNA decays to 5 per cent of its original value with a single rate constant favors the belief that only a single amino acid is formylated. The half life for the alkaline hydrolysis of formylmethionyl-sRNA under these conditions is 9.5 min, and of methionyl-sRNA about 2.1 min.   An increase in stability at high pH on blocking the amino group of aminoacyl-sRNA has also been found with polyphenylalanyl-sRNA.[10]   The decay rate constant for formyl-

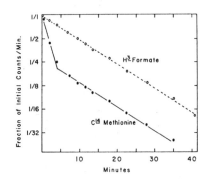

Fig. 1.—Alkaline hydrolysis of H[3]-formyl-C[14]- methionyl-sRNA. Double-labeled sRNA containing 60,000 cpm of H[3]-formate and 40,000 cpm of C[14]-methionine was adjusted to 0.1 M tris pH 8.6 and incubated at 37°C. At various times, aliquots were removed precipitated with 5% cold TCA onto Millipore filters which were counted on a Packard liquid scintillation counter.

methionyl-sRNA supports the conclusion that the formyl-methionine is bound to sRNA by the normal 2′ or 3′ ester bond. If it were bound to the N^6 of adenine, as suggested by Hall in analogy with compounds of this type found in yeast,[11] it would be much more stable. For example, the N^6-amide bond is reported to be stable for 24 hr at pH 8.5 and 37°C.[12]

Incorporation of H^3-formyl groups into protein: Incubation of a preincubated S-30 extract with R17 RNA[13] and H^3-formylmethionyl-sRNA results in the incorporation of H^3-formyl groups into an alkaline-resistant, TCA-precipitable product. Alkali was used to destroy the aminoacyl-sRNA rather than hot acid because of the lability of the formyl bond in hot acid. The kinetics of this incorporation in the presence and absence of R17 RNA is shown in Figure 2. One obtains very similar kinetics of R17 RNA-directed incorporation of H^3-formyl groups if one adds to the incubation mixture the formyl donor, labeled formyl-THFA, in place of the H^3-formylmethionyl-sRNA.

The magnesium ion dependence of the H^3-formyl group incorporation was measured and observed to be a typical magnesium profile for R17 RNA-directed protein synthesis with a sharp maximum at 11 mM Mg. This indicated that the formyl groups were being incorporated into newly synthesized polypeptides. In order to determine which phage-specific proteins contained the labeled formyl groups, the reaction mixture was analyzed on a sucrose gradient.

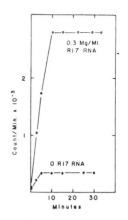

FIG. 2.—Kinetics of incorporation of H^3-formylmethionine in the presence and absence of R17 RNA. The 0.5-ml reaction mixtures contained 125 μl of preincubated S-30 and 62,000 cpm of H^3 - formylmethionyl-sRNA (170,000 cpm/ mg sRNA). At the designated times 50-μl aliquots were removed, incubated at pH 12.0 for 15 min at 0°C, and precipitated with cold 7% TCA onto Millipore filters.

Sedimentation properties of the in vitro product: Sucrose gradient analysis of the newly synthesized polypeptides in the reaction mixture, after incubation with R17 RNA and C^{14}-amino acids, reveals radioactivity peaks with sedimentation constants of 30S and 20S, in addition to the nascent chains bound to ribosomes and the much slower-sedimenting material (2–4S) at the top of the gradient. In an earlier paper it was shown that the protein which sediments in the 30S region consists of complete R17 coat protein molecules bound to R17 RNA and that the second R17 RNA-coded protein, sedimenting in the 20S region, was not coat protein.[13]

Figure 3 shows a sucrose gradient analysis of the alkaline-resistant, TCA-precipitable formyl counts after 15 min of incubation with R17 RNA and H^3-formyl-methionyl-sRNA. The bottom curve designated with x's shows the incorporation in the absence of R17 RNA. This experiment demonstrates that a large proportion of the formyl counts are incorporated into the *in vitro* synthesized coat protein (i.e., the 30S protein). Identical sucrose gradient profiles are obtained, whether the source of formyl groups is N-formylmethionyl-sRNA or N-formyl-THFA.

Two sources produce the shoulder in the 20S region: (1) incorporation of formyl groups into the 20S protein, and (2) partial degradation by endogenous RNase of the R17 RNA to which coat protein molecules are bound. This was shown by giving the reaction mixture a mild RNase treatment (1 μg/ml, 10 min at 0°C) prior to layering it onto a sucrose gradient. The material in the 30S peak was shifted to

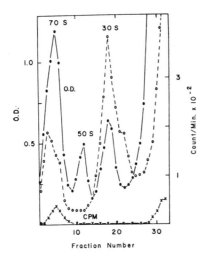

FIG. 3.—Sucrose gradient analysis of the total amino acid-incorporating system after 15 min incubation with H³-formylmethionyl-sRNA. The 300-μl reaction mixture contained 75 μl of preincubated *E. coli* S-30, 60,000 cpm of H³-formylmethionyl-sRNA, and either 0.3 mg/ml (O..O) or 0.0 mg/ml (x—x) R17 RNA. A 150-μl portion was layered onto a 5-ml (5–20% sucrose) gradient and centrifuged for 2.5 hr at 38,000 rev/min at 4°C. Each 0.16-ml fraction was first assayed for optical density at 260 mμ and then incubated at pH 12.0 for 15 min at 0°C to prepare samples for radioactive analysis. The optical density profile for the reaction mixture, which did not contain R17 RNA is not shown on the above figure. The 70*S* ribosomal O.D. peak, and the 50*S* and 30*S* subunit peaks were at the same position in the two gradients.

the top of the gradient revealing a small peak of radioactivity in the 20*S* region. The relatively small peak in the 20*S* region may reflect both the larger monomeric molecular weight of the 20*S* protein and the higher frequency of reading of the coat protein cistron.

The N-formyl bond is more sensitive than the peptide bond to mild acid hydrolysis. When double-labeled protein, containing the H³-formyl group and C¹⁴-amino acids, was heated at 100°C for 10 min in 1 *N* HCl, the H³-formyl counts were reduced by 95% while the C¹⁴ counts decreased less than 5 per cent. We conclude that the formyl group is incorporated into the phage protein as a N-formyl amino acid directly from the aminoacyl-sRNA.

Isolation of the N-formyl amino acid from in vitro synthesized coat protein: These results presented a paradox because the amino terminal amino acid of R17 coat protein is alanine.[3] A rather surprising solution was found by examining the nature of the *in vitro* coat protein. Double-labeled *in vitro* product was made by incubating the R17 RNA-dependent amino acid incorporating system with either C¹⁴-alanine and H³-formyl-THFA or C¹⁴-methionine and H³-formyl-THFA. The *in vitro* synthesized coat protein was separated from other newly synthesized polypeptides by sucrose gradient centrifugation. The coat protein was then digested with pronase and the hydrolysate analyzed by electrophoresis. The results with the C¹⁴-alanine-H³-formyl and the C¹⁴-methionine-H³-formyl-labeled products are shown in Figures 4 and 5, respectively. We observe that there are two peaks labeled with H³-formyl groups. Peak 1 contains C¹⁴-methionine as well as H³-formyl groups, whereas peak 2 contains H³-formyl counts, C¹⁴-methionine, and C¹⁴-alanine. The simplest interpretation of these results is that peak 1 is N-formyl methionine, and peak 2 is the dipeptide N-formylmethionylalanine. This interpretation is consistent with the markers *a*, *b*, and *c*, indicated in Figures 4 and 5. These standards are (*a*) N-formylalanine, (*b*) N-formylmethionine, and (*c*) N-formylmethionyl-alanine (prepared by formylation of the methionylalanine dipeptide[8]). Further evidence that peak 1 is N-formylmethionine is obtained by eluting peak 1 from the C¹⁴-alanine-H³-formyl-labeled product (which contains only tritium counts) and adding to the eluted material a S³⁵-formyl methionine standard. On electro-

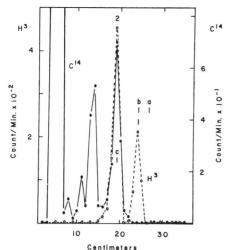

FIG. 4.—Electrophoretic analysis of the pronase digest of *in vitro* synthesized R17 coat protein labeled with H³-formyl groups and C¹⁴-alanine. The 1.0-ml reaction mixture contained 6 × 10⁶ cpm of H³-formyl-THFA (82 cpm/μμmole) and 8.8 × 10⁵ cpm of C¹⁴-alanine (18 cpm/μμmole). After 15 min incubation at 37°C with R17 RNA, the reaction mixture was put on a sucrose gradient to isolate the *in vitro* synthesized coat protein. The conditions for pronase digestion and electrophoretic separation of the protein hydrolyzate are described in *Materials and Methods*. The origin was at 6 cm. The letters *a*, *b*, and *c* designate the positions of the markers N-formylalanine, N-formylmethionine, and N-formylmethionylalanine, respectively.

phoresis and chromatography the S³⁵ and H³ counts superimposed exactly. During this process it was observed that a large proportion of the methionine auto-oxidized to methionine sulfoxide. To facilitate identification of the different peaks, both the samples and markers were oxidized with performic acid.[14] Similarly, it was found that the oxidized peak 2's of both the C¹⁴-alanine-H³-formyl and the C¹⁴-methionine-H³-formyl-labeled products ran in parallel on chromatography. Both peaks still contained C¹⁴ and H³ counts in the expected ratios and ran with the oxidized N-formylmethionylalanine marker.

Discussion.—We have shown that in an *in vitro* amino acid incorporating system programed with R17 RNA, N-formylmethionine is incorporated into R17 coat protein as well as into the phage-specific 20S protein. We have also observed that the pronase digest of the *in vitro* synthesized coat protein contains the dipeptide N-formylmethionylalanine. From these observations we propose that the *in vitro* sequence of the phage coat protein is:

N-formyl met ala ser aspNH₂ phe thr...

in contrast to the expected *in vivo* amino terminal sequence:

    ala ser aspNH₂ phe thr...

The latter sequence was determined by Konigsberg[15] for the very closely related bacteriophage f2. Comparison of the amino acid sequences of R17

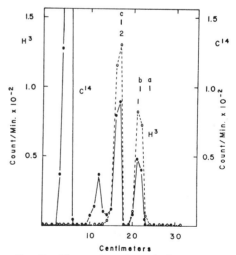

FIG. 5.—Electrophoretic analysis of the pronase digest of *in vitro* synthesized R17 coat protein labeled with H³-formyl groups and C¹⁴-methionine. The reaction mixture contained 6 × 10⁶ cpm of H³-formyl-THFA (82 cpm/μμmole) and 1.9 × 10⁶ cpm at C¹⁴-methionine (50 cpm/μμmole). The experimental conditions are described in Fig. 4.

and f2 coat proteins to date indicates greater than 90 per cent homology.[16]

We are currently attempting to isolate the amino-terminal chymotryptic peptide of the coat protein synthesized *in vitro*.   In an accompanying paper, Zinder and his co-workers describe the isolation of the incomplete polypeptide chain synthesized in a cell-free system directed by RNA with an amber mutation in the coat protein cistron.   Surprisingly, this small amino terminal peptide (6 amino acids) was found to contain one mole of formylmethionine per mole of phenylalanine.[17]   From the consistency of these two unexpected observations we would like to propose the following model for polypeptide initiation:

(1) There exists an initiation signal for protein synthesis.   This signal may be a codon for N-formylmethionyl-sRNA (for instance, AUG or UUG[18]) or be a longer sequence of six or nine bases containing an alanine or even a serine codon.   Pertinent to this is the recent demonstration by Marcker that only one of the two methionyl-sRNA's can be formylated.[7]

(2) *In vivo*, after completion of the polypeptide chain an enzyme removes the N-formylmethionine from the nascent R17 or f2 coat proteins.

$$\text{N-formyl met} \mid \text{ala ser aspNH}_2\text{phe}\ldots$$

For *E. coli* proteins, the same enzyme may cleave the formyl group or after methionine, etc.,

$$\text{N-formyl} \mid \text{met} \mid \text{ala} \mid \text{ser} \mid \ldots$$

which would account for the unexpectedly high values for methionine, alanine, and even serine as the amino terminal residues of *E. coli* proteins (about 45, 30, and 15%, respectively).[1]   The position of enzymatic cleavage could be controlled by the proteins' three-dimensional conformation.   Experiments are now in progress to test for this hypothetical peptidase activity.

Several reasons for employing N-formylated aminoacyl-sRNA as a polypeptide chain initiator can be envisioned.   The N-formylaminoacyl-sRNA may inherently define a direction for polypeptide growth on the ribosomal surface.   This could be accomplished, for example, if the N-formyl bond mimicked the peptide bond and thus aided in the selection of one of the two proposed sRNA binding sites on the ribosome.[19]   Also, blocking the amino group of the first amino acid may facilitate the formation of the initial peptide bond by eliminating the positive charge on the amino group.

The authors are grateful to Prof. J. D. Watson for his guidance and inspiration while pursuing this study and for his valuable criticism during the preparation of this manuscript. We are also happy to acknowledge the expert technical assistance of Mrs. Nancy Capecchi. This work was supported by grants from the National Institutes of Health RG-9541.

* Predoctoral trainee of the National Institutes of Health.

† Predoctoral fellow of the National Science Foundation.

[1] Waller, J. P., *J. Mol. Biol.*, **7**, 483 (1963); Waller, J. P., and J. I. Harris, these Proceedings, **47**, 18 (1961).

[2] Marcker, K., and F. Sanger, *J. Mol. Biol.*, **8**, 835 (1964).

[3] Guidotti, G., unpublished results (1965).   Dr. Guidotti, employing the cyanate procedure of Stark and Smyth for determining the amino terminal residue, found 0.89 mole of alanine, 0.10 mole of glycine, and less than 0.01 mole of the other amino acids per mole of coat protein.   Alanine also has been shown to be the amino terminal residue for the coat proteins of the very closely re-

lated bacteriophages f2 and MS2. Personal communications from W. Konigsberg and H. Fraenkel-Conrat.

[4] Goldthwait, D. A., and G. R. Greenberg, in *Methods in Enzymology*, ed. S. P. Colowick and N. O. Kaplan (New York: Academic Press, 1955), vol. 2, p. 516.

[5] Rabinowitz, J. C., in *The Enzymes*, ed. P. D. Boyer, H. Lardy, and K. Myrbäck (New York: Academic Press, 1960), vol. 2, p. 185.

[6] Huennekens, F. M., P. P. K. Ho, and K. G. Scrimgeour, in *Methods in Enzymology*, ed. S. P. Colowick and N. O. Kaplan (New York: Academic Press, 1963), vol. 6, p. 809.

[7] Marcker, K., *J. Mol. Biol.*, in press.

[8] Sheehan, J. C., and D. H. Yang, *J. Am. Chem. Soc.*, **80**, 1154 (1958).

[9] Sarin, P. S., and P. C. Zamecnik, *Biochim. Biophys. Acta*, **91**, 653 (1964).

[10] Gilbert, W., *J. Mol. Biol.*, **6**, 389 (1963).

[11] Hall, R. H., and G. B. Chheda, *J. Biol. Chem.*, **240**, PC 2754 (1965).

[12] Hall, R. H., *Biochemistry*, **3**, 769 (1964).

[13] Capecchi, M. R., and G. Gussin, *Science*, **149**, 417 (1965).

[14] Nathans, D., *J. Mol. Biol.*, **13**, 521 (1965).

[15] Konigsberg, W., personal communications (1965).

[16] Weber, K., and W. Konigsberg, personal communications (1965).

[17] Webster, W., D. Engelhardt, and N. Zinder, these PROCEEDINGS, **55**, 155 (1966).

[18] Clark, B. F. C., and K. A. Marcker, *Nature*, **207**, 1038 (1965).

[19] Watson, J. D., *Bull. Soc. Chim. Biol.*, **46**, 1399 (1965).

25

Reprinted from *Nature*, 224, 957–964 (Dec. 1969)

Polypeptide Chain Initiation: Nucleotide Sequences of the Three Ribosomal Binding Sites in Bacteriophage R17 RNA

by

JOAN ARGETSINGER STEITZ

MRC Laboratory of Molecular Biology
Hills Road, Cambridge

The initiator regions of the three cistrons of R17 bacteriophage RNA have been isolated and sequenced. All three regions contain a UGA triplet as well as the expected AUG and two contain the sequence GGUUUGA. The initiator regions thus represent untranslated sequences between termination and initiation codons.

In bacteria, formylmethionine transfer RNA has been shown to initiate polypeptide chains[1,2]. Studies *in vitro* indicate that an AUG or GUG triplet can code for this transfer RNA (refs. 2 and 3). The initiation process is believed to occur on 30*S* ribosomal particles[4] in the presence of GTP (refs. 5 and 6) and three initiation factors[7], one of which appears to be required for initiation with natural messengers only[7-9]. But the way in which the chain initiation machinery selects the AUG or GUG triplet at the beginning of a cistron and discriminates against the many internal AUG and GUG sequences in a natural RNA message has remained obscure.

It seems that the correct initiation triplet is not selected simply by its proximity to the 5' terminus of a messenger RNA. The 5' sequences of the several phage RNA molecules which have been elucidated do not begin with AUG or GUG[10-13]; and in at least one case, that of Qβ RNA, no AUG or GUG codon occurs within fifty-five residues of the 5' end[14]. Likewise, the messenger molecule is not required to thread through the 30*S* ribosomal particle until the AUG or GUG closest to the 5' terminus is reached; initiation can take place with high efficiency on a circular single-stranded DNA molecule[15]. Quite possibly ribosomes may not always start reading at one end of a polycistronic messenger and proceed to the other end. Rather, they may be able to initiate protein synthesis internally and simultaneously at several sites on a single polynucleotide chain[15-17].

It is hence of interest to consider the regions around true initiator codons in the messenger RNA molecule. Here, some distinguishing feature, such as a particular nucleotide sequence or a specific secondary or tertiary structure, might allow selective recognition of true initiator codons by the initiation factors in conjunction with the 30*S* ribosome.

For the direct study of such regions, the RNA genome of the closely related, small spherical RNA bacteriophages (R17, f2, MS2 and so on) provides an ideal system. This messenger molecule can be readily obtained in a pure form and appears to direct the synthesis of only three proteins[18-22]. These are: (1) The major coat protein of the phage, which is synthesized in largest amounts, both *in vivo* and *in vitro*. Its complete sequence of 129 amino-acids has been determined[23,24]. The nascent coat protein begins: fMet-Ala-Ser-Asn-Phe-Thr - - -. (2) The *A* protein or maturation protein, a minor constituent of the phage particle, which is produced in very small amounts *in vitro*[22], has a molecular weight of approximately 37,500 (ref. 25) and begins fMet-Arg[16]. (3) The synthetase or replicase protein, which is translated in an intermediate quantity *in vitro*[22] and has a molecular weight of about 50,000 (ref. 26) and an N-terminal sequence fMet-Ser[22,35]. The bacteriophage RNA has approximately 3,300 nucleotides[27,28] and is therefore large enough not only to accommodate the structural genes for these three proteins (requiring about 3,000 nucleotides) but also to contain substantial untranslated regions. These may be distributed between cistrons as well as at the ends of the RNA molecule and could be required for the initiation and termination of both protein and RNA synthesis.

Here I report the isolation and sequence analysis of

fragments from the R17 bacteriophage RNA corresponding to the initiator regions of the three cistrons. These have been obtained from R17 RNA–ribosome complexes which were formed in conditions of polypeptide chain initiation and were subsequently treated with pancreatic ribonuclease to trim away the ends of the messenger RNA not involved in the initiation complex. The sequences of the resulting radioactive RNA fragments were elucidated using the methods developed by Sanger and his colleagues[19-32]. They have been identified as authentic initiation regions, for their oligonucleotide sequences can in part code for the N-terminal amino-acids of the three R17 proteins.

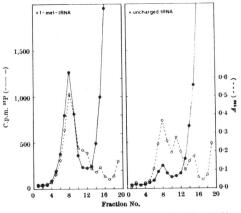

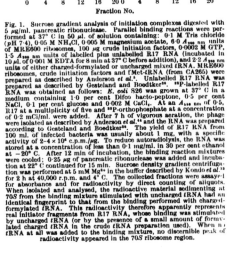

Fig. 1. Sucrose gradient analysis of initiation complexes digested with 5 μg/ml. pancreatic ribonuclease. Parallel binding reactions were performed at 37° C in 50 μl. of solution containing: 0·1 M Tris chloride (pH 7·4), 0·05 M NH₄Cl, 0·005 M magnesium acetate, 6·0 A_{260} nm units of MRE600 ribosomes, 100 μg crude initiation factors, 0·0002 M GTP, 1·5 A_{260} nm units of labelled plus unlabelled R17 RNA (incubated in 10 μl. of 0·001 M EDTA for 8 min at 37° C before addition), and 2·2 A_{260} nm units of either charged-formylated or uncharged mixed tRNA. MRE600 ribosomes, crude initiation factors and fMet-tRNA (from CA265) were prepared as described by Anderson *et al.*[9]. Unlabelled R17 RNA was prepared as described by Gesteland and Boedtker[33]. ³²P-labelled R17 RNA was obtained as follows: *E. coli* S26 was grown at 37° C in a medium containing 1·0 per cent Difco bacto-peptone, 0·5 per cent NaCl, 0·1 per cent glucose and 0·002 M CaCl₂. At an A_{650} nm of 0·5, R17 at a multiplicity of five and ³²P-orthophosphate at a concentration of 0·2 mCi/ml. were added. After 7 h of vigorous aeration, the phage were isolated as described by Anderson *et al.*[34] and the RNA was prepared according to Gesteland and Boedtker[33]. The yield of R17 RNA from 100 ml. of infected bacteria was usually about 1 mg, with a specific activity of 2–4 × 10⁶ c.p.m./μg. To reduce autoradiolysis, the RNA was stored at a concentration of less than 0·1 mg/ml. in 30 per cent ethanol at −20° C. After 12 min of incubation, the binding reaction mixtures were cooled; 0·25 μg of pancreatic ribonuclease was added and incubation at 22° C continued for 15 min. Sucrose density gradient centrifugation was performed at 5 mM Mg²⁺ in the buffer described by Kondo *et al.*[12] for 2 h at 40,000 r.p.m. and 4° C. The collected fractions were assayed for absorbance and for radioactivity by direct counting of aliquots. When isolated and analysed, the radioactive material sedimenting at 70*S* from the binding mixture stimulated with uncharged tRNA had an identical fingerprint to that from the binding performed with charged-formylated tRNA. This radioactivity therefore apparently represents real initiator fragments from R17 RNA, whose binding was stimulated by uncharged tRNA (or by the presence of a small amount of formylated charged tRNA in the crude tRNA preparation used). When no tRNA at all was added to the binding mixture, no discernible peak of radioactivity appeared in the 70*S* ribosome region.

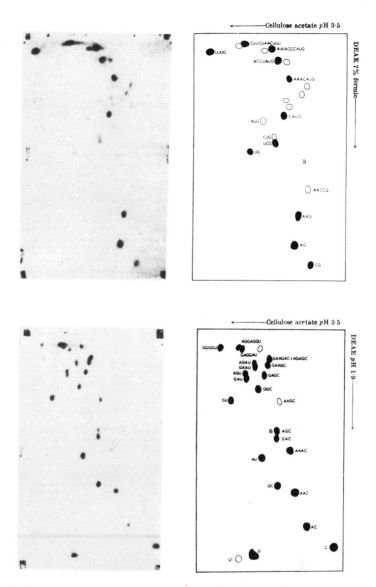

Fig. 2. Ribonuclease fingerprints of fragments from 70S ribosomes. Upper: T₁ ribonuclease; lower: pancreatic ribonuclease. Freshly prepared ³²P-labelled R17 RNA was bound to ribosomes in 500 μl. of reaction mixture as described in Fig. 1, except that the radioactive messenger was not diluted with unlabelled R17 RNA and the tRNA concentration was increased 2·5-fold. This brought the molar ratios of R17 RNA : ribosomes : fMet-tRNA to approximately 1 : 2 : 4, at which optimal ³²P binding was obtained. After treatment of the initiation complex with 5 μg/ml. pancreatic ribonuclease and sucrose gradient centrifugation (in three 5 ml. gradients as described, the fractions from the 70S ribosome peak were pooled and made 8 M in urea. They were then applied directly to a 0·5 ml. DEAE-cellulose column (prepared in a Pasteur pipette) which had been equilibrated with a buffer containing 8 M urea, 0·05 M Tris chloride (pH 7·4), 0·1 M NaCl, and 0·005 M MgSO₄. After thorough washing, approximately 90 per cent of the counts and 300 μg of RNA were eluted by addition of the above buffer made 0·5 M in NaCl, phenol extracted and ethanol precipitated twice. For fingerprint analysis, 60 μg portions of this material were digested with 3 μg of either T₁ ribonuclease or pancreatic ribonuclease in 5 μl. of a buffer containing 0·01 M Tris chloride (pH 7·4), 0·002 M EDTA for 30 min at 37° C. Fingerprinting was performed according to standard procedures, as was the primary sequence analysis of the major oligonucleotides (ref. 29 and J. A. S., in preparation). Note that the AUG-containing oligonucleotides are always among the most prominent in the T₁ fingerprint. Many of the minor spots, whose sequences are generally not shown, are derived from the ends of the initiator fragments. Because the fragments produced by pancreatic ribonuclease treatment of the initiation complex have frayed ends, these oligonucleotides appear in less than molar yield relative to the oligonucleotides from the interior of the ribosome-bound fragments.

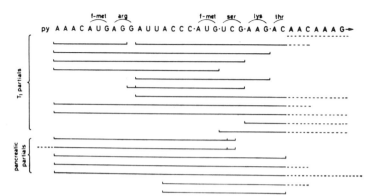

Fig. 3. Synthetase binding site partials. The T₁ ribonuclease partial products illustrated were obtained by partial digestion of the total radioactive RNA recovered from the 70S regions of sucrose gradients of binding reactions carried out as described in Fig. 2. In the two experiments which provided these data, the ³²P-labelled RNA used for binding was not freshly prepared and was only 10 per cent intact. 150 μg portions of RNA fragments isolated from the 70S regions of sucrose gradients were treated with 1/5,000 of their weight of T₁ ribonuclease in 7 μl. of 0·01 M Tris chloride (pH 7·4). After 10 min at 0° C, each digest was applied as a spot 15 cm from the end of a 3 × 85 cm cellulose acetate strip (Oxoid) and electrophoresed for 2·5 h at 6 kV in a 7 M urea solution buffered to pH 3·5 with pyridine acetate. The area between the pink and blue marker dyes was blotted through onto a 47 × 57 cm sheet of DEAE-paper, 5 cm from one of the short ends. Ascending homochromatography was performed in the second dimension at room temperature using a buffer containing 8 M urea and 5 per cent yeast RNA (B.D.H.) until the blue marker dye had travelled about 15 cm from the origin. After drying and autoradiography, the oligonucleotide spots were cut out and eluted from the paper with 30 per cent triethylamine carbonate (pH 9·5). They were then halved and further characterized by complete digestion with either T₁ ribonuclease (1·0 μg for 40 min) or with pancreatic ribonuclease (2·0 μg for 70 min) at 37° C. The T₁ digests were electrophoresed on DEAE-paper in 7 per cent formic acid, and the pancreatic digests on DEAE-paper in pH 1·9 buffer. The T₁ products were analysed by pancreatic ribonuclease digestion and the pancreatic products by T₁ digestion, both followed by ionophoresis at pH 3·5 on DEAE-paper (ref. 29 and J. A. S., in preparation). The pancreatic ribonuclease partial products illustrated have been obtained repeatedly on direct DEAE-cellulose thin layer homochromatography²⁷ (as described in Fig. 4) of the radioactive RNA isolated from the 70S ribosomes. Some of these synthetase site oligonucleotides can actually be seen as very faint spots in the lower centre of Fig. 4, the longer fragments being closer to the bottom of the chromatogram. It is not known whether these products all arise during the treatment of the binding complex with pancreatic ribonuclease or appear subsequent to the unfolding of the ribosomes due to the presence of small amounts of residual nuclease. After isolation from the thin layer plate, they were characterized exactly as described above for partial T₁ products fractionated by homochromatography on DEAE-paper (J. A. S., in preparation). The double ends on some of the partial products indicate that these were isolated as mixtures of oligonucleotides with the two termini shown. The tentative sequence (denoted by the dotted line) at the 3′ end of the synthetase site fragment was deduced by analysis of the various pancreatic ribonuclease partial products, rather than by prior sequence analysis of the complete T₁ oligonucleotide obtained from a fingerprint. Note that many T₁ partial products have termini arising from pancreatic cleavage; these are therefore ends of initiator fragments as isolated from the ribosomes.

Isolation of R17 RNA Initiator Fragments

Takanami, Yan and Jukes[45] were the first to show that regions of the bacteriophage RNA can be protected from nuclease digestion by association with ribosomes. For my study, it was essential to produce RNA–ribosome complexes with reasonable assurance that the interaction was occurring only at sites on the RNA actually specifying the initiation of protein synthesis. I have therefore taken advantage of an observation made by Kondo, Eggerston, Eisenstadt and Lengyel[33] that the binding of ³²P-labelled f2 RNA to 70S ribosomes from E. coli at 5 mM Mg²⁺ concentration is strictly dependent on the presence of fMet-tRNAf. Fig. 1 shows that this dependence on formylated charged methionine tRNA can also be observed when an initiation complex formed between MRE600 ribosomes and ³²P-labelled R17 RNA is treated with pancreatic ribonuclease before sucrose gradient analysis. A five-fold stimulation in the amount of radioactivity remaining bound to the 70S ribosomes is seen as compared with the control experiment, in which uncharged tRNA was substituted. Whereas 10 to 20 per cent of the R17 RNA was associated with the ribosomes before digestion, only about 0·5 per cent of the total radioactivity remains after ribonuclease treatment.

The binding experiment described here (see also legend to Fig. 1) was simply scaled up to obtain amounts of ³²P-labelled initiator fragments sufficient for sequence work. The ³²P R17 RNA utilized was of very high specific activity (about 3 × 10⁶ c.p.m./μg). Presumably because of autoradiolysis, it was not totally intact even immediately after preparation and degraded further on storage. The age and consequent state of fragmentation of the RNA appear to influence the relative affinity of

the ribosomes for the several binding sites in the R17 RNA molecule and will be discussed in detail below. After ribonuclease digestion, the initiation complex was fractionated by sucrose gradient centrifugation, and the ribosomes from the 70S region of the gradient were unfolded by the addition of 8 M urea. Passage over a small DEAE-cellulose column to remove the bulk of the unlabelled ribosomal RNA then rendered the R17 RNA fragments suitable for further study.

Sequence Analysis of the Initiator Fragments

T₁ and pancreatic ribonuclease fingerprints of the total fragments isolated from a pancreatic ribonuclease-digested initiation complex are shown in Fig. 2. Although the relative yields of the oligonucleotides vary somewhat, the fingerprints are generally consistent from one binding experiment to another. Moreover, they contain the same major oligonucleotides whether pancreatic or T₁ ribonuclease is used to trim the ends of the R17 RNA from the initiation complex. This result indicates that most of the labelled RNA being studied is actually protected in the ribosome–messenger complex before digestion and does not represent nuclease-released fragments which subsequently bind to the ribosomal particles. The RNA fragments obtained by pancreatic ribonuclease digestion of the initiation complex were chosen for further analysis since they have a somewhat simpler fingerprint and thus promised to be easier to sequence than the T₁-produced fragments.

The primary sequences of the major oligonucleotides in the T₁ and pancreatic ribonuclease fingerprints shown in Fig. 2 were determined by standard methods[19-21] and will be described in detail elsewhere (J. A. S., in prepara-

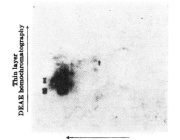

Cellulose acetate pH 3·5

AGAGC₁C₁U₁AACCGGGGUUUGAAGCAUGGCUUCUAACUUU

Fig. 4. Fractionation of fragments isolated from 70S ribosomes. Initiator fragments were produced and isolated as described in Fig. 2, using ³²P-labelled R17 RNA which was freshly prepared and therefore approximately 60 per cent intact. After isolation, the radioactive fragments were divided into portions of about 200 μg dissolved in 5 μl. of water. Each was applied as a spot 12 cm from the end of a 3 × 55 cm strip of cellulose acetate (Selectra-Elektrophoresefolien, Schleicher and Schüll) and electrophoresed at 5 kV for 1 h in 7 M urea buffered to pH 3·5 with pyridine acetate. The region around the pink marker dye was blotted through on to one end of a 20 × 20 cm thin layer of DEAE-cellulose (7·5 : 1, MN 300 and MN 300 DEAE, Machery, Nagel and Co.). After drying and equilibration, the fragments were chromatographed at 60° C, using a buffer containing 5 per cent yeast RNA (B.D.H.) and 8 M urea, pH 7·5, until the solvent front reached the top of the plate. After autoradiography, the spots were scraped from the plate, washed with ethanol to remove the urea and eluted from the DEAE-cellulose with approximately 0·3 ml. of 30 per cent triethylamine carbonate. Following evaporation of the buffer, each oligonucleotide was divided and further characterized by digestion with both pancreatic ribonuclease and T₁ ribonuclease as described in Fig. 3 (J. A. S., in preparation). The arrows in the sequence signify the points of pancreatic ribonuclease cleavage which gave rise to the fragments present in spots I and II. The tentative sequence (denoted by the dotted line) of the large AAC-containing T₁ oligonucleotide at the 5′ end of I was deduced not by direct analysis of the complete primary T₁ oligonucleotide from II, but by examination of the several partial products derived from it, each of which was present in low molar yield in I. Its sequence is uncertain in that one or two extra C residues may be present as indicated: CC(C ?)U(C ?)AACCG. Hence it may or may not contain the termination triplet UAA. Interestingly, the 3′ terminus of the coat protein binding site fragment does not appear to be variable. This could be explained by the existence of the secondary structure shown in Fig. 9; in the isolated fragment the base paired region would protect against further pancreatic ribonuclease attack.

tion). Only the sequence of the large oligonucleotide (CUUCUAACUUU) at the top of the T₁ ribonuclease fingerprint requires special comment. Its lack of a 3′ terminal G residue indicates that it arises from the 3′ end of at least one of the fragments in the mixture isolated from the 70S ribosomes. It also confirms that the coat protein initiation site is represented, for it would code for the N-terminus of the R17 coat protein in the following manner

Ala Ser Asn Phe
(G)CU UCU AAC UUU

(An identical sequence has been deduced by Robinson, Frist and Kaesberg, who have isolated a T₁ oligonucleotide

coding for the first six amino-acids of the R17 coat protein³⁴.) The probability of obtaining a nucleotide sequence corresponding to these four amino-acids is about 1 in 10⁵; hence it is highly unlikely that this oligonucleotide could arise from any other region in the 3,300 nucleotide-long RNA molecule.

The overlapping of several of the oligonucleotides from the T₁ and pancreatic ribonuclease fingerprints into longer sequences was achieved by analysis of partial T₁ digestion products of the total RNA fragments isolated from the 70S ribosomes. In this manner the complete set of T₁ partial products illustrated in Fig. 3 and some of those shown in Fig. 5 and Fig. 7 were obtained. There were, however, difficulties in isolating other partial products pure enough to overlap the many remaining primary oligonucleotides. In addition, from what information was obtained, it appeared that the primary oligonucleotides could be divided into several discrete linkage groups.

Fractionation by homochromatography³² of the total isolated fragments before further digestion revealed why this was so. This technique, which is a form of displacement chromatography performed on a thin layer of DEAE–cellulose, offers an excellent alternative to polyacrylamide gel electrophoresis for the separation of oligonucleotides up to fifty residues in length. The fragments isolated from the pancreatic-digested initiation complex resolved into a number of components as illustrated in Fig. 4. The two major spots in the autoradiogram are simply size variants of a sequence derived from the initiation region of the coat protein gene. The complete sequences of oligonucleotides I and II were deduced by analysis of T₁ partial digestion products (Fig. 5) from material isolated from such a homochromatogram.

In addition to the two strong oligonucleotides from the coat protein initiation site, other fragments are visible in Fig. 4. These are a group of spots in the lower centre, which correspond to the synthetase binding site (see Fig. 3), and those in the area above I and II, which are derived primarily from the A protein binding site. This particular distribution of oligonucleotides is observed only when fresh ³²P-labelled R17 RNA is used in the binding reaction. When older (and hence more degraded) RNA is used, the coat protein initiator fragments become much less prominent in the homochromatogram, while those derived from the synthetase and A protein binding sites become relatively more intense. The use of degraded RNA therefore provided valuable data for the analysis of the two latter sequences. None the less, the yield of A protein initiator fragments was barely sufficient to allow completion of the sequence of this region.

Initiation with *Bacillus stearothermophilus* Ribosomes

Lodish and Robertson¹⁶ have shown that when ribosomes from *B. stearothermophilus* rather than from *E. coli* are used for the *in vitro* translation of f2 RNA, only the A protein is initiated and synthesized. I have confirmed this observation in binding experiments with ³²P-labelled R17 RNA. Fig. 6 shows T₁ and pancreatic ribonuclease

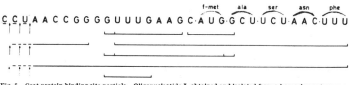

Fig. 5. Coat protein binding site partials. Oligonucleotide I, obtained and isolated from a homochromatogram as described in Fig. 4, was subjected to digestion with 1/10,000 of its weight of T₁ ribonuclease for 10 min. at 0° C in 7 μl. of 0·01 M Tris chloride (pH 7·4), 0·002 M EDTA. The partial degradation products were then fractionated by ionophoresis on cellulose acetate plus DEAE-paper homochromatography, isolated and further characterized as described in Fig. 3 (J. A. S., in preparation). Each partial product illustrated appeared and was analysed in two independent experiments. Again the double ends indicate that the product was isolated as a mixture of oligonucleotides possessing the two termini shown. Although relatively few T₁ ribonuclease partial products are obtained from the coat protein binding site fragment, note that they do unambiguously overlap the primary T₁ and pancreatic oligonucleotides into the given sequence.

217

fingerprints of unfractionated fragments isolated from 70S *B. stearothermophilus* ribosomes. Both fingerprints are considerably simpler than those shown in Fig. 2 and reveal only a subset of the oligonucleotides present when *E. coli* ribosomes are used for binding. All of these oligonucleotides had previously been identified as belong-

ing to the *A* protein initiation site, and many of them appear as minor spots in Fig. 2.

The use of *B. stearothermophilus* ribosomes increased the yield of *A* protein binding site fragments per mole of R17 RNA added by about ten-fold. This allowed analysis of a full set of T_1 ribonuclease partial digestion

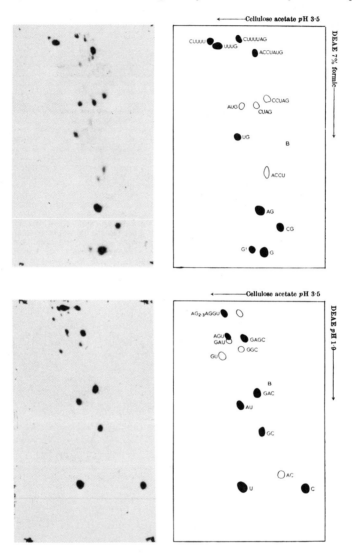

Fig. 6. Ribonuclease fingerprints of fragments isolated from *B. stearothermophilus* ribosomes. Upper: T_1 ribonuclease; lower: pancreatic ribonuclease. Freshly prepared ³²P-labelled R17 RNA was bound to *B. stearothermophilus* ribosomes in 500 μl. of reaction mixture containing: 0·1 M ammonium cacodylate (pH 7·1), 0·1 M NH₄Cl, 0·008 M magnesium acetate, 0·0002 M GTP, 55 A_{260} nm units of charged formylated mixed tRNA, 57 A_{260} nm units of *B. stearothermophilus* ribosomes (prepared by washing three times with buffer A (ref. 5) and 15 A_{260} nm units of ³²P-labelled R17 RNA. After incubation for 10 min at 61° C, the reaction mixture was cooled and treated with 2·5 μg of pancreatic ribonuclease for 15 min at 22° C. Sucrose gradient centrifugation, fragment isolation and fingerprinting were performed as described in Figs. 1 and 2. No large oligonucleotides derived from the coat or synthetase protein binding sites were ever detected in *B. stearothermophilus* ribosome fingerprints, but binding experiments performed in parallel with *E. coli* ribosomes yielded fragments derived from all three initiation sites of the R17 RNA. Note that several of the minor oligonucleotides in the T_1 ribonuclease fingerprint do not possess a 3′ terminal G residue. These arise either from the ends of ribosome-bound fragments or from contaminating pancreatic ribonuclease cleavage of the fragments during isolation.

products (Fig. 7), thereby completing the *A* protein initiation site sequence.

Identification of the R17 Initiator Sequences

The sequences of the three ribosomal binding sites in the R17 RNA are shown in Fig. 8. The identification of the first as that specifying the beginning of the coat protein cistron is unambiguous. The assignment of the third sequence to the R17 *A* protein is strongly supported both by the result obtained with *B. stearothermophilus* ribosomes and by the presence of an arginine codon following the AUG triplet.

By elimination the second sequence should encode the beginning of the R17 synthetase protein. This fragment contains two potential initiator triplets, both AUGs, one followed by an arginine codon and the other by a serine codon. The second AUG is implicated as the true initiator by the following considerations. First, if it is assumed that the ribosome shields approximately equivalent lengths of the messenger RNA at each binding site, then by comparison with the other two fragments the first AUG is unlikely to be the initiator codon; it lies too close to the 5′ end of the fragment. Second, the nucleotide sequence to the right of the second AUG predicts that the amino terminal sequence of the synthetase protein should be fMet-Ser-Lys-Thr- - -. This sequence has now been in part confirmed: a tryptic peptide, isolated from the R17-directed *in vitro* system and identified as the N-terminus of the synthetase protein, has been shown to be fMet-Ser-Lys (M. Osborn, K. Weber and H. Lodish, personal communication).

Ribosomal Recognition of Initiation Sites

The polypeptide chain is initiated at an AUG codon in each of the three cistrons of the R17 RNA. Any additional sequence elements which might distinguish these three AUGs from other AUGs are not immediately obvious, and at this point we do not know why the ribosome selects these particular initiator codons.

The difficulty is likely to arise from the fact that any general recognition by the ribosome may be influenced not only by the nucleotide sequence around the initiator AUG but also by the local secondary and tertiary structure of the RNA molecule. Obviously, a complete description

of the initiation event must explain both the fact that the three bacteriophage genes are translated with very different efficiencies (under one set of conditions *in vitro*, the molar amounts of coat protein : synthetase : *A* protein are 20 : 6 : 1 (ref. 22)) and the observation that *in vitro B. stearothermophilus* ribosomes recognize the *A* protein initiation site only. For ease of discussion, nucleotide sequence homologies and secondary and tertiary structure possibilities are considered separately, but in practice these two aspects cannot be divorced from one another.

Sequence Homology

The only extensive common sequence in the three R17 initiator fragments is GGUUUGA, which appears three nucleotides to the left of AUG in the coat and *A* protein binding sites. Because a portion of this sequence, UUUGA, also appears (two nucleotides) to the left of the initiator AUG in the coat protein cistron of Qβ RNA (J. Hindley, personal communication), it is extremely tempting to ascribe to it a role in initiation. A similar sequence does not, however, occur in the synthetase initiator fragment and therefore cannot be obligatory for ribosome recognition. Nor is this homology helpful in explaining the differing frequencies of translation of the three bacteriophage proteins or the fact that *B. stearothermophilus* ribosomes recognize the *A* protein initiation site only. The difference in the synthetase sequence, on the other hand, could possibly be related to the fact that initiation of translation at the synthetase cistron can be inhibited by the binding of coat protein to the phage RNA (refs. 22, 35). Because this is not true of translation of the other two genes, one might speculate that this region is the site of coat protein binding and repression.

Another curious sequence homology which occurs in all three fragments is a UGA triplet to the left and in phase with each initiator AUG. These UGA sequences may be elements in a general recognition site for the ribosome or reinforcement terminators strategically placed to ensure that reading from the previous cistron is not continued into the next. They are unlikely to be the normal termination signals at the ends of R17 genes, for either the one in the synthetase or that in the *A* protein fragment should then be preceded by a codon (UAPy) for tyrosine—the carboxy terminus of the coat protein[23,24]. This is evidently not the case.

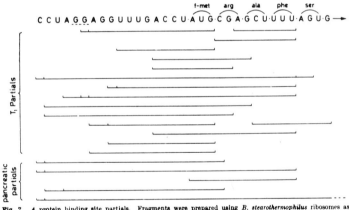

Fig. 7. *A* protein binding site partials. Fragments were prepared using *B. stearothermophilus* ribosomes as described in Fig. 6. After fractionation by DEAE-cellulose thin layer homochromatography (Fig. 4), the variably sized oligonucleotides were re-isolated, and two of the longer ones were chosen for T₁ partial digestion, which was performed as in Fig. 5 (J. A. S., in preparation). The remainder of the oligonucleotides from the homochromatogram and the T₁ ribonuclease partial products were analysed as described in Fig. 3 (J. A. S., in preparation). In addition to appearing in digests of fragments isolated from *B. stearothermophilus* ribosomes, at least one-fourth of the T₁ ribonuclease partial products illustrated had been previously obtained from partial digests of mixed initiator fragments (containing all three binding sites) isolated from *E. coli* ribosomes. Double ends again signify that the oligonucleotide was isolated as a mixture with several termini. The dotted line denotes an uncertainty as to the number of G residues present; the sequence could be either . . . AGGA . . . or . . . AGGGA . . .

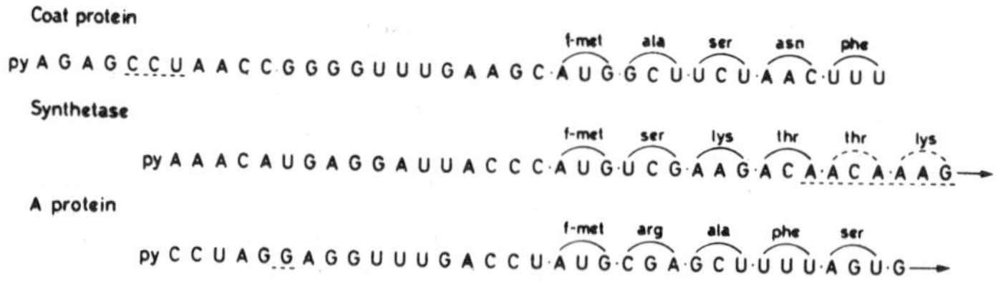

Coat protein

py A G A G C C U A A C C G G G G U U U G A A G C A U G G C U U C U A A C U U U
f-met ala ser asn phe

Synthetase

py A A A C A U G A G G A U U A C C C A U G U C G A A G A C A A C A A A G →
f-met ser lys thr thr lys

A protein

py C C U A G G A G G U U U G A C C U A U G C G A G C U U U U A G U G →
f-met arg ala phe ser

Fig. 8. Ribosome binding sites of R17 RNA. The initiation regions of the three cistrons in the R17 RNA and the predicted N-terminal amino-acid sequences of the corresponding proteins are shown. The dotted lines again denote regions of tentative sequence (see Figs. 3, 4 and 7 for details). Also note from Figs. 3, 4 and 7 that the fragments as isolated from the pancreatic ribonuclease treated initiation complex have frayed ends and that many of them are considerably shorter than the sequences shown above. The segment which appears most frequently in all three cases is slightly asymmetric about the AUG, extending about twelve nucleotides to the right and approximately seventeen to the left of the initiator triplet.

Further sequence homologies in the three initiator fragments which might represent general ribosome recognition sites can likewise be identified. For instance, even the synthetase sequence to the left of the initiator AUG appears to be slightly more similar to the coat protein sequence than a random polynucleotide would be expected to be. None of these homologies, however, is extensive enough to be convincing until supported by sequence data from other systems. It seems unlikely that the true recognition regions lie beyond the 5' ends of the isolated fragments and that ribosomes bind there before sliding along to the first available AUG. Were this so, protein synthesis would then be initiated at the wrong AUG codon in the synthetase fragment. The possibility thus remains that the ribosome or initiation factor population is heterogeneous, thereby allowing recognition of several different initiation site sequences.

Secondary and Tertiary Structure

The nucleotide sequence at the beginning of the R17 coat protein cistron may be folded as shown in Fig. 9. AUG is at the turn of a hairpin loop in which eleven out of twelve possible base pairs are made (counting G–U pairs). As demonstrated by Adams, Jeppesen, Sanger and Barrell[36], the R17 RNA does contain extensive regions of double helix. But whether a loop at the coat protein initiation site actually exists in the intact RNA molecule and whether it might have any physiological significance for the specificity of ribosomal binding is unknown.

The synthetase and A protein initiation site sequences cannot be fitted to similar satisfactorily hydrogen-bonded structures. Although it is not impossible that these sequences may lie at the turns of much longer loops, it nevertheless appears that hydrogen-bonding directly adjacent to the initiator codon is not required for ribosome recognition. On the other hand, at least two AUG codons within the R17 coat protein cistron (that specifying Met$_{88}$ and one out of phase between amino-acids 81 and 82) appear to lie buried in a hydrogen-bonded structure[36]. It is interesting to note that the sequence around this latter AUG is GCAUGGC, which is precisely the same as that found at the beginning of the coat protein cistron. Hence, it would seem that secondary structure might play a negative rather than a positive part in initiation; it may be important as a device for ensuring that internal AUG and GUG triplets are not available for ribosome attachment.

Secondary structure in the R17 RNA molecule has also been proposed to explain the polarity exerted on the synthetase cistron by amber mutants early in the coat protein gene[37-39,22]. We have therefore compared the sequence of the synthetase initiation site with that predicted for the coat protein structural gene hoping to identify regions of potential base pairing which might prevent the binding of ribosomes to the synthetase site until the coat protein cistron has been translated.

Although this search did not reveal any obvious complementarity, very short stretches of hydrogen bonding or secondary structure of a sort which may be described as tertiary structure may yet be the basis of the observed polarity.

The existence of tertiary structure, that is, the specific folding together of loops of the type analysed by Adams et al.[36], also offers a plausible interpretation of the binding results obtained with R17 RNA having different degrees of fragmentation. I find that the most intact ^{32}P-labelled preparations yield primarily coat protein initiator fragments (and some A protein). This is in agreement with the observations of Lodish and Robertson[16], who have shown by dipeptide synthesis that only two initiation sites, the coat and A protein, are available for ribosome attachment in the intact phage messenger. The synthetase site presumably lies buried by flexible structures present in the RNA molecule. It can be exposed by fragmentation of the RNA (refs. 22, 35, 40, 42) as well as by translation and consequent unfolding of hydrogen-bonded regions in the coat protein cistron. Hence, it would be expected that the more degraded the ^{32}P-RNA used in the binding reaction, the greater the relative yield of the synthetase initiation site. This is, in fact, what I observe. RNA fragmentation, with resulting alterations in secondary or tertiary structure, has therefore made the isolation and sequence analysis of the synthetase initiator fragment possible.

Although secondary and tertiary structure may play important parts in controlling the initiation of protein synthesis directed by bacteriophage RNA molecules, it may be incautious to assume that similar devices are used in other messengers. The phage RNA is a messenger

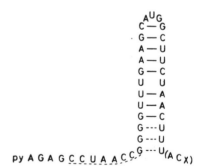

Fig. 9. Potential secondary structure at the R17 coat protein ribosomal binding site. The dotted line again denotes the tentative portion of this sequence (see Fig. 4). ACX, the codon for threonine, has been added at the 3' end of the fragment to show that the region of hydrogen bonding apparently does not extend beyond the determined sequence. Note that the unmade base pair in the loop is U–U, a non-complementarity which would not be expected seriously to disrupt the stability of the structure.

molecule which is highly resistant to degradation, as well as a genome which is capable of being tightly packed into a spherical coat protein shell. Clearly, generalizations must await the analysis of messengers from bacterial and other systems.

Gene Location in the R17 RNA

The position of termination triplets to the left of the three initiator AUGs in the isolated binding site fragments demonstrates that untranslated regions are present between cistrons in the R17 RNA. Hence, termination is not directly coupled to initiation by adjacent codons; and, in fact, the codons specifying these two events may be quite widely separated (J. Nichols, personal communication). This makes more plausible both the observation of independent translation of bacteriophage genes[40,42] and the idea that the efficiency of translation of the three cistrons need not be directly correlated with gene order.

No pppGp, the nucleotide tetraphosphate appearing at the 5' terminus of intact R17 RNA (ref. 41), has ever been detected in fingerprints of fragments produced by either pancreatic or T_1 ribonuclease digestion of an initiation complex. Moreover, comparison of the primary oligonucleotides from the ribosomal binding sites with those from an eighty to ninety nucleotide segment containing the 5' end of the R17 RNA (J. M. Adams, personal communication) indicates that there is no overlap in sequence. Thus no R17 cistron begins within about a hundred nucleotides of the 5' end of the RNA molecule.

Experiments designed to locate the three initiation sites and the fifty-seven long oligonucleotide fragment from the interior of the coat protein cistron[36] in well characterized large fragments of the R17 RNA (a 40 per cent fragment from the 5' end and a 60 per cent piece from the 3' end[35,42,43]) are now in progress. Preliminary results obtained in collaboration with R. Gesteland, P. Spahr and P. Jeppesen indicate that the gene order in the R17 RNA is 5'-A protein–coat protein-synthetase-3'.

I thank Dr M. S. Bretscher, whose interest and advice have been essential to the success of this investigation. I also thank B. G. Barrell for his help with sequencing techniques and Drs F. Crick, F. Sanger and R. Russell for discussions. This work was done while I was a post-doctoral fellow of the US National Science Foundation.

Note added in proof. On the basis of additional evidence, the sequence of the T_1 oligonucleotide in the coat protein binding site which was indicated as tentative (Figs. 4, 5, 8 and 9) now appears to be CCUCAACCG. The tentative regions in the synthetase (Figs. 3 and 8) and the

A protein (Figs. 7 and 8) binding site sequences are apparently correct as given.

Received October 9, 1969.

1 Adams, J. M., and Capecchi, M. R., *Proc. US Nat. Acad. Sci.*, **55**, 147 (1966).
2 Clark, B. F. C., and Marcker, K., *J. Mol. Biol.*, **17**, 394 (1966).
3 Ghosh, H. P., Söll, D., and Khorana, H. G., *J. Mol. Biol.*, **25**, 275 (1967).
4 Nomura, M., and Lowry, C. V., *Proc. US Nat. Acad. Sci.*, **58**, 946 (1967).
5 Anderson, J. S., Bretscher, M. S., Clark, B. F. C., and Marcker, K. A., *Nature*, **215**, 490 (1967).
6 Ohta, T., Sarkar, S., and Thach, R. E., *Proc. US Nat. Acad. Sci.*, **58**, 1638 (1967).
7 Iwasaki, K., Sabol, S., Wahba, A. J., and Ochoa, S., *Arch. Biochem. Biophys.*, **125**, 542 (1968).
8 Revel, M., and Gros, F., *Biochem. Biophys. Res. Commun.*, **25**, 124 (1966).
9 Brown, J. C., and Doty, P., *Biochem. Biophys. Res. Commun.*, **30**, 284 (1968).
10 DeWachter, R., Verhassel, J.-P., and Fiers, W., *FEBS Lett.*, **1**, 93 (1968).
11 Glitz, D. G., *Biochemistry*, **7**, 927 (1968).
12 Roblin, R., *J. Mol. Biol.*, **36**, 125 (1968).
13 DeWachter, R., and Fiers, W., *Nature*, **221**, 233 (1969).
14 Billeter, M., Dahlberg, J. E., Goodman, H. M., Hindley, J., and Weissmann, C., *Cold Spring Harbor Symp. Quant. Biol.*, **34** (in the press, 1969).
15 Bretscher, M. S., *Nature*, **220**, 1088 (1968).
16 Lodish, H. F., and Robertson, H. D., *Cold Spring Harbor Symp. Quant. Biol.*, **34** (in the press).
17 Voorma, H. O., *FEBS Abst.*, **6**, 114 (1969).
18 Gussin, G. N., *J. Mol. Biol.*, **21**, 435 (1966).
19 Horiuchi, K., Lodish, H. F., and Zinder, N. D., *Virology*, **28**, 438 (1966).
20 Viñuela, E., Algranati, I. D., and Ochoa, S., *Europ. J. Biochem.*, **1**, 1 (1967).
21 Nathans, D., Oeschger, M. P., Eggen, K., and Shimura, Y., *Proc. US Nat. Acad. Sci.*, **56**, 1844 (1966).
22 Lodish, H. F., *Nature*, **220**, 345 (1968).
23 Weber, K., and Konigsberg, W., *J. Biol. Chem.*, **242**, 3563 (1967).
24 Weber, K., *Biochemistry*, **6**, 3144 (1967).
25 Steitz, J. A., *J. Mol. Biol.*, **33**, 923 (1968).
26 Capecchi, M. R., *J. Mol. Biol.*, **21**, 173 (1966).
27 Strauss, J. H., and Sinsheimer, R. L., *J. Mol. Biol.*, **7**, 43 (1963).
28 Gesteland, R. F., and Boedtker, H., *J. Mol. Biol.*, **8**, 496 (1964).
29 Sanger, F., Brownlee, G. G., and Barrell, B. G., *J. Mol. Biol.*, **13**, 373 (1965).
30 Brownlee, G. G., and Sanger, F., *J. Mol. Biol.*, **23**, 337 (1967).
31 Brownlee, G. G., Sanger, F., and Barrell, B. G., *J. Mol. Biol.*, **34**, 379 (1968).
32 Brownlee, G. G., and Sanger, F., *Europ. J. Biochem.* (in the press).
33 Kondo, M., Eggerston, G., Eisenstadt, J., and Lengyel, P., *Nature*, **220**, 368 (1968).
34 Robinson, W. E., Frist, R. H., and Kaesberg, P., *Science* (in the press).
35 Gesteland, R. F., and Spahr, P. F., *Cold Spring Harbor Symp. Quant. Biol.*, **34** (in the press).
36 Adams, J. M., Jeppesen, P. G. N., Sanger, F., and Barrell, B. G., *Nature*, **223**, 1009 (1969).
37 Gussin, G. N., *J. Mol. Biol.*, **21**, 435 (1966).
38 Gussin, G. N., Capecchi, M. R., Adams, J. M., Argetsinger, J. E., Tooze, J., Weber, K., and Watson, J. D., *Cold Spring Harbor Symp. Quant. Biol.*, **31**, 257 (1966).
39 Zinder, N., Engelhardt, D. L., and Webster, R. E., *Cold Spring Harbor Symp. Quant. Biol.*, **31**, 251 (1966).
40 Lodish, H. F., *J. Mol. Biol.*, **32**, 681 (1968).
41 Roblin, R., *J. Mol. Biol.*, **31**, 51 (1968).
42 Spahr, P. F., and Gesteland, R. F., *Proc. US Nat. Acad. Sci.*, **59**, 876 (1968).
43 Spahr, P. F., Farber, M., and Gesteland, R. F., *Nature*, **222**, 455 (1969).
44 Anderson, J. S., Dahlberg, J. E., Bretscher, M. S., Revel, M., and Clark, B. F. C., *Nature*, **216**, 1072 (1967).
45 Takanami, M., Yan, Y., and Jukes, T. H., *J. Mol. Biol.*, **12**, 761 (1965).

26

Reprinted from *Biochemistry*, 8, 4289–4292 (1969)

Effect of Aminoacyl Transfer Ribonucleic Acid on Competition between Guanosine 5'-Triphosphate and Guanosine 5'-Diphosphate for Binding to a Polypeptide Chain Elongation Factor from *Escherichia coli*[*]

David Cooper† and Julian Gordon

ABSTRACT: The polypeptide chain elongation factor preparation referred to as T (an undissociated combination of T_u and T_s) exhibited preferential binding of guanosine 5'-diphosphate, even in the presence of a large excess of guanosine 5'-triphosphate, when tested for retention on Millipore filters. Previously published data on binding of guanosine 5'-triphosphate using this assay is now shown to be attributable to the preferential binding of the [³H]guanosine 5'-diphosphate present in the [³H]guanosine 5'-triphosphate preparation. Removal of this guanosine 5'-diphosphate with a guanosine 5'-triphosphate regenerating system considerably reduced the level of bound nucleoside phosphate. Nonradioactive guanosine 5'-triphosphate did not compete with Millipore-bindable [³H]guanosine 5'-diphosphate. Addition of aminoacyl transfer ribonucleic acid reversed this preference for guanosine 5'-diphosphate.

I t is now clear that the mechanism by which aminoacyl-tRNA binds to *Escherichia coli* ribosomes during polypeptide synthesis involves the polypeptide chain elongation factor T of Nishizuka and Lipmann (1966), which was subsequently resolved into two factors, T_u and T_s, by Lucas-Lenard and Lipmann (1966). This binding also requires GTP (Ravel *et al.*, 1967, 1968; Lucas-Lenard and Haenni, 1968; Ertel *et al.*, 1968b). The binding of aminoacyl-tRNA to ribosomes was originally proposed to proceed in two steps, on the basis of comparison of Millipore binding assays, which detected the first step (T–GTP); and gel filtration assays with Sephadex G-50 which detected a ternary complex (T–GTP–aminoacyl-tRNA), characterized by its inability to bind to the Millipore (Gordon, 1967, 1968; Ravel *et al.*, 1968; Ertel *et al.*, 1968a). More recent studies have suggested that the subfraction T_u is the acceptor for nucleoside phosphate

(Ertel *et al.*, 1968a; Ravel *et al.*, 1969). However, in the experiments described in this paper, T_u and T_s were not dealt with separately, so we retain the terminology T factor for convenience.

More detailed experiments have now shown that the Millipore assay selectively detects the binding of the trace of GDP preexisting in commercial preparations of GTP, and this preference was overcome by the addition of aminoacyl-tRNA. These experiments are the subject of this communication.

Materials and Methods

GTP was supplied by P-L Biochemicals, Inc., and repurified essentially by the method of Moffatt (1964). A 0–0.4 M triethylammonium bicarbonate gradient (pH 7.5) was used to elute the nucleotides from a DEAE-cellulose column (Whatman DE-11). Fractions containing GTP were identified by thin-layer chromatography as described below, pooled, and lyophilized.

[³H]GTP (lithium salt, specific activity 1.4 Ci/mmole) and [³H]GDP (lithium salt, specific activity 1.27 Ci/mmole) were

---

* From The Rockefeller University, New York, New York 10021. *Received June 19, 1969*. This work was supported by a grant to Dr. Fritz Lipmann from the National Institutes of Health, (No. GM-13972).

† Recipient of a Junior Fellowship for Medical Research from the Beit Memorial Fellowships Trust, London, England.

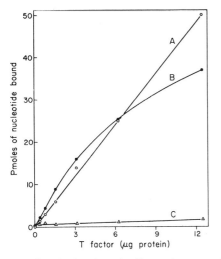

FIGURE 1: Binding of radioactive nucleotides to T factor assayed by Millipore binding. Each reaction mixture contained, in addition to the standard components: (A) 400 pmoles of [³H]GDP, (B) 700 pmoles of [³H]GTP, and (C) 740 pmoles of [γ-³²P]GTP. Reaction was continued for 2 min at 30°.

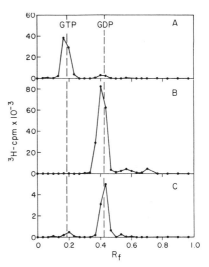

FIGURE 2: Chromatographic analysis of samples on polyethyleneimine thin-layer plates. (A) Commercial GTP, (B) commercial GDP, and (C) eluate from Millipore after binding assay with GTP sample as in part A used as substrate.

supplied by Schwarz BioResearch, Inc. [γ-³²P]GTP was from International Chemical and Nuclear Corp. The level of ³²P_i in the latter was reduced to 0.3% by stepwise chromatography on DEAE-cellulose, modified from the procedure of Moffatt (1964).

L-[¹⁴C]Phenylalanine (325 mCi/mmole), supplied by Schwarz BioResearch, Inc., was used to charge tRNA (General Biochemicals, Inc.) of *Escherichia coli* B by the method of Conway (1964). The product contained 500 pmoles of phenylalanine/mg of RNA. Radioactivity from [¹⁴C]-phenylalanine was counted in a Nuclear-Chicago low-background gas-flow counter at approximately 20% counting efficiency.

Nucleotides were chromatographed on polyethyleneimine cellulose impregnated MN 300 thin-layer plates (Brinkmann Instruments, Inc.) with 1 M potassium phosphate (pH 3.4) as developing solvent (Cashel et al., 1969). The spots were identified by viewing chromatograms in ultraviolet light, or by cutting them in 5-mm sections and determining the radioactivity of each section with 5 ml of toluene–Liquifluor (Pilot Chemicals, Inc.) in a Nuclear-Chicago scintillation counter. Radioactivity bound to the Millipore was transferred directly to a polyethyleneimine thin-layer plate by elution with 50 mM EDTA (pH 7.5) using a technique similar to that of Laskowski (1967). Over 90% of the radioactivity (assessed from a duplicate) was transferred in this way. Unlabeled GDP and GTP (0.01 μmole of each) were added before development of the chromatogram.

E. coli B used for the preparation of polypeptide chain elongation factors T and G (Nishizuka and Lipmann, 1966) was grown at 37° with vigorous aeration in a Biogen (American Sterilizer Co.). The medium contained, per liter, 10 g of glucose, 8 g of Difco nutrient broth, and 5 g of Difco yeast extract supplemented with the salts solution of Schaechter

et al. (1958). Cells were continuously harvested in logarithmic phase (optical density of 3.5 at 450 mμ in a Zeiss Model PHQ II spectrophotometer) and were not frozen before use.

Details of the methods used to purify the ribosomes and factors T and G are to be published separately (Gordon, 1969). In summary, the steps were: centrifugation of the cell homogenate at 30,000g, and extraction of nucleic acids and ribosomes from the supernatant fraction by a polyethylene glycol–Dextran phase system (Albertsson, 1960; Alberts, 1967); separation of T factor from G factor by ammonium sulfate fractionation; DEAE-cellulose column chromatography; hydroxylapatite column chromatography (Parmeggiani, 1968); and preparative polyacrylamide gel electrophoresis (Parmeggiani, 1968). Both factors T and G showed single bands on analytical polyacrylamide gel electrophoresis. The factors T_u and T_s (Lucas-Lenard and Lipmann, 1966) were not studied separately in the present work.

The Millipore assay for binding of nucleotide to T factor was essentially as before (Gordon, 1968). The reaction mixtures contained in a total volume of 0.105 ml: 50 mM Tris-HCl (pH 7.4), 160 mM ammonium chloride, 10 mM magnesium chloride, 12 mM dithiothreitol, nucleotides, tRNA, and factor as specified. Reactions were started by the addition of 5 μl of T factor solution, and terminated after 30 sec at 0° (when the reaction had already gone to completion) or as otherwise specified. Reaction mixtures were filtered and washed as previously described, and radioactivity was assayed in the filters by scintillation counting. The Millipore filters used in this assay (HA 0.45 μ pore size) were washed with NaEDTA (pH 8) and then with water before use. All buffer solutions were Millipore filtered to remove particles of dust which caused variable blanks.

GTP hydrolysis and poly U directed phenylalanine poly-

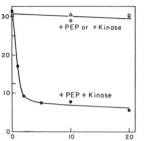

FIGURE 3: Effect of preincubation of [³H]GTP with triphosphate-regenerating system on the amount of nucleotide subsequently bound to T factor. The standard reaction mixture (before addition of enzyme) was supplemented with either phosphoenolpyruvate (5 mM) or 50 μg of pyruvate kinase (upper curve), or both together (lower curve) during preincubation. Each tube contained 700 pmoles of [³H]GTP. After preincubation, the reaction mixtures were cooled in an ice bath, and 12.5 μg of T factor was added. After 30 sec they were filtered and washed.

merization were assayed by the method of Nishizuka and Lipmann (1966) using dithiothreitol instead of mercaptoethanol.

Phosphoenolpyruvate (tricyclohexylammonium salt) and phosphoenolpyruvate kinase (EC 2.7.1.40) were supplied by Boehringer Mannheim Corp.

Results

The T factor used in this study was active in complementing factor G for phenylalanine polymerization. In the standard assay, 1 μg of T resulted in the polymerization of approximately 30 pmoles of phenylalanine. It therefore contained both factors T_u and T_s which are necessary for amino acid polymerization. The T factor was also free of GTPase activity (less than 0.2 pmole of P_i was released from 1000 pmoles of GTP by 10 μg of T factor in 1 min at 30°).

Figure 1 illustrates an experiment in which labeled GDP or GTP was incubated with various amounts of T factor. The binding of tritium-labeled GDP by Millipore filter was linear with protein concentration, but when [³H]GTP was used there was a detectable leveling of the amount of tritium bound. Very little ³²P was retained when [γ-³²P]GTP was used, suggesting that much of the nucleotide bound to T factor was not GTP.

To identify the nucleotide retained on the Millipore filter, [³H]GTP was incubated with T factor and the reaction mixture was filtered and washed in the usual way. The bound radioactivity was then analyzed by thin-layer chromatography after elution from the Millipore filter (see Methods section). Figure 2 shows that although [³H]GTP was added to the reaction mixture, 90% of the radioactivity bound chromatographed as GDP and only 10% as GTP. Figure 2A shows that the commercial [³H]GTP contained [³H]GDP as a contaminant. Although GDP comprised only 10% of the nucleotide present in tritiated GTP added to the binding reaction mixture, the T factor showed a remarkably high selectivity for GDP. When reaction mixtures containing T factor and [³H]GTP and no aminoacyl-tRNA were passed through Sephadex

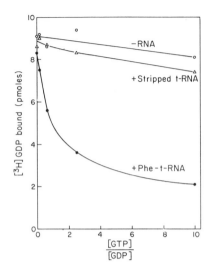

FIGURE 4: Effect of phenylalanyl-tRNA and GTP on the binding of GDP to T factor. Each reaction mixture contained 100 pmoles of [³H]GDP and purified unlabeled GTP as shown by the abscissa. Where indicated, 75 μg of stripped tRNA or 73 μg of [¹⁴C]phenylalanyl-tRNA (containing 37 pmoles of phenylalanine) was also added. The binding was started by the addition of 3 μg of T factor and continued for 1.5 min at 30° before filtration.

G-50 columns, relatively small amounts of bound radioactivity were obtained (for direct comparison of the Millipore and Sephadex assays, see Gordon, 1968). However, an analysis similar to that of Figure 2 of the complex that survived passage through Sephadex G-50 also showed a strong preferential binding of the contaminating GDP.

Experiments were also performed in which the tritiated GTP was preincubated with a nucleoside triphosphate regenerating system before the addition of T factor. Figure 3 shows that this procedure reduced by 80% the amount of tritiated nucleotide bound to the T factor in the Millipore assay. Chromatography showed that the tritium binding, which could not be eliminated by preincubation with the regenerating system, was largely GTP. The lower level of [γ-³²P]GTP bound in the experiment of Figure 1 was probably due to the presence of nonradioactive GDP. The T factor is therefore highly selective in preferentially binding GDP and little GTP binding can be detected by the Millipore filtration assay. This experiment also eliminates the possibility that GTP, which might have initially bound to the T factor, was being hydrolyzed as the complex was filtered on the Millipore.

It was found with the electrophoretically homogeneous T factor preparation used here, as reported earlier by Gordon (1968), Ravel et al. (1968), and Ertel et al. (1968a,b), that addition of aminoacyl-tRNA to reaction mixtures containing tritiated GTP resulted in a reduction of label retained on the Millipore filter. The requirement for GTP in this process is shown in Figure 4. The GTP was repurified by DEAE-cellulose chromatography to remove contaminating GDP. The graph shows that unlabeled GTP was not effective in displacing labeled GDP bound to T factor, whereas both GTP and

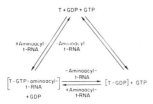

FIGURE 5: Equilibrium suggested from Millipore assays. The two complexes are thermodynamically highly favored over a mixture of separate components, but the relative amounts of each complex are determined by the concentrations of GTP, GDP, and of aminoacyl-tRNA (Gordon, 1968).

amino acid charged tRNA were necessary to displace the GDP. Approximately 70% of the GDP could be displaced in this way, indicating that, in the presence of aminoacyl-tRNA, GTP competes more effectively with GDP for binding to T factor, with formation of the T–GTP–aminoacyl-tRNA complex previously characterized (Gordon, 1967). Thus, the addition of aminoacyl-tRNA allows the selectivity for GDP binding to be reversed.

Discussion

The experiments described show that T factor from *E. coli* will bind GDP strongly and with high selectivity, even in the presence of a large excess of GTP. For example, from Figure 2, when GTP and GDP are present in the ratio 9:1, 90% of the nucleotide bound is GDP and 10% GTP. Ravel *et al.* (1968), using the Millipore assay, also found preferential binding of the tritium label of GTP compared with GTP labeled in the terminal phosphate, but they attributed this to hydrolysis. When aminoacyl-tRNA was added, the equilibrium moved in favor of GTP binding, and nonradioactive GTP was able to compete effectively with the [³H]GDP (Figure 4 above). Earlier data showing that the addition of aminoacyl-tRNA caused the disappearance of the bound GTP in the Millipore assay (Gordon, 1968; Ravel *et al.*, 1968; Ertel *et al.*, 1968a) must now be reinterpreted. In our setup, the radioactivity bound was actually the trace of GDP preexisting in the [³H]GTP; addition of aminoacyl-tRNA not only resulted in the formation of a ternary complex that was Millipore filtrable, but also reversed the preference for GDP. The present finding of only low levels of T–GTP suggests that this complex is much more labile than T–GDP. The equilibria postulated on the basis of our findings are summarized in Figure 5.

The data presented in this paper were all based on the Millipore assay, which measures the radioactivity bound to material adsorbable to the nitrocellulose membrane. Earlier results (Gordon, 1967) showed that GTP, labeled with either ³H or ³²P, bound to high molecular weight material in the excluded volume of Sephadex columns, but only in the presence of aminoacyl-tRNA. The reason why the T–GDP complex described here did not appear in the earlier Sephadex analyses was attributed to its relative instability (Gordon, 1968). However, when the small amount of radioactivity that was excluded from Sephadex in the absence of aminoacyl-tRNA was analyzed, a preferential binding of GDP was found in that case too. The reason for the apparent differences in stability of the same complex obtained with the two assays is not clear. It may be that the Millipore itself has some stabilizing effect. This is consistent with the fact that the groups that are involved in the interaction with tRNA are also involved in the binding to the Millipore (concluded from the reversal of the Millipore binding by aminoacyl-tRNA; Gordon, 1968).

Acknowledgment

We are deeply grateful to Dr. Fritz Lipmann, in whose laboratory this work was done, for his valuable discussions and criticism.

References

Alberts, B. M. (1967), *Methods Enzymol. 12A*, 579.
Albertsson, P.-E. (1960), Partition of Cell Particles and Macromolecules, New York, N. Y., Wiley.
Cashel, M., Lazzarini, R. A., and Kalbacher, B. (1969), *J. Chromatog. 40*, 103.
Conway, T. W. (1964), *Proc. Natl. Acad. Sci. U. S. 51*, 1216.
Ertel, R., Brot, N., Redfield, B., Allende, J. E., and Weissbach, H. (1968b), *Proc. Natl. Acad. Sci. U. S. 59*, 861.
Ertel, R., Redfield, B., Brot, N., and Weissbach, H. (1968a), *Arch. Biochem. Biophys. 128*, 331.
Gordon, J. (1967), *Proc. Natl. Acad. Sci. U. S. 58*, 1574.
Gordon, J. (1968), *Proc. Natl. Acad. Sci. U. S. 59*, 179.
Gordon, J. (1969), *J. Biol. Chem.* (in press).
Laskowski, M. (1967), *Methods Enzymol. 12A*, 281.
Lucas-Lenard, J., and Haenni, A.-L. (1968), *Proc. Natl. Acad. Sci. U. S. 59*, 554.
Lucas-Lenard, J., and Lipmann, F. (1966), *Proc. Natl. Acad. Sci. U. S. 55*, 1562.
Moffatt, J. G. (1964), *Can. J. Chem. 42*, 599.
Nishizuka, Y., and Lipmann, F. (1966), *Proc. Natl. Acad. Sci. U. S. 55*, 212.
Parmeggiani, A. (1968), *Biochem. Biophys. Res. Commun. 30*, 613.
Ravel, J. M., Shorey, R. L., Froehner, S., and Shive, W. (1968), *Arch. Biochem. Biophys. 125*, 514.
Ravel, J. M., Shorey, R. L., Garner, C., and Shive, W. (1969), *Fed. Proc. 28*, 598.
Ravel, J. M., Shorey, R. L., and Shive, W. (1967), *Biochem. Biophys. Res. Commun. 29*, 68.
Schaechter, M., Maaløe, O., and Kjeldgaard, N. O. (1958), *J. Gen. Microbiol. 19*, 592.

27

Reprinted from *Arch. Biochem. Biophys.*, **137**(1), 262–269 (1970)

Studies on the Role of Factor Ts in Polypeptide Synthesis

HERBERT WEISSBACH, DAVID LEE MILLER, AND JOHN HACHMANN

Roche Institute of Molecular Biology, Nutley, New Jersey 07110

Received December 19, 1969

Three forms of Tu, namely, Tu-GDP, Tu-GTP, and Tu free of nucleotide, have been examined with respect to the requirement of Ts for their reaction with GDP, GTP, and aminoacyl-tRNA (AA-tRNA). Ts does not appear to be required for the binding of GTP to Tu or the reaction of Tu-GTP with AA-tRNA. However, Ts stimulates the exchange between Tu-GDP and GDP or GTP. The possibility is discussed that Ts is required to dissociate the Tu-GDP complex formed during the binding of AA-tRNA to ribosomes.

Recent studies on the function of the soluble transfer factors, Ts[1] and Tu, in protein synthesis have concerned the interaction of these factors with GTP and their role in the messenger dependent binding of AA-tRNA to ribosomes (2–14). The initial observation that the protein-nucleotide complex was retained by a nitrocellulose filter (13) has provided the basis for a rapid assay for Tu and Ts, which has been helpful in the purification of these factors from *E. coli* extracts (5, 11). Previous data (8, 11, 12) have indicated that an initial step leading to AA-tRNA binding to ribosomes is the formation of Tu-GTP (reaction 1), and that the rate of formation of the Tu-nucleotide complex is stimulated by Ts (11).

$$Tu + GTP \rightarrow Tu\text{-}GTP \qquad (1)$$

An interaction of the transfer factors and GTP with AA-tRNA in the bacterial system was first suggested by Ravel (2), who showed

[1] Since there is not a uniform nomenclature for the soluble transfer factors, we are using the abbreviations reported by Lucas-Lenard and Lipmann (1). Ts and Tu correspond to factors FIs and FIu described by Ravel and co-workers (12) and factors S₁ and S₃ reported by Lengyel and co-workers (8). Other abbreviations used in this paper include: AA-tRNA, aminoacyl-tRNA; DTT, dithiothreitol; GTP, guanosine 5′-triphosphate; GDP, guanosine 5′-diphosphate; PEP, phosphoenolypyruvate.

that a soluble factor and GTP stimulated AA-tRNA binding to ribosomes. More recent studies have shown that both transfer factors Ts and Tu are required for the enzymatic binding of AA-tRNA to ribosomes (5, 7, 8, 12). In addition, Gordon (10) provided more direct evidence for an interaction of AA-tRNA with transfer factors and GTP when he showed that AA-tRNA inhibited the retention of the transfer factor-GTP complex on a nitrocellulose filter. This finding has been confirmed in other laboratories (4, 11, 12, 14) and has provided a simple means with which to study the interaction of AA-tRNA with the Tu-GTP complex. The most direct evidence for the presence of an AA-tRNA–protein-GTP complex has come from the studies of Ravel *et al.* (4, 12), Skoultchi *et al.* (8), and Weissbach *et al.* (15). It was possible in these studies to show a shift in the chromatographic profile of AA-tRNA on Sephadex G-100 after incubating the acylated tRNA with GTP and two transfer factors corresponding to Ts and Tu. The above observations would be compatible with a reaction sequence in which the Tu-GTP complex formed in reaction 1 reacted with AA-tRNA to form a ternary complex (reaction 2).

$$Tu\text{-}GTP + AA\text{-}tRNA \rightarrow$$
$$AA\text{-}tRNA\text{-}Tu\text{-}GTP \qquad (2)$$

More recently, several groups have shown that the AA-tRNA–Tu-GTP complex is involved in the messenger-dependent binding of AA-tRNA to ribosomes (6, 12, 16). Hydrolysis of GTP accompanies this reaction (reaction 3), and Tu-GDP is formed as a product of this reaction (12, 16, 17).

$$\text{AA-tRNA-Tu-GTP} \xrightarrow[\text{Messenger}]{\text{Ribosomes}}$$

$$\text{AA-tRNA-Messenger-Ribosomes} + \text{Tu-GDP} + \text{Pi} \quad (3)$$

The availability of a crystalline and homogenous preparation of Tu (15, 18) has enabled further studies on the role of Ts in the formation of the Tu-nucleotide complex. We have confirmed the observation of Cooper and Gordon (19) that Tu preferentially binds GDP, not GTP. In addition, Tu preparations relatively free of nucleotide, hereafter referred to as Tu, or containing bound GDP (Tu-GDP) or GTP (Tu-GTP) have been studied. Ts does not appear to be required for the reaction of Tu with either GTP or GDP but does stimulate the exchange of Tu-GDP with exogenous GTP or GDP. It is concluded that Ts may be involved in the regeneration of Tu-GTP from Tu-GDP.

MATERIALS AND METHODS

E. coli B cells grown on minimal media and harvested at midlog were obtained from Grain Processing Corporation. ^{3}H-GTP (1.1–1.4 mCi/μmole) and ^{3}H-GDP (1.12 mCi/μmole) were purchased from Schwarz Biochemicals, and γ-^{32}P-GTP was obtained from International Chemical and Nuclear Corporation (sp. activity ranged between 200–600 cpm/$\mu\mu$mole). ^{14}C-Phenylalanine (360 μCi/μmole) was purchased from New England Nuclear Corp., and *E. coli* tRNA (640 A_{260} = 1 μmole) was obtained from General Biochemicals and acylated with 19 unlabeled amino acids and ^{14}C-phenylalanine (about 1% charging with phenylalanine) according to a modification of the procedure of Conway (20). All radioactive determinations were done in a Beckman Model LS100 liquid scintillation counter using a counting fluid described by Bray (21).

The *E. coli* transfer factor Tu (sp. act. 18,000 units/mg) used in these studies was purified by a modification (18) of a previously described procedure (5, 11). The Tu preparations contained one equivalent of bound GDP [the nucleotide had been

added during the purification procedure (18)] and will be referred to as Tu-GDP.

The bound nucleotide could be removed from the Tu-GDP by dialyzing the enzyme solution (1–2 ml) for 3–5 days against 4 liters of a buffer containing 0.05 M Tris pH 8.0, 0.75 × 10^{-4} M EDTA, and 2.5 × 10^{-3} M DTT. The buffer was changed every 24 hr. The enzyme was incubated with ^{3}H-GDP prior to dialysis to form a Tu-^{3}H-GDP complex which could be assayed by retention of the complex on a nitrocellulose filter (see below). The degree of removal of the nucleotide could be determined from the loss of radioactivity bound to Tu during dialysis.

A Tu-GTP complex could be formed by incubating 0.1 ml of Tu-GDP (600 units of Tu) with phosphoenolpyruvate (PEP), 0.75 μmoles, and pyruvate kinase (10 μg cryst. enzyme from Boeringer Corp.) for 10 min at 37°. Further data on the formation of Tu and Tu-GTP are presented in the legend of Table III. Ts was purified by a modification of a reported procedure (5, 11) and had a specific activity of 2.6 × 10^5 units/mg.

The assay for both Ts and Tu was based upon retention of radioactive GDP on a nitrocellulose filter (Millipore Corp. 0.45 μ, 25 mm) in the presence of both factors as previously described (11). The incubation mixtures contained 0.05 M Tris buffer pH 7.5, 0.05 M NH$_4$Cl, 0.01 M MgCl$_2$, Tu, Ts, 2.5 × 10^{-6} M nucleotide, and H$_2$O to a total volume of 0.2 ml. The incubations were usually done at 0° for 5 min and terminated by the addition of 0.8 ml of a wash buffer containing 0.01 M MgCl$_2$, 0.01 M NH$_4$Cl, and 0.01 M Tris–Cl buffer pH 7.4. After passing the diluted incubation mixture through a nitrocellulose filter (0.45 μ, 25 mm), the filter was washed with 15 ml of cold buffer. The filter was then dissolved in a counting fluid described by Bray (21), and the radioactivity determined. One unit of either Tu or Ts is defined as the amount of enzyme needed to complex 1 $\mu\mu$mole of GDP after 5 min at 0°. For the Tu assay, sufficient Ts was present so that the amount of GDP complexed represented the total extent of binding. For the Ts assay a standard amount of Tu-GDP was used (35 units), and an amount of Ts was added so that less than 20 $\mu\mu$moles of Tu-GDP exchanged with ^{3}H-GDP after 5 min at 0°. Under these conditions the amount of GDP complexed to Tu was proportional to the Ts concentration (5, 11). When either ^{3}H or γ-^{32}P-GTP was used (except for the experiments in Table I), trace levels of contaminating GDP were removed by preincubating the GTP solution with PEP (3.75 × 10^{-3} M) and pyruvate kinase (10 μg of purified protein) for 10 min at 37° before the addition of Tu. This useful procedure was suggested to us by Dr. J. Gordon (19).

Assay for the formation of the AA-tRNA–Tu-GTP complex. Although a Tu-nucleotide complex is retained by a nitrocellulose filter, the AA-tRNA–Tu-GTP complex passes through the filter (4, 10, 11, 12, 14). Therefore, the decrease in radioactivity retained by the filter due to the addition of AA-tRNA was used as a measure of ternary complex formation.

GTP hydrolysis using γ-^{32}P-GTP was done according to the procedure of Conway and Lipmann (22).

RESULTS

Reaction of GTP and GDP with Tu-GDP. Previous studies (5, 11) on the formation of a Tu-nucleotide complex were complicated by the high level of phosphatase activity in the Tu and Ts preparations. Thus, the finding that Tu-GDP was always formed after incubating with GTP was believed to be due to hydrolysis of Tu-GTP to Tu-GDP by a contaminating phosphohydrolase activity. It was apparent that until Tu and Ts preparations were obtained which were relatively free of contaminating hydrolytic activity it would not be possible to examine the individual reactions of Tu with GTP and GDP. The availability of a highly purified preparation of Tu (containing one equivalent of bound GDP and referred to as Tu-GDP) with only a limited ability to hydrolyze GTP prompted the present studies. Since this enzyme preparation contained a bound nucleotide, the studies presented here on the binding of radioactive GDP or GTP to Tu represent an exchange between free and protein-bound nucleotide. In initial experiments (Table I) with this Tu-GDP preparation, it was surprising to find that although ^{3}H-GTP could efficiently exchange

TABLE I

COMPARISON OF BINDING OF ^{3}H-GTP WITH γ-^{32}P-GTP TO Tu AND ^{32}P-HYDROLYSIS[a]

^{3}H-GTP $\mu\mu$moles bound	γ-^{32}P-GTP $\mu\mu$moles bound	^{32}P $\mu\mu$moles hydrolyzed
24	1.3	9.5

[a] Incubations were performed at 37° for 5 min. Thirty units of Tu and 2.5×10^{-6} M GTP were present in the incubation mixture in the absence of Ts. Details of the filter assay are described in Methods. GTP hydrolysis was performed according to the procedure of Conway and Lipmann (22).

TABLE II

EFFECT OF PHOSPHOENOLYPYRUVATE AND PYRUVATE KINASE ON THE FORMATION OF Tu-GTP[a]

Nucleotide	PEP and pyruvate kinase	^{3}H or ^{32}P $\mu\mu$moles bound
^{3}H-GTP	−	29.0
^{3}H-GTP	+	21.4
γ-^{32}P-GTP	−	0.9
γ-^{32}P-GTP	+	22.1

[a] The incubation mixtures contained 30 units of Tu, 2.5×10^{-6} M nucleotide and were performed at 37° for 5 min. The Tu-nucleotide complex was assayed by the filter technique (see Methods).

with the bound nucleotide, γ-^{32}P-GTP did not. It should be noted that Ts was not added in the experiments in Table I since it has been shown previously (11) that at 37° the formation of a Tu-nucleotide complex can occur at a significant rate in the absence of Ts. Although some hydrolysis did occur during these incubations, it is noted in Table I that under the conditions employed, there was less inorganic phosphate (^{32}P) released than ^{3}H-GTP bound. Thus, the inability of γ-^{32}P-GTP to exchange with GDP bound to Tu could not be explained by hydrolysis of GTP to GDP during the reaction. A possible explanation for these anomalous results was that the ^{3}H-GTP and γ-^{32}P-GTP solutions contained significant levels of ^{3}H-GDP and unlabeled GDP, respectively, and that in a mixture of the two nucleotides (GTP and GDP), GDP is the preferred substrate (i.e., is more tightly bound to Tu). A similar conclusion concerning the preference of a T factor (containing both Ts and Tu) for GDP was reached recently by Cooper and Gordon (19).

It was possible to show that GTP could bind to Tu in the absence of GDP by using γ-^{32}P-GTP in the presence of a generating system (i.e., PEP and pyruvate kinase) to remove any contaminating GDP. Although the commercial preparation of γ-^{32}P-GTP yielded very little ^{32}P bound to Tu (see Table I), the incubation of Tu-GDP with ^{32}P nucleotide and, in addition, PEP and pyruvate kinase resulted in the formation of a Tu-γ-^{32}P-GTP complex (Table II). As seen in Table II the amount of γ-^{32}P-GTP bound compared favorably with the results

TABLE III

Ts Dependency for Nucleotide Binding Using Various Tu Fractions[a]

Fraction	γ-^{32}P-GTP $\mu\mu$moles bound		^{3}H-GDP $\mu\mu$moles bound	
	+Ts	−Ts	+Ts	−Ts
Tu-GDP[b]	5.2	0.6	29.0	1.6
Tu[c]	11.8	14.6	15.5	13.4
Tu-GTP[d]	25.3	20.4	21.8	17.2

[a] The incubations were performed at 0° for either 2 min (with γ-^{32}P-GTP) or 3 min (with ^{3}H-GDP). About 15 units of Tu and 30 units of Tu-GDP and Tu-GTP were used in the incubations. Details of the incubations are described in the text.

[b] Tu-GDP was routinely checked by forming Tu-^{3}H-GDP and removal of the excess GDP by dialysis as follows: Tu-GDP (6000 units in 100 μl) was incubated at 37° for 15 min with ^{3}H-GDP (0.7 mμmoles, 850 cpm/$\mu\mu$mole) to form a Tu-^{3}H-GDP complex which could be assayed by retention of the complex on a nitrocellulose filter. Total radioactivity was also determined. The Tu-preparation was then dialyzed 16 hr against 4 liters of a buffer (0.01 M Tris buffer pH 7.5, 0.01 M MgCl$_2$, 0.01 M NH$_4$Cl, and 0.001 M DTT) to remove any unbound ^{3}H-GDP. Tu-bound radioactivity did not decrease during dialysis, and after dialysis total radioactivity was equal to Tu-bound radioactivity.

[c] Tu was prepared from Tu-GDP in the following way: 1 ml of Tu-^{3}H-GDP was dialyzed for 72 hr against 4 liters of a buffer containing 7.5×10^{-5} M EDTA, 0.0025 M DTT, and 0.05 M Tris–Cl buffer pH 8.0. Each 24 hr the buffer was changed and the enzyme assayed for bound radioactivity. After 16, 32, and 72 hr the percentage of free Tu in a typical preparation was 52, 70, and 92%, respectively.

[d] Tu-GTP was prepared from Tu-GDP using PEP and pyruvate kinase as described in the Methods section. Evidence for the formation of Tu-GTP from Tu-GDP was obtained as follows: Tu-^{3}H-GDP was prepared by exchanging Tu-GDP (100 units of Tu) with ^{3}H-GDP as described above in part A. After dialysis to remove any unbound ^{3}H-GDP, 1 ml of the Tu-GDP solution contained 80 units of Tu and 68,000 cpm of bound ^{3}H-GDP. Two 0.1-ml aliquots were removed; one aliquot was incubated with PEP (0.75 μmoles) and pyruvate kinase (10 μg) for 10 min at 37°, while the other was kept at 37° for 10 min but was not incubated with PEP and pyruvate kinase. Both samples were chromatographed

obtained using ^{3}H-GTP as substrate. The above data confirmed the view that GDP contaminating the radioactive GTP solutions is the substrate of choice in the reaction with Tu, and that previous studies on the binding of GTP to Tu were very likely measuring GDP binding. In addition, the data in Table II also show that GTP, in the absence of GDP, can exchange with a nucleotide bound to Tu to form a stable Tu-GTP complex.

Studies on the role of Ts using Tu, Tu-GDP, and Tu-GTP. The experiments in Tables I and II were performed at 37°; at this temperature even with the most purified Tu preparations, the formation of a Tu-nucleotide complex reaches completion in 5–10 min without the addition of exogenous Ts (11). Although Ts has been shown to stimulate the rate of the reaction at both 0° and 37°, a better dependency is observed at the lower temperature (11). The experiments reported in Tables I ad II also represent an exchange reaction since the enzyme contained a bound nucleotide (GDP). Therefore the effect of Ts was reexamined at 0° using three different enzyme preparations; Tu (free of bound nucleotide), Tu-GDP, and Tu-GTP. The removal of bound nucleotide and the conversion of Tu-GDP to Tu-GTP are described in Methods and the legend of Table III.

Typical results showing the effects of Ts on the reaction of γ-^{32}P-GTP or ^{3}H-GDP with the three different Tu preparations are seen in Table III. When Tu was used, the reaction essentially reached completion with γ-^{32}P-GTP or ^{3}H-GDP within 3 min in the absence of Ts. However, with Tu-GDP and

on DEAE-cellulose paper using formic acid: NH$_4$OH:H$_2$O (20:3.3:180) as the solvent. In this system GTP has an R_f value of 0.18 and GDP an R_f of 0.42. Essentially all of the radioactivity migrated with GTP in the sample of Tu-GDP that was treated with PEP and pyruvate kinase, whereas GDP was identified as the nucleotide present in Tu-GDP (not exposed to PEP and pyruvate kinase).

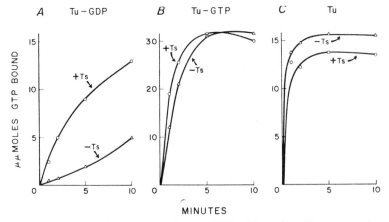

FIG. 1. Effect of Ts on the reaction of γ-^{32}P-GTP with various Tu preparations. Tu (15 units), Tu-GDP (30 units), and Tu-GTP (30 units) were prepared as described in the legend of Table III. The incubation were performed at 0° in the presence of 2.5×10^{-6} M γ-^{32}P-GTP, which had been preincubated for 10 min at 37° with PEP and pyruvate kinase as described in Methods. Where indicated, 60 units of Ts were added.

to a much less degree with Tu-GTP,[2] a stimulation of the exchange reaction with γ-^{32}P-GTP or ^{3}H-GDP was obtained in the presence of Ts. It should also be noted that at 0° much less γ-^{32}P-GTP exchanged with Tu-GDP than ^{3}H-GDP presumably because of the higher affinity of Tu for GDP than for GTP. The high binding of γ-^{32}P-GTP observed in the experiments in Table II could be explained by the fact that the incubations were performed at 37° with PEP and pyruvate kinase. Under these conditions the Tu-GDP was very likely converted to Tu-GTP which readily exchanges with γ-^{32}P-GTP. Figures 1 and 2 show the kinetics of the reactions of the various forms of Tu with either γ-^{32}P-GTP or ^{3}H-GDP. It is clearly seen that with both GTP and GDP marked dependencies on Ts are observed only when Tu-GDP is used. Further evidence showing that the loss of Ts dependency with Tu was due to removal of bound GDP from Tu-GDP is seen in Fig. 2. Preincubation of Tu with GDP to reform Tu-GDP converted the Tu to a species which again required Ts in

[2] Part of the effect of Ts seen with Tu-GTP may be due to the failure of the GTP generating system (PEP, pyruvate kinase) to completely convert all the Tu-GDP to Tu-GTP. The "nucleotide free" preparation also usually contained 5–15% Tu-GDP which could account for a slight Ts stimulation of nucleotide binding that was observed with this preparation occasionally.

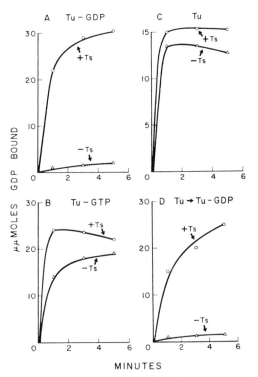

FIG. 2. The effect of Ts on the reaction of ^{3}H-GDP with various Tu preparations. The amount of Ts, Tu, and nucleotide (^{3}H-GDP) added are the same as in Fig. 1. In Fig. D, 120 units of Tu used in Fig. C were converted to Tu-GDP by incubating the Tu for 10 min at 37° with 3 mμmoles of GDP (unlabeled) in a total volume of 130 μl. 20 μl of this mixture were added as the Tu-GDP preparation in Fig. D.

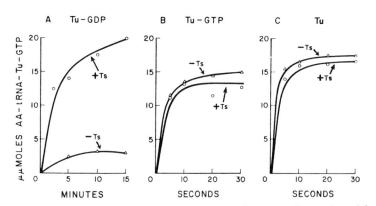

FIG. 3. The effect on Ts on the formation of an AA-tRNA-Tu-GTP complex with various Tu preparations. Tu and Tu-GTP were prepared as described in the legend of Table III from Tu-GDP. The incubations with AA-tRNA (0.23 A_{260} of an unfractionated tRNA preparation acylated with ^{14}C-L-phenylalanine and 19 unlabeled amino acids) and ^{3}H-GTP were performed at 0°. Where indicated, 80 units of Ts and about 30 units of either Tu, Tu-GDP, or Tu-GTP were employed. The formation of the AA-tRNA-Tu-GTP complex was assayed by the decrease in radioactivity retained on a nitrocellulose filter.

order to obtain binding with ^{3}H-GDP (Fig. 2D).

Interaction of AA-tRNA with Tu. The above experiments were extended to studies on the role of Ts in the formation of the ternary complex, AA-tRNA–Tu-GTP. As seen in Fig. 3, Ts was required for the formation of the ternary complex, AA-tRNA–Tu-GTP, only when the reaction was initiated with Tu-GDP. With Tu or Tu-GTP the rate of ternary complex formation was not affected by the presence of Ts. Thus, the results obtained support the conclusion that the requirement of Ts for Tu-GTP and AA-tRNA–Tu-GTP formation is observed only when Tu-GDP is the initial form of Tu employed.

Studies with Tu isolated in the absence of added nucleotide. The characteristics of Tu-GDP were similar to the Tu preparations described previously (11), in that Ts stimulated the formation of a Tu-nucleotide complex. The one basic difference in the purification procedure for the isolation of the Tu preparation used in the present studies was the presence of GDP in the solutions used in the chromatographic separations (18). This offered a ready explanation for the isolation of Tu-containing bound GDP. Since a homogenous preparation of Tu was not obtained previously, the possibility existed that the Tu initially isolated from *E. coli* extracts contained bound GDP. In order to investigate this, Tu was again

purified (in the absence of added nucleotide) through the Sephadex G-100 chromatography as described by Ertel *et al.* (5). Although this Tu preparation also has not been isolated in a homogeneous state which would permit direct assay of a bound nucleotide, preparations about 50% pure (sp. act. 10,000 units/mg) have been obtained. Results with this partially purified preparation of Tu have provided indirect evidence that it contains bound GDP. This Tu preparation had a 280/260 absorbancy ratio of 1.3 (a Tu-GDP complex has a 280/260 absorbancy ratio of 1.2) and behaved like Tu-GDP[3] as shown in Table IV.

[3] Further evidence indicating that this Tu preparation contained bound GDP was obtained as follows: One milliliter of a Tu solution containing 28 mμmoles of Tu was heated at 56° for 20 min to denature the Tu and release any bound nucleotide. The insoluble protein was removed, and the solution was lyophilized. The residue was dissolved in 0.1 ml of H_2O, and additional protein was removed by the addition of 5 vol. of ethanol and centrifugation. The 280/260 mμ absorbancy ratio of the supernatant solution was 0.8, and its uv spectrum resembled that of guanosine. The solution was evaporated, and the residue was chromatographed on DEAE-cellulose paper using a formic acid–ammonium formate buffer (see Table III). A nondetectable (by uv) amount of ^{3}H-GDP (0.1 mμmole, 80,000 cpm) was added as a tracer. When examined under uv light, the chromatogram showed a distinct spot that migrated with the ^{3}H-GDP marker.

It is seen that the binding of ^{3}H-GDP to this Tu preparation is stimulated 20–30-fold by Ts after 1 min at 0°. However, when the Tu preparation was preincubated with PEP and pyruvate kinase or dialyzed against EDTA for 5 days (see legend of Table III),

TABLE IV

EFFECT OF Ts ON THE REACTION OF Tu PURIFIED
IN THE ABSENCE OF ADDED GDP[a]

Fraction	$\mu\mu$Moles ^{3}H-GDP bound +Ts	$\mu\mu$Moles ^{3}H-GDP bound −Ts	Stimulation by Ts (fold)
A. Tu, not treated[b]	32	1.5	21
B. Tu, PEP + Pyr. Kin.[c]	34	7	4.8
C. Tu, Dialyzed[d]	31	17	1.8
D. Tu, Dialyzed + GDP[e]	34	1	34

[a] All incubations contained between 30–35 units of Tu which was determined by the ability of each preparation to bind ^{3}H-GDP in the presence of excess Ts (11). The binding reaction was performed with the different Tu preparations at 0° for 1 min in the presence of ^{3}H-GDP ($\pm$Ts) as described in Methods.

[b] A. Tu (not treated and presumed to be Tu-GDP) was prepared essentially according to the procedure of Ertel *et al.* (5). It had a specific activity of about 10,000 units/mg which is about one-half that obtained with a homogenous preparation.

[c] B. The Tu in A was preincubated with PEP and pyruvate kinase as described in Table III for the formation of Tu-GTP.

[d] C. The Tu in A was dialyzed for 3 days in the presence of EDTA as described in Table III for the formation of Tu. Since the Tu preparation did not contain a radioactive nucleotide bound to it, the extent of the loss of bound GDP could not be determined. In control experiments using Tu-^{3}H-GDP, 80–95% of the bound GDP would have dissociated under these conditions.

[e] D. The Tu in C was exposed to GDP to form Tu-GDP as described in Fig. 2.

the Ts stimulation was markedly diminished. In addition, as seen in Table IV, conversion of Tu to Tu-GDP restored the Ts dependency for the binding of the radioactive nucleotide.

DISCUSSION

The present results necessitate a reevaluation of the previous conclusion that Ts is required for the reaction Tu + GTP → Tu-GTP. The evidence suggests that Ts is required for an exchange between free nucleotide and nucleotide bound to Tu. Since Tu-GDP is a product of the binding of AA-tRNA to ribosomes (12, 16, 17), it seems probable that Ts is involved in the regeneration of Tu-GTP from Tu-GDP. Thus Tu-GDP can be pictured as an inactive form of Tu which must be dissociated before Tu can again function in AA-tRNA binding. A less but significant dependency on Ts was observed for the exchange of GTP or GDP with Tu-GTP (Figs. 1 and 2). It is therefore simplest to picture Ts as being involved in the exhange reaction with either Tu-GDP or Tu-GTP, and that the reaction of Tu (free of nucleotide) with GTP or GDP does not require Ts (or requires so little Ts that a significant Ts dependency could not be observed in the present studies).

A summary of the postulated reaction sequences leading to the binding of AA-tRNA to ribosomes is shown in Fig. 4. It should be noted in Fig. 4 that the conversion of Tu-GDP to Tu-GTP requires Ts, and this reaction is pictured as proceeding via the intermediate formation of a Ts-Tu complex. Evidence for such a complex has been obtained (23), and recent experiments have shown that Ts can readily displace GDP from Tu-GDP (24). Experiments on the binding of AA-tRNA to ribosomes are currently in progress with the hope of obtaining

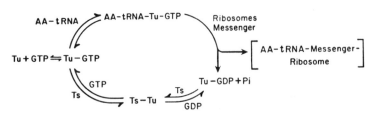

FIG. 4. Scheme showing the postulated reactions leading to the binding of AA-tRNA to ribosomes.

further support for the postulated role of Ts in the regeneration of Tu-GTP.

ACKNOWLEDGMENTS

The authors thank Dr. Nathan Brot for many helpful discussions, and Dr. J. Gordon for making available preprints of his studies and for his advice.

REFERENCES

1. LUCAS-LENARD, J., AND LIPMANN, F., *Proc. Nat. Acad. Sci. U. S.* **55**, 1562 (1966).
2. RAVEL, J. M., *Proc. Nat. Acad. Sci. U. S.* **57**, 1811 (1967).
3. RAVEL, J. M., SHOREY, R. L., AND SHIVE, W., *Biochem. Biophys. Res. Commun.* **29**, 68 (1967).
4. RAVEL, J. M., SHOREY, R. L., FROEHNER, S., AND SHIVE, W., *Arch. Biochem. Biophys.* **125**, 514 (1968).
5. ERTEL, R., BROT, N., REDFIELD, B. ALLENDE, J. E., AND WEISSBACH, H., *Proc. Nat. Acad. Sci. U. S.* **59**, 861 (1968).
6. LUCAS-LENARD, J., AND HAENNI, A. L., *Proc. Nat. Acad. Sci. U. S.* **59**, 554 (1968).
7. HAENNI, A. L., AND LUCAS-LENARD, J., *Proc. Nat. Acad. Sci. U. S.* **61**, 1363 (1968).
8. SKOULTCHI, A., OHO, Y., MOON, H. M., AND LENGYEL, P., *Proc. Nat. Acad. Sci. U. S.* **60**, 675 (1968).
9. GORDON, J., *Proc. Nat. Acad. Sci. U. S.* **58**, 1574 (1967).
10. GORDON, J., *Proc. Nat. Aczd. Sci. U. S.* **59**, 179 (1968).
11. ERTEL, R., REDFIELD, B., BROT, N., AND WEISSBACH, H., *Arch. Biochem. Biophys.* **128**, 331 (1968).
12. SHOREY, R. L., RAVEL, J. M., GARNER, C. W., AND SHIVE, W., *J. Biol. Chem.* **244**, 4555 (1969).
13. ALLENDE, J. E., AND WEISSBACH, H., *Biochem. Biophys. Res. Commun.* **28**, 32 (1967).
14. JEREZ, C., SANDOVAL, A., ALLENDE, J., HENES, C., AND OFENGAND, J., *Biochem.* **8**, 3006 (1967).
15. WEISSBACH, H., BROT, N., MILLER, D. L., ROSMAN, M., AND ERTEL, R., *Cold Spring Harbor Symp. Quant. Biol.*, (1969).
16. ONO, Y., SKOULTCHI, A., WATERSON, J., AND LENGYEL, P. *Nature* **222**, 645 (1969).
17. GORDON, J., *J. Biol. Chem.* **244**, 5680 (1969).
18. MILLER, D. L., AND WEISSBACH, H., To be published.
19. COOPER, D., AND GORDON, J., *Biochemistry* **8**, 4289 (1969).
20. CONWAY, T. W., *Proc. Nat. Acad. Sci. U. S.* **51**, 1216 (1964).
21. BRAY, G., *Anal. Biochem.* **1**, 279 (1960).
22. CONWAY, J. W., AND LIPMANN, F., *Proc. Nat. Acad. Sci. U. S.* **52**, 1462 (1964).
23. MILLER, D. L., AND WEISSBACH, H., *Arch. Biochem. Biophys.* **132**, 146 (1969).
24. MILLER, D. L., AND WEISSBACH, H., To be published.

28

Reprinted from *Cold Spring Harbor Symp. Quant. Biol.*, **34**, 39–48 (1969)

On The Catalytic Center of Peptidyl Transfer:
A Part of the 50 S Ribosome Structure

T. Staehelin, D. Maglott, and R. E. Monro*

*Department of Zoology, University of Michigan, and * Instituto de Biología Celular, Madrid, Spain*

The enormous complexity of ribosome structure even in simple organisms such as bacteria has become fully evident during the last few years. Most of the detailed information on the chemical aspects of ribosomal proteins has been obtained from the 30 S ribosome of *Escherichia coli* (Traut et al., 1967; Moore et al., 1968; Fogel and Sypherd, 1968; Delius and Traut, 1969; Traut et al., this volume; Kurland et al., this volume). The functional aspects and how they relate to the structural complexity of the 30 S ribosome have been studied to a most remarkable degree of detail in Nomura's laboratory (this volume and Traub et al., 1967, 1968b, 1969; Ozaki et al., 1969). Although some protein components are of particular importance for specific functions of the 30 S ribosome (chain initiation, codon specific aminoacyl-transfer RNA (tRNA) binding, fidelity of coding) it is quite clear that the functions of many proteins are interrelated and interdependent. Thus a genetic alteration of one component may influence the function of another.

Fewer details on the structural-functional relationship are known for the 50 S ribosome. The complexity of its structure, consisting of two RNA molecules and approximately 35 different protein components, makes it extremely difficult to analyze this particle in the manner that Nomura and his co-workers investigated the 30 S ribosome. Until very recently, reconstitution of 50 S particles active in polypeptide synthesis from protein deficient cores and proteins was possible only with a core lacking 5 or 6 protein components (Staehelin and Meselson, 1966; Hosokawa et al., 1966; Traub and Nomura, 1968a). Very little information about specific functions of the 50 S ribosome had been gained by this earlier work.

One of the key functions of the 50 S ribosome in protein synthesis is the catalysis of an enzymatic reaction, namely the peptide bond formation (Monro, 1967). Thus the 50 S ribosome or part of its structure can be considered an enzyme which we shall call peptidyl transferase as suggested by Monro et al. (1967).

In this study we attempted to characterize further the catalytic center of peptide bond formation as part of the molecular fine structure of the 50 S ribosome of *E. coli*. Our approach was to remove in steps defined groups of proteins from the 50 S ribosomes and test the resulting core particles for function in polypeptide synthesis with poly U as messenger RNA (mRNA) and for the specific function of peptide bond formation assayed by the "fragment reaction" as described by Monro and Marcker (1967). In the assay system of the present study we used N-acetyl-(^{3}H)leucyl-ACCAC (the 3'-terminal pentanucleotide fragment of Leu-tRNA) and puromycin as substrates. The product of the peptidyl transferase catalyzed reaction is N-acetyl-(^{3}H)leucyl-puromycin.

By the stepwise removal of proteins from the 50 S ribosome we wanted to determine which proteins are essential and which are not essential in the catalysis of peptide bond formation as well as in the whole process of polypeptide synthesis. We finally wanted to reconstitute the catalytic activity for peptidyl transfer from a completely inactive ribonucleoprotein particle (core) and the smallest possible number of missing proteins.

Preparation of 50 S and 30 S Ribosomal Subunits Active in Polypeptide Synthesis

Whereas the isolation of ribosomal subunits behaving as 50 S and 30 S particles in the ultracentrifuge has been a routine procedure in dozens of laboratories for many years, the performance of such isolated subunits in polypeptide synthesis in vitro with synthetic or natural messenger is usually much less than would be expected if all ribosomes participated in the synthetic process. Good estimates of the fraction of ribosomes which is active are in the most cases impossible to get from the published data. Usually one can calculate the number of amino acid residues polymerized, on the average, per 50 S + 30 S ribosome couple present in the assay. This number ranges from 1 or 2 up to 20 or 25 in the most frequently used poly U system. Gilbert (1963) determined both the number and average length of the polyphenylalanine chains synthesized in his poly U system. This allows one to calculate that usually 20% to 30% of the ribosomes participated in the synthetic process. In his best experiments, up to 50% of the ribosomes were active (Gilbert, pers. commun.).

When looking at the 50 S ribosome as an enzyme (peptidyl transferase) we considered it important to prepare this "enzyme" in an active form: that is, we wanted to be certain that most of the presumed enzyme particles were competent in catalyzing the chemical reaction. By showing that all or most 50 S ribosomes can participate in polypeptide synthesis, we may conclude that these same 50 S ribosomes must have an active peptidyl transferase site.

All ribosomes were prepared from the RNase I deficient *E. coli* strain from Gesteland (1966), most frequently referred to as D_{10}, which is a derivative of *E. coli* strain AB301 (K-12, Hfr, *met*⁻, from Adelberg).

The cells were grown by the New England Enzyme Center, Tufts University, Boston (in 1966) in a 50 gallon fermenter under forced aeration in a tryptone broth medium at 37°C to late logarithmic growth phase (about 10^9 cells per ml). At this time, aeration was stopped and the cells were harvested by continuous flow centrifugation at room temperature. The cell paste was divided into 30 to 40 g aliquots which were wrapped in cellophane and kept frozen at −80°C until used during the last three years. The fact that during several hours the cells went through a period of slow cooling without aeration before finally being chilled to about 0°C might be relevant for the yield of highly active 50 S and 30 S ribosomes. The cells seemed to have no or very few polysomes.

All manipulations during ribosome isolation were done at 0° to 4°C. The cells were broken by grinding with 2 to 2.5 g per g cell paste of levigated alumina (Alcoa). The ground paste was extracted with buffered salt solution (2.0 to 2.5 ml per g of wet cells) containing 0.1 M NH_4Cl, 0.01 M Mg acetate, 0.02 M Tris-HCl, pH 7.5, 0.0005 M EDTA and 0.003 M beta-mercaptoethanol. Electrophoretically purified DNase (Worthington), 3 to 5 μg per ml, was added during the extraction. A low speed centrifugation to remove the alumina was followed by 30 min centrifugation at 30,000 × g average. The resulting supernatant fluid (S-30) was layered in about 40 ml aliquots over 35 ml of 1.1 M sucrose containing 0.5 M NH_4Cl, 0.01 M (or 0.02 M) Mg acetate, 0.02 M Tris, pH 7.5 and 0.0005 M EDTA in 75 ml IEC polycarbonate centrifuge bottles and centrifuged for 16 to 18 hr at 40,000 rpm in the A170 type angle rotor in the IEC B60 ultracentrifuge at 2° to 4°C. Approximately the upper two-thirds of the supernatant fluid was withdrawn and dialyzed for 16 to 18 hr against 0.1 M NH_4Cl 0.01 M Mg acetate, 0.02 M Tris, pH 7.5, 0.0005 M EDTA and 0.002 M β-mercaptoethanol. After

dialysis this high speed supernatant (S-100) was frozen in small aliquots at −80°C and used as enzyme source for in vitro polypeptide synthesis. The crystal clear pellets were rinsed gently with distilled H_2O to remove some loose flocculant material. They were then dissolved in about 3 ml per tube of 0.5 M NH_4Cl, 5 × 10^{-4} M Mg acetate, 0.02 M Tris, pH 7.5, which assures complete and instant dissociation of remaining 70 S ribosomes into 50 S and 30 S subunits. This solution was immediately dialyzed for 4 to 6 hr against 0.1 M NH_4Cl, 0.001 M Mg acetate and 0.02 M Tris, pH 7.5. Isolation and purification of 50 S and 30 S subunits was achieved by two successive sucrose gradient centrifugations in the same ionic conditions as the last dialysis solution. Subunits recovered from sucrose gradients can be quickly concentrated by adding 0.6 to 0.65 volumes of cold ethanol (after having raised the Mg acetate concentration of the ribosome solution to between 7 and 10 mM). The ribosome precipitate is pelleted in 10 min at about 20,000 × g and redissolved in the desired volume of 0.1 M NH_4Cl, 0.001 M Mg acetate and 0.02 M Tris, pH 7.5. The purified 50 S ribosomes containing less than 3% 30 S, and the 30 S containing not more than 1% 50 S are stored in 0.1 M NH_4Cl, 0.005 M Mg acetate, 0.02 M Tris, 0.00025 M EDTA (200 to 250 OD_{260} units of 30 S per ml; 400 to 500 OD_{260} units of 50 S per ml) at −80°C. (This method will be published in detail in Method. Enzymol., Nucleic Acids, Vol. 12, Part C.)

POLY U DIRECTED POLYPHENYLALANINE SYNTHESIS WITH 50 S AND 30 S RIBOSOMES

Before attempting to analyze and characterize the molecular structure of the enzymatic site in the 50 S ribosome catalyzing peptide bond formation, it was important to establish first that the "enzyme," i.e., the 50 S ribosomes, consists mostly of active particles. This point was established by showing that in the presence of 30 S units most of the 50 S ribosomes are capable of participating in polypeptide synthesis. Figure 1 shows sucrose density gradient analyses of poly U directed phenylalanine incorporation mixtures after various times of incubation. Thirty S ribosomes were present in an excess concentration over 50 S (see legend to Fig. 1). After only 5 min incubation at 35°C, about 60% of the 50 S particles initially present have combined with 30 S ribosomes to form 70 S couples and polysomes carrying nascent polyphenylalanine chains. The gradual decrease of free 50 S and 30 S particles, presumably due to polyphenylalanine chain initiation, is virtually complete by 20 min. At that time about 90% of the 50 S ribosomes have been converted into 70 S particles and polysomes. Polypeptide chain

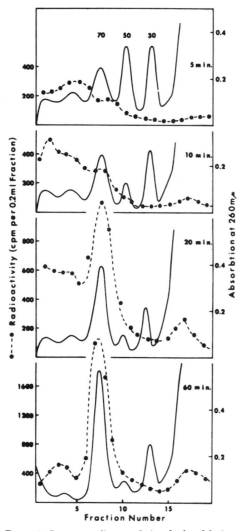

FIGURE 1. Sucrose gradient analysis of phenylalanine incorporations.

Aliquots (0.15 ml) of reaction mixture, incubated at 35°C and taken at the times indicated, were layered onto 4 ml exponential gradients and spun at 60,000 rpm for 75 min in the SB 405 rotor of the IEC B-60 ultracentrifuge. The gradients were monitored continuously at 260 mμ with a Gilford 2400 spectrophotometer using an LKB flow cell. The reaction mixture contained, per 1.0 ml: 1 μmole ATP; 0.4 μmoles GTP; 12 μmoles phosphoenolpyruvate; 10 μg pyruvate kinase; 40 mμmoles of each of 20 ^{12}C-amino acids, including Phe; 300 μg *E. coli* tRNA; 0.2 ml S-100; 100 μg poly U; 104 μμmoles 30 S, 66 μμmoles 50 S (30 S:50 S = 1.5); 0.45 μc ^{14}C-Phe, (final sp. act. = 11 mc/mm); 160 μmoles NH$_4$Cl; 18 μmoles MgAc$_2$; 20 μmoles Tris-Cl, pH 7.5; and 3 μmoles β-mercaptoethanol.

Hot TCA precipitable counts from 50 μliter aliquots of the reaction mixture showed that after 60 min of incu-

extension usually ceases after 90 to 150 min of incubation. It should be pointed out that a considerable amount of the ribosomal and of the hot TCA insoluble radioactivity input sedimented with large ribosome aggregates to the bottom of the centrifuge tubes. After 5 min incubation the pellet contained about 25% of the ribosome OD$_{260}$ and about 45% of the hot TCA insoluble radioactivity input, at 10 and 20 min 40% to 45% of the OD$_{260}$ and 60% to 65% of the radioactivity had pelleted, and at 60 min, about 35% of the OD$_{260}$ and about 50% of the radioactivity were in the pellet. This aggregation of ribosomes is not, or only to a minor extent due to the formation of true polysomes, since pancreatic ribonuclease (2 μg per ml) would not convert the aggregates into 70 S ribosomes. Furthermore, we can exclude unspecific aggregation of free 50 S and 30 S subunits because of the stability of the 70 S couples and larger aggregates in 0.5 M KCl, (at 0.01 M Mg acetate and 0.02 M Tris), under which conditions no such unspecific aggregates could exist. There is a striking correlation between length of the polyphenylalanine chains and tendency to ribosome aggregation. In all experiments the 70 S couples have the lowest specific activity, that is, the shortest average polyphenylalanine chains. In the experiment shown in Fig. 1, after 60 min incubation, the average chain length per 70 S unit in the monomer peak is about 70 to 80 phenylalanine residues; in the dimer region it is between 200 and 300; and for the pellet this number can be calculated to be about 400. These calculations and the failure by conventional means (high salt concentration; low Mg ion concentration; ribonuclease) to desaggregate or dissociate the ribosomes lead us to the following conclusions: (1) The formation of 70 S ribosomes is the result of polypeptide synthesis. (2) The aggregation of 70 S ribosomes into larger conglomerates is most likely due to the long, extremely hydrophobic, and "sticky" polyphenylalanine chains of the active ribosomes. This is not too surprising in view of the fact that the amount of polyphenylalanine per ribosome corresponds to 5 to 6% of the 70 S ribosomal protein mass or to 8 to 9% of the 50 S ribosomal protein.

Although we have not excluded the possibility that some inactive 50 S and (or) 30 S particles might aggregate with the active 70 S complexes, we are convinced that the majority, and probably about 90%, of the 50 S ribosomes are competent in polypeptide synthesis and must therefore have an active center for catalysis of peptide bond formation.

bation, 280 Phe had been polymerized per 50 S subunit present in the mixture. Note change of radioactivity scale at 60 min.

CHARACTERIZATION OF THREE PROTEIN-DEFICIENT
NUCLEOPROTEIN CORES DERIVED FROM 50 S
RIBOSOMES

1. Preparation of core particles. With the use of
high performance swinging bucket or angle rotors
which allow shortening of the centrifugation time
of CsCl equilibrium density gradients, we observed
in the presence of high Mg ion concentration (0.04
to 0.05 M Mg acetate) that 50 S ribosomes would
initially form three bands with approximate
buoyant densities of 1.62 g/cm³, 1.65 g/cm³ and
1.68 g/cm³. Particles recovered from the three
bands are designated α-core ($\rho = \sim 1.62$), β-core
($\rho = \sim 1.65$) and γ-core ($\rho = \sim 1.68$). The scheme
of conversion of 50 S ribosomes into its various
derived cores is as follows:

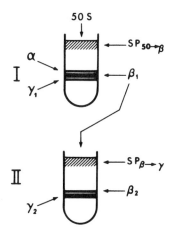

FIGURE 2. Stepwise dissociation of 50 S ribosomes in
CsCl density gradients into α, β, and γ-cores and split
proteins.

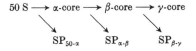

SP stands for split proteins, representing specific
groups of proteins dissociated upon the conversion
of one particle into another.

The α-cores are the least stable in CsCl, even in
high Mg ion concentration. After about 30 hr
centrifugation at 4°C they are completely converted into β-cores. Beta-cores are quite stable in
CsCl density gradients at 0.04 to 0.05 M Mg ion
concentration; conversion into γ-cores is very slow.
In order to achieve efficient conversion into
γ-cores, 0.008 to 0.01 M Mg is used in the CsCl
density gradients. Since for the present study
β-cores and γ-cores were the most useful and
interesting particles, we chose a dissociation scheme
of 50 S ribosomes in CsCl density gradients in two
steps as outlined in Fig. 2. The method will be
published in detail elsewhere (Staehelin and
Maglott, in prep.). In a first centrifugation at 0.04
M Mg ions 50 S ribosomes are mostly (about 70 %)
converted into β-cores and a split protein fraction
designated SP$_{50-\beta}$. Only a minor portion (about 15
to 20 %) of the particles are remaining in the α-core,
form; and also only a minor portion (about 10 to
15 %) are converted into γ-cores. These γ-cores,
designated γ$_1$, are possibly not quite identical with
regard to their protein content and composition to
the subsequently produced γ-cores which are prepared by centrifuging the β-cores isolated in the
first CsCl centrifugation for 35 to 40 hr in CsCl containing 0.008 to 0.01 M Mg ions. The split protein
resulting from this conversion is designated
SP$_{\beta-\gamma}$. (The detailed method of 50 S dissociation
and reconstitution will be published in Method.
Enzymol., Nucleic Acids, Vol. 12, C.)

2. Characterization of proteins. Proteins were
extracted from 50 S ribosomes and the derived
cores for analysis by acrylamide gel electrophoresis
by the LiCl-urea method of Spitnik-Elson (1965).
Electrophoresis at pH 4.5 in 6 M urea was carried
out according to Leboy et al. (1964) as described
previously (Gesteland and Staehelin, 1967) and in
the legend to Fig. 3. Acrylamide gel electrophoresis
patterns of proteins from 50 S ribosomes, α-cores,
β-cores and γ-cores are shown in Fig. 3. The α-core
has lost one major basic protein (the *4th* strong band
from the top in the 50 S proteins) and the slowest
moving acidic protein. (The designation acidic
means adsorbed on DEAE-cellulose, and basic
means not adsorbed on DEAE-cellulose at pH
7.4.) The β-core has lost all acidic proteins plus the
basic one already missing in the α-core; i.e., the
β-core results from the loss of at least 5 proteins.
This can be more readily seen in Fig. 4, where the
split proteins from converting 50 S ribosomes into
mainly β-cores (SP$_{50-\beta}$) are displayed. The γ-core
protein pattern seems to differ from the β-core
proteins only by the lack of one component,
namely the lowest band in the upper group
(Fig. 3 and 4). However, the split proteins from
converting β- into γ-cores (SP$_{\beta-\gamma}$, Fig. 4) show at
least four additional bands which in their electrophoretic mobility coincide with components still
present in the γ-core. This may be interpreted in
three alternative ways: 1. The conversion from
β- into γ-cores can occur by "multiple choice," i.e.,
different β-particles lose different proteins leading
to chemically heterogeneous γ-cores with very
similar buoyant densities. Heterogeneity could, of
course, exist already in the 50 S and β-cores. 2. All

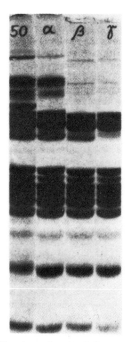

FIGURE 3. Disc electrophoresis of the proteins from the 50 S subunit and its derivative α-, β-, and γ-cores.

Electrophoresis was done at pH 4.5 with 2.5 mamp per tube for 5–6 hr. The separating gel was 10% acrylamide; the spacer and sample were 1.25%. The proteins were dialyzed against the upper gel buffer in 6 M urea before being mixed with the sample gel.

β-core. It should be noted in Fig. 4 that the band at the position of the arrow in the SP$_{\beta-\gamma}$ fraction is a basic protein and is often resolved into two weakly stained components, whereas the band in the SP$_{50-\beta}$ at the same position is of acidic nature, as judged from its affinity for DEAE-cellulose.

3. *Physical properties of cores and* RNA. Core particles isolated from CsCl density gradients are usually dialyzed against 0.1 M NH$_4$Cl, 0.01 M Mg acetate, and 0.02 M Tris pH 7.5. Sedimentation analysis in sucrose density gradients at 0.001 M Mg ion concentration and 0.02 M Tris, pH 7.5, showed much to our surprise that all three cores cosediment with 50 S ribosomes. The γ-cores might sediment slightly slower as indicated by a broadening of the "50 S" peak when 50 S and γ-cores are analyzed in the same gradient. This indicates that in these conditions of analysis all cores are in a very compact shape, similar to the shape of the 50 S particles. From the loss of mass (protein) which we estimate for the γ-core to be about 30% to 35% of the total protein or about 10% of the total 50 S mass, one would indeed expect only a very small decrease in sedimentation velocity unless there is a significant

or some of the proteins removed may not only have identical electrophoretic mobilities to those remaining, but may also be chemically identical and thus exist in more than one copy in the 50 S (or β-core) particle. This would mean that one out of two or more identical proteins dissociates during β- to γ-core conversion. 3. All or some components in the SP$_{\beta-\gamma}$ fraction, although having the same electrophoretic mobility as core proteins, differ chemically, i.e., in their primary structure from those core proteins. This would mean that the split proteins represent a specific and chemically unique group missing in all γ-cores. Not until a thorough chemical analysis of all single components has been done and the molecular weights of the split proteins determined and their sum compared with the mass difference between the β- and γ-cores can this question be decided. However, since the buoyant density difference between the β- and γ-cores is similar to that between the α- and β-cores, it seems unlikely that only one or two protein components per β-particle are removed in converting to γ, assuming the ribosomal RNA has the same degree of hydration in CsCl in the γ-core as it has in the

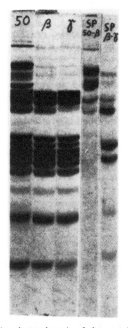

FIGURE 4. Disc electrophoresis of the proteins from the 50 S subunit, its derivative β- and γ-cores, and the split proteins SP$_{50-\beta}$ and SP$_{\beta-\gamma}$ dissociated during the conversion of 50 S into β-cores, and of β- into γ-cores, respectively.

The bands in SP$_{50-\beta}$ + SP$_{\beta-\gamma}$, indicated by the arrow, are different proteins. One, SP$_{50-\beta}$, is acidic, the other, SP$_{\beta-\gamma}$, is basic, as defined by DEAE-chromatography.

shape change. The cores in the earlier work (Staehelin and Meselson, 1966; Hosokawa et al., 1966; Traub and Nomura, 1968a) derived from 50 S ribosomes correspond to our β-cores. Therefore sedimentation coefficients reported in that work ranging from 40 S to 43 S must have been obtained from unfolded particles.

Ribonucleic acid extracted from fresh preparations of α-, β-, and γ-cores consists of at least 90 % intact 23 S RNA. Upon prolonged storage in solution at 4°C the 23 S RNA in γ-cores and to a lesser degree β-cores breaks down progressively, presumably due to traces of ribonuclease, into 16 S half molecules and most likely quarter molecules of about 12 S. These breakdown products sediment in remarkably sharp bands which indicates that the breaks occur within specific and relatively short stretches of the RNA. Under the same storage conditions the 23 S RNA in α-cores and 50 S ribosomes is very stable and protected. Upon reconstitution of γ-cores with the combined split proteins the 23 S RNA of the particles becomes protected again against ribonucleolytic attack.

Both β- and γ-cores, at least in their compact form, contain the same amount of 5 S RNA as 50 S ribosomes (Brownle and Staehelin, unpubl.).

THE FUNCTION OF CORES AND RECONSTITUTED PARTICLES IN CATALYSIS OF PEPTIDE BOND FORMATION AND POLYPEPTIDE SYNTHESIS

In order to learn how the specific function of peptidyl transfer and its molecular basis are related to other less well defined 50 S functions and structural components relevant to the overall performance in polypeptide synthesis, we tested the various 50 S derived particles for peptidyl transferase activity in the fragment assay as well as for activity in the overall process of poly U directed

polypeptide synthesis in the presence of 30 S ribosomes.

Figure 5 shows the results of the peptidyl transferase assay with 50 S and the three core particles. The formation of N-acetyl-leucyl-puromycin is shown as a function of incubation time at 0°C. Each point represents the average of duplicate samples. The three minute values allow a fair estimate of initial rates of the reaction. It is quite clear that both α- and β-cores are active, whereas the γ-cores are completely inactive. Freshly prepared β-cores are usually just as active as 50 S ribosomes. However, prolonged storage without freezing or rebanding in CsCl gradients with subsequent dialysis seems to inactivate β-cores partially. Two assays with such "aged" and slightly inactivated β-core preparations are included in Fig. 5. We do not know at this time whether or not this inactivation of peptidyl transferase is related to the observed breaks in 23 S RNA occurring during prolonged storage without freezing. Disregarding its relative instability we conclude that the β-core still has a fully active catalytic site for peptide bond formation, and that this site is either physically removed or inactivated by conformational changes as a result of the dissociation of the approximately five proteins during conversion of β-cores into γ-cores. Repeated attempts have failed to show any peptidyl transferase activity in the split protein fractions (both $SP_{50-\gamma}$ and $SP_{\beta-\gamma}$). Thus, the protein or proteins forming the active site need the supporting structure of ribosomal RNA and other proteins for allowing and maintaining the correct conformation. Since the fully active β-core is converted into two completely inactive fractions, namely the γ-nucleoprotein core and a group of about five proteins, we do not know at this time in which fraction the molecular

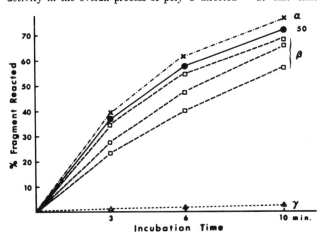

FIGURE 5. Peptidyl transferase activity of the 50 S subunit and the α-, β-, and γ-cores as measured by the fragment assay.

The reaction mixture for the assay contained, per 1.0 ml: 0.33 ml methanol; 260 μmoles KCl; 13 μmoles MgAc$_2$; 33 μmoles Tris-Cl, pH 7.5; 0.67 μmoles puromycin, and about 10,000 count/min of CACCA-N-acetyl-leucyl-^{3}H. The reaction, started by addition of the fragment, was carried out at 0°C. At the times indicated, 0.1 ml of the mixture was added to 0.1 ml of BeCl$_2$-NaAc-MgSO$_4$ salt solution to stop the reaction. The N-acetyl-leucyl-puromycin product was extracted with ethyl acetate and counted according to Leder and Bursztyn (1966).

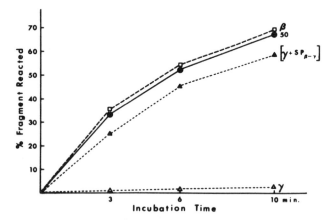

FIGURE 6. Comparison of the peptidyl transferase activity of 50 S subunits, β- and γ-cores, and reconstituted $[\gamma + SP_{\beta-\gamma}]$ particles.

Conditions for the fragment assay were as described in the legend to Fig. 5. Reconstitution was as outlined in the legend to Table 1.

structure making up the catalytic center resides. For this reason it was most important to achieve the reconstitution of the active site in the subribosomal particle from its inactive components.

Even before the beautiful work on reconstitution of active 30 S ribosomes from 16 S RNA and proteins by Traub and Nomura (1968b), we had tried in vain to activate γ-cores dialyzed with split proteins by about 10 min incubation at 35°C at ionic conditions close to the ones used by Traub. Also elaborate transitions from high to low ionic strength, 0.8 M to 0.2 M NH$_4$Cl or KCl at various Mg ion concentrations, gave only marginal and rather erratic results. After the appearance of Traub and Nomura's work (1968b), even their exact conditions for 30 S reconstitution applied to our γ-cores with the relevant split proteins gave no better results than many of our previous attempts.

Only at 50° to 55°C for 30 to 40 min was efficient functional reconstitution of γ-cores with SP$_{\beta-\gamma}$ or SP$_{50-\gamma}$ finally achieved. The optimal or near optimal ionic conditions are stated in the legend to Fig. 6, which compares peptidyl transferase activity of γ-cores reconstituted with SP$_{\beta-\gamma}$ with that of β-cores and 50 S ribosomes. The results show that a nearly complete reconstitution of the catalytic center for peptidyl transfer with about 5 protein components and γ-cores is possible.

A preliminary study of the time dependence of γ-core reconstitution indicates that the reconstitution rate at 50°C is slightly slower than the rate of 30 S reconstitution from 16 S RNA and 30 S proteins observed at 35°C by Traub and Nomura (1969). Thus, the energy barrier for reconstitution of 50 S derived γ-cores seems to be considerably higher than for total reconstitution of 30 S ribosomes from RNA and proteins.

Table 1 shows an experiment in which the activity

of 50 S derived cores and various reconstituted particles in the overall mechanism of polypeptide synthesis was tested. Table 2 gives a comparison

TABLE 1. POLYPHENYLALANINE SYNTHESIS BY 50 S AND ITS DERIVATIVES IN THE PRESENCE OF 30 S RIBOSOMES

Ribosome input	count/min (−150 for 30 S alone)	Phe-residues polymerized per ribosome	% of 50 S
30 S + 0	0		
50 S	11520	210	100
α	4080	74	35
β	2320	42	20
$[\beta + SP_{50-\beta}]$	8450	153	73
$[\beta + SP_{\beta-\gamma}]$	2220	40	19
γ	27	0	0
$[\gamma + SP_{50-\gamma}]$	2560	47	22
$[\gamma + SP_{\beta-\gamma}]$	1200	22	10

Conditions for the assay of polyphenylalanine (poly-Phe) synthesis were as in Fig. 1. Symbols in brackets such as $[\beta + SP_{50-\beta}]$ signify the reconstituted particle obtained by heating the designated core and split protein fractions at 50°C for 30 min in the following ionic conditions: 0.25 M NH$_4$Cl; 0.015 M MgAc; 0.02 M Tris, pH 7.5; and 0.003 M β-mercaptoethanol.

TABLE 2. ACTIVITY OF 50 S RIBOSOME DERIVATIVES IN POLYPEPTIDE SYNTHESIS AND CATALYSIS OF PEPTIDE BOND FORMATION

Particle	Poly-Phe synthesis % activity	Peptidyl transferase % activity
50 S	100	100
α	35–50	100
β	20–30	70–100
$[\beta + SP_{50-\beta}]$	70–85	not tested
γ	0	<3
$[\gamma + SP_{50-\gamma}]$	20–30	50–80
$[\gamma + SP_{\beta-\gamma}]$	~10	30–70
$[\gamma + SP_{50-\beta}]$	~5	~10

Conditions for the polyphenylalanine synthesis, the peptidyl transferase assay, and reconstitution were as described in Fig. 1 and Fig. 5, and Table 1.

between the competence of the particles in the integrated protein synthesizing system and their ability to serve only as catalysts in peptide bond formation in the fragment assay. In overall polypeptide synthesis the lack of probably only two proteins in the α-core causes a severe impairment of function. The β-core lacking at least 5 proteins is very inefficient, though still significantly active in protein synthesis. Beta-cores in the presence of 30 S consistently polymerize between 40 and 60 phenylalanine residues per core particle present in the incubation mixture. Reconstitution of β-cores for protein synthesis with $SP_{50-\beta}$ is quite efficient even in the cold, in agreement with the earlier work done on 50 S reconstitution (Staehelin and Meselson, 1966; Hosokawa et al., 1966; Traub and Nomura, 1968a). Gamma-core reconstitution, on the other hand, depends on heating at 50° to 55°C. The reconstitution of γ-cores with either $SP_{50-\gamma}$ or $SP_{\beta-\gamma}$ is rather poor for activity in polypeptide synthesis but excellent for activity in peptide bond formation (Table 1 and 2). It is also evident from Table 1 and 2 that $SP_{50-\beta}$ and $SP_{\beta-\gamma}$ have different functional significance. Thus $SP_{50-\beta}$ cannot substitute for $SP_{\beta-\gamma}$ in reconstituting peptidyl transferase activity with γ-cores; and $SP_{\beta-\gamma}$ cannot substitute for $SP_{50-\beta}$ in stimulating severalfold the activity of β-cores in polypeptide synthesis.

Table 2 summarizes the differences between the activities of the various particles in the complete protein synthesizing system on one hand and in the peptidyl transferase assay on the other hand. The β-core, which lacks 5 to 6 proteins, still has a fully active peptidyl transferase site but is very inefficient in protein synthesis. This catalytic site can be restored to almost full function from the inactive γ-core and the $SP_{\beta-\gamma}$ proteins, without restoring much of the protein synthesizing activity. Thus the efficiency of the 50 S derived particles in complete protein synthesis depends not only on the catalytic center for peptidyl transfer but also quite strongly on the presence and correct conformation of the $SP_{50-\beta}$ proteins which are not essential for peptide bond formation.

DISCUSSION

We have shown in this study that the function of the catalytic center for peptide bond formation can be related to the presence of a defined group of about 5 protein components ($SP_{\beta-\gamma}$) in a 50 S derived subunit. The removal of these proteins completely abolishes this enzymatic function, whereas the removal of a different group of some 5 or 6 proteins ($SP_{50-\beta}$) does not affect the active site. The crucial component or components in the $SP_{\beta-\gamma}$ fraction have an all-or-nothing effect on this

TABLE 3. ^{14}C-CHLORAMPHENICOL BINDING AND PEPTIDYL TRANSFERASE ACTIVITY BY 50 S RIBOSOMES AND DERIVED PARTICLES

Ribosomal input	^{14}C-CM bound per 4 OD_{260} count/min	% CM binding activity	% peptidyl transferase activity
50 S	2880	100	100
β	1310	45	70–100
γ	90	3	<3
[γ + $SP_{\beta-\gamma}$]	1340	46	30–70

[14]C-chloramphenicol (CM) (Amersham, sp. act. = 18.9 mc/mm) binding was done by equilibrium dialysis at 3×10^{-6} M chloramphenicol outside concentration, in 0.10 M NH_4Cl, 0.01 M Mg acetate, 0.02 M Tris pH 7.5, 0.003 M β-mercaptoethanol. Ribosomal particles at about 80 OD_{260} per ml were dialyzed in 0.15 ml aliquots overnight at 4°C. Duplicates of 50 μliter samples were counted in Bray's solution in a liquid scintillation spectrometer. Counts per minute (5090) corresponding to 50 μl outside fluid were subtracted from the samples.
Reconstitution was done as described in Table 1. Peptidyl transferase activities were taken from Table 2.

function and might actually contain the molecular structure of the active site. This complete inactivation of peptide bond formation occurring upon the transition from the β- to the γ-core indicates strongly that the catalytic center is a common and identical structural feature of all 50 S ribosomes. In other words, the 50 S ribosomes are homogenous with regard to the structure of this important enzymatic site. This assumption is also strongly supported by the fact that various antibiotics, e.g., chloramphenicol, of defined chemical structure affecting peptide bond formation, bind specifically with the 50 S ribosome, one molecule per ribosome at saturation of binding. Table 3 shows an experiment which illustrates, at least qualitatively, that the ability to bind chloramphenicol is closely related to the ability to catalyze peptide bond formation. The same group of proteins ($SP_{\beta-\gamma}$) whose removal abolishes peptide bond formation also abolishes binding of chloramphenicol. The same proteins, with the same requirements for reconstitution (50°C for 30 to 40 min) are sufficient to restore both activities. For a more detailed discussion of antibiotics relating to peptidyl transfer see Monro and Vazquez (1967), Monro et al. (1969, and this volume), Vazquez et al. (1969).

In contrast to the activity for peptidyl transfer, the performance of 50 S derived particles in the overall mechanism of polypeptide synthesis is drastically impaired upon the removal of proteins, which leaves the catalytic site for peptide bond formation still intact. The question arises whether all particles are inactivated to the same degree, or whether some ribosomes are selectively inactivated. Almost all α-cores, which lack 2 proteins as judged from the acrylamide electrophoresis pattern, seem to be competent in poly U directed polypeptide synthesis. Sucrose gradient analysis shows most of them to combine with 30 S subunits to form stable 70 S particles and polysomes. However, chain extension seems to be slower, resulting in shorter polyphenylalanine chains. Sucrose gradients of phenylalanine incorporation mixtures with β-cores always show at least half of the cores as free particles. The possibility that the β-cores are heterogenous has to be carefully tested. On the other hand, we consider it not unlikely that the β-cores at the moment of initiation of a polyphenylalanine chain form such an unstable complex with the 30 S partner that the chance of catalyzing the first peptide bond is very small. Once a di- or oligophenylalanyl-tRNA is present, the stability of the 70 S couple should increase with the length of the polypeptide chain. Experiments are now in progress to distinguish between explanation on the basis of inherent heterogeneity of β-cores, some being functional and some incompetent, or on probability as determinant of participation in polypeptide synthesis.

In our initial work on 50 S reconstitution we, as well as Nomura's group, found the cores to be inactive in polyphenylalanine synthesis. Yet the fact that they can be reconstituted in the cold, and the acrylamide gel electrophoresis patterns, clearly identify them as β-cores. There are several possible reasons for our failure to detect activity in polypeptide synthesis of those cores. (1) They may have been inactivated by partial unfolding and not able to refold under the assay conditions, in the absence of the split proteins. (2) Our present assay system contains 0.16 M NH_4Cl as opposed to the 0.08 M previously used. This may be more favorable for maintaining the active conformation of β-cores, whereas 50 S ribosomes perform about equally well under both conditions. (3) In those early experiments our so-called 100% activities ascribed to 50 S + 30 S ribosomes were more than an order of magnitude lower than with our present ribosome preparations, indicating that a large fraction of particles was somehow damaged to begin with. This may have been the reason for the formation in CsCl of truly inactive cores which needed for restoration of any activity at least part of the split proteins, namely the acidic group, as demonstrated by Traub and Nomura (1968a).

We are now in the process of studying in more detail the kinetics and energy requirements of γ-core reconstitution. Still a wide open and quite puzzling question is how, if at all, these in vitro reconstitutions resemble the in vivo assembly process. So far, complete self assembly of 30 S ribosomes (Traub and Nomura, 1968b, 1969) and partial assembly of 50 S ribosomes from subcomponents has been achieved in vitro in the absence of any non-ribosomal cell constituents.

Thus, all the information for correct assembly is contained in the structure of the components. However, for both ribosome subunits, some steps of the in vitro assembly involve thermodynamic energy barriers which would be incompatible with a spontaneous and efficient rate of ribosome assembly in vivo. Therefore the assembly pathway in vivo must either be different, such as to avoid the formation of such energetically insurmountable dead-end particles as the γ-cores or Traub's RI intermediates of the in vitro 30 S assembly; or else there must be some kind of enzymatic helper mechanisms to overcome these energy barriers.

The actual structural component or components of the catalytic center for peptidyl transfer have not yet been identified. This site could still be in the γ-core but need for its active conformation the presence of one or more components of the $SP_{\beta-\gamma}$ proteins. Alternatively, one or more components of the $SP_{\beta-\gamma}$ fraction may contain the structure for substrate binding and catalysis of peptide bond formation, but be unable to maintain the active conformation when separated from the nucleoprotein core structure. Or thirdly, the active site may be composed of parts from both the γ-core and the $SP_{\beta-\gamma}$ proteins. There are no easy experimental ways of distinguishing between these possibilities. But we are now trying to identify and isolate the crucial protein or proteins in the $SP_{\beta-\gamma}$ fraction necessary for reconstitution of the enzymatic activity.

ACKNOWLEDGMENTS

This investigation was supported by a research grant GB-7259 from the National Science Foundation and in part by an Institutional Research Grant IN-40-H from the American Cancer Society to the University of Michigan Cancer Institute. One of us, D. M., is the recipient of a National Science Foundation predoctoral fellowship.

REFERENCES

DELIUS, H., and R. R. TRAUT. 1969. Characterization of ribosomal proteins of *Escherichia coli*. VI. (Meeting Fed. Europ. Biochem. Soc.) *In* S. Ochoa, C. Asensio, and C. F. Heredia [ed.] Symposium on biosynthesis of macromolecules. (In press)

FOGEL, S., and P. SYPHERD. 1968. Chemical basis for heterogeneity of ribosomal proteins. Proc. Nat. Acad. Sci. *59*: 1329.

GESTELAND, R. F. 1966. Unfolding of *E. coli* ribosomes by removal of magnesium. J. Mol. Biol. *18*: 356.

GESTELAND, R. F., and T. STAEHELIN. 1967. Electrophoretic analysis of proteins from normal and CsCl-treated *Escherichia coli* ribosomes. J. Mol. Biol. *24*: 149.

GILBERT, W. 1963. Polypeptide synthesis in *Escherichia coli*. II. The polypeptide chain and sRNA. J. Mol. Biol. *6*: 389.

HOSOKAWA, K., R. FUJIMURA, and M. NOMURA. 1966

Reconstitution of functionally active ribosomes from inactive subparticles and proteins. Proc. Nat. Acad. Sci. *55:* 198.

LEBOY, P. S., E. C. COX, and J. G. FLAKS. 1964. The chromosomal site specifying a ribosomal protein in *Escherichia coli*. Proc. Nat. Acad. Sci. *52:* 1367.

LEDER, P., and H. BURSZTYN. 1966. Initiation of protein synthesis. II. A convenient assay for the ribosome dependent synthesis of N-formyl-^{14}C-methionylpuromycin. Biochem. Biophys. Res. Commun. *25:* 233.

MONRO, R. E. 1967. Catalysis of peptide bond formation by 50 S ribosomal subunits from *E. coli*. J. Mol. Biol. *26:* 147.

MONRO, R. E., M. L. CELMA, and D. VAZQUEZ. 1969. Action of sparsomycin on ribosome-catalyzed peptidyl transfer. Nature *222:* 356.

MONRO, R. E., B. E. H. MADEN, and R. R. TRAUT. 1967. The mechanism of peptide bond formation in protein synthesis, p. 179. *In* D. Shugar [ed.] Symp. Fed. Europ. Biochem. Soc. Academic Press, New York.

MONRO, R. E., and K. A. MARCKER. 1967. Ribosome-catalyzed reaction of puromycin with a formylmethionine-containing oligonucleotide. J. Mol. Biol. *25:* 347.

MONRO, R. E., and D. VAZQUEZ. 1967. Ribosome-catalyzed peptidyl transfer: effects of some inhibitors of protein synthesis. J. Mol. Biol. *28:* 161.

MOORE, P. B., R. R. TRAUT, H. NOLLER, P. PEARSON, and H. DELIUS. 1968. Ribosomal proteins from *Escherichia coli*. II. Proteins from the 30 S subunit. J. Mol. Biol. *31:* 441.

OZAKI, M., S. MIZUSHIMA, and M. NOMURA. 1969. Identification and functional characterization of the protein controlled by the streptomycin-resistant locus in *E. coli*. Nature *222:* 333.

SPITNIK-ELSON, P. 1965. The preparation of ribosomal protein from *Escherichia coli* with lithium chloride and urea. Biochem. Biophys. Res. Commun. *18:* 557.

STAEHELIN, T., and M. MESELSON. 1966. In vitro recovery of ribosomes and of synthetic activity from synthetically inactive ribosomal subunits. J. Mol. Biol. *16:* 245.

TRAUB, P., K. HOSOKAWA, G. R. CRAVEN, and M. NOMURA. 1967. Structure and function of *E. coli* ribosomes. IV. Isolation and characterization of functionally active proteins. Proc. Nat. Acad. Sci. *58:* 2430.

TRAUB, P., and M. NOMURA. 1968a. Structure and function of *Escherichia coli* ribosomes. I. Partial fractionation of the functionally active ribosomal proteins and reconstitution of artificial subribosomal particles. J. Mol. Biol. *34:* 575.

——, ——. 1968b. Structure and function of *Escherichia coli* ribosomes. V. Reconstitution of functionally active 30 S ribosomal particles from RNA and proteins. Proc. Nat. Acad. Sci. *59:* 777.

——, ——. 1969. Structure and function of *Escherichia coli* ribosomes. VI. Mechanism of assembly of 30 S ribosomes studied in vitro. J. Mol. Biol. *40:* 391.

TRAUT, R. R., P. B. MOORE, H. DELIUS, H. NOLLER, and A. TISSIÈRES. 1967. Ribosomal proteins of *Escherichia coli*. I. Demonstration of different primary structures.

VAZQUEZ, D., T. STAEHELIN, M. L. CELMA, E. BATTANER, R. FERNANDEZ-MUNOZ, and R. E. MONRO. 1969. Inhibitors as tools in elucidating ribosomal function. *In* 20th Mosbacher Kolloquium der Gesellschaft für Biologische Chemie, *in* Inhibitoren als Werkzeuge der Zellforschung.

29

Reprinted from *J. Mol. Biol.*, **39**(3), 441–460 (1969)

Translation and Translocation of Defined RNA Messengers

Richard W. Erbe†, Marion M. Nau and Philip Leder

Laboratory of Biochemistry, National Cancer Institute
National Institutes of Health, Bethesda, Maryland 20019, U.S.A.

(*Received 9 September 1968, and in revised form 4 November 1968*)

A purified system for protein synthesis, derived from *Esherichia coli*, makes use of small, synthetic mRNA's initiated by the fMet‡ codon, AUG, followed by a sequence of 3, 6 or 9 uridylic acid residues. These di-, tri- and tetra-codons direct the binding of fMet- and Phe-tRNA to ribosomes and the synthesis of the corresponding fMet-initiated di-, tri- and tetrapeptides. This system, directing the synthesis of unique and conveniently detected oligopeptide products, has permitted us to correlate the soluble elements required for protein synthesis with specific steps in the translation of initial and succeeding codons. Complete translation of these small mRNA's has certain of the stringent requirements necessary for the accurate cell-free translation of naturally occurring RNA messengers. Thus, initiation factors, fMet-tRNA and GTP must be provided. After recognition of the first codon and formation of the initial complex, one of two transfer factors, in the presence of GTP, catalyzes a *binding reaction* in which the second (first internal) codon is recognized. This permits formation of the dipeptide, fMet-Phe, but does not permit translation of succeeding codons. A second transfer factor, probably an enzyme, participates in a *translocation reaction* in which the third codon is made available for translation, apparently displacing the second codon at the recognition site on the ribosome. In addition, the complex formed between peptidyl-tRNA, mRNA and ribosome is stabilized, possibly by translocation of the peptidyl-tRNA from a site of lesser affinity to one of greater affinity on the ribosome. The role of GTP in the translocation reaction will be certain only when it has been uncoupled from its participation in the binding reaction. Once requirements for translation of the third codon have been met no further additions are necessary for elongation of the peptide chain.

1. Introduction

A variety of enzymes and factors is required for protein biosynthesis and for the unique events which serve in its accurate initiation and termination (Keller & Zamecnik, 1956; Hardesty, Arlinghaus, Shaeffer & Schweet, 1963; Fessenden & Moldave, 1963; Nakamoto, Conway, Allende, Spyrides & Lipmann, 1963; Eisenstadt & Brawerman, 1966; Revel & Gros, 1966; Stanley, Salas, Wahba & Ochoa, 1966; Capecchi, 1967*a*; Ganoza, 1967; Ravel, 1967). In order to describe this obviously complex sequence of reactions, a number of useful models has been advanced (Watson,

---

† Present address: Biophysics Research Laboratory, Department of Biological Chemistry, Harvard Medical School, Boston, Mass. 02115, U.S.A.

‡ Abbreviations used: fMet, *N*-formylmethionine. As far as possible all other abbreviations follow the recommendations of IUPAC-IUB as published in *J. Biol. Chem.* (1966) **241**, 527. AUGU$_3$, AUGU$_6$ and AUGU$_9$ indicate the trinucleoside diphosphate ApUpG followed by 3, 6 or 9 uridylic acid residues; GMP · PCP, 5′-guanylylmethylene diphosphonate.

1964; Monro, Maden & Traut, 1966; Nakamoto & Kolakofsky, 1966; Nishizuka & Lipmann, 1966a; Noll, 1966; Igarashi & Kaji, 1967; Schlessinger, Mangiarotti & Apirion, 1967; Bretscher, 1968; Guthrie & Nomura, 1968; Ohta & Thach, 1968). Perhaps the most general of these is that outlined by Watson in 1964. He proposed a cycle of three synthetic reactions involving two functionally distinct sites for binding tRNA to the ribosome. The reactions included the codon-dependent *binding* of aminoacyl-tRNA to a recognition site on the ribosome, *transfer* of nascent peptide from its tRNA to newly bound aminoacyl-tRNA with formation of a peptide bond, and, finally, *translocation* of the elongated peptidyl-tRNA from the recognition site to a holding site on the ribosome and *translocation* of mRNA, exposing the succeeding codon at the recognition site for translation.

The codon recognition or aminoacyl-tRNA binding step, resulting in the formation of an aminoacyl-tRNA–mRNA–ribosome complex, has been well documented (Arlinghaus, Favelukes & Schweet, 1963; Kaji & Kaji, 1963; Nakamoto et al., 1963; Nirenberg & Leder, 1964; Pestka & Nirenberg, 1966). Despite earlier evidence in higher organisms (Arlinghaus, Shaeffer & Schweet, 1964; Bont, Huizinga, Bloemendal, Van Weenen & Bosch, 1965), the enzyme and co-factor requirements for this reaction in bacterial systems have only recently been recognized (Ravel, 1967; Ertel, Brot, Redfield, Allende & Weissbach, 1968; Lucas-Lenard & Haenni, 1968; Erbe & Leder, 1968). Peptide bond synthesis as measured by the chemically analogous puromycin reaction or the unit elongation of peptidyl-tRNA is now recognized to be a function associated with the ribosome itself (Traut & Monro, 1964; Bretscher & Marcker, 1966; Hultin, 1966; Rychlík, 1966; Zamir, Leder & Elson, 1966; Goldberg & Mitsugi, 1967; Gottesman, 1967; Pestka, 1968) and, more specifically, with the 50 s subunit (Traut & Monro, 1964; Monro, 1967). Translocation, the last of these reactions, has proved the most difficult to examine experimentally. Nishizuka & Lipmann (1966a) suggested that GTP and a transfer factor associated with ribosome-dependent GTPase might be involved in this reaction. Their argument was by analogy to the contraction-linked ATPases. A number of studies involving the transfer of peptide to puromycin have suggested a role for GTP in translocation (Brot, Ertel & Weissbach, 1968; Tanaka, Kinoshita & Masukawa, 1968; Skogerson & Moldave, 1968). However, the translocation model itself provides at least one clear prediction which distinguishes the aminoacyl- or peptidyl-tRNA–mRNA–ribosome complex which has undergone translocation from that which has not. It requires that the codon adjacent to that most recently translated will be unavailable for aminoacyl-tRNA recognition until it is translocated into the appropriate recognition site on the ribosome. Making use of this prediction, we have been able to provide more direct evidence for the translocation mechanism and to identify the element, apparently an enzyme, required to accomplish this step. A preliminary report of these results has appeared (Erbe & Leder, 1968).

Here we describe experiments carried out in a purified protein-synthesizing system which makes use of a series of very short synthetic mRNA's initiated by the fMet codon, AUG, and followed by 3, 6 or 9 uridylic acid residues. These di-, tri- and tetra-codons direct the ribosomal binding of fMet- and Phe-tRNA's and the synthesis of the corresponding, easily identifiable fMet-initiated di-, tri- and tetrapeptides. Thus, the small mRNA's provide an experimental system for correlating the soluble elements required for protein synthesis with specific steps in translation of initial and succeeding codons. They also permit us to define the point beyond which translation of additional codons requires no additional enzymes or co-factors.

2. Materials and Methods

(a) *Materials*

Tris base was purchased from Sigma Chemical Co.; radioactive amino acids from New England Nuclear Corp.; Triton X-100 and dimethyl POPOP from Packard Instrument Co.; puromycin from Nutritional Biochemicals; GDP, UDP, GTP and [^{14}C]GTP from Schwartz BioResearch; [γ-^{32}P]GTP from International Chemical and Nuclear Co.; GMP · PCP, ApU and poly U from Miles Chemical; unfractionated *E. coli* strain B tRNA from General Biochemicals; DNase and bacterial alkaline phosphatase from Worthington Biochemicals; T_1 RNase from Calbiochem; DEAE cellulose DE52 from Whatman; DEAE Sephadex A50 and Sephadex G25 from Pharmacia; and nitrocellulose filters type HAWP, pore size 0·45 μ, 25 mm diameter, from Millipore. 10-formyl tetrahydrafolic acid was generously supplied by Lederle Laboratories, and diphenylalanine by Dr Sidney Pestka to whom we are most grateful.

(b) *Preparation of ribosomes and initiation and transfer factors*

(i) *Supernatant, ribosomes and ribosome wash*

All procedures were carried out at 4 to 6°C. 50 g of frozen, early log-phase *E. coli* MRE600 (Cammack & Wade, 1965) were thawed and suspended in 150 ml. of buffer containing 0·01 M-Tris acetate, pH 7·2; 0·05 M-NH$_4$Cl; and 10 mM-magnesium acetate. Following disruption in a French pressure cell at 10,000 lb./in^2, the lysate was incubated at 4°C with 1 μg DNase/ml. for 30 min and centrifuged at 22,400 g for 30 min and the precipitate discarded. Centrifugation was resumed at 106,000 g for 2 hr and the supernatant used to prepare the transfer factors. The ribosomal pellet was suspended in ribosomal wash solution containing 0·01 M-Tris acetate, pH 7·2, 1 M-NH$_4$Cl, 10 mM-magnesium acetate, and 10^{-4} M-dithiothreitol by gentle rocking overnight and centrifuged at 106,000 g for 2 hr. This washing procedure was repeated 6 times. Protein concentrations were determined by the method of Lowry, Rosebrough, Farr & Randall, 1951.

(ii) *Preparation of initiation factors*

The procedure is modified from the method of Stanley *et al.* (1966). The first 2 pooled ribosome-wash supernatants were brought to 70% saturation with solid (NH$_4$)$_2$SO$_4$ (43·6 g/100 ml.). After 30 min of stirring the solution was centrifuged at 40,000 g for 30 min. The precipitate was dissolved in 10 ml. of ribosome-wash solution and dialyzed 6 hr against 4 l. of buffer containing 0·01 M-Tris–HCl, pH 8·0, 0·05 M-KCl and 10^{-4}M-dithiothreitol. The dialysate was applied to a 1 cm × 50 cm DEAE cellulose column previously equilibrated with the same buffer and eluted with a linear salt–pH gradient: 0·05 to 0·50 M in KCl, pH 8·0 to 7·0 in 0·01 M-Tris–HCl containing 10^{-4} M-dithiothreitol; 400 ml. per chamber. Five-ml. fractions were collected every 20 min. The fractionation is shown in Fig. 1, wherein F1 and F2 are separated from transfer factors T and G (G not shown).

(iii) *Preparation of transfer factors*

Transfer factors T and G (Nishizuka & Lipmann, 1966*b*) were purified by a modification of the method of Lucas-Lenard & Lipmann (1966). The 106,000 g supernatant was brought to 40% saturation with solid (NH$_4$)$_2$SO$_4$ (22·6 g/100 ml.) while the pH was maintained at 7·4 with 0·1 N-KOH. After 20 min the solution was centrifuged at 40,000 g for 20 min and the precipitate discarded. Additional (NH$_4$)$_2$SO$_4$ was added to the supernatant to bring it to 65% saturation (17·2 g/100 ml.). After 20 min, the precipitate was collected by centrifugation at 40,000 g for 20 min and dissolved in 6 ml. of buffer containing 0·01 M-Tris–HCl, pH 8·0, 0·15 M-KCl and 10^{-4} M-dithiothreitol. Following a 12-hr dialysis against 4 l. of the same solution, the dialysate was applied to a 1 cm × 100 cm DEAE Sephadex A50 column previously equilibrated with this buffer and eluted with a linear salt–pH gradient: 0·15 to 0·65 M-in KCl; pH 8·0 to 7·0 in 0·01 M-Tris–HCl containing 10^{-4}M-dithiothreitol; 185 ml. per chamber. Two ml. fractions were collected every 11 min. Under these conditions, transfer factor T is well separated from transfer factor G, but not resolved into its sub-components, T_u and T_s.

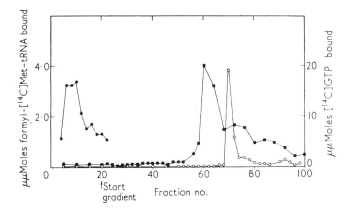

FIG. 1. Separation of initiation and transfer factors by DEAE cellulose column chromatography. The chromatographic procedure and factor assays are described in Materials and Methods. Initiation factor F1 (—●—●—) was detected by its effect on the ribosomal binding of fMet-tRNA at 5 mM-Mg^{2+} in the presence of F2. F2 (—■—■—) was detected by its effect in the presence of F1. Transfer factor T (—○—○—) was detected by [^{14}C]GTP binding.

(c) *Preparation of oligoribonucleotides*

The triplet AUG was prepared by the method of Leder, Singer & Brimacombe (1965) as modified (Leder, 1968). AUG was used as a primer for the further addition of a limited number of 5′-uridylic acid residues by polynucleotide phosphorylase. The reaction mixture contained 0·05 M-Tris–HCl, pH 9·0, 0·05 mM-EDTA, 10 mM-$MgCl_2$, 2·1 mM-ApUpG; and 20 mM-UDP, in a total volume of 15 ml. The mixture was heated to 95°C for 45 sec, then cooled to 37°C and 22·5 mg partially purified polynucleotide phosphorylase (fraction VI of Singer & Guss, 1962) were added. Following incubation for 4·25 hr at 37°C, the reaction was heated to 90°C for 3 min, cooled and diluted to 400 ml. with 0·01 M-triethylammonium bicarbonate. The diluted reaction mixture was applied to a 2·5 cm × 30 cm DEAE cellulose column previously equilibrated with 0·05 M-triethylammonium bicarbonate and eluted with a linear 0·05 to 1·0 M gradient of this solution, 3 l. per chamber. Ten-ml. fractions were collected every 6 min. pU_3, pU_5 and pU_6 were isolated from poly U which had been partially digested with an endonuclease purified from pig liver nuclei (Heppel, 1966). Selected fractions were treated with bacterial alkaline phosphatase to remove the 5′-monophosphate and were isolated by DEAE cellulose column chromatography in triethylammonium bicarbonate and characterized as described previously (Leder, 1968).

(d) *Characterization of oligonucleotides*

A 20-mμmole sample of each purified oligonucleotide fraction was digested with 10 units of T_1 RNase and 0·3 unit of bacterial alkaline phosphatase and co-chromatographed with an undigested oligonucleotide sample and authentic standards on Whatman 3 MM paper in descending n-propanol–NH_4OH–water (55 : 10 : 35) solvent. The results of this analysis, indicating the purity of each oligonucleotide, are diagramatically represented in Fig. 2.

(e) *Preparation of aminoacyl-tRNA*

Preparation of singly-labeled [^{14}C]Phe-tRNA (sp. act. 393 μc/μmole) in the presence of the other 19 unlabeled amino acids, including fMet, and of doubly-labeled formyl-[^{3}H]Met-tRNA (sp. act. 3·1 mc/μmole) and [^{14}C]Phe-tRNA (sp. act. 393 μc/μmole) from unfractioned *E. coli* B tRNA has been described (Leder & Bursztyn, 1966a). Preparations in which only formyl-[^{3}H]Met- or [^{14}C]Phe-tRNA's were active for binding were prepared using specific aminoacyl-tRNA synthetases partially purified by the DEAE cellulose column chromatographic method of Muench & Berg (1966). Aminoacylation reaction mixtures differed from those previously described in containing only Met and/or Phe, and their corresponding aminoacyl-tRNA synthetases which had been separated from one

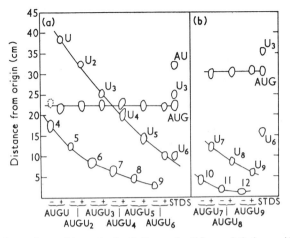

FIG. 2. Paper chromatographic analysis of oligonucleotides and their specific degradation products.

The chromatographic and degradative procedures are given in Materials and Methods. The untreated samples are indicated by the minus signs (−); the T_1 RNase, alkaline phosphatase-treated samples, by plus signs (+); (a) represents elution for 16 hr; (b) for 40 hr. Stds, standards.

another. After acylation and phenol extraction, the tRNA's were isolated by Sephadex G25 column chromatography and incubated in 5×10^{-4} M-NaIO$_4$ at 30°C for 30 min to destroy any unacylated tRNA (Capecchi, 1967a). The mixture was made 5×10^{-3}M in glucose to consume the unused IO$_4^-$ and passed over a Sephadex G25 column, reacylated with the specific aminoacyl-tRNA synthetases and purified again as above. Samples of the formyl-[³H]Met-tRNA preparation did not accept [¹⁴C]Phe after NaIO$_4$ treatment, nor would samples of the [¹⁴C]Phe-tRNA preparation accept [³H]Met.

(f) Binding assays

(i) Binding of tRNA to ribosomes

The assay is that of Nirenberg & Leder (1964). Each complete 50-μl. reaction mixture contained 0·04 mM-oligoribonucleotide; 1 mM-GTP; 1·5 μg initiation factor F1; 3·7 μg initiation factor F2; 12 μg transfer factor T; 11 μg transfer factor G; 0·3 A_{260} unit of ribosomes; 20 and 9 μμmoles, respectively, of formyl-[³H]Met- and [¹⁴C]Phe-tRNA's, or 7 μμmoles [¹⁴C]Phe-tRNA; 0·05 M-Tris acetate, pH 7·2; 0·05 M-NH$_4$Cl, and 5 mM-magnesium acetate. Where indicated, the [Mg²⁺] was increased to 10 mM, or 1 mM-GMP · PCP substituted for GTP. Incubation was at 23°C for 30 min. Ribosome-bound radioactivity retained by the nitrocellulose filter was measured in a liquid-scintillation counter in a toluene–PPO–POPOP solution. All determinations were in duplicate.

(ii) Detection of initiation factors

Initiation factors were identified by their ability to complement in stimulating the AUG-directed ribosomal binding of fMet-tRNA at 5 mM-Mg²⁺.

(iii) Detection of transfer factors

Transfer factor T was identified by retention of the [¹⁴C]GTP–T factor complex on a nitrocellulose filter (Allende & Weissbach, 1967; Ertel et al., 1968) as well as in the complementation of polyphenylalanine synthesis as described below.

(g) Synthesis assays

(i) Blocked oligopeptides

The assay is that of Erbe & Leder (1968). Reaction components and conditions were the same as the aminoacyl-tRNA binding assay except that the tRNA had been aminoacylated with all 20 amino acids, including fMet, of which only [¹⁴C]Phe-tRNA (7 μμmoles) was radioactive. The reaction, carried out in conical 12-ml. centrifuge tubes, was stopped by the addition of 1·2 ml. of 0·1 N-KOH and incubation resumed for 30 min at 37°C to hydrolyze aminoacyl- and peptidyl-tRNA's completely. Following acidification with 0·1 ml. of 12 N-HCl, 1·5 ml. of ethyl acetate were added and the mixture agitated vigorously on a Vortex Jr. mixer for 10 sec to extract the fMet-blocked ^{14}C-labeled Phe peptides. Tubes were centrifuged at 2000 rev./min for 1 min and 1 ml. of the upper ethyl acetate phase transferred to a vial containing 10 ml. of Bray's (Bray, 1960) or Triton X–toluene– PPO–POPOP (Patterson & Greene, 1965) solution for liquid-scintillation counting.

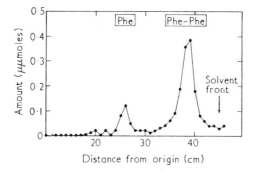

FIG. 3. Chromatographic analysis of the cyanogen bromide degradation products of an AUGU$_6$-directed synthetic reaction. Components of twofold reaction mixtures, conditions and analytic procedure are described in Materials and Methods.

As shown in Table 1, the unblocked amino acids Met and Phe are essentially insoluble in ethyl acetate. In contrast, 34% of the blocked fMet originally added to the aqueous phase is recovered in the 1·0-ml. ethyl acetate portion. The expected translation products fMet-Phe and fMet-Phe-Phe are very efficiently extracted, 63 and 70%, respectively, whereas only 3% of the unblocked dipeptide Met-Phe is recovered in the ethyl acetate portion. It should be noted that polyphenylalanine is quite soluble in ethyl acetate over a wide range of pH values. The extracted products of synthetic reactions were characterized directly by electrophoresis (Figs 8 and 9) and further by cyanogen bromide degradation. The cyanogen bromide reaction of the extracted products of an AUGU$_6$-directed synthetic reaction yielded the degradation products expected of the parent compounds, fMet-Phe-Phe and fMet-Phe, namely the dipeptide, Phe-Phe and the amino acid, Phe (Fig. 3). Thus, using these small mRNA's and under appropriately stringent conditions of synthesis, labeled Phe enters the ethyl acetate phase only when incorporated into blocked peptide.

TABLE 1

Recovery of free and blocked amino acids and peptides by acidified ethyl acetate extraction

Compound	Input recovered (%)
Met	1
Phe	1
fMet	34
Met-Phe	3
fMet-Phe	63
fMet-Phe-Phe	70

10 μμmoles of each compound were dissolved in alkaline solution, acidified and extracted with 1·5 ml. ethyl acetate as noted in Materials and Methods. The percentage of the added compound recovered in a 1·0-ml. ethyl acetate sample is indicated.

(ii) Blocked puromycin peptides

The assay is that of Leder & Bursztyn (1966b). Reaction components and conditions were as noted above for aminoacyl-tRNA binding. Following incubation, the reaction mixture was made 0·9 mM in puromycin and incubation resumed at 23°C for 30 min. The reaction was stopped by adding 1·2 ml. 0·1 M-sodium acetate, pH 5·5. fMet-[^{14}C]Phe-puromycin was then extracted with 1·5 ml. ethyl acetate of which 1·0 ml. was transferred and counted as above.

(iii) Poly U-directed polyphenylalanine synthesis

Components and conditions are the same as for the aminoacyl-tRNA binding assay, except that poly U served as mRNA, initiation factors were omitted and [Mg^{2+}] was 10 mM. The reaction was stopped by addition of 1 ml. 10% trichloroacetic acid, heated to 95°C for 20 min, cooled to 0°C for 5 min, poured over a nitrocellulose filter and washed with 10 ml. cold 5% trichloroacetic acid (Nirenberg, 1963). The dried filters were counted in 10 ml. toluene–PPO–POPOP solution.

(h) Detection of transfer factors T and G by complementation for polypeptide synthesis

Using reaction components and conditions as for poly U-directed polyphenylalanine synthesis, transfer factor T was identified by the ability of 5-μl. column fraction samples to stimulate polyphenylalanine synthesis in the presence of 11 μg of factor G. Transfer factor G was similarly identified using 12 μg of factor T. G factor had been previously identified by the associated ribosome-dependent GTPase activity (see below) and T factor by the associated [^{14}C]GTP binding activity.

(i) Ribosome-dependent GTPase activity

The ribosome-dependent GTPase activity associated with transfer factor G (Nishizuka & Lipmann, 1966b) was detected by incubating a 2-μl. portion of each fraction for 10 min at 23°C in a 50-μl. reaction mixture containing 2 mM-[γ-^{32}P]GTP (10^4 cts/min) in 0·01 M-Tris acetate, pH 7·2, 0·05 M-NH$_4$Cl, 10 mM-magnesium acetate in the presence and absence of 0·7 A_{260} unit of ribosomes. The reaction was stopped by adding 1 ml. of 5% trichloroacetic acid containing 1% (w/v) activated charcoal to absorb the unreacted [γ-^{32}P]GTP. The suspension was mixed and poured over a nitrocellulose filter, washed with an additional 1 ml. of water and the 2-ml. filtrate collected in scintillation vials to which were added 10 ml. Bray's or Triton X–toluene–PPO–POPOP solution.

(j) Stability of the binding complex

Components were as described above for the aminoacyl-tRNA binding assay. Following incubation at 23°C for 30 min, the reaction mixture was diluted to 1 ml. with ice-cold 0·01 M-Tris acetate, pH 7·2, 0·05 M-NH$_4$Cl and 5 mM-magnesium acetate, poured over a nitrocellulose filter and washed with an additional 1 ml. of the above solution. The ribosome-free filtrate of each reaction mixture was made alkaline with 0·02 ml. 10 N-KOH and incubated for 30 min at 37°C to hydrolyze aminoacyl- and peptidyl-tRNA completely. The mixture was then acidified and extracted with ethyl acetate as described in the assay for synthesis of blocked oligopeptides.

(k) Characterization of fMet initiated oligopeptide products

Twofold oligopeptide synthesis reaction mixtures were hydrolyzed for 30 min at 37°C in 0·1 N-KOH, acidified to pH 1 and extracted in ethyl acetate as above. The extract was evaporated to dryness, taken up in 0·1 ml. ethyl acetate and applied to Whatman no. 1 paper for electrophoresis in 0·05 M-pyridine acetate, pH 3·5, at 53 v/cm for 75 min in the presence of authentic standards. 1 cm × 3 cm strips were counted in toluene–PPO–POPOP scintillation solution. In a second method of analysis, the extracted products were degraded with cyanogen bromide (Gross & Witkop, 1962) and characterized by descending co-chromatography with authentic fMet, Phe and Phe-Phe on Whatman no. 3 paper in n-butanol–acetic acid–water (4 : 1 : 1) for 15 hr (Pestka, 1968).

3. Results

(a) *Translation of the di-codon $AUGU_3^-$ at 5 mM-Mg^{2+}*

The factors required for translation of each codon of the di-codon $AUGU_3$ are shown in Table 2. Since the binding assay does not distinguish bound fMet- or Phe-tRNA's from that bound as the dipeptidyl-tRNA, fMet-Phe-tRNA, results of the aminoacyl-

TABLE 2

$AUGU_3$-directed binding of fMet-, Phe- and fMet-Phe-tRNA's to ribosomes, and synthesis of fMet-Phe at 5 mM-Mg^{2+}

Reaction system	Formyl[^{3}H]Met [^{14}C]Phe in ribosome-bound product ($\mu\mu$moles)		fMet[^{14}C]Phe synthesized
Complete	3·20	2·08	2·48
minus initiation factors	0·31	0·35	0·22
minus T factor	2·94	0·18	0·11
minus G factor	2·20	1·60	2·75
minus GTP	1·03	0·69	0·47
minus GTP, plus GMP · PCP	1·03	0·19	0·26
minus mRNA	1·37	0·24	0·24
minus ribosomes	0·11	0·12	0·07

Reaction mixture components, conditions of incubation and details of the binding and synthesis assays are given in Materials and Methods. The omitted component or addition is indicated in the Table. The Mg^{2+} concentration was 5 mM.

tRNA binding assay are given as fMet and Phe contents of the product bound to ribosomes (cf. Fig. 5). fMet binding reflects recognition of the first codon, AUG, which requires initiation factors as indicated by the marked reduction in fMet binding when these are omitted. fMet binding is not, however, significantly affected by omission of transfer factor T but is somewhat reduced in the absence of transfer factor G. The effect of G is commented upon below. GTP is required and is not replaced by an equimolar concentration of GMP · PCP, the β, γ-methylene analog of GTP. The low level of fMet binding which occurs in the absence of added mRNA reflects a non-specific but functional interaction with ribosomes described previously (Leder & Nau, 1967). Phe binding, reflecting recognition of the second or internal codon, UUU, requires initiation factors and GTP, but also requires the transfer factor T. As in the case of fMet, binding of Phe is moderately reduced when G is omitted. GMP · PCP does not replace GTP for binding in response to the internal codon and both mRNA and ribosomes are required.

fMet-Phe synthesis represents a complex series of reactions involving recognition of initial and internal codons and formation of the first peptide bond. It, therefore, requires binding of both fMet- and Phe-tRNA's and, as expected, requires all the elements necessary for recognition of each component codon. Thus, initiation and T factors, GTP, mRNA and ribosomes are required and, as noted by Thach, Dewey & Mykolajewycz (1967), GMP · PCP does not replace GTP for dipeptide synthesis. In contrast to its effect on binding, omission of G does not significantly affect dipeptide synthesis. This differential effect of G factor on binding and dipeptide synthesis is the subject of experiments reported later in this and the following sections. It is important to note, however, that the dipeptide, fMet-Phe, is not released from tRNA during the synthetic reaction, for the blocked dipeptide cannot be extracted by ethyl acetate prior to alkaline hydrolysis.

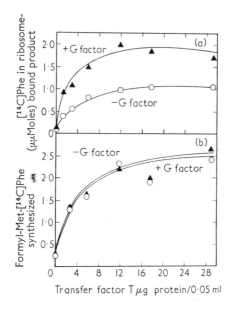

Fig. 4. The effect of transfer factor T concentration on the extent of Phe binding and fMet-Phe synthesis in the presence and absence of transfer factor G.
(a) Phe binding; (b), fMet-Phe synthesis. —▲—▲—, binding or synthesis in presence of transfer factor G; —○—○—, in its absence.

The dependence of AUGU$_3$-directed binding and synthesis upon the concentration of factor T and the effect of factor G are shown in Figure 4. At lower concentrations of T, the extent of both binding and synthesis reactions is dependent upon the concentration of this factor. Both reactions are essentially complete in 30 minutes when 12 μg of T are present. In this experiment as in that shown in Table 2, omission of factor G results in a reduction in the amount of Phe bound to the ribosome, but is without effect upon the synthesis of the dipeptide, fMet-Phe (see below).

(b) *Analysis of the product bound to ribosomes at 5 mM-Mg^{2+}*

Since the binding assay does not distinguish bound aminoacyl-tRNA from bound peptidyl-tRNA, it is of interest to determine the nature of the product bound to ribosomes in a complete reaction mixture at 5 mM-Mg^{2+}. The material retained on a nitrocellulose filter was subjected to alkaline hydrolysis and analyzed by paper electrophoresis at pH 3·5 (Fig. 5). Over 95% of the Phe bound to ribosomes in the presence of AUGU$_3$ (Fig. 5(a)) is in the form of the dipeptide, fMet-Phe, whereas only a small amount is bound as unincorporated Phe. With AUGU$_6$ (Fig. 5(b)), over 95% of the bound Phe is again incorporated into peptide, in this case largely into the tri-peptide, fMet-Phe-Phe.

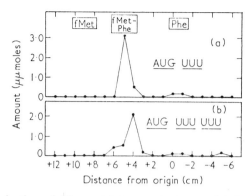

FIG. 5. Characterization of Phe-containing products associated with the ribosome.
Binding reactions, directed by $AUGU_3$ (a) and $AUGU_6$ (b), were carried out as described in
Materials and Methods. The ribosome-bound product, isolated on a nitrocellulose filter, was
hydrolyzed in 2 ml. of 0·1 N-NH_4OH for 60 min at 37°C and the filter removed. The products were
electrophoresed at pH 3·5 as described in Materials and Methods. The bars indicate the positions
of authentic standard compounds.

(c) Role of transfer factor T

Since almost all of the Phe bound to ribosomes at 5 mM-Mg^{2+} is incorporated into
peptide, it is not possible to distinguish an effect of transfer factor T directly upon the
formation of the peptide bond from that only upon the binding of Phe-tRNA to the
ribosome. At 10 mM-Mg^{2+} when initiation factors are omitted and little fMet-tRNA
is bound to ribosomes (Table 3, line 2), 3·29 $\mu\mu$moles of Phe-tRNA are bound. Of this,

TABLE 3

*$AUGU_3$-directed binding of fMet-, Phe- and fMet-Phe-tRNA's to ribosomes,
and synthesis of fMet-Phe at 10 mM-Mg^{2+}*

Reaction system	Formyl[³H]Met	[¹⁴C]Phe	fMet[¹⁴C]Phe synthesized
	in ribosome-bound product		
	($\mu\mu$moles)		
Complete	3·71	4·39	2·63
minus initiation factors	0·61	3·29	0·54
minus T factor	4·81	0·34	0·11
minus GTP	0·65	2·57	0·46
minus GTP, plus GMP · PCP	0·50	0·29	0·31

Reaction mixture components, conditions of incubation and details of binding and synthesis
assays are given in Materials and Methods. The omitted component or addition is indicated in the
Table. The Mg^{2+} concentration was 10 mM.

very little is incorporated into dipeptide, despite the presence of both transfer factors
T and G. The Phe-tRNA binding is clearly dependent upon the addition of T factor,
less dependent upon added GTP at this [Mg^{2+}], but is markedly inhibited by the
addition of GMP · PCP. The ability of T and GTP to stimulate binding of aminoacyl-
tRNA in the absence of peptide bond formation implicates this transfer factor T and
GTP in catalysis of the binding reaction affecting recognition of internal, as contrasted
to initial, codons. These observations confirm the previous studies of Ravel (1967);
Lucas-Lenard & Haenni (1968); Erbe & Leder (1968); and Ertel et al. (1968). Although
involvement of T factor in peptide bond formation is not ruled out by these experiments, this activity appears to be a function of the 50 s ribosomal subunit and, under

appropriate conditions, has no further requirements for soluble factors (Traut & Monro, 1964; Hultin, 1966; Rychlík, 1966; Goldberg & Mitsugi, 1967; Gottesman, 1967; Monro & Marcker, 1967; Pestka, 1968; Pulkrábek & Rychlík, 1968).

(d) *Effect of Mg²⁺ concentration on translation of the di-codon AUGU₃*

The effect of [Mg²⁺] on AUGU₃-directed binding of fMet and Phe and the synthesis of fMet-Phe are shown in Figure 6. The optimum [Mg²⁺] for the ribosomal binding of

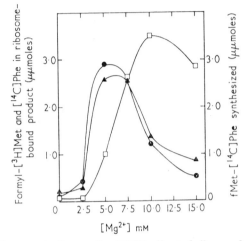

FIG. 6. Effect of Mg²⁺ concentration on the AUGU₃-directed ribosomal binding of fMet and Phe and the synthesis of fMet-Phe. Otherwise complete reactions, *lacking transfer factor G*, were carried out as described in Materials and Methods. The ribosomal binding of fMet is indicated by —●—●—; the binding of Phe, by —□—□—; and the synthesis of fMet-Phe, by—▲—▲—.

fMet is in the range of 5·0 to 7·5 mM, whereas the optimum for Phe binding is higher, at approximately 10 mM. The [Mg²⁺] dependence for synthesis of the dipeptide, fMet-Phe, parallels precisely that observed for fMet binding. Although Phe binding is more extensive at higher [Mg²⁺], in-frame translation of AUGU₃ requires a low Mg²⁺ concentration favoring recognition of its initial codon. Thus, cell-free protein synthesis carried out at relatively high levels of Mg²⁺ would be expected to result in random initiation events and frequent incomplete and out-of-frame translation despite enhanced polypeptide synthesis. Studies using synthetic oligoribonucleotides (Leder & Bursztyn, 1966c; Sundaraarjan & Thach, 1966) and R17 viral mRNA (Capecchi, 1967b) confirm these expectations.

(e) *Requirement of fMet-tRNA for translation of the second codon*

The requirement for initiation factors for recognition of both the initial and second codons of AUGU₃ at low [Mg²⁺] was explored further to determine whether the effect represented a requirement for initiation factors themselves or was due to their effect upon the binding of fMet-tRNA (Table 4). When both fMet- and Phe-tRNA species were available for binding, the Phe was readily bound to ribosomes to the extent of 1·18 μμmoles. When only Phe-tRNA was available, no significant binding of Phe occurred despite the presence of initiation factors. Recognition of the second codon thus requires the tRNA corresponding to the first codon, as well as both initiation and T factors. Under these conditions, the complete initiation system might be required

TABLE 4

Effect of fMet-tRNA on the AUGU₃-directed binding of Phe-tRNA to ribosomes

Aminoacyl-tRNA available	[¹⁴C]Phe in ribosome-bound product (μμmoles)
fMet- and [¹⁴C]Phe-tRNA	1·18
[¹⁴C] Phe-tRNA	0·17

Reaction mixture components, including initiation and transfer factors, conditions of incubation and assay are given in Materials and Methods. The Mg^{2+} concentration was 5 mM.

for stable mRNA–ribosome interaction in appropriate frame. On the other hand, the effect of the initiation system might be related to the incorporation of Phe into peptidyl-tRNA and reflect a more stable interaction between peptidyl-tRNA and ribosome than that between unincorporated Phe-tRNA and ribosome.

(f) *Continuity of mRNA*

In terms of a model for protein synthesis which specifies two ribosomal binding sites, binding at one site might in some way alter the other so that binding occurs more readily or with greater stability. Such a co-operative effect upon the binding of the in-coming aminoacyl-tRNA might be unrelated to the covalent continuity of the two codons being recognized. To test this, separate oligonucleotides in which the AUG and UUU sequences were *not* covalently joined were added to otherwise complete reaction mixtures (Table 5). fMet was bound efficiently whenever the AUG codon was present, whether as triplet or covalently linked to a sequence of U's. In contrast, Phe was bound only when the UUU codon was covalently joined to AUG as in the di-codon, AUGU₃. Despite the ability of these oligouridylic acid mRNA's to induce Phe-tRNA

TABLE 5

Effect of covalent continuity of AUG and UUU codons upon the binding of fMet-, Phe- and fMet-Phe-tRNA's to ribosomes at 5 mм-Mg^{2+}

mRNA added	Formyl-[³H]Met in ribosome-bound product (μμmoles)	[¹⁴C]Phe
None	2·21	0·14
AUG	4·97	0·18
AUGU₃	6·23	3·72
AUG + U₃	5·60	0·21
AUG + U₅	5·19	0·17
AUG + pU₆	5·38	0·17

Reaction mixture components, including initiation and transfer factors, conditions of incubation and assay are given in Materials and Methods. Oligonucleotide mRNA's where present were in equimolar amounts. The Mg^{2+} concentration was 5 mM.

binding to ribosomes at 20 mM-Mg^{2+} (Nirenberg & Leder, 1964), at 5 mM-Mg^{2+} none was active in inducing Phe binding unless covalently joined with the fMet codon, AUG. This requirement for continuity with the initiating codon for recognition of subsequent codons may be related to the role of the initiation system in facilitating or stabilizing the interaction between mRNA and the ribosome (Brown & Doty, 1968; Iwasaki, Sabol, Wahba & Ochoa, 1968; Maitra & Dubnoff, 1968). The requirement for fMet-tRNA (Table 4), however, suggests a more complicated effect.

(g) Effect of G factor: translocation

(i) Transfer of fMet-Phe to puromycin

Although G factor is not required for formation of the first peptide bond, it has an apparent effect on transfer of the dipeptide to puromycin (Table 6) involving formation of a second peptide bond. Thus, the synthesis of fMet-Phe-puromycin is reduced from 3·21 to 0·79 $\mu\mu$moles when G is omitted. Synthesis of peptidyl-puromycin is also greatly reduced when GTP is omitted or replaced by GMP · PCP, an effect which reflects the decreased aminoacyl-tRNA binding and dipeptide synthesis under these conditions.

TABLE 6

Effect of G factor upon fMet-Phe-puromycin synthesis

Reaction system	fMet[^{14}C]Phe-puromycin ($\mu\mu$moles)
Complete	3·21
minus G factor	0·79
minus T factor	0·25
minus initiation factors	0·09
minus GTP	0·38
minus GTP, plus GMP · PCP	0·04

Reaction mixture components, conditions of incubation and assay are given in Materials and Methods. The component omitted or added is indicated in the Table. The Mg^{2+} concentration was 5 mM.

(ii) Stability of the binding complex

The ability of a blocked amino acid or peptide to transfer to puromycin has been assumed to define its position in the peptidyl-tRNA binding site on the ribosome (Traut & Monro, 1964; Bretscher & Marcker, 1966; Heintz, McAllister, Arlinghaus & Schweet, 1966; Matthaei & Voigt, 1967; Brot et al., 1968; Ohta & Thach, 1968; Skogerson & Moldave, 1968; Tanaka et al., 1968). In this connection we examined the effect of G factor on the stability of the mRNA–tRNA–ribosome binding complex (Table 7) as an alternative mechanism which might account for the effect of G on the puromycin reaction. For if dipeptidyl-tRNA is released from the ribosome in the absence of G, it is likely that it will not be transferred to puromycin. Although transfer factor G is not required for synthesis of the initial dipeptide, in the absence of G the amount of fMet and Phe bound to the ribosome is reduced (Table 2 and Fig. 4). Loss of both fMet and Phe from the ribosome complex can be accounted for by an increase in the amount of fMet-Phe-tRNA recovered in the ribosome-free binding reaction filtrate, having been synthesized on the ribosome but released from it (Table 7). The released product remains associated with tRNA for it is not extracted in ethyl acetate prior to alkaline hydrolysis. G factor apparently stabilizes the mRNA–peptidyl-tRNA–ribosome complex. This is compatible, among other possibilities, with translocation of the peptidyl-tRNA from a site of lesser to a site of greater affinity on the ribosome, conceivably from the aminoacyl-tRNA or recognition site to the peptidyl-tRNA site.

TABLE 7

Effect of G factor upon the stability of the complex formed between fMet-Phe-tRNA and ribosome

Reaction system	$[^{14}C]$Phe in ribosome-bound product	fMet-$[^{14}C]$Phe-tRNA released
Complete	2·21	1·06
minus G factor	1·24	1·88

Reaction mixture components and conditions of incubation are given in Materials and Methods. fMet-$[^{14}C]$Phe-tRNA released from the ribosome represents that present in the ribosome-free filtrate and is detected as described in Materials and Methods. The Mg^{2+} concentration was 5 mM.

(iii) *Translocation*

The role of G factor in the translocation process was examined more directly by comparing its effect on translation of the di- and tri-codons, $AUGU_3$ and $AUGU_6$. The peptide products corresponding to these mRNA's were extracted from acidified reaction mixtures with ethyl acetate and analyzed by paper electrophoresis at pH 3·5 (Fig. 7). Where indicated, factor G alone was omitted from otherwise complete reaction mixtures in which the concentration of ribosomes limited the extent of the synthesis. Both in the presence (Fig. 7(b)) and absence (Fig. 7(a)) of transfer factor G, the product of the $AUGU_3$-directed reaction is the dipeptide, fMet-Phe. However, translation of the tri-codon, $AUGU_6$, progresses only as far as the second codon in the absence of G and the major product of the reaction is also the dipeptide, fMet-Phe (Fig. 7(c)). The third codon of the tri-codon message $AUGU_6$ is available for translation only upon addition of G factor and this product is the tripeptide, fMet-Phe-Phe (Fig. 7(d)). This result is consistent with the participation of G factor in the translocation of the small mRNA, exposing the next succeeding codon for translation. The cycle of reactions can be repeated in a propagative fashion in the presence of transfer factors T and G and GTP. Whereas in the absence of factor G the product directed by the tetra-codon, $AUGU_9$, remains the dipeptide, fMet-Phe (Fig. 8(a)), that directed by $AUGU_9$ in the complete system is primarily the tetra-peptide, fMet-Phe-Phe-Phe (Fig. 8(b)). Thus, excepting termination, translation of successive codons beyond the third requires no additional factors.

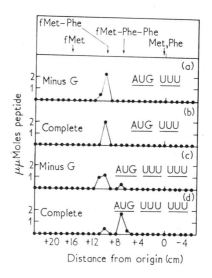

FIG. 7. Electrophoretic analysis of AUGU$_3$- and AUGU$_6$-directed peptide synthesis.
The components of double reaction mixtures, conditions and analytic procedure are given in Materials and Methods.
(a) AUGU$_3$-directed synthesis in the absence of transfer factor G; (b) in the presence of the complete system; (c) AUGU$_6$-directed synthesis in the absence of transfer factor G; (d) in the presence of the complete system.

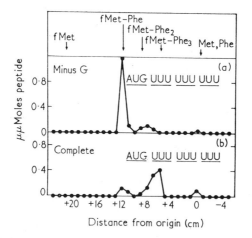

FIG. 8. Electrophoretic analysis of AUGU$_9$-directed peptide synthesis.
The components of double reaction mixtures, conditions and analytic procedure are given in Materials and Methods.
(a) Synthesis in the absence of transfer factor G; (b) in the presence of the complete system.

4. Discussion

Complete translation of the small synthetic mRNA's used in these studies has certain of the stringent requirements necessary for the accurate cell-free translation of naturally occurring messengers (Stanley et al., 1966; Capecchi, 1967b; Anderson Dahlberg, Bretscher, Revel & Clark, 1967). The synthetic reaction must be carried out at a suitably low level of Mg^{2+} to ensure initiation in appropriate frame including the codon corresponding to the N-terminal amino acid of the encoded protein. This is also necessary to avoid ambiguous interactions between aminoacyl-tRNA, mRNA and ribosomes which occur at higher concentrations of this cation (Davies, Gilbert & Gorini,1964; Szer & Ochoa, 1964). In addition, the fMet codon AUG must be incorporated covalently into the short messenger sequence and the initiation system, including fMet-tRNA, initiation factors and GTP, must be provided. The small mRNA's, directing the synthesis of unique and conveniently detected oligopeptide products, permit us to correlate the soluble elements required for protein synthesis with specific steps in the translation of initial and succeeding internal codons.

(a) Formation of the initial complex

A number of studies indicate that the binding complex is formed from ribosomal subunits, first by interaction of fMet-tRNA and mRNA with the smaller 30 s subunit, which is joined subsequently by the 50 s component (Schlessinger et al., 1967; Nomura & Lowry, 1967; Hille, Miller, Iwasaki & Wahba, 1967; Mukandan, Hershey, Dewey & Thach, 1968). Formation of the larger complex is thought to make available an additional site on the ribosome for the binding of unblocked, internal aminoacyl-tRNA's. There is evidence, however, that under certain conditions recognition of unblocked aminoacyl-tRNA's can take place on the isolated 30 s subunit as well (Matthaei, Amelunxen, Eckert & Heller, 1964; Kaji, Suzuka & Kaji, 1966; Pestka & Nirenberg, 1966). Experiments by Sarkar & Thach (1968) suggest that fMet-tRNA is bound first to the recognition or aminoacyl-tRNA site on the ribosome and is translocated subsequently to an initiator or peptidyl-tRNA site. An alternative is direct access to the peptidyl site. Our own studies do not distinguish between these possibilities. They do indicate, however, that translocation of fMet-tRNA, if required, occurs without addition of the enzyme required for translocation of internal codons (Table 2). This raises the possiblity that one or a combination of initiation factors is responsible for an initial translocation reaction. These studies also emphasize the stringent framing mechanism imposed by the presence of the AUG codon (Leder & Bursztyn, 1966c; Thach, Sundaraarjan, Dewey, Brown & Doty, 1966; Ghosh, Söll & Khorana, 1967; Salas et al., 1967) and the necessity of its being present in order that translation occur. Thus, at 5 mM-Mg^{2+} internal codons not covalently joined to the fMet codon, AUG, do not direct ribosomal binding of their corresponding aminoacyl-tRNA's (Table 5). It is possible that interactions with internal codons are comparatively unstable and that prior formation of the more stable initiation complex in appropriate frame is required for translation of succeeding codons. The effect, however, is not restricted to the interaction of mRNA and ribosome, for it requires all the elements involved in the formation of the initiation complex (Table 4). Although additional sequences may be required for inter- or intra-cistronic initiation, the AUG codon apparently suffices for formation of the initial complex and complete in-frame translation of these small RNA messengers.

(b) *Elongation of the peptide*

In addition to the unique initiation event, subsequent propagative steps can be related conveniently to a general model for protein synthesis. This is represented diagrammatically in Figure 9 (adapted from Watson, 1964). In this case, the model specifically represents the experimental system used in these studies wherein the tricodon, $AUGU_6$, directs the synthesis of the tripeptide, fMet-Phe-Phe. Details of the ribosomal subunit structure are omitted, for these studies do not allow us to make meaningful distinctions between events occurring on alternative subunits. The experiments which we have described allow us to correlate elements required for protein synthesis with specific reactions indicated by the model. The first postulated intermediate is the initial binding complex (Fig. 9–1), the formation of which we have discussed above. In this structure, the second codon, UUU, is positioned at the recognition site on the ribosome and, therefore, available to direct ribosomal binding of Phe-tRNA. In accord with previous observations (Ravel, 1967; Brot *et al.*, 1968; Erbe & Leder, 1968; Lucas-Lenard & Haenni, 1968), we have found that the binding of aminoacyl-tRNA requires the transfer factor T, apparently a binding enzyme, and GTP (Tables 2 and 3; Fig. 4). This reaction is inferred to result in the formation of a

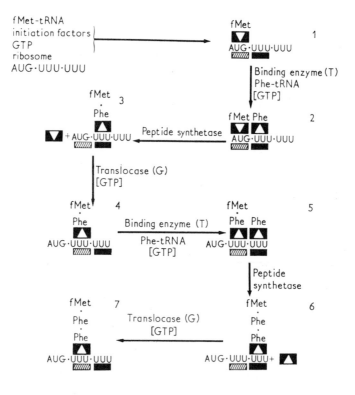

Fig. 9. Diagrammatic representation of postulated reactions and intermediates in the $AUGU_6$-directed synthesis of the tripeptide, fMet-Phe-Phe.
Symbols represent the following: ▼, tRNA corresponding to fMet; ▲, tRNA corresponding to Phe; ▨, peptidyl or holding site on the ribosome; ■, aminoacyl- or recognition site on the ribosome.

short-lived intermediate (Fig. 9–2) in which fMet-tRNA resides at the peptidyl-tRNA binding site and Phe-tRNA at the recognition site on the ribosome. Our studies indicate that very little Phe-tRNA remains free and unincorporated into dipeptide if fMet-tRNA is available for peptide bond formation (Fig. 5). The peptide bond is readily formed (Fig. 9–3), catalyzed by an activity associated with the ribosome (Traut & Monro, 1964; Bretscher & Marcker, 1966; Hultin, 1966; Rychlík, 1966; Zamir et al., 1966; Goldberg & Mitsugi, 1967; Gottesman, 1967; Pestka, 1968), resulting in the formation of fMet-Phe-tRNA. Because the peptide synthetase reaction is very rapid, the binding intermediate (Fig. 9–2), in which unincorporated aminoacyl-tRNA's are associated with the ribosome, can be identified only in the presence of antibiotics which prevent peptide formation (Goldberg & Mitsugi, 1967; Lucas-Lenard & Haenni, 1968) or, as in certain of these studies (Table 3) and those of Ravel (1967), under conditions which do not require the prior binding of blocked aminoacyl-tRNA. As measured in these studies, Phe is bound to ribosomes at 5 mM-Mg^{2+} almost entirely as fMet-Phe-tRNA. According to the model, both fMet-Phe-tRNA and the corresponding second codon, UUU, remain at the recognition site in a pre-translocation intermediate (Fig. 9–3). The next codon, in this case the third, is not available for translation until translocated into the recognition site on the ribosome displacing the second, previously translated codon (Fig. 9–4). We have shown this to be accomplished only in the presence of the transfer factor G (Figs 7 and 8). Recent evidence suggests that this translocating element is an enzyme, which we have recently isolated in crystalline form (Leder, Skogerson & Nau (manuscript in preparation)), confirming by independent methods the results of Parmeggiani (1968). The translocation reaction may involve hydrolysis of GTP, though its role in this reaction will become clear only when it has been uncoupled from that in the previous binding step. The postulated reaction would result in translocation of the third codon to the recognition site, where it would be available for Phe-tRNA binding and in translocation of fMet-Phe-tRNA from the recognition to the peptidyl-tRNA site (Fig. 9–4). The increased stability of the fMet-Phe-tRNA–mRNA–ribosome complex induced by transfer factor G (Table 7) suggests an increased binding affinity at the peptidyl-tRNA site as compared to that at the recognition site on the ribosome. Alternative explanations for this observation, however, are not excluded. After translation of the third codon, there are no further requirements for elongation of the peptide chain (Fig. 8). Subsequent reactions would be expected to repeat the earlier sequence (Fig. 9–5,–6 and –7), resulting in the synthesis of fMet-Phe-Phe-tRNA and its translocation to the peptidyl site on the ribosome. The observations are consistent, therefore, with a two-ribosomal binding site model for protein synthesis, involving enzymically catalyzed *binding, peptide bond formation* and *translocation* reactions.

We are grateful to Mr David Rogerson and his staff for preparing the bacteria used in these studies and to Mrs Elizabeth Stotler for her expert assistance in the preparation of this manuscript.

REFERENCES

Allende, J. E. & Weissbach, H. (1967). *Biochem. Biophys. Res. Comm.* **28**, 82.
Anderson, J. S., Dahlberg, J. E., Bretscher, M. S., Revel, M. & Clark, B. F. C. (1967). *Nature*, **216**, 1072.
Arlinghaus, R., Favelukes, G. & Schweet, R. (1963). *Biochem. Biophys. Res. Comm.* **11**, 92.
Arlinghaus, R., Shaeffer, J. & Schweet, R. (1964). *Proc. Nat. Acad. Sci., Wash.* **5**, 1291.

Bont, W. S., Huizinga, F., Bloemendal, H., Van Weenen, M. F. & Bosch, L. (1965). *Arch. Biochem. Biophys.* **109**, 207.

Bray, G. A. (1960). *Analyt. Biochem.* **1**, 279.

Bretscher, M. S. (1968). *Nature,* **218**, 675.

Bretscher, M. S. & Marcker, K. A. (1966). *Nature,* **211**, 380.

Brot, N., Ertel, R. & Weissbach, H. (1968). *Biochem. Biophys. Res. Comm.* **31**, 563.

Brown, J. C. & Doty, P. (1968). *Biochem. Biophys. Res. Comm.* **30**, 284.

Cammack, K. A. & Wade, H. E. (1965). *Biochem. J.* **96**, 671.

Capecchi, M. R. (1967a). *Proc. Nat. Acad. Sci., Wash.* **58**, 1144.

Capecchi, M. R. (1967b). *J. Mol. Biol.* **30**, 213.

Davies, J., Gilbert, W. & Gorini, L. (1964). *Proc. Nat. Acad. Sci., Wash.* **51**, 883.

Eisenstadt, J. & Brawerman, G. (1966). *Biochemistry,* **5**, 2777.

Erbe, R. W. & Leder, P. (1968). *Biochem. Biophys. Res. Comm.* **31**, 798.

Ertel, R., Brot, N., Redfield, B., Allende, J. E. & Weissbach, H. (1968). *Proc. Nat. Acad. Sci., Wash.* **59**, 861.

Fessenden, J. M. & Moldave, K. (1963). *J. Biol. Chem.* **238**, 1479.

Ganoza, M. C. (1967). *Cold Spr. Harb. Symp. Quant. Biol.* **31**, 273.

Ghosh, H. P., Söll, D. & Khorana, H. G. (1967). *J. Mol. Biol.* **25**, 275.

Goldberg, I. H. & Mitsugi, K. (1967). *Biochemistry,* **6**, 383.

Gottesman, M. E. (1967). *J. Biol. Chem.* **242**, 5564.

Gross, E. & Witkop, B. (1962). *J. Biol. Chem.* **237**, 1856.

Guthrie, C. & Nomura, M. (1968). *Nature,* **219**, 232.

Hardesty, B., Arlinghaus, R., Shaeffer, J. & Schweet, R. (1963). *Cold Spr. Harb. Symp. Quant. Biol.* **28**, 215.

Heintz, R., McAllister, H., Arlinghaus, R. & Schweet, R. (1966). *Cold Spr. Harb. Symp. Quant. Biol.* **31**, 633.

Heppel, L. A. (1966). In *Procedures in Nucleic Acid Research,* ed. by G. L. Cantoni & D. R. Davies, p. 31. New York: Harper & Row.

Hille, M. B., Miller, M. J., Iwasaki, K. & Wahba, A. J. (1967). *Proc. Nat. Acad. Sci., Wash.* **58**, 1652.

Hultin, T. (1966). *Biochim. biophys. Acta,* **123**, 561.

Igarashi, K. & Kaji, A. (1967). *Proc. Nat. Acad. Sci., Wash.* **58**, 1971.

Iwasaki, K., Sabol, S., Wahba, A. J. & Ochoa, S. (1968). *Arch. Biochem. Biophys.* **125**, 542.

Kaji, A. & Kaji, H. (1963). *Biochem. Biophys. Res. Comm.* **13**, 186.

Kaji, H., Suzuka, I. & Kaji, A. (1966). *J. Biol. Chem.* **241**, 1251.

Keller, E. B. & Zamecnik, P. C. (1956). *J. Biol. Chem.* **221**, 45.

Leder, P. (1968). In *Methods in Enzymology,* ed. by L. Grossman & K. Moldave, vol. 12B, p. 837. New York: Academic Press.

Leder, P. & Bursztyn, H. (1966a). *Proc. Nat. Acad. Sci., Wash.* **56**, 1579.

Leder, P. & Bursztyn, H. (1966b). *Biochem. Biophys. Res. Comm.* **25**, 233.

Leder, P. & Bursztyn, H. (1966c). *Cold Spr. Harb. Symp. Quant. Biol.* **31**, 297.

Leder, P. & Nau, M. M. (1967). *Proc. Nat. Acad. Sci., Wash.* **58**, 774.

Leder, P., Singer, M. F. & Brimacombe, R. L. C. (1965). *Biochemistry,* **4**, 1561.

Lowry, O. H., Rosebrough, N. J., Farr, A. L. & Randall, R. J. (1951). *J. Biol. Chem.* **193**, 265.

Lucas-Lenard, J. & Haenni, A. L. (1968). *Proc. Nat. Acad. Sci., Wash.* **59**, 554.

Lucas-Lenard, J. & Lipmann, F. (1966). *Proc. Nat. Acad. Sci., Wash.* **55**, 1562.

Maitra, U. & Dubnoff, J. (1968). *Fed. Proc.* **27**, 398.

Matthaei, H., Amelunxen, F., Eckert, K. & Heller, G. (1964). *Ber. Bunsen.* **68**, 735.

Matthaei, H. & Voigt, H. P. (1967). *Biochem. Biophys. Res. Comm.* **28**, 730.

Monro, R. E. (1967). *J. Mol. Biol.* **26**, 147.

Monro, R. E., Maden, B. E. H. & Traut, R. R. (1966). *Symp. Fed. European Biochem. Soc.* (April, 1966), ed. by D. Shugar, p. 179. London: Academic Press.

Monro, R. E. & Marcker, K. A. (1967). *J. Mol. Biol.* **25**, 347.

Muench, K. H. & Berg, P. (1966). In *Procedures in Nucleic Acid Research,* ed. by G. L. Cantoni & D. R. Davies, p. 375. New York: Harper & Row.

Mukandan, M. A., Hershey, J. W. B., Dewey, K. F. & Thach, R. E. (1968). *Nature,* **217**, 1013.

Nakamoto, T., Conway, T. W., Allende, J. E., Spyrides, G. J. & Lipmann, F. (1963). *Cold Spr. Harb. Symp. Quant. Biol.* **28**, 227.

Nakamoto, T. & Kolakofsky, D. (1966). *Proc. Nat. Acad. Sci., Wash.* **55**, 606.

Nirenberg, M. W. (1963). In *Methods in Enzymology*, ed. by S. P. Colowick & N. O. Kaplan, vol. 6, p. 17. New York: Academic Press.

Nirenberg, M. & Leder, P. (1964). *Science*, **145**, 1399.

Nishizuka, Y. & Lipmann, F. (1966a). *Arch. Biochem. Biophys.* **116**, 344.

Nishizuka, Y. & Lipmann, F. (1966b). *Proc. Nat. Acad. Sci., Wash.* **55**, 212.

Noll, H. (1966). *Science*, **151**, 1241.

Nomura, M. & Lowry, C. V. (1967). *Proc. Nat. Acad. Sci., Wash.* **58**, 946.

Ohta, T. & Thach, R. E. (1968). *Nature*, **219**, 238.

Parmeggiani, A. (1968). *Biochem. Biophys. Res. Comm.* **30**, 613.

Patterson, M. S. & Greene, R. C. (1965). *Analyt. Chem.* **37**, 854.

Pestka, S. (1968). *J. Biol. Chem.* **243**, 2810.

Pestka, S. & Nirenberg, M. W. (1966). *J. Mol. Biol.* **21**, 145.

Pulkrábek, R. & Rychlík, I. (1968). *Biochim. biophys. Acta*, **155**, 219.

Ravel, J. M. (1967). *Proc. Nat. Acad. Sci., Wash.* **57**, 1811.

Revel, M. & Gros, F. (1966). *Biochem. Biophys. Res. Comm.* **25**, 124.

Rychlík, I. (1966). *Biochim. biophys. Acta*, **114**, 425.

Salas, M., Miller, M. J., Wahba, A. J. & Ochoa, S. (1967). *Proc. Nat. Acad. Sci., Wash.* **57**, 1865.

Sarkar, S. & Thach, R. E. (1968). *Proc. Nat. Acad. Sci., Wash.* **60**, 1479.

Schlessinger, D., Mangiarotti, G. & Apirion, D. (1967). *Proc. Nat. Acad. Sci., Wash.* **58**, 1782.

Singer, M. F. & Guss, J. K. (1962). *J. Biol. Chem.* **237**, 182.

Skogerson, L. & Moldave, K. (1968). *Arch. Biochem. Biophys.* **125**, 497.

Stanley, W. M., Jr., Salas, M., Wahba, A. J. & Ochoa, S. (1966). *Proc. Nat. Acad. Sci., Wash.* **56**, 290.

Sundararajan, T. A. & Thach, R. E. (1966). *J. Mol. Biol.* **19**, 74.

Szer, W. & Ochoa, S. (1964). *J. Mol. Biol.* **8**, 823.

Tanaka, N., Kinoshita, T. & Masukawa, H. (1968). *Biochem. Biophys. Res. Comm.* **30**, 278.

Thach, R. E., Dewey, K. F. & Mykolajewycz, N. (1967). *Proc. Nat. Acad. Sci., Wash.* **57**, 1103.

Thach, R. E., Sundararajan, T. A., Dewey, K. F., Brown, J. C. & Doty, P. (1966). *Cold Spr. Harb. Symp. Quant. Biol.* **31**, 85.

Traut, R. R. & Monro, R. E. (1964). *J. Mol. Biol.* **10**, 63.

Watson, J. D. (1964). *Bull. Soc. chim. Biol.* **46**, 1399.

Zamir, A., Leder, P. & Elson, D. (1966). *Proc. Nat. Acad. Sci., Wash.* **56**, 1794.

30

Reprinted from *Cold Spring Harbor Symp. Quant. Biol.*, 34, 479–488 (1969)

Peptide Chain Termination, Codon, Protein Factor, and Ribosomal Requirements

T. Caskey, E. Scolnick, R. Tompkins, J. Goldstein, and G. Milman

Laboratory of Biochemical Genetics, National Heart Institute, National Institutes of Health, Bethesda, Maryland

Results of biochemical studies (Last et al., 1967; Capecchi, 1967; Bretscher, 1968) and genetic studies (Weigert and Garen, 1965; Brenner, Stretton, and Kaplan, 1965) show that the synthesis of a peptide chain is terminated upon translation of messenger RNA codons UAA or UAG. Genetic evidence also indicates UGA functions as a barrier or terminator codon (Brenner, Barnett, Katz, and Crick, 1967; Weigert, Lanka, and Garen, 1967).

The mechanism of peptide chain termination has been investigated by stimulating cell-free protein synthesis with randomly ordered or sequenced polyribonucleotides (Takanami and Y. H. Yan, 1965; Bretscher et al., 1965; Ganoza and Nakamoto, 1966; Morgan, Wells, and Khorana, 1966; Kössel, 1968), or viral RNA (Capecchi, 1967; Bretscher, 1968). Translation of UAA or UAG results in the release of free peptides from ribosomes. Capecchi (1967) partially purified a factor required for the release of nascent peptide chains from ribosomes upon the translation of the amber codon, UAG.

We have developed an alternate method for in vitro study of peptide chain termination permitting us to describe the requirements for peptide chain termination and to propose a mechanism for the event. The following headings will be used in this discussion: (1) Trinucleotide directed formyl-methionine release, (2) Release factors R1 and R2, (3) Stimulatory protein, (4) Chain termination ribosomal requirements and, (5) Mechanism of peptide chain termination.

Trinucleotide Directed Formyl-Methionine Release

Trinucleotide codons are used, as shown below, in a sequential manner to initially bind N-formyl-[³H]methionyl-transfer RNA (F[³H]met-tRNA) with AUG to *E. coli* ribosomes and subsequently to release F[³H]methionine with terminator codons (Caskey et al., 1968).

F[³H]met-tRNA + ribosomes + AUG $\rightleftharpoons$

             F[³H]met-tRNA·AUG·ribosome

F[³H]met-tRNA·AUG·ribosome + R Factor

    + Terminator Codon $\rightarrow$ F[³H]methionine

The release of F[³H]methionine from ribosomal bound F[³H]met-tRNA requires both release factor and specific terminator codons and is analogous to the release of nascent peptides upon peptide chain termination. Singular omission of ribosomes, Mg++, K+ or NH4+, or AUG (F-methionine codon) prevents release, suggesting that release occurs from a F[³H]met-tRNA·AUG·ribosomal intermediate rather than from dissociated components. The product, F[³H]methionine, is quantitated by ethyl acetate extraction of the reaction mixture at pH 1.0 (Leder and Bursztyn, 1966).

The time course of F-methionine formation or release is shown in Fig. 1. In the presence of partially purified R factor UAA markedly stimulates the release of F[³H]methionine. This release is proportional to time during the first 40 min, with 72% of the ribosomal bound F[³H]met-tRNA present at zero time released at 60 min. The initial rate, but not extent of F-methionine formation, is directly proportional to R factor concentration, indicating the usefulness of the assay for assessment of R specific activity.

Other trinucleotides were examined for their

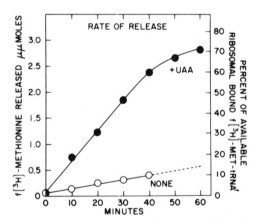

FIGURE 1. Each reaction contained in 0.05 ml: 3.8 pmole F[³H]met-tRNA·AUG·ribosome; 0.78 µg of R protein; 7.3 nmole UAA where indicated; and buffers as previously described. The amount of F[³H]methionine present at zero time (0.30 pmole) is subtracted from each value (Caskey et al., 1968).

264

TABLE 1. PEPTIDE CHAIN TERMINATION CODONS

Addition	Δpmoles F[³H]methionine
UAA	0.66
UAG	0.67
UGA	0.53
CUU, AAU, or AGU	0.09
UUU, UUA, CUC,	0.05
AAA, AGA, GUA,	
GAU, AUA, UGU,	
UGC, and UGG	
None	(0.40)

Specificity of codons for termination. Each reaction contained 5.1 pmoles F[³H]met-tRNA·AUG·ribosome; 11.2 μg R protein; 7.5 nmole of trinucleotide as indicated. The Δpmoles F[³H]methionine is the amount released due to added trinucleotide, and is above the value enclosed in parenthesis. The amount of F[³H]methionine present at zero time (0.35 pmole) was subtracted from each value (Caskey et al., 1968).

TABLE 2. PURIFICATION OF RELEASE FACTORS

Fraction	Specific activity		Purification
	R1 with UAG	R2 with UGA	
II Ammonium sulfate	0.7	1.2	2
III G-100 Sephadex	2.5	4.6	7.5
IV DEAE-Sephadex	46.1	0	136
V DEAE-Sephadex	0	160	274
VI CM-Sephadex	480	—	1,410
VII Hydroxylapatite	—	670	1,150

Purification of release factors. The details of R1 and R2 purification are described elsewhere (Milman et al., 1969). Fractions were assayed for R1 with 2.5 nmoles UAG and R2 with 2.5 nmoles UGA, at 24°C. Specific activity is the number of micro-units of R per mg protein as previously described (Scolnick et al., 1968). Purification is the ratio of the specific activity at the end of each purification step to that of the S-100 supernatant, Fraction I.

ability to release F-methionine in the presence of unfractionated R factor. As is shown in Table 1, only 3 (UAA, UAG, and UGA) of 19 codons tested stimulate the rate of F-methionine formation. The other trinucleotides, including isopleths of terminator trinucleotides, have little detectable effect. Therefore, E. coli appears to utilize the codons UAA, UAG, and UGA for peptide chain termination, each of which requires R factor for its expression.

RELEASE FACTORS R1 AND R2

Release factor is a component of the supernatant fraction of E. coli extracts (Capecchi, 1967), and is required for peptide chain termination directed by UAA, UAG, or UGA (Caskey et al., 1968). We have identified two proteins from the supernatant fraction (Scolnick, Tompkins, Caskey, and Nirenberg, 1968), which release F[³H]methionine from the F[³H]met-tRNA·AUG ribosomal intermediate in reactions containing one of the terminator codons. Release factors are fractionated by DEAE-Sephadex column chromatography, as shown in

Fig. 2, into two well-separated peaks, R1 active in the presence of UAA and UAG, and R2 active in the presence of UAA and UGA. Approximately 60% of the total UAA dependent R activity is found with R1, and 40% with R2. R1 is approximately as active with UAG as with UAA; whereas R2 is more active with UGA than UAA. These proteins do not correspond to any of the previously described protein factors required for peptide chain initiation or elongation. Further purification procedures for R1 and R2 were developed and the results are summarized in Table 2 (Milman, Goldstein, Scolnick, and Caskey, 1969). Although early fractions (I, II, and III) contain both R factors, the activity of each is easily appraised by assay with their specific codons (R1, UAG and R2, UGA). These procedures yield a 1400-fold purification for R1 with 64% recovery of activity; and 1100-fold purification of R2 with 42% recovery. Purified R1 (Fraction VI) and R2 (Fraction VII) have no detectable transfer factor T and contain less than 0.3% transfer factor G activity. Each of these fractions has been additionally purified by Ampholine electrofocusing, a

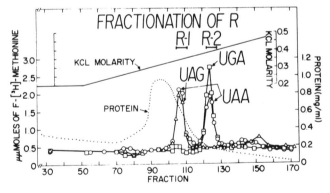

FIGURE 2. Fractionation of R factors. The 0–55% (NH₄)₂SO₄ fraction (2.2 g protein) of E. coli B supernatant fraction (S-100) was eluted from a DEAE-Sephadex A-50 column (112 × 2.8 cm) with 3200 ml of buffer containing a linear potassium chloride gradient (0.15–0.70 M). Each reaction for R assay contained 4.4 pmoles F[³H]met-tRNA·AUG·ribosome; 7.5 nmole of indicated trinucleotide; and 0.010 ml of each column fraction. Reactions were incubated for 25 min. The amount of F[³H]methionine present at zero time (0.30 pmole) is subtracted from each value (Scolnick et al., 1968).

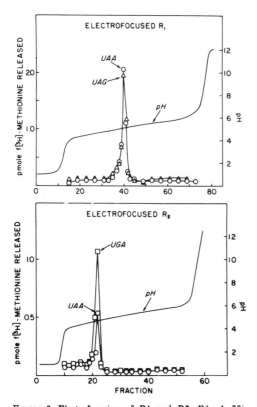

FIGURE 3. Electrofocusing of R1 and R2. R1: A 2% Ampholine solution (pH 4.0–6.0) containing 440 µg R1 (Fraction VI) was electrofocused 44 hr at 700 volts in an LKB 8100 column. Each reaction for R activity was incubated 20 min at 24°C and contained 4.9 pmoles F[³H]met-tRNA·AUG·ribosome; 0.002 ml of each 1.27 ml fraction; and 2.5 nmoles of UAA or UAG. The amount of F[³H]methionine present at zero time (0.31 pmole) was subtracted from each value. R2: Fraction VII of R2 (320 µg) was electrofocused under conditions described above. Each reaction for R activity was incubated 20 min, 24°C and contained 5.3 pmoles F[³H]met·tRNA·AUG·ribosome; 0.002 ml of each 1.50 ml fraction; and 2.5 nmoles of UAA or UGA. The amount of F[³H]methionine present at zero time (0.15 pmole) is subtracted from each value.

technique which separates molecules on the basis of their isoelectric values (Fig. 3). Both R1 and R2 electrofocus to a single peak of release activity at pH 5.1 and 4.7, respectively. Throughout 5000-fold purification the ratio of UAA/UAG activity for R1 and UAA/UGA activity for R2 has remained constant. Acrylamide gel electrophoretic analysis of the Ampholine R1 fraction reveals an apparent homogeneous protein. These studies indicate R1 and R2 are single molecular species, each active with either of two terminator codons (R1, UAA or UAG; R2, UAA or UGA). On the basis of the purification data we estimate both R1 and R2 exist in low cellular content, 10^2 molecules per cell. This number is comparable to estimates for the cellular level of a minor tRNA species (Anderson, 1969).

R factors have the chemical and physical properties of proteins. Analysis of purified R2 for nucleic acid content indicates <0.1 Up and Ap residue per R molecule. Release factors are inactivated as shown in Table 3 by incubation with ribonuclease-free trypsin, N-ethylmaleimide (SH inhibitor), or by incubation at 100°C. However, incubation with T_1 or pancreatic ribonuclease, or di-isopropylfluorophosphate (esterase inhibitor) has little effect on R activity. The results of both purification and selective inactivation procedures indicate that R1 and R2 are acidic protein molecules with essential free sulfhydral groups.

Although the codon specificity of R1 and R2 suggests that they recognize terminator trinucleotide codons, we have investigated the possibility that tRNA may be involved in this codon recognition. Since all in vitro release studies require added F-met-tRNA and ribosomes, these components were purified. Essentially pure F-met-tRNA (prepared at Oak Ridge National Laboratory under AEC-NIGMS), ribosomes, and purified R factors without detectable tRNA release F-methionine in the presence of terminator trinucleotides. The addition of unfractionated aminoacyl-tRNA to release reactions containing these purified components does not stimulate the rate of F-methionine release (Fig. 4), thus indicating no requirement for tRNA in peptide chain termination. Similar conclusions were reached by Bretscher (1968); Fox and Ganoza (1968) by alternate approaches.

The recognition of terminator trinucleotides by the protein release factors is suggested by their codon specificity, and demonstrated directly by

TABLE 3. SENSITIVITY OF R TO INACTIVATION

Experiment	Incubation of R	UAA Dependent F[³H]methionine (pmole)
1	Control	1.08
	+ trypsin	0.05
2	Control	0.96
	+ N-ethylmaleimide	0.13
3	Control	0.73
	+ T_1 ribonuclease	0.66
4	Control	0.56
	100°C	−0.60
	Control + boiled R	0.41
	Control + di-isopropyl-fluorophosphate	0.56

The experimental details of the inactivation process are described elsewhere (Caskey et al., 1968).

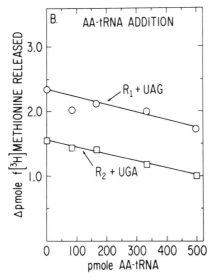

FIGURE 4. Aminoacyl-tRNA addition to purified assay components. Each termination assay was incubated 7.5 min at 24°C and contained 6.1 pmole F[³H]met-tRNA·AUG·ribosome complex, aminoacyl-tRNA as indicated, and either 0.4 μg R1 (○) and 2.5 nmole UAG, or 0.14 μg R2 (□) and 2.5 nmole UGA. The F[³H]methionine released (0.31 pmole) in the absence of terminator trinucleotide is subtracted from each value. Ribosomal bound F[³H]met-tRNA decreases in parallel with the rate of F[³H]methionine release with added aminoacyl-tRNA (Milman et al., 1969).

the trinucleotide-directed ribosomal binding of R factors (Table 4). Reactions containing ribosomes, 10% ethanol, and R1 or R2, incubated at 4°C in the presence of each terminator trinucleotide were separated into ribosomal and supernatant fractions by centrifugation. The R activity of each fraction was determined in the presence of UAA. The codons UAA and UAG, but not UGA, direct the binding of R1 to ribosomes; UAA and UGA, but not UAG, direct the binding of R2 to ribosomes. Furthermore R1, ribosomal bound with UAG, is active in the presence of UAA; similarly R2 bound with UGA, is active in the presence of UAA. Therefore, a single molecule of R1 or R2 each recognize two terminator codons. Thus, R1 and R2 are codon recognition molecules for peptide chain termination. These protein molecules recognize trinucleotide codons on ribosomes. Further study will be required to define the precise mechanism of codon recognition by these proteins.

Recently we have been able to investigate the requirements for release of F-methionine independent of codon recognition. In the presence of ethanol, but absence of codon, both R1 and R2 release F-methionine from a F[³H]met-tRNA·AUG·ribosomal intermediate or in reactions con-

TABLE 4. CODON SPECIFIC R1 AND R2 BINDING TO RIBOSOMES

R Binding reaction additions		UAA-Dependent R activity pmole F[³H]methionine fraction	
R Factor	Codon	Supernatant	Ribosomal
R1	None	1.45	0.55
	UAG	0.07	1.34
	UAA	0.12	1.50
	UGA	1.40	0.62
R2	None	1.54	0.49
	UAA	0.12	1.24
	UGA	0.45	1.02
	UAG	1.52	0.47

Ribosomal binding of R1 and R2. Ribosomal binding reaction: Each reaction was incubated 5 min at 4°C and contained in 0.10 ml: 6.0 A²⁶⁰ NH₄Cl washed ribosomes; 0.25 M ammonium acetate; 0.03 M magnesium acetate; 0.05 M Tris acetate pH 7.4; 0.006 M β-mercaptoethanol; 10% ethanol; 4.9 μg R1 (Fraction VI) or 6.0 μg R2 (Fraction VII); and 2.5 nmole of the indicated trinucleotide codon. Ribosomal and supernatant fractions of each reaction were separated by 60 min 200,000 k × g centrifugation of each reaction in a Spinco #50 rotor with tube adaptors. The supernatant was aspirated and the visible pellet suspended in 0.10 ml of buffer containing 0.25 M ammonium acetate; 0.05 M Tris acetate pH 7.2; and 0.006 M β-mercaptoethanol.

Assay of supernatant and ribosomal bound fraction for release activity: Each release assay was incubated 15 min at 24°C and contained in 0.05 ml: 4.5 pmoles F[³H]met-tRNA·AUG·ribosome; 2.5 nmole UAA; 0.05 M ammonium acetate; and 0.005 ml of the indicated ribosomal or supernatant fractions. The F[³H]methionine released (0.22 pmole) in the absence of either fraction is subtracted from each value.

taining F[³H]met-tRNA and ribosomes. These conditions are similar to those reported by Monro and Marcker (1967) for the study of peptidyl synthetase. R factor and ribosomes exist as a complex under these conditions. The time course of this release is shown in Fig. 5. Reactions containing 70 S ribosomes, F[³H]met-tRNA, and R1

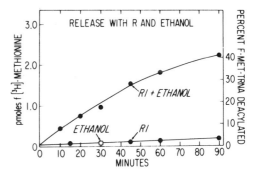

FIGURE 5. *R Factor Release with Ethanol.* Each reaction was incubated at 0°C for the indicated time and contained in 0.05 ml: 5.6 pmoles F[³H]met-tRNA; 1.4 A²⁶⁰ units of NH₄Cl washed ribosomes; 4.4 μg of R1 (Fraction VI); 20% ethanol; 0.25 M NH₄Cl; 0.03 M magnesium acetate; and 0.050 M Tris acetate pH 7.2. Plotted below is a reaction which contains R1 without ethanol (●), or ethanol in the absence of R1 (□) (at 30 min).

TABLE 5. ENZYMIC SPECIFICITY FOR F-METHIONINE
RELEASE WITH ETHANOL

Addition	Δpmole F[³H]-methionine
R1	0.78
R2	0.92
Ts	−0.01
Tu	0.02
G	−0.04
S	0.01
None	(0.16)

Each release assay was incubated 10 min, 24°C and contained in 0.05 ml 3.2 pmoles F[³H]-met-tRNA·ribosome; 0.05 M NH₄Cl; 0.03 M MgAc₂; 0.05 Tris-Ac pH 7.2; 20% ethanol; and as indicated 0.5 μg R1; 0.5 μg R2; 4.4 μg Ts; 21 μg Tu; 0.5 μg G; or 2.0 μg S. The Δpmoles F[³H]-methionine represents release due to added protein above the value enclosed in parenthesis.

(Fraction VI) in the presence of ethanol release F-methionine linearly for 90 min. N-Formyl-methionine is not released if either ribosomes, ethanol, or R factor is omitted, suggesting release must occur from an R-ribosome·F[³H]met-tRNA intermediate. Release occurs only in the presence of the R factors and not with the proteins Tu, Ts, G, or S (Table 5). In other experiments, the ethanol dependent release activity cochromatographed on DEAE- and CM-Sephadex with codon-specific R2 and R1 activity. These studies indicate that R factors participate in the release of F-methionine by their trinucleotide or ethanol-directed binding to ribosomes.

STIMULATORY PROTEIN

The purification of components for the release assay enabled identification of another protein, S, which stimulates the rate of F-methionine release (Milman et al., 1969). This protein is devoid of R activity and stimulates release only in the presence of R factor and terminator codon, as illustrated in Fig. 6. The addition of S to reactions containing R2 and UAA increases the rate of F-methionine appearance 4- to 5-fold without affecting the extent of the reaction. Apparently, S protein acts at a rate limiting event in F-methionine release. Since S stimulates F-methionine release in the presence of either R1 or R2 and each of their appropriate codons (Table 6), S affects a common event in peptide chain termination. The stimulation by S is totally inhibited by 10^{-4} M GTP, GDP, and GOPOPCP, but not GMP. These nucleotides have no effect on the rate or extent of F-methionine release in reactions without S (Table 6), indicating S is not a component of purified ribosomes or R factors.

The S factor has properties of a heat labile protein and is a component of the supernatant fraction (S-100) of E. coli extracts. The purification of S by DEAE-Sephadex column chromatography

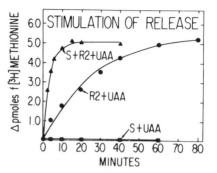

FIGURE 6. S protein stimulation. Each release assay was incubated 30°C for the indicated time and contained in 0.05 ml: 5.8 pmoles F[³H]met-tRNA·AUG·ribosome; 2.5 nmole UAA; 0.05 M potassium acetate; 0.03 M magnesium acetate; 0.05 M Tris-acetate, pH 7.2; and as indicated, 1.6 μg R2; and 15.2 μg S-protein (partially purified by DEAE-Sephadex). Release in the absence of UAA is subtracted from all values.

is shown in Fig. 7. A single peak of S activity cochromatographs with a minor peak of T activity (determined both by GTP binding and T-dependent polyphenylalanine polymerization). The major peaks of T and G activity have no effect on F-methionine release. Following the initial separation of S activity from the major T activity by DEAE-Sephadex, S is further purified by CM-Sephadex, hydroxylapatite, and G-100 column chromatography. The highest purity S preparation contains 2% Tu protein. Capecchi reported at these meetings Tu, purified to apparent homogeneity by SDS acrylamide gel analysis, stimulates F-methionine release. Our studies indicate S differs from Tu because: (1) The bulk of cellular T activity (Fig. 7) does not effect F-methionine release. (2) We have purified Tu from this major T fraction to apparent homogeneity by acrylamide gel analysis (Weissbach and Miller, pers. commun.), and find

TABLE 6. STIMULATION OF TERMINATION

Factors	UAA Dependent release pmoles F[³H]methionine additions	
	None	GTP
S	0.05	0.01
R1	0.51	0.50
S + R1	1.13	0.53
R2	0.42	0.45
S + R2	0.80	0.42

GTP inhibition of stimulation. Each release assay was incubated 5 min at 30°C and contained in 0.05 ml: 4.9 pmole F[³H]met·RNA·AUG·ribosome; 2.5 nmole UAA; 2.0 μg R1 (Fraction VI), or 1.0 μg R2 (Fraction VII); 54 μg S; 0.05 M potassium acetate; and as indicated 10^{-4} M GTP. Release in the absence of UAA (0.41 pmoles) is subtracted from all values (Milman et al., 1969).

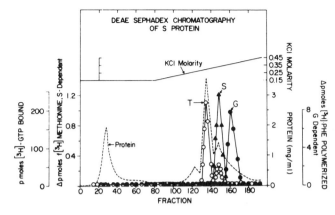

FIGURE 7. S protein purification. The 55–80% $(NH_4)_2SO_4$ fraction (1.17 g protein) of *E. coli* B supernatant fractions (S-100) was eluted from a DEAE-Sephadex column (65 × 2.5 cm) with 1500 ml of buffer containing a linear potassium chloride gradient (0.15–0.45 M) from pH 8.0 to 7.0. Each reaction for S activity was incubated 25 min at 30°C and contained in 0.05 ml: 4.1 pmole F[³H]met-tRNA·AUG·ribosome; 0.5 nmole UAA; 0.66 μg R2 (Fraction VII); 0.005 ml of each column fraction; and 0.05 M potassium acetate. The amount of F[³H]methionine released in the absence of column addition (0.41 pmole) is subtracted from each value. The transfer activity T and G was determined as previously described (Allende, Seeds, Conway, and Weissbach, 1967; Milman et al., 1969).

this fraction has no effect on F-methionine release (3) We have also purified Tu by an alternate method (Lucas-Lenard, pers. commun.) which also has no effect on F-methionine release.

While additional purification is needed to determine the significance of the 2% Tu in S, it is clear that functionally S acts at a rate-limiting event in F-methionine release. Earlier studies suggested codon recognition by R1 and R2 is rate limiting, for at a fixed concentration of R factor, the rate of F-methionine release varies with terminator trinucleotide concentration (Smrt, Kemper, Caskey, and Nirenberg; Template activity of modified terminator codons, in prep.). This rate

limiting effect of trinucleotide codon is altered by S protein (Fig. 8). From 0 to 5.0×10^{-5} M UAA, the rate of F-methionine becomes maximal in the presence of S but is severely limited in its absence. Analysis of similar data by Lineweaver and Burk plots (1934) indicates S lowers the K_m of trinucleotide codons 7–10-fold without altering V_{max} (Fig. 9). It appears S has its site of action at codon recognition. One possible mechanism is that S factor facilitates codon directed R factor binding to ribosomes in a fashion similar to the facilitation of codon directed binding of aminoacyl-tRNA to ribosomes by GTP and T factor (Ravel, 1967).

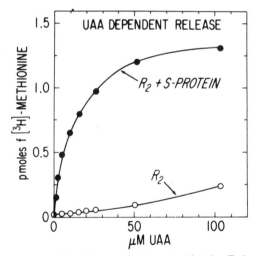

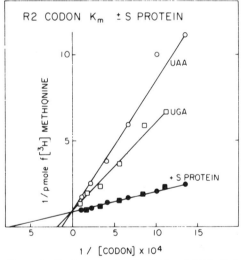

FIGURE 8. Stimulation at limiting trinucleotide. Each reaction was incubated 10 min at 30°C and contained in 0.050 ml: 5.3 pmole F[³H]met-tRNA·AUG·ribosome; 0.66 μg R2 (Fraction VII); 16.0 μg S protein; 0.05 M potassium acetate; and UAA as indicated. The amount of F[³H]-methionine present at zero time (0.15 pmole) is subtracted from each value.

FIGURE 9. K_m determination for UAA and UGA. Each release assay was incubated 10 min at 24°C and contained in 0.050 ml: 5.9 pmole F[³H]met-tRNA·AUG·ribosome; 0.05 M potassium acetate; 0.66 μg R2 (Fraction VII); 6.3 μg S protein (filled symbols); UAA (○) or UGA (□) as indicated. Release in the absence of trinucleotide codon (0.14 pmole) is subtracted from all values.

CHAIN TERMINATION RIBOSOMAL REQUIREMENTS

In order to examine the ribosomal subunit requirements for codon directed peptide chain termination, ribosomal intermediates of two types were prepared. In the first case, F[³H]met-tRNA was bound to 30 S or 30 S + 50 S ribosomal subunits in the presence of AUG. These ribosomal intermediates were compared for release of F-methionine in the presence of R1 and UAG (Fig. 10). Release of F-methionine occurs only when 30 S and 50 S subunits are present. Since there was no significant binding of F-met-tRNA to the 50 S subunits we isolated poly[¹⁴C]Phe-tRNA·50 S ribosomal intermediates in order to determine if polyphenylalanine could be released with R1 and UAG (Table 7). Release of polyphenylalanine occurred only when 30 S subunits were added to reactions containing R1 and UAG. The poly[¹⁴C]Phe-tRNA·50 S ribosomal intermediates were reactive however with puromycin (Traut and Monro, 1964). These studies indicate that both 30 S and 50 S ribosomal subunits are required for peptide chain termination. These requirements are similar to those for peptide chain elongation (Gilbert, 1963).

The ribosomal site requirements for peptide chain termination have been investigated by an

TABLE 7. RELEASE REQUIREMENTS FOR POLY-Phe-tRNA·50 S INTERMEDIATES

Additions	Poly-Phe released pmoles
None	1.37
R1	1.00
R1, UAG	1.11
30 S, R1	1.39
30 S, R1, UAG	2.68

Ribosomal subunit requirement. The poly-Phe-tRNA·50 S ribosome complex was prepared according to Maden, Traut, and Monro, 1968. Each reaction was incubated 60 min at 24°C and contained in 0.05 ml: 8.6 pmole poly [¹⁴C]-Phe-tRNA·50 S ribosome; 0.1 M ammonium acetate; 0.03 M magnesium acetate; 0.05 M Tris-acetate pH 7.2; and where indicated 4.0 μg R1 (Fraction VI); 7.5 nmole UAG; and 2.3 pmole 30 S ribosomal subunits. Released [¹⁴C]-polyphenylalanine was determined by ethyl acetate extraction at pH 8.0. Values given are corrected for total release.

alternate approach. F-met-tRNA bound to ribosomes at 5 mM Mg⁺⁺ in the presence of AUG, initiation factors and GTP is puromycin reactive (P-site) while that bound with GOPOPCP is not reactive (A-site) (Ohta, Sarkar, and Thach, 1967). The ability of R1 and UAG to release F-methionine from F[³H]met-tRNA·AUG·ribosomal intermediates formed with GTP (P-site) or GOPOPCP (A-site) is illustrated in Fig. 11. Release occurs from ribosomal intermediates formed with GTP but not with GOPOPCP. Control studies assured that GOPOPCP does not inhibit peptide chain termination. The effect of puromycin on the two ribosomal intermediates confirms earlier observations (Ohta et al., 1967). These studies suggest that peptidyl-tRNA must be translocated from the A- to P-site before peptide chain termination can occur.

MECHANISM OF PEPTIDE CHAIN TERMINATION

The process of peptide chain termination has been experimentally separated into a minimum of two events, codon recognition and release of peptides. Although this oversimplification (Fig. 12) does not take into account other possible intermediate events in peptide chain termination, it serves as framework for discussion of the available information.

The codons for peptide chain termination in *E. coli* are UAA, UAG, or UGA. Although the occurrence of a single terminator codon in mRNA as a result of mutation produces premature chain termination (Weigert and Garen, 1965; Kaplan, Stretten, and Brenner, 1965; and Brenner et al., 1967), the arrangement of naturally occurring chain termination signals is not known. Premature chain termination mutations (nonsense) of UAG and UGA type are corrected with high efficiency in bacteria carrying UAG or UGA suppressor

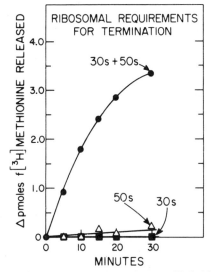

FIGURE 10. Ribosomal subunit requirements. Each 0.050 ml release assay was incubated at 24°C for the indicated time intervals and contained in 0.05 ml: 4.1 pmole F[³H]met-tRNA·AUG·30 S ribosome, 4.1 pmole of F[³H]-met-tRNA·AUG·(30 S + 50 S) ribosome, or 5.0 pmole F[³H]met-tRNA and 50 S ribosome; 0.88 μg R1; 2.5 nmole UAG; 0.15 M magnesium acetate; 0.1 M ammonium acetate; and 0.05 M Tris-acetate pH 7.2. The amount of F[³H]methionine present at zero time is subtracted from each value.

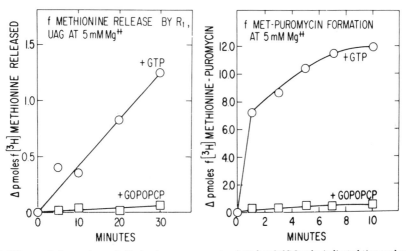

Figure 11. Ribosomal site requirement. Each release assay was incubated at 24°C for the indicated time and contained in 0.05 ml: 3.4 pmole F[³H]met-tRNA·AUG·ribosome formed in the presence of partially purified F1, F2, and GTP or GOPOPCP; 0.66 μg R1; 2.5 nmole UAG; 0.1 M ammonium acetate; 0.005 M magnesium acetate; and 0.05 M Tris-acetate pH 7.2. Reactions for the formation of F-met-puromycin contained 10^{-3} M puromycin; R1 and UAG were omitted.

genes (Weigert et al., 1965; Sambrook, Fan, and Brenner, 1967), while the cells have apparently normal growth characteristics. Bacteria containing UAA suppressor genes grow poorly (Smith et al., 1966; Gallucci and Garen, 1966). It has been suggested on the basis of this data, UAA is a stronger candidate than UAG or UGA for the naturally occurring chain termination signal (Garen. 1968). However, tRNA suppressors of UAG and UGA do not suppress UAA whereas tRNA suppressors of UAA can also suppress a second terminator codon, UAG. Therefore, ochre suppressors affect two terminator codons. Naturally occurring chain termination signals may be arranged as multiple or tandem codons which are of mixed or identical type. Further studies are required to elucidate the exact composition of the natural termination signal.

The protein molecules R1 and R2 recognize different sets of the three terminator codons (R1, UAA or UAG; R2, UAA or UGA). The recognition of a trinucleotide codon with respect to both base composition and sequence by R protein molecules is an interaction of extremely high fidelity. While the precise mechanism of this recognition has not been elucidated, the recognition can be physically demonstrated. Release factor in the presence of its appropriate terminator codon forms an intermediate with ribosomes, which can be isolated. This ribosomal intermediate formed in response to trinucleotide codon or in vitro with ethanol is an essential intermediate for peptide release.

Protein factor S increases affinity for trinucleotide codon in termination reactions and therefore affects the process of terminator codon recognition. Since codon recognition results in the formation of a R·terminator codon·ribosome intermediate, S may act to increase the rate of formation or stability of this complex.

Peptide release by R factor, as shown in the lower portion of Fig. 12, occurs only when peptidyl-tRNA and R factor occur in a ribosomal bound state, induced by either codon or ethanol. R factor-dependent termination yields deacylated tRNA, the identical by-product of the transfer of

$$\begin{bmatrix} \text{R1} + \begin{array}{c} \text{UAA} \\ \text{or} \\ \text{UAG} \end{array} \end{bmatrix} \text{ or } \begin{bmatrix} \text{R2} + \begin{array}{c} \text{UAA} \\ \text{or} \\ \text{UGA} \end{array} \end{bmatrix} + \text{Ribosome·AUG·F[³H]met-tRNA (P-Site)}$$

S Protein

(A-Site) R·Terminator codon·ribosome·AUG·F[³H]met-tRNA (P-Site)

N-Formyl[³H]methionine

Figure 12

271

TABLE 8. EFFECT OF ANTIBIOTICS

Addition	Termination assay UAA-dependent F[³H]methionine pmoles	Binding assay Ribosomal bound F[³H]met-tRNA pmoles
None	0.74	5.36
+ Tetracycline	0.02	5.08
+ Streptomycin	0.28	5.36
+ Sparsomycin	0.28	5.36
+ Chloramphenicol	0.59	5.26
+ Fusidic acid	0.74	5.45

Each termination reaction contained in 0.05 ml: 5.5 pmole F[³H]met-tRNA·AUG·ribosome intermediate, 1.5 μg R1 (Fraction V), 7.5 nmole UAA, and 3×10^{-4} M indicated antibiotic. The amount of F[³H]methionine present at zero time (0.33 pmole) was subtracted from each value. The effect of antibiotics upon the dissociation of the ribosomal intermediate after incubation of reactions identical to those described above (except that R1 was omitted) by the aminoacyl-tRNA binding assay.

peptides from peptidyl-tRNA occurring in protein synthesis. This latter reaction is catalyzed by peptidyl transferase, a 50 S ribosomal subunit enzyme (Traut and Monro, 1964). Therefore, the removal of peptides from peptidyl-tRNA during peptide chain termination may not be a simple hydrolysis by R factor but rather may involve the peptidyl transferase activity. The latter mechanism is suggested by two studies. One, peptide chain termination is inhibited by chloramphenicol and sparsomycin (Table 8), antibiotics known to inhibit peptidyl transferase (Monro and Marcker, 1967; Monro, Celma, and Vazquez, 1969). Two, release by R1 and UAG occurs only when F-met-tRNA is located in the P-site. Similarly, peptidyl transferase transfers peptidyl groups to yield tRNA only when peptidyl-tRNA is located in the P-site. The precise requirements and intermediate events of peptide chain release will require a more thorough understanding of the mechanism of R factor action and the role of peptidyl transferase.

ACKNOWLEDGMENTS

The authors would like to thank Dr. M. Nirenberg for his helpful advice during the course of these studies; Mrs. T. Caryk for her devoted efforts throughout the studies, and Mrs. E. Murray, whose cooperative attitude made the preparation of this manuscript possible. G. M. is a fellow of the Helen Hay Whitney Foundation.

REFERENCES

ALLENDE, J., N. W. SEEDS, T. W. CONWAY, and H. WEISSBACH. 1967. Guanosine triphosphate interaction with an amino acid polymerization factor from *E. coli*. Proc. Nat. Acad. Sci. *58:* 1566.

ANDERSON, F. 1969. The effect of tRNA concentration on the rate of protein synthesis. Proc. Nat. Acad. Sci. *62:* 566.

BRENNER, S., L. BARNETT, E. R. KATZ, and F. H. C. CRICK. 1967. UGA: A third nonsense triplet in the genetic code. Nature *213:* 449.

BRENNER, S., A. O. W. STRETTON, and S. KAPLAN. 1965. Genetic code: The 'nonsense' triplets for chain termination and their suppression. Nature *206:* 994.

BRETSCHER, M. S. 1968. Polypeptide chain termination: An active process. J. Mol. Biol. *34:* 131.

BRETSCHER, M. S., H. M. GOODMAN, J. R. MENNINGER, and J. D. SMITH. 1965. Polypeptide chain termination using synthetic polynucleotides. J. Mol. Biol. *14:* 634.

CASKEY, T., R. TOMPKINS, E. SCOLNICK, T. CARYK, and M. NIRENBERG. 1968. Sequential translation of trinucleotide codons for the initiation and termination of protein synthesis. Science *162:* 135.

CAPECCHI, M. R. 1967. Polypeptide chain termination in vitro: Isolation of release factor. Proc. Nat. Acad. Sci. *58:* 1144.

FOX, J. L., and M. C. GANOZA. 1968. Chain termination in vitro. Studies on the specificity of amber and ochre triplets. Biochem. Biophys. Res. Commun. *32:* 1064.

GALLUCCI, E., and A. GAREN. 1966. Suppressor genes for nonsense mutations. II. The Su-4 and Su-5 suppressor genes of *Escherichia coli*. J. Mol. Biol. *15:* 193.

GANOZA, M. C., and T. NAKAMOTO. 1966. Studies of the mechanism of polypeptide chain termination in cell-free extracts of *E. coli*. Proc. Nat. Acad. Sci. *55:* 162.

GAREN, A. 1968. Sense and nonsense in the genetic code. Science *160:* 149.

GILBERT, W. 1963. Polypeptide synthesis in *Escherichia coli*. I. Ribosomes and the active complex. J. Mol. Biol. *6:* 374.

KAPLAN, S., A. O. W. STRETTON, and S. BRENNER. 1965. Amber suppressors: Efficiency of chain propagation and suppressor specific amino acids. J. Mol. Biol. *14:* 528.

KÖSSEL, H. 1968. Studies on polynucleotides. LXXXIII. Synthesis in vitro of the tripeptide valyl-seryl-lysine directed poly r(G–U–A–A). Biochim. Biophys. Acta *157:* 91.

LAST, J. A., M. WENDEL, J. R. STANLEY, M. SALAS, M. B. HILLE, A. J. WAHBA, and S. OCHOA. 1967. Translation of the genetic message. IV. UAA as a chain termination codon. Proc. Nat. Acad. Sci. *57:* 1062.

LEDER, P., and H. BURSZTYN. 1966. Initiation of protein synthesis. II. A convenient assay for the ribosome-dependent synthesis of N-formyl-¹⁴C-methionyl puromycin. Biochem. Biophys. Res. Commun. *25:* 233.

LINEWEAVER, H., and D. BURK. 1934. The determination of enzyme dissociation constants. J. Am. Chem. Soc. *56:* 658.

MADEN, B. E. H., R. R. TRAUT, and R. E. MONRO. 1968. Ribosome-catalyzed peptidyl transfer: The polyphenylalanine system. J. Mol. Biol. *35:* 333.

MILMAN, G., J. GOLDSTEIN, E. SCOLNICK, and T. CASKEY. 1969. Peptide chain termination. III. Stimulation of in vitro termination. Proc. Nat. Acad. Sci. *63:* 183.

MONRO, R. E., M. L. CELMA, and D. VAZQUEZ. 1969. Action of sparsomycin on ribosome-catalyzed peptidyl transfer. Nature *222:* 356.

MONRO, R. E., and K. A. MARCKER. 1967. Ribosome-catalyzed reaction of puromycin with a formyl methionine-containing oligonucleotide. J. Mol. Biol. *25:* 347.

MORGAN, A. R., R. D. WELLS, and H. G. KHORANA. 1966. Studies on polynucleotides. LIX. Further codon assignments from amino acid incorporation directed by ribopolynucleotides containing repeating trinucleotide sequences. Proc. Nat. Acad. Sci. *56:* 1899.

OHTA, T., S. SARKAR, and R. E. THACH. 1967. The role of

guanosine-5′-triphosphate in the initiation of formyl-methionyl-tRNA to ribosomes. Proc. Nat. Acad. Sci. *58:* 1638.

RAVEL, J. M. 1967. Demonstration of a guanosine triphosphate-dependent enzymatic binding of aminoacyl-ribonucleic acid to *Escherichia coli* ribosomes. Proc. Nat. Acad. Sci. *57:* 1811.

SAMBROOK, J. F., D. P. FAN, and S. BRENNER. 1967. A strong suppressor specific for UGA. Nature *214:* 452.

SCOLNICK, E., R. TOMPKINS, T. CASKEY, and M. NIRENBERG, 1968. Release factors differing in specificity for terminator codons. Proc. Nat. Acad. Sci. *61:* 768.

SMITH, J. D., J. N. ABELSON, B. F. C. CLARK, H. M. GOODMAN, and S. BRENNER. 1966. Studies on amber suppressor tRNA. Cold Spring Harbor Symp. Quant. Biol. *31:* 479.

TAKANAMI, M., and Y. H. YAN. 1965. The release of polypeptide chain from ribosomes in cell-free amino

acid-incorporating system by specific combinations of bases in synthetic polyribonucleotides. Proc. Nat. Acad. Sci. *54:* 1450.

TRAUT, R. R., and R. E. MONRO. 1964. The puromycin reaction and its relation to protein synthesis. J. Mol. Biol. *10:* 63.

WATSON, J. D. 1964. The synthesis of protein upon ribosomes. Bull. Soc. Chim. Biol. *46:* 1399.

WEIGERT, M. G., E. LANKA, and A. GAREN. 1965. Amino acid substitutions resulting from suppression of nonsense mutations. II. Glutamine insertion by the Su-3 gene. J. Mol. Biol. *14:* 522.

WEIGERT, M. G., E. LANKA, and A. GAREN. 1967. Base composition of nonsense codons in *Escherichia coli*. II. The N_2 codon UAA. J. Mol. Biol. *23:* 391.

WEIGERT, M. G., and A. GAREN. Base composition of nonsense codons in *E. coli*. Nature *206:* 992.

Part V

TRANSLATION: THE GENETIC CODE AND TRANSFER RNA

Editor's Comments
on Papers 31 Through 37

This volume will conclude with a section on the genetic code and the role of tRNA in the translation process. The existence of a code for the transfer of genetic information was suggested by the Watson–Crick model for DNA structure (Paper 1). A colinearity was envisioned between the base sequence of a DNA chain and the amino acid sequence of the resulting polypeptide gene product (reviewed by Crick, 1966). The discovery of mRNA and the role that it plays in this transfer of genetic information is detailed in Part II of this volume. Stereochemical and hydrogen-bonding considerations suggested that amino acids are not assembled into proteins directly on the messenger template. Instead, it was reasoned that a polynucleotide adaptor species mediates the translation process (Crick, 1957). It was soon demonstrated that such an adaptor function is provided by tRNA (Hoagland et al., 1958).

The studies of Yanofsky et al. (1964) established that the base sequence of DNA maps colinearly into corresponding polypeptides. It was found that the ordering of amino acid replacements in the tryptophan synthetase A protein corresponded exactly to the linear order of their associated mutations determined by genetic mapping. This was corroborated in the studies of Sarabhai et al. (1964), who demonstrated that the position of chain termination mutations of the phage T4 genome corresponded to the fraction of head protein that was produced. These methods together with the studies on suppression of chain termination mutations (Part IV) and the frameshift mutations in the phage T4 lysozyme gene investigated by Streisinger et al. (1966) eventually provided *in vivo* verification of RNA code word assignments made *in vitro*.

A nonoverlapping triplet code read from a fixed starting point was first suggested on the basis of genetic studies (Crick et al., 1961). Proof for a triplet code and the fidelity of its reading was provided by Khorana et al. (Paper 31). Using a cell-free amino acid incorporating system from *Escherichia coli* (see below) and RNAs that they had synthesized with known repeating sequences, the following was found: repeating dinucleotide sequences resulted in the formation of polypeptides of repeating dipeptide sequence (Nishimura et al., 1965a); polymers with repeating trinucleotide sequence gave rise to homopeptides (Nishimura et al., 1965b); from repeating tetranucleotide sequences, polypeptides containing alternating tetrapeptide sequences were produced (Paper 31). These studies not only established codon size but also determined the direction in which the messenger is read. They also provided a method for determining several of the codon assignments and confirmed many of those made by Nirenberg and his collaborators (Paper 32).

The earliest of the 64 codon assignments were provided by the cell-free system discovered by Nirenberg and Matthaei (1961) in which the capacity to stimulate the incorporation of amino acids into protein was dependent upon the presence of a polyribonucleotide messenger. By using synthetic RNA homopolymers and random copolymers, the nucleotide composition of about 45 code words was assigned (Nirenberg et al., 1965); however, it was possible to assign the base sequence of only a few of the codons in this manner. A more dynamic technique was necessary for making the remaining codon assignments, and this was provided by the triplet binding assay devised by Nirenberg and Leder (1964). In this method, interactions between ^{14}C-aminoacyl-

277

tRNA and the ribosome were measured by the retention of the complex on nitrocellulose filters. The interactions were specifically stimulated with the appropriate trinucleotide codons that were enzymatically synthesized (Leder et al., 1965; Bernfield, 1966). Most of the unknown codon sequences were assigned in this manner (Nirenberg et al., Paper 32).* Some of these codon assignments were independently confirmed by Matthaei and his collaborators (1966) using an independent technique.

That the genetic code is universal was suggested by the ability of charged tRNAs of microbial origin to insert amino acids into hemoglobin synthesized on rabbit reticulocyte ribosomes (Weisblum et al., 1965; Gonano, 1967). Marshall et al. (1967) demonstrated by the filter binding technique that aminoacyl tRNAs of amphibian and mammalian origin recognized essentially the same triplet codons as the charged tRNAs of *E. coli*. Thus, an identical codon catalog appears to be used by all organisms, with the exception of a differential use of the three *nonsense* codons. Woese (1969) has provided a review on various aspects of the genetic code, including the evolutionary possibilities that it has suggested.

The elucidation of the role of tRNA as the adaptor species bridging the individual codon of the mRNA and its cognate amino acid dates from the studies of Hoagland et al. (1958) and Berg et al. (Paper 33). It was found that aminoacyl-tRNA formation is catalyzed by aminoacyl-tRNA synthetase and that this reaction can be divided into two steps: an activation reaction in which 1 mole of ATP is consumed, which results in the formation of an aminoacyl-AMP-enzyme intermediate and pyrophosphate, and a charging reaction in which the amino acid is transferred to tRNA, and free AMP and enzyme are released. It was demonstrated by Hoagland et al. (1958) that amino acids in tRNA linkage are the immediate precursors of those in the growing polypeptide chain. The experiments of Bergmann et al. (1961) first suggested a single aminoacyl-tRNA synthetase for each of the amino acids, and it is now known that individual synthetases are capable of acylating different iso-accepting tRNA species (Baldwin and Berg, 1966). However, multiple forms of leucine-tRNA synthetase have been found that transfer this amino acid to different tRNA[leu] species (Yu, 1966). It is thought that the overall exterior conformation of tRNA is an important determinant permitting its recognition by the appropriate aminoacyl-tRNA synthetase (Gauss et al., 1971).

*With regard to the uncertain codon assignments in Table 3 (Paper 32), UUA and UUG have been designated as leucine code words; UAA, UAG, and UGA serve as terminator code words. GUG apparently codes for formylmethionine when it initiates polypeptide chains, whereas it codes for valine when placing amino acids into the interior of a chain. A more complete codon catalog is presented in Paper 31 (Table 8).

Beginning with the pioneering investigations of Holley and his collaborators (Paper 34), many structural details of tRNA have been elucidated. They reported the complete primary structure of tRNA[ala] from yeast, the first nucleic acid to be entirely sequenced. It was found that tRNA is composed of a single chain of approximately 80 nucleotides with its 3'-end terminating in CCA.* The amino acid is attached to the 3'-terminal A through an ester linkage during the charging reaction (Preiss et al., 1961). The aminoacylation occurs exclusively at the 2'-hydroxyl group of the adenosine (Sprinzl and Cramer, 1973), and the attached amino acid is thought to undergo a transacylation to the 3'-hydroxyl during protein synthesis. The lack of exact correspondence between the content of A and U and that of G and C in tRNA together with possible models for the secondary structure of the molecule suggested that several unpaired bases form a series of hairpin loops, and that the remaining bases are capable of hydrogen bonding to form double-helical stem regions. Since mixtures of different tRNAs form regular three-dimensional crystals (Watson, 1970), a single basic secondary structure in the form of a cloverleaf has been suggested for tRNA (see the conformation in the upper right portion of Figure 2, Paper 34). This structure also has as its basis the several regions of similar primary structure found in a number of tRNAs whose complete nucleotide sequences have been determined (Madison, 1968; Cramer, 1971). The cloverleaf arrangement of six tRNAs is presented by Madison (1968). The basic features of the cloverleaf structure are the CCA stem and a series of three to four additional stems connected by looped structures: the DHU loop; the anticodon loop; the extra loop of variable size (4–21 nucleotides; Cramer, 1971), with or without a stem; and a TψC loop. The anticodon (the three tRNA bases with which the codon interacts) is placed in a position opposite the CCA stem; its high sensitivity to nuclease digestion (Penswick and Holley, 1965; Nishimura and Novelli, 1965) suggests that it is accessible for the hydrogen bonding essential to the decoding process.

As there are 64 possible trinucleotide codons, but only 20 amino acids found in protein, many of the amino acids have more than one code word. This overabundance of codons for a given amino acid is referred to as degeneracy (Bernfield and Nirenberg, 1965). Although there are specific tRNAs for each of the amino acids, insufficient tRNAs have been found to account for all the assigned code words if standard base pairing in the codon–anticodon interaction is assumed (Söll et al., 1967). The solution to this paradox was provided by Crick's clever "wobble" hypothesis (Paper 35), which accounts for the degeneracy in the decoding process. It was postulated that the base in the III' (5') position of the anticodon is capable of moving slightly out of orientation to accommodate base pairings other than A–U and

---

*For the abbreviations used in these comments, see Paper 37 (note 5).

G–C in the III (3′) position of the codon. Essentially standard base pairings are strictly employed in the other two anticodon positions.

A molecular basis for the "wobble" hypothesis is presented by Fuller and Hodgson in Paper 36. They chose a conformation for the anticodon loop that maximized single-stranded base stacking such that it assumed a secondary structure equivalent to a single strand of a double helix. The base in the III′ ("wobble") position is stacked in such a manner as to afford it maximum conformational flexibility. This structure accounts for all the base movement necessary to accommodate codon degeneracy patterns, and it may be said that the Fuller–Hodgson model is truly stacked in favor of the "wobble" hypothesis.

The final paper in this collection, will, like the first, deal with an outstanding contribution of X-ray diffraction studies to molecular biology. This contribution was recently made by Kim and his collaborators (Paper 37) and concerns the elucidation of the three-dimensional structure of tRNA. They obtained a 4.0-Å resolution electron-density map of heavy atom derivatives of crystalline yeast phenylalanine tRNA. The polynucleotide chain was seen as a series of electron-dense masses whose peaks had an average separation of 5.8 Å; these peaks of electron density were associated with adjoining phosphate groups in the polynucleotide chain. Less electron-dense areas were associated with areas of hydrogen bonding. A perspective diagram was deduced from the electron-density map (see Figures 4 and 5, Paper 37), and although the basic stem and loop features of the cloverleaf model are retained, the folding shown in these diagrams is very different. The CCA stem is connected to the TψC loop by a continuous double helix comprising both the CCA and TψC stems. The TψC loop is placed at the corner of the L-shaped molecule adjoining the DHU loop whose stem is connected to the CCA stem. Both the DHU and anticodon stems form the other continuous double helix, which is oriented approximately at right angles to that formed by the TψC and CCA stems. The anticodon and the 3′-hydroxyl terminal A are at opposite ends of the molecule, separated by approximately 82 Å. The molecule is 20 Å thick, which corresponds to the thickness of an RNA double helix.

More recently, electron-density maps were obtained at 3.0-Å resolution for the yeast phenylalanine tRNA molecule (Kim et al., 1974; Robertus et al., 1974). These data confirm the general folding of the previous model; remarkably, individual ribose, phosphate, and base moieties are clearly visualized. The 3.0-Å maps also provide many additional details of the tertiary structure and remove much of the uncertainty in the 4.0-Å map. Accordingly, appropriate assignments for many tertiary interactions have been made, and essentially identical models for the three-dimensional folding of the tRNA molecule have been constructed by each research group (Kim et al., 1974; Robertus et al., 1974). There is an excellent correlation between chemical reactivity of the bases and their accessibility in the proposed structure. Most of the proposed tertiary interactions involve nucleotides common to all tRNAs, which suggests that the model may aid in elucidating the three-dimensional folding of many tRNA molecules.

REFERENCES

Baldwin, A. N., and P. Berg (1966). Purification and properties of isoleucyl ribonucleic acid synthetase from *E. coli. J. Biol. Chem., 241,* 831–838.

Bergmann, F. H., P. Berg, and M. Dieckmann (1961). The enzymic synthesis of aminoacyl derivatives of ribonucleic acid: II. The preparation of leucyl-, valyl-, isoleucyl-, and methionyl-ribonucleic acid synthetases from *E. coli. J. Biol. Chem., 236,* 1735–1740.

Bernfield, M. (1966). Ribonuclease and oligoribonucleotide synthesis: II. Synthesis of oligonucleotides of specific sequence. *J. Biol. Chem., 241,* 2014–2023.

——, and M. W. Nirenberg (1965). RNA codewords and protein synthesis: the nucleotide sequences of multiple codewords for phenylalanine, serine, leucine, and proline. *Science, 147,* 479–484.

Cramer, F. (1971). Three-dimensional structure of tRNA. *Progr. Nucleic Acid Res. Mol. Biol., 11,* 391–421.

Crick, F. H. C. (1957). Discussion. In *The Structure of Nucleic Acids and Their Role in Protein Synthesis,* Biochemical Society Symposium, No. 14. Cambridge University Press, New York, pp. 25–26.

—— (1966). The genetic code—yesterday, today, and tomorrow. *Cold Spring Harbor Symp. Quant. Biol., 31,* 3–9.

——, L. Barnett, S. Brenner, and R. J. Watts-Tobin (1961). General nature of the genetic code for proteins. *Nature (London), 192,* 1227–1232.

Gauss, D. H., F. von der Haar, A. Maelicke, and F. Cramer (1971). Recent results of tRNA research. *Ann. Rev. Biochem., 40,* 1045–1078.

Gonano, F. (1967). Specificity of serine transfer ribonucleic acids in the synthesis of hemoglobin. *Biochemistry, 6,* 977–983.

Hoagland, M. B., M. L. Stephenson, J. F. Scott, L. I. Hecht, and P. C. Zamecnik (1958). A soluble ribonucleic acid intermediate in protein synthesis. *J. Biol. Chem., 231,* 241–257.

Kim, S. H., F. L. Suddath, G. J. Quigley, A. McPherson, J. L. Sussman, A. H. J. Wang, N. C. Seeman, and A. Rich (1974). Three-dimensional tertiary structure of yeast phenylalanine transfer RNA. *Science, 185,* 435–440.

Leder, P., M. F. Singer, and R. L. C. Brimacombe (1965). Synthesis of trinucleoside diphosphates with polynucleotide phosphorylase. *Biochemistry, 4,* 1561–1567.

Madison, J. T. (1968). Primary structure of RNA. *Ann. Rev. Biochem., 37,* 131–148.

Marshall, R. E., C. T. Caskey, and M. W. Nirenberg (1967). Fine structure of RNA codewords recognized by bacterial, amphibian, and mammalian transfer RNA. *Science, 155,* 820–826.

Matthaei, J. H., H.-P. Voigt, G. Heller, R. Neth, G. Schöch, H. Kübler, F. Amelunxen, G. Sander, and A. Parmeggiani (1966). Specific interactions of ribosomes in decoding. *Cold Spring Harbor Symp. Quant. Biol., 31,* 25–38.

Nirenberg, M., and P. Leder (1964). RNA codewords and protein synthesis: the effect of trinucleotides upon the binding of sRNA to ribosomes. *Science, 145,* 1399–1407.

——, and J. H. Matthaei (1961). The dependence of cell-free protein synthesis in *E. coli* upon naturally occurring or synthetic polyribonucleotides. *Proc. Natl. Acad. Sci. U.S.A., 47,* 1588–1602.

——, P. Leder, M. Bernfield, R. Brimacombe, J. Trupin, F. Rottman, and C. O'Neal (1965). RNA codewords and protein synthesis: VII. On the general nature of the RNA code. *Proc. Natl. Acad. Sci. U.S.A., 53,* 1161–1168.

Nishimura, S., and G. D. Novelli (1965). Dissociation of amino acid acceptor function of s-RNA from its transfer function. *Proc. Natl. Acad. Sci. U.S.A., 53,* 178–184.

——, D. S. Jones, and H. G. Khorana (1965a). The *in vitro* synthesis of a co-polypeptide containing two amino acids in alternating sequence dependent upon a DNA-like polymer containing two nucleotides in alternating sequence. *J. Mol. Biol., 13,* 302–324.

——, D. S. Jones, E. Ohtsuka, A. Hayatsu, T. M. Jacob, and H. G. Khorana (1965b). Studies on polynucleotides: XLVII. The *in vitro* synthesis of homopeptides as directed by a ribopolynucleotide containing a repeating trinucleotide sequence. New codon sequences for lysine, glutamic acid and arginine. *J. Mol. Biol., 13,* 283–301.

Penswick, J. R., and R. W. Holley (1965). Specific cleavage of the yeast alanine RNA into two large fragments. *Proc. Natl. Acad. Sci. U.S.A., 53,* 543–546.

Preiss, J., M. Dieckmann, and P. Berg (1961). The enzymic synthesis of amino acyl derivatives of ribonucleic acid: IV. The formation of the 3′-hydroxyl terminal trinucleotide sequences of amino acid-acceptor ribonucleic acid. *J. Biol. Chem., 236,* 1748–1757.

Robertus, J. D., J. E. Ladner, J. T. Finch, D. Rhodes, R. S. Brown, B. F. C. Clark, and A. Klug (1974). Structure of yeast phenylalanine tRNA at 3 Å resolution. *Nature (London), 250,* 546–551.

Sarabhai, A. S., A. O. W. Stretton, S. Brenner, and A. Bolle (1964). Co-linearity of the gene with the polypeptide chain. *Nature (London), 201,* 13–17.

Söll, D., J. D. Cherayil, and R. M. Bock (1967). Studies on polynucleotides: LXXV. Specificity of tRNA for codon recognition as studied by the ribosomal binding technique. *J. Mol. Biol., 29,* 97–122.

Sprinzl, M., and F. Cramer (1973). Accepting site for aminoacylation of tRNA[phe] from yeast. *Nature New Biol., 245,* 3–5.

Streisinger, G., Y. Okada, J. Emrich, J. Newton, A. Tsugita, E. Terzaghi, and M. Inouye (1966). Frameshift mutations and the genetic code. *Cold Spring Harbor Symp. Quant. Biol., 31,* 77–84.

Watson, J. D. (1970). *Molecular Biology of the Gene.* W. A. Benjamin, Inc., Menlo Park, Calif.

Weisblum, B., G. von Ehrenstein, and S. Benzer (1965). A demonstration of coding degeneracy for leucine in the synthesis of protein. *Proc. Natl. Acad. Sci. U.S.A., 53,* 329–334.

Woese, C. (1969). The biological significance of the genetic code. *Progr. Mol. Subcell. Biol., 1,* 5–46.

Yanofsky, C., B. C. Carlton, J. R. Guest, D. R. Helinski, and U. Henning (1964). On the colinearity of gene structure and protein structure. *Proc. Natl. Acad. Sci. U.S.A., 51,* 266–272.

Yu, C.-T. (1966). Multiple forms of leucyl sRNA synthetase of *E. coli. Cold Spring Harbor Symp. Quant. Biol., 31,* 571–580.

31

Reprinted from *Cold Spring Harbor Symp. Quant. Biol.*, 31, 39–49 (1966)

Polynucleotide Synthesis and the Genetic Code

H. G. Khorana, H. Büchi, H. Ghosh, N. Gupta, T. M. Jacob, H. Kössel,
R. Morgan, S. A. Narang, E. Ohtsuka, and R. D. Wells

Institute for Enzyme Research, University of Wisconsin, Madison, Wisconsin

The approach that forms the basis of the greater portion of our work on the genetic code involves the preparation of long ribopolynucleotide messengers containing known repeating nucleotide sequences. The first step in the preparation of these defined messengers consists in the preparation of short deoxyribopolynucleotides of known sequences by entirely chemical methods, and the second step involves the use of the synthetic polynucleotides as templates for RNA polymerase or DNA polymerase. Intensive studies with both these enzymes have shown that the route involving use of DNA polymerase for "amplification" of the synthetic short deoxyribopolynucleotides is much the more satisfactory and, consequently, the preparation of high molecular weight DNA-like polymers with repeating nucleotide sequences has been a principal concern of our recent work. In the following, the work on the chemical synthesis of deoxyribopolynucleotides and that on the use of these as templates for DNA polymerase is reviewed first, and this is followed by the preparation and characterization of RNA messengers and the use of these in a cell-free amino acid incorporating system for the purpose of codon assignments. Subsequently, studies on different aspects of the genetic code, all of which were made possible by the availability of polynucleotides of defined sequences, will be presented.

1. Chemical Synthesis of Deoxyribopolynucleotides

In earlier work, syntheses of short deoxyribopolynucleotides with repeating dinucleotide sequences (Ohtsuka, Moon, and Khorana, 1965) and of a few series of deoxyribopolynucleotides containing repeating trinucleotide sequences (Jacob and Khorana, 1965; Narang and Khorana, 1965; Narang, Jacob, and Khorana, 1965) were accomplished. Studies in vitro on the genetic code, made possible by the availability of these deoxypolynucleotides, demonstrated the reliability of this general approach for rigid assignments of codon sequences. It was therefore clearly desirable to try to *prove* the total structure of the genetic code by this method. For a comprehensive attack on the codon assignments and for comprehensive physical studies of DNA-like polymers, the synthesis of a

TABLE 1. Synthetic Deoxyribopolynucleotides with Repeating Sequences

Repeating Trinucleotide Sequences

d(TTC)$_4$ d(AAG)$_4$	d(CCT)$_{3-5}$ d(CCA)$_{3-5}$	d(TAC)$_{4-6}$ d(TAG)$_{4-6}$
d(TTG)$_{4-6}$ d(CAA)$_{4-6}$	d(CGA)$_{3-5}$ d(CGT)$_{3-5}$	d(ATC)$_{3-5}$ d(ATG)$_{3-5}$
		d(GGA)$_{3-5}$ d(GGT)$_{3-5}$

Repeating Tetranucleotide Sequences

d(TTTC)$_3$ d(AAAG)$_{3-4}$	d(TATC)$_3$ d(TAGA)$_2$	d(TTAC)$_4$ d(TAAG)$_2$

variety of deoxyribopolynucleotides containing repeating tri- and tetranucleotide sequences was recently undertaken. Chemical work necessitated the development of satisfactory methods for (a) the polymerization of suitably protected deoxyribotrinucleotides and the isolation and characterization of the resulting oligomers and (b) the synthesis of specific deoxyribopolynucleotides by stepwise condensation of preformed di-, tri-, and tetranucleotides. Satisfactory progress has been made in both these areas, and the syntheses of the deoxyribopolynucleotides listed in Table 1 have been accomplished. The first general point about the list in this table is that the synthetic polynucleotides all comprise sets which are complementary in the antiparallel Watson-Crick base-pairing sense. The requirement of synthesis of segments of both strands was forced upon us by the failure of DNA-polymerase to respond to deoxyribopolynucleotides corresponding to one strand only. The second general point about the polynucleotides synthesized is that the theoretical number of double-stranded DNA-like polymers containing all possible repeating trinucleotide sequences derived *from more than one type of base* is ten. Of these, seven DNA-like polymers can be generated from the list given in Table 1.

Justification for the selection of the repeating tetranucleotide sequences of Table 1 is given later.

2. DNA Polymerase-catalyzed Synthesis of DNA-like Polymers Containing Repeating Tri- and Tetranucleotide Sequences

Table 2 lists the types of reactions so far elicited from DNA-polymerase. Reaction (1) leading to the

TABLE 2. TYPES OF REACTIONS CATALYZED BY DNA-POLYMERASE

(1) $d(TG)_6 + d(AC)_6 + \begin{Bmatrix} dTTP \\ dATP \\ dCTP \\ dGTP \end{Bmatrix} \rightarrow$ Poly d-TG:CA*

(2) $d(TTC)_4 + d(AAG)_3 + \begin{Bmatrix} dTTP \\ dATP \\ dCTP \\ dGTP \end{Bmatrix} \rightarrow$ Poly d-TTC:GAA*

(3) $d(TATC)_3 + d(TAGA)_2 + \begin{Bmatrix} dTTP \\ dATP \\ dCTP \\ dGTP \end{Bmatrix} \rightarrow$ Poly d-TATC:GATA*

* All of the DNA-like polymers are written so that the colon separates the two complementary strands. The complementary sequences in the individual strands are written so that antiparallel base-pairing is evident.

synthesis of DNA-like polymers with repeating dinucleotide sequences has been fully documented previously (Byrd et al., 1965; Wells, Ohtsuka, and Khorana, 1965) and attention is devoted here to reactions (2) and (3). As an example of the reaction of type (2), we may cite the kinetics of the synthesis of poly d-TTC:GAA (Fig. 1). The most direct means

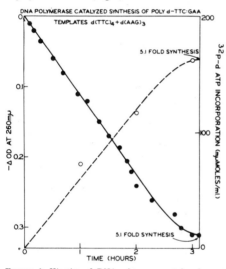

FIGURE 1. Kinetics of DNA-polymerase catalyzed synthesis of poly d-TTC:GAA

for showing the chemical structure of the polymer is by nearest neighbor analysis using one α-P[32]-labeled deoxynucleoside triphosphate at a time, and degradation of the isolated polymer to deoxyribonucleoside 3'-phosphates. The results which are given in Table 3, show *complete* accord with theoretical expectation for a polymer with strictly repeating trinucleotide sequences in the complementary strands. As an example of a DNA-like polymer with a repeating tetranucleotide sequence, we show the kinetics of the incorporation of isotope from α-P[32]-labeled TTP into acid-insoluble product using $d(TATC)_3 + d(TAGA)_2$ as templates (Fig. 2). Here again, the reaction starts off without any lag period and leads to net synthesis. Nearest neighbor frequency analysis as seen in Table 4 gives, within experimental error, excellent agreement between experimentally observed and theoretically expected results for repeating tetranucleotide sequence. It should be further added that with every repeating DNA-like polymer that has been prepared, the individual strands have been copied by RNA polymerase (see below) and the ribopolynucleotide products, when subjected to the nearest neighbor analysis technique, give results in *complete* accord with theory. A double check is thus provided for the conclusion that all of the DNA-like polymers prepared have the strictly repeating patterns of nucleotide sequences originally provided in the chemically synthesized templates.

A property of the DNA-like polymers that is of paramount significance is their ability to reseed

TABLE 3. NEAREST NEIGHBOR FREQUENCY ANALYSIS OF POLY d-TTC:GAA
Templates: $d(TTC)_4 + d(AAG)_3$

| α-P[32]-labeled Triphosphate | Radioactivity in deoxynucleoside 3'-phosphates | | | | | | | |
| | dAp | | dGp | | dCp | | dTp | |
	count/min	%	count/min	%	count/min	%	count/min	%
dATP	12,836	50.0	12,851	50.0	0	0	0	0
dGTP	13,684	100	0	0	0	0	0	0
dCTP	0	0	0	0	0	0	9,623	100
dTTP	0	0	0	0	12,860	50.6	12,565	49.4

TABLE 4. NEAREST NEIGHBOR FREQUENCY ANALYSIS OF POLY d-TATC:GATA
Templates: d(TATC)$_3$ + d(TAGA)$_2$

| α-P^{32}-labeled Triphosphate | Radioactivity in deoxynucleoside 3'-phosphates | | | | | | | |
| | dAp | | dGp | | dCp | | dTp | |
	count/min	%	count/min	%	count/min	%	count/min	%
dATP	0	0	6,425	35.0	0	0	11,938	65.0
dGTP	22,467	100	0	0	0	0	0	0
dCTP	34	0.2	0	0	0	0	16,682	99.8
dTTP	15,920	66.6	0	0	7,985	33.4	0	0

the synthesis by the DNA polymerase of more of the same product. An example of this type of reaction leading to more than 10-fold net synthesis is shown in Figure 3. The same feature has previously been established for repeating dinucleotide polymers (Wells, Ohtsuka, and Khorana, 1965). The importance of this finding can be hardly overstressed: once the specific sequences have been put together by well-defined and unambiguous chemical synthesis, DNA-polymerase ensures their permanent availability, an expected but nevertheless dramatic feature of DNA-structure, namely, its ability to guide its own replication.

The total DNA-like polymers prepared so far are listed in Table 5 and the total characteristics of the DNA polymerase-catalyzed reactions can be summarized as follows: (1) Chemically synthesized segments corresponding to both strands are required for reactions to proceed; (2) Minimal size of the two complementary segments used as primers varies between 8 and 12 nucleotide units; (3) Synthesis is extensive; (4) Products are high molecular weight and are double-stranded with sharp melting transi-

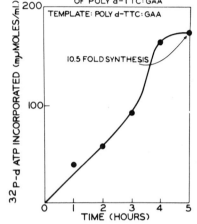

FIGURE 3. Re-utilization of poly d-TTCG:AA as template for DNA-polymerase.

tions; (5) Nearest neighbor analyses invariably show the individual strands to contain appropriate repeating tri- or tetranucleotide sequences; (6) High molecular weight products can be re-utilized as primers for more synthesis.

3. SINGLE-STRANDED RIBOPOLYNUCLEOTIDES WITH REPEATING TRI- AND TETRANUCLEOTIDE SEQUENCES

All of the DNA-like polymers prepared contain a maximum of three different bases in every strand. It is therefore possible, by providing only three appropriate ribonucleoside triphosphates in the reaction mixture, to restrict the action of RNA

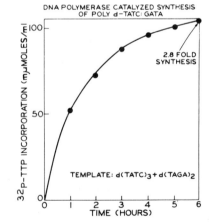

FIGURE 2. Kinetics of DNA-polymerase catalyzed synthesis of poly d-TATC:GATA.

TABLE 5. NEW DNA-LIKE POLYMERS
WITH REPEATING SEQUENCES

Repeating trinucleotide sequences	Repeating tetranucleotide sequences
Poly d-TTC:GAA	
Poly d-TTG:CAA	Poly d-TTAC:GTAA
Poly d-TAC:GTA	Poly d-TATC:GATA
Poly d-ATC:GAT	

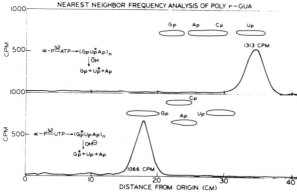

FIGURE 4. Nearest neighbor frequency analysis of poly GUA.

polymerase so as to transcribe only one strand at a time. The same principle was used previously in the preparation of single-stranded ribopoly-nucleotides containing two nucleotides in alternating sequence (Nishimura, Jones, and Khorana, 1965). This expectation has been fully realized in the case of all of the DNA-like polymers listed in Table 5. Thus, for example, poly d-TAC:GTA (Chart 1) when exposed to RNA polymerase in the presence of UTP + ATP + GTP gives poly GUA and when this enzymatic reaction is carried out in the presence of UTP + ATP + CTP, the result is the formation of poly UAC. All of the single-stranded ribopolynucleotides thus prepared have been shown to contain the expected repeating trinucleotide sequences by the technique of nearest neighbor analysis, the results being exemplified with respect to poly GUA in Fig. 4. As seen from these data, the transfer of the label in each experiment is exactly as would be theoretically expected. There is also evident in these results a further confirmation of the antiparallel base-pairing mechanism in the transcription by RNA polymerase.

The above observations apply exactly to the DNA-like polymers containing repeating tetra-nucleotide sequences. Thus, for example, using

CHART 1. PREPARATION OF SINGLE-STRANDED RIBO-POLYNUCLEOTIDES FROM DNA-LIKE POLYMERS CONTAINING REPEATING TRI- AND TETRANUCLEOTIDE SEQUENCES.

poly d-TAC:GTA ─┬─ UTP + ATP + GTP ──→ Poly GUA
                └── RNA Polymerase
                    UTP + ATP + CTP ──→ Poly UAC

poly d-TATC:GATA ─┬─ UTP + ATP + GTP ──→ Poly GAUA
                  └── RNA Polymerase
                      UTP + ATP + CTP ──→ Poly UAUC

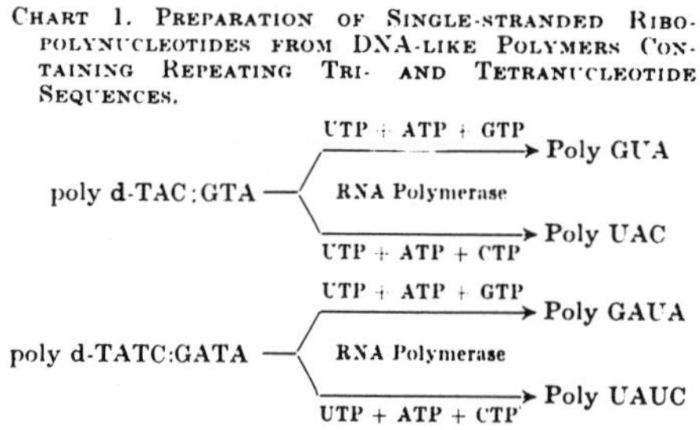

poly d-TATC:GATA as template for RNA polymerase, either of the two ribopolynucleotides containing the repeating tetranucleotide sequences, poly GAUA and poly UAUC, may be prepared (Chart 1). Again, only an example of results of nearest-neighbor frequency analysis is given in Fig. 5 for poly GAUA. A further point of practical interest is that the synthesis of the ribopoly-nucleotides is usually extensive (3–10-fold that of the DNA-template), the amount of synthesis showing an increase with an increase in the amount of the enzyme.

4. CELL-FREE AMINO ACID INCORPORATIONS AND THE CODON ASSIGNMENTS

From a combination of the results obtained with poly AAG (formation of polylysine, polyarginine, and polyglutamate) and with polynucleotides containing repeating dinucleotide sequences (formation of four co-polypeptides) direct proof for the three-letter and nonoverlapping nature of the genetic code has previously been provided (Nishimura et al., 1965; Nishimura, Jones, and Khorana, 1965; Jones, Nishimura, and Khorana, 1966). For assignment of codon sequences to different amino acids, the binding technique (Nirenberg and Leder, 1964) was extensively used by Nirenberg and co-workers (1965) and by Söll et al., (1965). The total structure of the code which emerged from (a) cell-free poly-peptide synthesis of defined sequence, (b) tri-nucleotide-stimulated binding of aminoacyl-sRNA to ribosomes and (c) other genetical and mutagenic evidence, has been presented and discussed recently (Söll et al., 1965; Khorana, 1965). The simple and elegant binding technique which could, in principle, have given information on all of the code, proved, in fact, not to be completely reliable. Often the effects observed were too small and, in certain cases, the results were ambiguous. In other

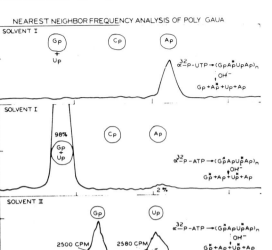

FIGURE 5. Nearest neighbor frequency analysis of poly GAUA.

situations, authentic codons were shown not to give any detectable effect in the binding assay (Söll et al., 1965; Brimacombe et al., 1965). It was therefore concluded that further information on the codon assignments should be sought by the use of repeating polymers. The amino acid incorporations, directed by the messengers that are afforded by the DNA-like polymers of Table 5, are presented in Table 6, and a summary of the total results obtained with all of the repeating polymers appears in Table 7. The codon assignments that are thus derived, when incorporated into the previous data (Khorana, 1965), lead to the codon catalogue of Table 8. The following comments on Tables 7 and 8 are worthy of note in particular. (1) Poly UAG stimulates, in the incorporating system prepared from *E. coli* B, a nonpermissive strain, the incorporation of valine (GUA) and serine (AGU); AGU is clearly a codon for serine, the only reliable previous evidence on this point being that of Streisinger et al. (see elsewhere in this volume). (2) UUA, CUA, and CUU are all now shown to be codons for leucine; the binding experiments in these cases again failed to give any detectable effects. (3) The assignment of AUA as a codon for isoleucine is deduced from the recent incorporation data (see elsewhere in this volume). (4) Poly GAU stimulated the incorporation of methionine and of aspartic acid. These findings confirm the codon assignments AUG and GAU for methionine and aspartic acid respectively. A homopolypeptide corresponding to the third repeating codon (UGA) present in poly GAU should also have been observed. Incorporation of no other amino acid has, however, been detected so far. Since AUG is also a codon for peptide chain initiation (Clark and Marcker, 1966) and may be expected to fix the reading frame (cf. Thach et al., 1966), poly GAU might have stimulated the incorporation of methionine only. However, as mentioned above, under the conditions used in the present experiment, aspartic acid (0.01 M Mg⁺⁺) was also incorporated and the product of this incorporation has been characterized as poly-aspartic acid. The failure to detect so far the incorporation of a third amino acid in the presence of poly GAU is, therefore, probably not due to the phasing effect of AUG. (5) The repeating tetranucleotide polymers, such as poly UAUC, stimulated the incorporation of only four amino acids, in agreement with the theoretical expectation that a repeating tetrapeptide should be formed (see also below in Section 5).

TABLE 6. AMINO ACID INCORPORATIONS USING POLYMERS WITH
REPEATING TRI- AND TETRANUCLEOTIDE SEQUENCES

RNA	Amino Acid	Incorporation (mμmoles)		Codon assignment
		−template	+ template	
Poly r-UUC	Phe	0.010	0.15	UUC
Poly r-UUC	Ser	0.044	0.16	UCU
Poly r-UUC	Leu	0.042	0.40	CUU
Poly r-CAA	Gln	0.028	0.11	CAA
Poly r-CAA	Thr	0.11	2.4	ACA
Poly r-CAA	Asn	(deamidation to		AAC
			aspartic acid)	
Poly r-UUG	Cys	0.25	0.79	UGU
Poly r-UUG	Leu	0.068	2.0	UUG
Poly r-UUG	Val	0.050	3.3	GUU
Poly r-GUA	Val	0.024	9.3	GUA
Poly r-GUA	Ser	0.034	0.15	AGU
Poly r-UAC	Tyr	0.052	0.44	UAC
Poly r-UAC	Thr	0.035	4.7	ACU
Poly r-UAC	Leu	0.043	0.20	CUA
Poly r-AUC	Ileu	0.027	3.6	AUC
Poly r-AUC	Ser	0.037	0.17	UCA
Poly r-AUC	His	0.045	0.35	CAU
Poly r-GAU	Met	0.46	11.0	AUG
Poly r-GAU	Asp	0.07	1.770	GAU
Poly r-UAUC	Tyr	0.010	0.230	UAU
Poly r-UAUC	Ileu	0.012	0.296	AUC
Poly r-UAUC	Ser	0.036	0.445	UCU
Poly r-UAUC	Leu	0.034	0.358	CUA
Poly r-UUAC	Leu	0.099	0.242	UUA, CUU
Poly r-UUAC	Thr	0.091	0.143	ACU
Poly r-UUAC	Tyr	0.092	0.150	UAC

Except perhaps for a few codon assignments of Table 8, most of the code can now be regarded as established with certainty. As described below in Section 5, support from the work with repeating polymers is available for the assignments of UAA and UAG as chain-terminating triplets. Perhaps the only trinucleotide for which an assignment is still completely uncertain is UGA. As mentioned above, poly AUG is potentially capable of providing a definitive answer on this point.

TABLE 7. AMINO ACID INCORPORATIONS STIMULATED BY MESSENGERS CONTAINING REPEATING NUCLEOTIDE SEQUENCES

(System, E. coli B)

Messenger	Amino acids incorporated	Messenger	Amino acids incorporated
Repeating Dinucleotides		Poly GUA	val, ser
Poly UC	ser-leu	Poly UAC	tyr, thr, leu
Poly AG	arg-glu	Poly AUC	ileu, ser, his
Poly UG	val-cys	Poly GAU	met, asp
Poly AC	thr-his	*Repeating Tetranucleotides*	
Repeating Trinucleotides		Poly UAUC	tyr, leu, ileu, ser
Poly UUC	phe, ser, leu	Poly GAUA	none
Poly AAG	lys, glu, arg	Poly UUAC	leu, thr, tyr
Poly UUG	cys, leu, val	Poly GUAA	none
Poly CAA	gln, thr, asn*?		

* The expected incorporation of asparagine has not been realized so far because of the presence of a powerful enzyme which deaminates asparagine in the amino acid incorporating system (cf. Schwartz, 1965).

TABLE 8. CODON ASSIGNMENTS FROM POLYPEPTIDE SYNTHESIS AND/OR STIMULATION OF AMINOACYL-sRNA BINDING TO RIBOSOMES

1st	2nd				3rd
	U	C	A	G	
U	Phe	Ser	Tyr	Cys	U
	Phe	Ser	Tyr	Cys	C
	Leu	Ser	Ochre	?	A
	Leu	Ser	Amber	Trp	G
C	Leu	Pro	His	Arg	U
	Leu	Pro	His	Arg	C
	Leu	Pro	Gln	Arg	A
	Leu	Pro	Gln	Arg	G
A	Ileu	Thr	Asn	Ser	U
	Ileu	Thr	Asn	Ser	C
	Ileu	Thr	Lys	Arg	A
	Met	Thr	Lys	Arg	G
G	Val	Ala	Asp	Gly	U
	Val	Ala	Asp	Gly	C
	Val	Ala	Glu	Gly	A
	Val	Ala	Glu	Gly	G

The assignments not underlined are on the basis of binding experiments only. The assignments singly underlined are on the basis of copolypeptide and/or homopolypeptide syntheses, and gave essentially no binding. The assignments doubly underlined are derived from both polypeptide synthesis and binding experiments.

5. Polypeptide Synthesis Using Polynucleotide-Containing Repeating Tetranucleotide Sequences

(Direction of Reading of Messenger- and Chain-terminating Triplets)

The DNA-like polymers containing repeating tetranucleotide sequences, poly d-TATC:GATA and poly d-TTAC:GTAA, were prepared with two specific questions in mind, in addition to the information on codon assignments which they would be expected to provide. These two questions were: the direction of reading of the messenger, and the sequence of the polypeptide chain-terminating triplets.

As shown in Chart 2, the ribopolynucleotide messengers, poly UAUC and poly UUAC, result from the transcription of one strand each of poly d-TATC:GATA and of poly d-TTAC:GTAA. In the amino acid incorporating system, these messengers should give polypeptidic products containing repeating tetrapeptide sequences (Chart 2). Determination of the sequence of either of these polypeptides should give definitive and new independent evidence regarding the direction of the reading of the messenger. Thus, for the repeating tetrapeptide obtained using the messenger poly UAUC (Chart 2), if the direction of reading is 5′ → 3′, then the carboxyl of tyrosine will be found linked to the α-amino group of leucine. If the direction is 3′ → 5′, then the carboxyl group of tyrosine will be found linked to the α-amino group of isoleucine. The experiment, thus, is to cleave the polypeptidic product selectively at the carboxyl group of tyrosine by using chymotrypsin, and to look for the free α-amino-group-carrying amino acid. It is expected that the answer to this question will be forthcoming shortly.

The transcription of the strands poly d-TATC and poly d-TTAC, in the above two DNA-like polymers, gives ribopolymers, poly UAGA and poly GUAA. These two polymers contain in every

fourth position the triplets UAG and UAA, respectively. These two triplets have recently been concluded to cause polypeptide chain-termination in "nonpermissive" strains of *E. coli*, such as *E. coli* B. All attempts to demonstrate polypeptide synthesis (assays used acid-insolubility and paper chromatography) using poly UAGA and poly GUAA as messengers in the *E. coli*. B amino acid incorporating system have given negative results, and this failure to demonstrate any polypeptide synthesis is restricted only to these polymers, in contrast with the positive results with all of the other repeating polymers (Tables 6 and 7). We feel confident that our failure to obtain any polypeptide synthesis using the above two messengers *is* due to these polymers having nonsense triplets in them and, therefore, in this indirect way, support is given to the assignment of the sequences UAG and UAA as the nonsense triplets in *E. coli* B.

6. Stimulation of the Incorporation of Formate and Methionine into Polypeptide by Poly UG

We observed last year (Jones, Nishimura, and Khorana, 1966; Khorana, 1965) that among poly UC, poly AG, poly AC, and poly UG the last-mentioned polymer was the most efficient in directing polypeptide synthesis. With the accumulating evidence of the involvement of formylmethionine in polypeptide chain-initiation in *E. coli*, we re-examined the above messengers, containing repeating dinucleotide sequences, for the possible utilization of formylmethionine, and the results summarized below do in fact show that poly UG is unique among the repeating dinucleotide polymers in stimulating the incorporation of formate and methionine, under appropriate conditions, into polypeptidic material.

Tables 9 and 10 show the requirements for the incorporation of H³-formate and C¹⁴-methionine into polypeptidic material. The experiment of

CHART 2. CELL-FREE POLYPEPTIDE SYNTHESIS USING POLYMERS CONTAINING REPEATING TETRANUCLEOTIDE SEQUENCES

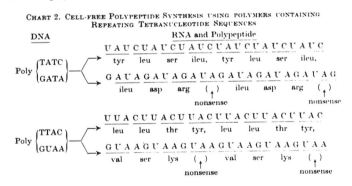

TABLE 9. INCORPORATION OF H³-FORMATE IN
COPOLYPEPTIDES

	H³-Formate incorporated ($\mu\mu$moles/ml)			
	poly UG	poly AC	poly UC	poly AG
Complete	202	19	17	19
—Methionine		19	21	20
—Template	22	18	19	22

Table 9, which was carried out using the 0.01 M Mg⁺⁺ concentration used in earlier experiments (Nishimura, Jones, and Khorana, 1965; Jones, Nishimura, and Khorana, 1966) shows that the incorporation of H³-formate is specifically stimulated by poly UG, and not by any of the other three polynucleotides. Table 10 describes further the requirements for the incorporation of formyl-methionine under the direction of poly UG. Experiment A was done using unfractionated *E. coli* sRNA and it is clear (1) that poly UG stimulates the incorporation of both formate and methionine; (2) that the incorporation of formate requires the presence of methionine; and (3) the incorporation of formate or methionine requires the presence of valine. The formation of valyl-cysteine copolypeptide is thus a requirement for the incorporation of formylmethionine. The results of Experiment B of Table 10 using the appropriately charged met-sRNA fractions I and II (obtained by countercurrent distribution of total sRNA) show that only the charged methionyl-sRNA of peak I, which can be formylated (Marcker, 1965), serves as the donor of formylmethionine, and that C¹⁴-methionyl-sRNA-II is inactive. Clark and Marcker (1966)

have already shown that, in the presence of randomly linked polynucleotides (poly U, G, and poly A, U, G), the incorporation of methionine is stimulated largely into the terminal position only by the methionyl-sRNA fraction that can be formylated.

In further experiments the effect of Mg⁺⁺ concentration, and of formyl-methionine on the synthesis of copolypeptides under the direction of poly UG, poly AC, poly UC, and poly AG was investigated. The results given in Table 11 show that at 0.011 M concentration polypeptide syntheses proceed well with all of the polynucleotides, but at reduced Mg⁺⁺ ion concentrations polypeptide syntheses due to poly UC, poly AC, and poly AG are strikingly inhibited. In contrast, poly UG-stimulated incorporation of valine proceeds rather well even at 0.004 Mg⁺⁺ concentration, and the presence of formylmethionine is clearly required to support this synthesis. It is likely that the appreciable incorporation of C¹⁴-valine, even without added formylmethionine, is due to the endogeneous presence of the latter compound. Nakamoto and Kolakofsky (1966) have similarly described the dependence of polypeptide synthesis as directed by randomly linked poly A, G, U on formylmethionine at low Mg⁺⁺ ion concentration.

It is the objective of current work to prove by peptide sequential analysis that poly UG directed incorporation of methionine is in the amino terminal position, and to learn whether the peptide sequence is then H₂N-met-cys-val-cys-val----. If, as seems likely, this is the case, then we believe that the present experiments provide direct proof for the first time from cell-free polypeptide synthesis

TABLE 10. POLY UG DIRECTED INCORPORATION OF FORMATE AND METHIONINE

	A. Total sRNA		B. Fractionated Met-sRNA	
Conditions	H³-Formate ($\mu\mu$moles/ml)	C¹⁴-Met ($\mu\mu$moles/ml)	H³-Formyl-C¹⁴-Met-sRNA-I count/min	C¹⁴-Met-sRNA-II count/min
Complete	501	162	16,700	960
—Template	74	44	1,560	540
—Methionine	100	—		
—Valine	79	56	1,360	980

TABLE 11. EFFECT OF Mg⁺⁺ CONCENTRATION AND OF FORMYLMETHIONINE ON
COPOLYPEPTIDE SYNTHESIS

Mg⁺⁺ (M)	Poly UG (C¹⁴-val, $\mu\mu$moles)			Poly AC (C¹⁴-his, $\mu\mu$moles)		Poly UG (C¹⁴-ser, $\mu\mu$moles)		Poly AG (C¹⁴-arg, $\mu\mu$moles)	
	—UG	+UG	+UG +For-Met	—AC	+AC	—UC	+UC	—AG	+AG
0.004	30	470	1850	10	10	10	11	60	68
0.006	60	5160	6560	16	290	26	168	124	590
0.011	110	3700	4220	25	3820	58	3320	186	666

that GUG is a chain-initiating triplet as well as a codon for valine. That AUG also seems to be a chain initiation triplet has been mentioned above, in Section 4.

7. MISSENSE TO SENSE SUPPRESSION: sRNA INVOLVEMENT

A further example of the use of ribopolynucleotides, containing repeating dinucleotide sequences, as messengers has been in the study of the mechanism of genetic suppression (missense to sense) in some of the suppressed strains of *E. coli* which normally make only a defective (CRM) protein A of tryptophan synthetase. From the work of Yanofsky and co-workers it is known that in one case (mutant A-78) one glycine residue in A protein is replaced by cysteine. A suppressed mutant (A-78-Su-78) restores, to a small extent, the original glycine in place of cysteine. It has now been shown (Gupta and Khorana, 1966) that the amino acid incorporating system prepared from *E. coli* B, when supplemented with sRNA from A-78-Su-78, brings about the synthesis of valine-glycine copolypeptide under the direction of poly UG (Fig. 6). (The latter polymer normally directs the synthesis of valine-cysteine copolypeptide.) Valine-glycine copolypeptide formed *specifically* in the presence of sRNA from A-78-Su-78 strain has been thoroughly charcter-

ized. Similarly, Carbon, Berg, and Yanofsky (this volume) have shown that another missense suppressor of glycine to arginine mutation in protein A also acts at the level of sRNA. As reported by Dr. Gupta elsewhere in this volume, we are currently investigating the nature of the structural change involved in the elaboration of the suppressed sRNA in A-78-Su-78 strain. Previously, sRNA has been shown to be responsible for suppression of an *amber* codon in the genetic material specifying a bacteriophage coat protein (Capecchi and Gussin, 1965; Engelhardt et al., 1965).

8. DIRECT TRANSLATION OF SINGLE-STRANDED DEOXYRIBOPOLYNUCLEOTIDES IN THE PRESENCE OF AMINOGLYCOSIDE ANTIBIOTICS

The double-stranded poly d-TG:CA and poly d-A:T separate very well upon equilibrium centrifugation in alkaline cesium chloride density gradient (see, for example, Doerfler and Hogness, 1965). We have recently prepared adequate amounts of the four single-stranded deoxyribopolynucleotides from the above polymers and, stimulated by the discovery of McCarthy, Holland and Buck (see this volume) that denatured DNA can directly serve as messenger in the presence of certain antibiotics, have tested the messenger activity of these polymers. An example of the experimental results is given in Fig. 7. Thus, neomycin B is very effective in eliciting messenger response and, in contrast with the positive results of McCarthy and Holland with streptomycin + denatured DNAs, streptomycin is ineffective for poly d-AC; so are many of the other aminoglycoside antibiotics tested (humatin, hygromycin, kanamycin, and viocin), though some effect was elicited by gentamycin.

The results of the tests of specificity in translation are summarized in Table 12. For comparison, the normal incorporations (in the absence of neomycin B) for the corresponding ribopolynucleotides are also shown. The main points that emerge are: (1) Neomycin B so alters the ribosomes, or at least the messenger RNA-ribosomes-sRNA complex, that often deoxyribopolynucleotides can be read efficiently; the failure of poly d-A to give any response is surprising. (2) Whereas the combination of a ribopolynucleotide messenger and neomycin B leads to a variety of rather unspecific mistakes (Davies, Jones, and Khorana, 1966), the mistakes in the direct translation of deoxyribopolynucleotides are very restricted.

9. CONCLUDING REMARKS

The combined use of the methods of organic chemistry and of enzymology has made available DNA-like polymers containing most of the possible repeating di- and trinucleotide sequences, as well

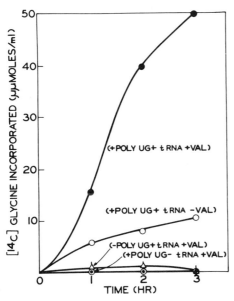

FIGURE 6. The incorporation of C^{14}-glycine into valine-glycine copolypeptide as stimulated by poly UG in the presence of tRNA from A-78-Su-78 strain of *E. coli*.

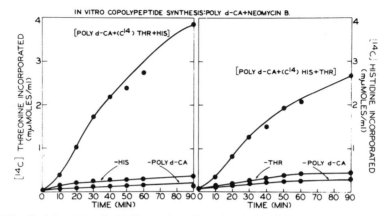

FIGURE 7. The stimulation of threonine-histidine copolypeptide synthesis in the cell-free amino acid incorporating system in the presence of poly d-AC + neomycin B.

TABLE 12. TRANSLATION OF DEOXYPOLYNUCLEOTIDES IN PRESENCE OF NEOMYCIN B

Ribopoly-nucleotide	Poly-peptide	Deoxypolynucleotide + Neomycin B	Amino acids incorporated	Base-misreading
Poly AC	(thr-his)$_n$	d-Poly AC	thr-his copolypeptide	None detected
Poly UG	(val-cys)$_n$	d-Poly TG	val-cys	
			ala-cys (10x)*	GTG → GCG
			glu-cys (2x)	GTG → GAG
			gly-cys (2x)	GTG → GGG
Poly U	(phe)$_n$	d-Poly T	(phe)$_n$	TTT
			leucine (2x)	TTT → TTA
				TTT → TTG
Poly A	(lys)$_n$	d-Poly A	no reading	

* The numbers in brackets show the stimulation above background.

as a few selected polymers containing repeating tetranucleotide sequences. Transcription of all of the DNA-polymers can be restricted at will to either of the two strands, and single-stranded ribopoly-nucleotide messengers are thus afforded. The use of the latter in a cell-free amino acid incorporating system has given information such that most of the genetic code can now be regarded as established with complete certainty. Continued work using the same approach is expected to clarify the few remaining uncertainties in the codon assignments in the very near future.

The availability of a variety of DNA-like polymers containing limited but specified information has opened up opportunities for further studies of different aspects of protein biosynthesis and of genetic suppression. In addition, the DNA-like polymers are expected to facilitate further studies of the chemistry and enzymology of DNA and, in particular, of the mechanism of the transcription process.

ACKNOWLEDGMENTS

This work has been supported by grants from the National Science Foundation, (Grant No. GB-3342), the National Cancer Institute of the National Institutes of Health, U.S. Public Health Service (Grant No. CA-05178) and the Life Insurance Medical Research Fund (Grant No. G-62-54).

H. Kössel wishes to thank the Deutscher Akademischer Austauschdienst, Germany for a fellowship.

REFERENCES

BRIMACOMBE, R., J. TRUPIN, M. NIRENBERG, P. LEDER, M. BERNFIELD, and T. JAOUNI. 1965. RNA codewords and protein synthesis, VIII. Nucleotide sequences of synonym codons for arginine, valine, cysteine, and alanine. Proc. Natl. Acad. Sci. 54: 954–960.
BYRD, C., E. OHTSUKA, M. W. MOON, and H. G. KHORANA. 1965. Synthetic deoxyribo-oligonucleotides as templates for the DNA polymerase of Escherichia coli. New DNA-like polymers containing repeating nucleotide sequences. Proc. Natl. Acad. Sci. 53: 79–86.
CAPECCHI, M. R. and G. N. GUSSIN. 1965. Suppression

in vitro: Identification of a serine-sRNA as a"Nonsense" suppressor. Science *149*: 417–422.

CLARK, B. and K. MARCKER. 1966. The role of N-formyl-methionyl-sRNA in protein biosynthesis. J. Mol. Biol., in press.

DAVIES, J., D. S. JONES, and H. G. KHORANA. 1966. A further study of misreading of codons induced by streptomycin and neomycin using ribopolynucleotides containing two nucleotides in alternating sequence as templates. J. Mol. Biol., in press.

DOERFLER, W. and D. S. HOGNESS. 1965. Separation of the strands of poly d-TG:AC in alkaline CsCl. J. Mol. Biol. *14*: 237–240.

ENGELHARDT, D. L., ROBERT E. WEBSTER, ROBERT C. WILHELM, and NORTON D. ZINDER. 1965. *In vitro* studies on the mechanism of suppression of a nonsense mutation. Proc. Natl. Acad. Sci. *54*: 1791–1797.

GUPTA, N. K., and H. G. KHORANA. 1966. Missense suppression of the tryptophan synthetase A-protein mutant A78. Proc. Natl. Acad. Sci. *56*: 772.

JACOB, T. M., and H. G. KHORANA. 1965. Studies on Polynucleotides. XLIV. The synthesis of dodecanucleotide containing the repeating trinucleotide sequence, thymidylyl - (3' → 5') - thymidylyl - (3' → 5') - deoxycytidine. J. Amer. Chem. Soc. *87*: 2971–2981.

JONES, D. S., S. NISHIMURA, and H. G. KHORANA. 1966. Studies on polynucleotides. LVI. Further syntheses, *in vitro*, of copolypeptides containing two amino acids in alternating sequence dependent upon DNA-like polymers containing two nucleotides in alternating sequence. J. Mol. Biol. *16*: 454–472.

KHORANA, H. G. 1965. Polynucleotide synthesis and the genetic code. Fed. Proc. *24*: 1473–1487.

MARCKER, K. 1965. The formation of N-formylmethionyl-sRNA. J. Mol. Biol. *14*: 63–70.

NAKAMOTO, T. and D. KOLAKOFSKY. 1966. A possible mechanism for initiation of protein synthesis. Proc. Natl. Acad. Sci. *55*: 606–613.

NARANG, S. A. and H. G. KHORANA. 1965. Studies on polynucleotides. XLV. The synthesis of dodecanucleotide containing the repeating trinucleotide sequence, thymidylyl-(3'→5')-thymidylyl-(3'→5')-deoxyinosine. J. Amer. Chem. Soc. *87*: 2981–2988.

NARANG, S. A., T. M. JACOB and H. G. KHORANA. 1965. Studies on polynucleotides. XLVI. The synthesis of the hexanucleotides containing the repeating trinucleotide sequences, deoxycytidylyl-(3' → 5')-deoxyadenylyl-(3' → 5')-deoxyadenosine and deoxyguanylyl-(3' → 5')-deoxyadenylyl-(3' → 5')-deoxyadenosine. J. Amer. Chem. Soc. *87*: 2988–2995.

NIRENBERG, M. W. and P. LEDER. 1964. The effect of trinucleotides upon the binding of sRNA to ribosomes. Science *145*: 1399–1407.

NIRENBERG, M. W., P. LEDER, M. BERNFIELD, R. BRIMACOMBE, J. TRUPIN, F. ROTTMAN, and C. O'NEAL. 1965. RNA codewords and protein synthesis, VII. On the general nature of the RNA code. Proc. Natl. Acad. Sci. *53*: 1161–1168.

NISHIMURA, S., D. S. JONES, and H. G. KHORANA. 1965. The *in vitro* synthesis of a co-polypeptide containing two amino acids in alternating sequence dependent upon a DNA-like polymer containing two nucleotides in alternating sequence. J. Mol. Biol. *13*: 302–324.

NISHIMURA, S., D. S. JONES, E. OHTSUKA, H. HAYATSU, T. M. JACOB, and H. G. KHORANA. 1965. Studies on polynucleotides. XLVII. The *in vitro* synthesis of homopeptides as directed by a ribopolynucleotide containing a repeating trinucleotide sequence. New codon sequences for lysine, glutamic acid and arginine. J. Mol. Biol. *13*: 283–301.

OHTSUKA, E., M. W. MOON, and H. G. KHORANA. 1965. Studies on deoxyribopolynucleotides containing repeating dinucleotide sequences. J. Amer. Chem. Soc. *87*: 2956–2970.

SCHWARTZ, J. H. 1965. An effect of streptomycin on the biosynthesis of the coat protein of coliphage F2 by extracts of *E. coli*. Proc. Natl. Acad. Sci. *53*: 1133–1140.

SÖLL, D., E. OHTSUKA, D. S. JONES, R. LOHRMANN, H. HAYATSU, S. NISHIMURA, and H. G. KHORANA. 1965. Studies on polynucleotides. XLIX. Stimulation of the binding of aminoacyl-sRNA's to ribosomes by ribotrinucleotides and a survey of codon assignments for twenty amino acids. Proc. Natl. Acad. Sci. *54*: 1378–1385.

THACH, R. E., K. F. DEWEY, J. C. BROWN, and P. DOTY. 1966. Formylmethionine codon AUG as an initiator of polypeptide synthesis. Science, in press.

WELLS, R. D., E. OHTSUKA and H. G. KHORANA. 1965. Studies on polynucleotides. L. Synthetic deoxyribopolynucleotides as templates for the RNA polymerase of *Escherichia coli*. A new double-stranded DNA-like polymer containing repeating dinucleotide sequences. J. Mol. Biol. *14*: 221–240.

32

Reprinted from *Cold Spring Harbor Symp. Quant. Biol.*, 31, 11–24 (1966)

The RNA Code and Protein Synthesis

M. Nirenberg, T. Caskey, R. Marshall, R. Brimacombe, D. Kellogg, B. Doctor†,
D. Hatfield, J. Levin, F. Rottman, S. Pestka, M. Wilcox, and F. Anderson

Laboratory of Biochemical Genetics, National Heart Institute, National Institutes of Health, Bethesda, Maryland and
† *Division of Biochemistry, Walter Reed Army Institute of Research, Walter Reed Army Medical Center, Washington, D.C.*

Many properties of the RNA code which were discussed at the 1963 Cold Spring Harbor meeting were based on information obtained with randomly ordered synthetic polynucleotides. Most questions concerning the code which were raised at that time related to its fine structure, that is, the order of the bases within RNA codons. After the 1963 meetings a relatively simple means of determining nucleotide sequences of RNA codons was devised which depends upon the ability of trinucleotides of known sequence to stimulate AA-sRNA binding to ribosomes (Nirenberg and Leder, 1964). In this paper, information obtained since 1963 relating to the following topics will be discussed:

(1) The fine structure of the RNA code
(2) Factors affecting the formation of codon-ribosome-AA-sRNA complexes
(3) Patterns of synonym codons for amino acids and purified sRNA fractions
(4) Mechanism of codon recognition
(5) Universality
(6) Unusual aspects of codon recognition as potential indicators of special codon functions
(7) Modification of codon recognition due to phage-infection.

FINE STRUCTURE OF THE RNA CODE

FORMATION OF CODON-RIBOSOME-AA-sRNA COMPLEXES

The assay for base sequences of RNA codons depends, first upon the ability of trinucleotides to serve as templates for AA-sRNA binding to

ABBREVIATIONS

The following abbreviations are used: Ala-, alanine-; Arg-, arginine-; Asn-, asparagine-; Asp-, aspartic acid-; Cys-, cysteine-; Glu-, glutamic acid-, Gln-, glutamine-, Gly-, glycine, His-, histidine-, Ile-, isoleucine-, Leu-, leucine-, Lys-, lysine, Met-, methionine-, Phe-, phenyl-alanine-, Pro-, proline-, Ser-, serine-, Thr-, threonine-, Trp-, tryptophan-, Tyr-, tyrosine-, and Val-, valine-sRNA; sRNA, transfer RNA; AA-sRNA, aminoacyl-sRNA; sRNA^{Phe}, deacylated phenylalanine-acceptor sRNA; Ala-sRNA^{Yeast}, acylated alanine- acceptor sRNA from yeast. U, uridine; C, cytidine; A, adenosine; G, guanosine; I, inosine; rT, ribothymidine; ψ, pseudouridine; DiHU, dihydro-uridine; MAK, methylated albumin kieselguhr; F-Met, N-formyl-methionine. For brevity, trinucleoside diphosphates are referred to as trinucleotides. Internal phosphates of trinucleotides are (3′,5′)-phosphodiester linkages.

TABLE 1. CHARACTERISTICS OF AA-sRNA BINDING TO RIBOSOMES

Modifications	C^{14}-Phe-sRNA bound to ribosomes ($\mu\mu$mole)
Complete	5.99
– Poly U	0.12
– Ribosomes	0.00
– Mg^{++}	0.09
+ deacylated sRNA at 50 min	
0.50 A^{260} units	5.69
2.50 A^{260} units	5.39
+ deacylated sRNA at zero time	
0.50 A^{260} units	4.49
2.50 A^{260} units	2.08

Complete reactions in a volume of 0.05 ml contained the following: 0.1 M Tris acetate (pH 7.2) (in other experiments described in this paper 0.05 M Tris acetate, pH 7.2 was used), 0.02 M magnesium acetate, 0.05 M potassium chloride (standard buffer); 2.0 A^{260} units of *E. coli* W3100 70 S ribosomes (washed by centrifugation 3 times); 15 mμmoles of uridylic acid residues of poly U; and 20.6 $\mu\mu$moles C^{14}-Phe-sRNA (0.71 A^{260} units). All components were added to tubes at 0°C. C^{14}-Phe-sRNA was added last to initiate binding reactions.

Incubation was at 0°C for 60 min (in all other experiments described in this paper, reactions were incubated at 24° for 15 min). Deacylated sRNA was added either at zero time or after 50 min of incubation, as indicated. After incubation, tubes were placed in ice and each reaction was immediately diluted with 3 ml of standard buffer at 0° to 3°C. A cellulose nitrate filter (HA type, Millipore Filter Corp., 25 mm diameter, 0.45 μ pore size) in a stainless steel holder was washed with gentle suction with 5 ml of the cold standard buffer. The diluted reaction mixture was immediately poured on the filter under suction and washed to remove unbound C^{14}-Phe-sRNA with three 3-ml and one 15-ml portions of standard buffer at 3°. Ribosomes and bound sRNA remained on the filter (Nirenberg and Leder, 1964). The filters were then dried, placed in vials containing 10 ml of a scintillation fluid (containing 4 gm 2,5-diphenyloxazole and 0.05 gm 1,4-bis-2-(5-phenyloxazolyl)-benzene per liter of toluene) and counted in a scintillation spectrometer.

ribosomes prior to peptide bond formation, and second, upon the observation that codon-ribosome-AA-sRNA complexes are retained by cellulose nitrate filters (Nirenberg and Leder, 1964). Results shown in Table 1 illustrate characteristics of codon-ribosome-sRNA complex formation. Ribosomes, Mg^{++}, and poly U are required for the binding of C^{14}-Phe-sRNA to ribosomes. The addition of deacylated sRNA to reactions at zero time greatly reduces the binding of C^{14}-Phe-sRNA (Table 1), since poly U specifically stimulates the binding of both deacylated sRNA^{Phe} and C^{14}-Phe-sRNA to ribosomes. Ribosomal bound

C[14]-Phe-sRNA is not readily exchangeable with unbound Phe-sRNA or deacylated sRNA[Phe] except at low Mg[++] concentrations (Levin and Nirenberg, in prep.). Later in this volume Dr. Dolph Hatfield discusses the characteristics of exchange of ribosomal bound with unbound AA-sRNA when trinucleotides are present.

Two enzymatic methods were devised for oligonucleotide synthesis, since most trinucleotide sequences had not been isolated or synthesized earlier. One procedure employed polynucleotide phosphorylase to catalyze the synthesis of oligonucleotides from dinucleoside monophosphate primers and nucleoside diphosphates (Leder, Singer, and Brimacombe, 1965; Thach and Doty, 1965); the other approach (Bernfield, 1966) was based upon the demonstration (Heppel, Whitfeld, and Markham, 1955) that pancreatic RNase catalyzes the synthesis of oligonucleotides from uridine- or cytidine-2′,3′-cyclic phosphate and acceptor moieties. Elegant chemical procedures for oligonucleotide synthesis devised by Khorana and his associates (see Khorana et al., this volume) also are available.

TEMPLATE ACTIVITY OF OLIGONUCLEOTIDES WITH TERMINAL AND INTERNAL SUBSTITUTIONS

The trinucleotides, UpUpU and ApApA, but not the corresponding dinucleotides, stimulate markedly the binding of C[14]-Phe- and C[14]-Lys-sRNA, respectively. Such data directly demonstrate a triplet code and also show that codons contain three *sequential* bases. The template activity of triplets with 5′-terminal phosphate, pUpUpU, equals that of the corresponding tetra- and penta-nucleotides; whereas, oligo U preparations with 2′,3′-terminal phosphate are much less active. Hexa-A preparations, with and without 3′-terminal phosphate, are considerably more active as templates than the corresponding pentamers; thus, one molecule of hexa-A may be recognized by two Lys-sRNA molecules bound to adjacent ribosomal sites (Rottman and Nirenberg, 1966).

An extensively purified doublet with 5′-terminal phosphate, pUpC, serves as a template for Ser-sRNA (but not for Leu- or Ile-sRNA), whereas a doublet without terminal phosphate, UpC, is inactive (see Figs. 1a and b). However, the template activity of pUpC is considerably lower than that of the triplet, UpCpU. The relation between Mg[++] concentration and template activity is shown in Fig. 1b. pUpC and UpCpU stimulate Ser-sRNA binding in reactions containing 0.02–0.08 M Mg[++]. These results demonstrate that a doublet with 5′-terminal phosphate can serve as a specific, although relatively weak, template for AA-sRNA. It is particularly intriguing to relate recognition of a doublet to the

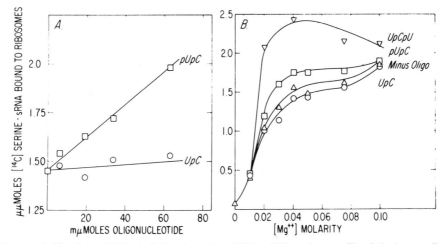

FIGURE 1a, b. The effects of UpC and pUpC on the binding of C[14]-Ser-sRNA to ribosomes. The relation between oligonucleotide concentration and C[14]-Ser-sRNA binding to ribosomes at 0.03 M Mg[++] is shown in Fig. 1a. It should be noted that the ordinate begins at 1.25 µµmoles of C[14]-Ser-sRNA. The relation between Mg[++] concentration and C[14]-Ser-sRNA binding to ribosomes is shown in Fig. 1b. As indicated, 50 mµmoles of UpC or pUpC, or 15 mµmoles of UpCpU, were added to each reaction. Each point in parts a and b represents a 50 µl reaction containing the components described in the legend to Table 1 except for the following: 14.3 µµmoles C[14]-Ser-sRNA (0.42 A[260] units); 1.1 A[260] units of ribosomes. Incubations were for 15 min at 24°C. (Data from Rottman and Nirenberg, 1966.)

TABLE 2. RELATIVE TEMPLATE ACTIVITY OF SUBSTITUTED OLIGONUCLEOTIDES

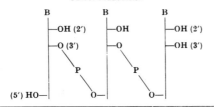

Oligonucleotide	Relative template activity
p-5′-UpUpU	510
UpUpU	100
CH₃O-pUpUpU	74
UpUpU-3′3-p	48
UpUpUp-OCH₃	18
UpUpU-2′,3′-cyclic p	17
(2′-5′)-UpUpU	0
Oligodeoxy T	0
p-5′-ApApA	181
ApApA	100
ApApA-3′-p	57
ApApA-2′-p	15
(2′-5′)-ApApA	0
Oligodeoxy A	0

Relative template activities are approximations obtained by comparing the amount of AA-sRNA bound to ribosomes in the presence of limiting concentrations of oligonucleotides (0.50 or 0.12 mμmoles of oligonucleotides containing U or A, respectively) compared to either UpUpU, for C¹⁴-Phe-sRNA; or ApApA, for C¹⁴-Lys-sRNA (each assumed to be 100%). Data are from Rottman and Nirenberg (1966) except results with oligodeoxynucleotides which are from Nirenberg and Leder (1964).

possibility that only two out of three bases in a triplet may be recognized occasionally during protein synthesis, and also to the possibility that a triplet code evolved from a more primitive doublet code.

Further studies on template activities of oligonucleotides with terminal and internal modifications are summarized in Table 2. At limiting oligonucleotide concentrations, the relative template activities of oligo U preparations are as follows: p-5′-UpUpU > UpUpU > CH₃O-p-5′-UpUpU > UpUpU-3′-p > UpUpU-3′-p-OCH₃ > UpUpU-2′,3′-cyclic phosphate. Trimers with (2′-5′) phosphodiester linkages, (2′-5′)-UpUpU and (2′-5′)-ApApA, do not serve as templates for Phe- or Lys-, sRNA respectively. The relative template efficiencies of oligo A preparations are as follows: p-5′-ApApA > ApApA > ApApA-3′-p > ApApA-2′-p.

These studies led to the proposal that RNA and DNA contain three classes of codons, differing in structure; 5′-terminal, 3′-terminal, and internal codons (Nirenberg and Leder, 1964). Certainly

the first base of a 5′-terminal codon and the third base of a 3′-terminal codon may be recognized with less fidelity than an internal codon, for in the absence of a nucleotide neighbor a terminal base may have a greater freedom of movement on the ribosome. Substitution of 5′- or 3′-terminal hydroxyl groups may impose restrictions upon the orientation of terminal bases during codon recognition. 5′-Terminal and perhaps also 3′-terminal codons possibly serve, together with neighboring codons, as operator regions.

Since many enzymes have been described which catalyze the transfer of nucleotides, amino acids, phosphate, and other molecules to or from terminal ribose or deoxyribose of nucleic acids, modification of sugar hydroxyl groups was proposed as a possible mechanism for regulating the reading of RNA or DNA (Nirenberg and Leder, 1964).

NUCLEOTIDE SEQUENCES OF RNA CODONS

A summary of nucleotide sequences of RNA codons by *E. coli* AA-sRNA is shown in Table 3

TABLE 3. NUCLEOTIDE SEQUENCES OF RNA CODONS

1st Base	2nd Base U	C	A	G	3rd Base
U	PHE*	SER*	TYR*	CYS*	U
	PHE*	SER*	TYR*	CYS	C
	leu*?	SER	TERM?	cys?	A
	leu*, f-met	SER*	TERM?	TRP*	G
C	leu*	pro*	HIS*	ARG*	U
	leu*	pro*	HIS*	ARG*	C
	leu	PRO*	GLN*	ARG*	A
	LEU	PRO	gln*	arg	G
A	ILE*	THR*	ASN*	SER	U
	ILE*	THR*	ASN*	SER*	C
	ile*	THR*	LYS*	arg*	A
	MET*, F-MET	THR	lys	arg	G
G	VAL*	ALA*	ASP*	GLY*	U
	VAL	ALA*	ASP*	GLY*	C
	VAL*	ALA*	GLU*	GLY*	A
	VAL	ALA	glu	GLY	G

Nucleotide sequences of RNA codons were determined by stimulating binding of *E. coli* AA-sRNA to *E. coli* ribosomes with trinucleotide templates. Amino acids shown in capitals represent trinucleotides with relatively high template activities compared to other trinucleotide codons corresponding to the same amino acid. Asterisks (*) represent base compositions of codons which were determined previously by directing protein synthesis in *E. coli* extracts with synthetic randomly-ordered polynucleotides (Speyer et al., 1963; Nirenberg et al., 1963). F-Met, represents N-formyl-Met-sRNA which may recognize initiator codons. TERM represents possible terminator codons. Question marks (?) indicate uncertain codon function. Data are from Nirenberg et al., 1965; Brimacombe et al., 1965; also see articles by Khorana et al., Söll et al., and Matthaei et al., in this volume.

TABLE 4. PATTERNS OF DEGENERATE CODONS FOR AMINO ACIDS

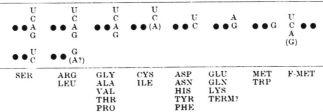

SER	ARG	GLY	CYS	ASP	GLU	MET	F-MET
	LEU	ALA	ILE	ASN	GLN	TRP	
		VAL		HIS	LYS		
		THR		TYR	TERM?		
		PRO		PHE			

Solid circles represent the first and second bases of trinucleotides; U, C, A, and G indicate bases which may occupy the remaining position of degenerate codons. In the case of F-Met (N-formylmethionine), circles represent the second and third bases. Parentheses indicate codons with relatively low template activities.

and patterns of degeneracy in Table 4. Almost every trinucleotide was assayed for template specificity with 20 AA-sRNA preparations (unfractionated sRNA acylated with one labeled and 19 unlabeled amino acids). It is important to test trinucleotide template specificity with 20 AA-sRNA preparations, since relative responses of AA-sRNA are then quite apparent. In surveying trinucleotide specificity, unfractionated AA-sRNA should be used initially because altering ratios of sRNA species often influences the fidelity of codon recognition.

Almost all triplets correspond to amino acids; furthermore, patterns of codon degeneracy are logical. Six degenerate codons correspond to serine, five or six to arginine and also to leucine, and from one to four to each of the remaining amino acids. Alternate bases often occupy the third positions of triplets comprising degenerate codon sets. In all cases triplet pairs with 3'-terminal pyrimidines (XYU and XYC, where X and Y represent the first and second bases, respectively, in the triplet) correspond to the same amino acid; often XYA and XYG correspond to the same amino acid; sometimes XYG alone corresponds to an amino acid. For eight amino acids, U, C, A, or G may occupy the third position of synonym codons. Alternate bases also may occupy the first position of synonyms, as for N-formyl-methionine.

One consequence of logical degeneracy is that many single base replacements in DNA may be silent and thus not result in amino acid replacement in protein (cf. Sonneborn, 1965). Also, the code is arranged so that the effects of some errors may be minimized, since amino acids which are structurally or metabolically related often correspond to similiar RNA codons (for example, Asp-codons, GAU, and GAC, are similar to Glu-codons, GAA, and GAG). When various amino acids are grouped according to common biosynthetic precursors, close relationships among their synonym codons

sometimes are observed. For example, codons for amino acids derived from aspartic acid begin with A: Asp, GAU, GAC; Asn, AAU, AAC; Lys, AAA, AAG; Thr, ACU, ACC, ACA, ACG; Ile, AUU, AUC, AUA; Met, AUG. Likewise, aromatic amino acids have codons beginning with U; Phe, UUU, UUC; Tyr, UAU, UAC; Trp, UGG. Such relationships may reflect either the evolution of the code or direct interactions between amino acids and bases in codons (see Woese et al., this volume).

At the time of the 1963 meeting at Cold Spring Harbor, 53 base compositions of RNA codons had been estimated (14 tentatively) in studies with randomly-ordered synthetic polynucleotides and a cell-free protein synthesizing system derived from *E. coli* (Speyer et al., 1963; Nirenberg et al., 1963). Forty-six base composition assignments now are confirmed by base sequence studies with trinucleotides (shown in Table 3). Thus, codon base compositions and base sequence assignments, obtained by assaying protein synthesis and AA-sRNA binding, respectively, agree well with one another. In addition, codon base sequences are confirmed by most amino acid replacement data obtained in vivo (see Yanofsky et al.; Wittman et al., this volume).

PATTERNS OF SYNONYM CODONS RECOGNIZED BY PURIFIED sRNA FRACTIONS

Table 5 contains a summary of synonym codons recognized by purified sRNA fractions obtained either by countercurrent distribution or by MAK column chromatography. The following patterns of codon recognition involving alternate bases in the third positions of synonym codons were found; $C = U$; $A = G$; G; $U = C = A$; $A = G = (U)$. For example, Val-sRNA$_3$ recognizes GUU and GUC, whereas the major peak of Val-sRNA (fractions 1 and 2) recognizes GUA, GUG and, to a lesser extent, GUU. The possibility that the latter Val-sRNA fraction contains two or more Val-sRNA components has not been excluded. Met-sRNA$_1$

TABLE 5. CODON PATTERNS RECOGNIZED BY PURIFIED sRNA FRACTIONS

Alternate acceptable bases in 3rd or 1st positions of triplet					
$\dfrac{C}{U}$	$\dfrac{A}{G}$	G	$\begin{matrix}U\\C\\A\end{matrix}$	$\begin{matrix}A\\G\\(U)\end{matrix}$	Possibly only 2 bases recognized
TYR$_{1,2}$ UA$_U^C$	LYS AA$_G^A$	LEU$_2$ CUG	ALA^{yeast} $\begin{matrix}U\\GCC\\A\end{matrix}$	ALA$_1$ GCG $\begin{matrix}A\\(U)\end{matrix}$	LEU$_9$ CU$_{(C)}^{(U)}$
VAL$_3$ GU$_U^C$		LEU$_5$ UUG	SER$_{2,3}^{yeast}$ $\begin{matrix}U\\UCC\\A\end{matrix}$	VAL$_{1,2}$ GUG $\begin{matrix}A\\(U)\end{matrix}$	LEU$_{4a,b}$ UU$_{(C)}^{(U)}$
		MET$_2$ AUG	F-MET$_1$ $\begin{matrix}U\\C\\UG\\A\end{matrix}$		LEU$_1$ (U)UG
			TRP$_2$ $\begin{matrix}U\\CGG\\(A)\end{matrix}$		

Patterns of degenerate codons recognized by purified AA-sRNA fractions. sRNA fractions are from *E. coli* B, unless otherwise specified. At the top of the table are shown the alternate bases which may occupy the third or first positions of degenerate codon sets. Purified sRNA fractions and corresponding codons are shown below. Parentheses indicate codons with relatively low template activity. sRNA fractions were obtained by counter-current distribution (Kellogg et al., 1966), unless otherwise specified. Yeast Ser-sRNA fractions 2 and 3 (Connelly and Doctor, 1966) are thought to be equivalent to yeast Ser-sRNA fractions 1 and 2, respectively, discussed by Zachau et al. in this volume. Yeast Ala-sRNA was the gift of R. W. Holley; results are from Leder and Nirenberg (unpubl.). Results obtained with Val-, Met-, and Ala-sRNA*E. coli* fractions are from Kellogg et al. (1966). For additional results with Tyr-sRNA fractions, see Doctor, Loebel and Kellogg, this volume. Leu-sRNA fractions (see Fig. 6 and Sueoka et al., this volume) and Lys-sRNA (Kellogg, Doctor, and Nirenberg, unpubl.) were obtained by MAK column chromatography. Three Leu-sRNA fractions also were obtained by counter-current distribution (Nirenberg and Leder, 1964). Reactions contained the usual components (see legend to Table 1) and 0.01 or 0.02 M Mg⁺⁺. Incubation was at 24° for 15 min.

responds to UUG, CUG, AUG and, to a lesser extent, GUG, and can be converted enzymatically to N-formyl-Met-sRNA, whereas, Met-sRNA$_2$ responds primarily to AUG and does not accept formyl moieties (see later discussion). Unfractionated Trp-sRNA responds only to UGG; however one fraction of Trp-sRNA, after extensive purification, responds to UGG, CGG and AGG. Possibly the latter responses depend upon the removal of sRNA for other amino acids (e.g., Arg-sRNA) which also may recognize CGG or AGG. Yeast Ala- and Ser-sRNA$_{2,3}$ fractions recognize synonyms containing U, C, or A in the third position. Leu-sRNA$_{1,3,4}$ bind to ribosomes in response to polynucleotide templates but not to trinucleotides. Possibly, only two of the three bases are recognized by these Leu-sRNA fractions.

MECHANISM OF CODON RECOGNITION

Crick (1966; also this volume) has suggested that certain bases in anticodons may form alternate hydrogen bonds, via a wobble mechanism, with corresponding bases in mRNA codons. This hypothesis and further experimental findings are discussed below.

Yeast Ala-sRNA of known base sequence and of high purity (>95%) was the generous gift of Dr. Robert Holley. In Figs. 2 and 3 are shown the responses of purified yeast and unfractionated

E. coli C¹⁴-Ala-sRNA, respectively, to synonym Ala-codons as a function of Mg⁺⁺ concentration. Purified yeast C¹⁴-Ala-sRNA responds well to GCU, GCC, and GCA, but only slightly to GCG. Similar results were obtained with unfractionated Ala-sRNA^{yeast}. In contrast, unfractionated *E. coli* C¹⁴-Ala-sRNA responds best to GCG and GCA, less well to GCU, and only slightly to GCC.

In Fig. 4a and b, the relation between concentration of yeast or *E. coli* C¹⁴-Ala-sRNA and response to synonym Ala-codons is shown. At *limiting* concentrations of purified yeast C¹⁴-Ala-sRNA, at least 59, 45, 45, and 3% of the available C¹⁴-Ala-sRNA molecules bind to ribosomes in response to GCU, GCC, GCA, and GCG, respectively. The response of unfractionated *E. coli* C¹⁴-Ala-sRNA to each codon was 18, 2, 38, and 64%, respectively. Similar results have been obtained by Keller and Ferger (1966) and Söll et al. (this volume). Since the purity of the yeast Ala-sRNA was greater than 95%, the extent of binding at limiting Ala-sRNA concentrations indicates that one molecule of Ala-sRNA recognizes 3, possibly 4, synonym codons. In addition, the data demonstrate marked differences between the relative responses of yeast and *E. coli* Ala-sRNA to synonym codons.

Correlating the base sequences of yeast Ala-sRNA with corresponding mRNA codons also provides insight into the structure of the Ala-sRNA

298

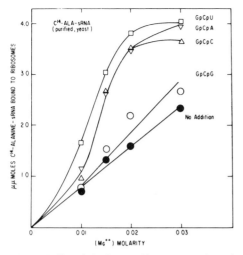

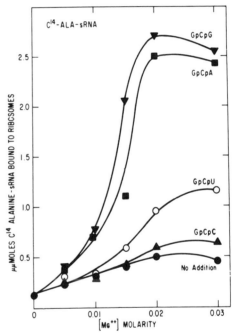

FIGURE 2. The relation between Mg^{++} concentration and binding to ribosomes of purified yeast C^{14}-Ala-sRNA of known base sequence (Holley et al., 1965) in response to trinucleotides. Each point represents a 50 μl reaction containing the components described in the legend to Table 1 except for the following: 1.5 A^{260} units of *E. coli* ribosomes, 11.2 $\mu\mu$moles of purified yeast C^{14}-Ala-sRNA (0.038 A^{260} units); and 0.1 A^{260} units of trinucleotide as specified. Reactions were incubated at 24° for 15 min (Leder and Nirenberg, unpubl.).

anti-codon and the mechanism of codon recognition. Possible anticodon or enzyme recognition sequences in Ala-sRNA^{Yeast} are –IGC MeI– and DiHU–CGG–DiHU (Fig. 5; Holley et al., 1965). Each site potentially comprises a single-stranded loop region at the end of a hairpin-like double-stranded segment. If CGG were the anticodon, *parallel* hydrogen bonding with GCU, GCC, GCA codons would be expected. If IGC were the anticodon, *antiparallel* Watson-Crick hydrogen bonding between GC in the anticodon and GC in the first and second positions of codons, and alternate pairing of inosine in the anticodon with U, C, or A, but not G, in the third position of Ala-codons, would be expected. All of the available evidence is consistent with an IGC Ala-anticodon. Zachau has shown that Ser-sRNA$^{Yeast}_{1\ and\ 2}$ contain, in appropriate positions, IGA sequences (Zachau, Dütting, and Feldmann, 1966), and we find that Ser-sRNA^{Yeast} fractions 2 and 3 (believed to correspond to fractions 1 and 2 of Zachau) recognize UCU, UCC, and UCA, but not UCG (see Table 5). A purified Val-sRNA^{Yeast} fraction contains the sequence IAC which corresponds to three Val-codons, GUU, GUC, and GUA (Ingram and Sjöqvist, 1963). In addition, the sequence, GψA, is found at the postulated anticodon site of Tyr-

sRNA^{Yeast} which corresponds to the Tyr-codons, UAU and UAC (Madison, Everett, and Kung, 1966).

Crick's wobble hypothesis and patterns of synonym codons found experimentally are in full agreement. In Table 6 are shown bases in anticodons which form alternate hydrogen bonds, via the wobble mechanism, with bases usually occupying the third positions of mRNA codons. U in the sRNA anticodon may pair alternately with A or G in mRNA codons; C may pair with G; A with U; G with C or U; and I with U, C, or A. In addition, we suggest that ribo T in the anticodon may hydrogen bond more strongly with A, and perhaps with G also, than U; and ψ in the anticodon may hydrogen bond alternately with A, G or, less well, U.

Dihydro U in an anticodon may be unable to hydrogen bond with a base in mRNA but may be repelled less by pyrimidines than by purines.

FIGURE 3. Relation between Mg^{++} concentration and binding of unfractionated *E. coli* C^{14}-Ala-sRNA to ribosomes in response to trinucleotides. Each point represents a 50 μl reaction containing the components described in the legend to Table 1, 2.0 A^{260} units of ribosomes; 18.8 $\mu\mu$moles of unfractionated *E. coli* C^{14}-Ala-sRNA (0.54 A^{260} units); and 0.1 A^{260} unit of trinucleotide, as specified (Leder and Nirenberg, unpubl.).

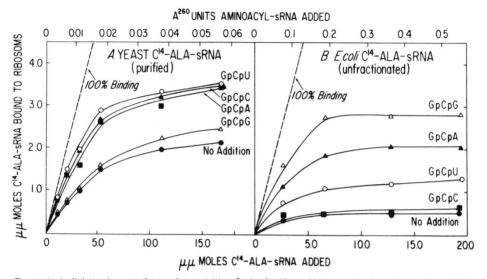

FIGURE 4a, b. Relation between the template activities of trinucleotides and the concentrations of purified yeast C14-Ala-sRNA (part a) and unfractionated *E. coli* C14-Ala-sRNA (part b). Each point represents a 50 μl reaction containing the components described in the legend of Table 1, and the following components: 0.02 M magnesium acetate; 0.1 A260 unit of trinucleotide as specified; 1.1 A260 units of *E. coli* ribosomes (part a) and 2.0 A260 units of *E. coli* ribosomes (part b); and C14-Ala-sRNA as indicated on the abscissa (Leder and Nirenberg, unpubl.).

Possibly, hydrogen bonds then form between the two remaining bases of the codon (bases 1 and 2, or 2 and 3) and the corresponding bases in the anticodon. Only two out of three bases in a codon would then be recognized. This possibility is supported by the studies of Rottman and Cerutti (1966) and Cerutti, Miles, and Frazier. (1966). Possibly, some synonym codon patterns may be due to the formation of two rather than three base pairs per triplet, particularly if both are

RECOGNITION OF ALA-CODONS BY YEAST ALA-sRNA

	ME	Di Di
		H H
sRNA	CUUIGCIψGG	UAGUCGGUAGC
	↓↓↓	↓↓↓
mRNA	GCU	GCU
	GCC	GCC
	GCA	GCA
	(GCG)	(GCG)

FIGURE 5. Base sequences from yeast Ala-sRNA shown in the upper portion of the figure represent possible anticodons. Base sequences of synonym RNA Ala-codons are shown in the lower portion of the figure. The first and second bases of Ala-codons on the left would form antiparallel Watson-Crick hydrogen bonds with the anticodon, while those on the right would form parallel hydrogen bonds. See text for further details.

TABLE 6. ALTERNATE BASE PAIRING

sRNA Anticodon	mRNA Codon
U	A
	G
C	G
A	U
G	C
	U
I	U
	C
	A
rT	A
	G
ψ	A
	G
	(U)
DiHU	No base pairing

The base in an sRNA anticodon shown in the left-hand column forms antiparallel hydrogen bonds with the base(s) shown in the right-hand column, which usually occupy the third position of degenerate mRNA codons. Relationships for U, C, A, G, and I of anticodons are "wobble" hydrogen bonds suggested by Crick (1966; also this volume). See text for further details.

TABLE 7. NUCLEOTIDE SEQUENCES OF RNA CODONS RECOGNIZED BY AA-sRNA FROM BACTERIA AND AMPHIBIAN AND MAMMALIAN LIVER

	U	C	A	G	
	PHE	SER	TYR	cys	U
	PHE	SER	TYR	cys	C
U	leu?	SER	TERM?	[cys]	A
	leu, F-MET	[SER]	TERM?	trp	G
	leu	PRO	HIS	ARG	U
	leu	PRO	HIS	ARG	C
C	leu	PRO	gln	ARG	A
	leu	PRO	gln	[ARG]	G
	ILE	THR	asn	[SER]	U
	ILE	THR	asn	[SER]	C
A	[ILE]	THR	LYS	[ARG*]	A
	MET, F-MET?	THR	[LYS]	[ARG]	G
	VAL	ALA	ASP	GLY	U
	VAL	ALA	ASP	GLY	C
G	VAL	ALA	GLU	gly	A
	VAL	[ALA]	GLU	gly	G

Universality of the RNA code. Nucleotide sequences and relative template activities of RNA codons determined with trinucleotides and AA-sRNA from E. coli, Xenopus laevis and guinea pig liver. Rectangles represent trinucleotides which are active templates for AA-sRNA from one organism, but not from another. Assignments in capitals indicate that the trinucleotide was assayed with AA-sRNAs from E. coli, Xenopus laevis liver, and guinea pig liver. Assignments in lower case indicate that the trinucleotide was assayed only with E. coli AA-sRNA (with the exception of cys-codons which were assayed with both E. coli and guinea pig liver Cys-sRNA).

⁺Söll et al. (1965) reported that both AGA and AGG stimulate yeast Arg-sRNA binding to ribosomes. The trinucleotide, AGA, however, has little or no effect upon the binding of E. coli, Xenopus laevis or guinea pig Arg-sRNA to ribosomes.

Reactions contained components described in the legend to Table 1, 0.01 or 0.02 M Mg⁺⁺, E. coli ribosomes, and 0.150 A²⁶⁰ units of trinucleotides (data from Marshall, Caskey, and Nirenberg, in prep.).

(C) · (G) pairs (also see earlier discussion concerning template activity of pUpC).

In summary, patterns for amino acids often represent the sum of two or more codon patterns recognized by different sRNA species. Specific sRNA patterns, in turn, often result from alternate pairing between bases in the codon and anticodon or, possibly, from the formation of only two base pairs if the remaining bases do not greatly repel one another.

UNIVERSALITY

The results of many studies indicate that the RNA code is largely universal. However, translation of the RNA code can be altered in vivo by extragenic suppressors and in vitro by altering components of reactions or conditions of incubation. Thus, cells sometimes differ in specificity of codon translation.

To investigate the fine structure of the code recognized by AA-sRNA from different organisms, nucleotide sequences and relative template activities of RNA codons recognized by bacterial, amphibian, and mammalian AA-sRNA (E. coli, Xenopus laevis and guinea pig liver, respectively) were determined (Marshall, Caskey, and Nirenberg, submitted for publication). Acylation of sRNA was catalyzed in all cases by aminoacyl-sRNA synthetases from corresponding organisms and tissues. E. coli ribosomes were used for binding studies. Therefore, the specificities of sRNA and AA-sRNA synthetases were investigated.

The results are shown in Table 7. Almost identical translations of nucleotide sequences to amino acids were found with bacterial, amphibian, and mammalian AA-sRNA. In addition, similar sets of synonym codons usually were recognized by AA-sRNA from each organism. However, E. coli AA-sRNA sometimes differed strikingly from Xenopus and guinea pig liver AA-sRNA in relative response to synonym codons. Differences in codon recognition are shown in Table 8. The following

TABLE 8. SPECIES DEPENDENT DIFFERENCES IN RESPONSE OF AA-sRNA TO TRINUCLEOTIDE CODONS

		sRNA		
Codon		Bacterial (E. coli)	Amphibian (Xenopus laevis)	Mammalian (Guinea pig liver)
ARG	AGG	±	+ · · ·	+ · · ·
	CGG	±	· · · ·	+ · ·±
MET	UUG	+ +	±	±
ALA	GCG	+ + · ·	±	-- · --
ILE	AUA	±	+ · ·	+ + ·
LYS	AAG	±	· · -- + ·	± ± -- --
SER	UCG	± + + +	±	-- +
	AGU	±	· + ·	· + ·
	AGC	±	+ + +	· + ·
CYS	UGA	±		-- + +

Possible differences: ACG, THR; AUC, ILE; CAC, HIS; GUC, VAL; and GCC, ALA.

No differences found: ASP, GLY, GLU, PHE, PRO, and TYR.

The following scale indicates the approximate response of AA-sRNA to a trinucleotide relative to the responses of the same AA-sRNA preparation to all other trinucleotides for that amino acid (except Gly-sRNA which was assayed only with GGU and GGC).

+ + + +	70–100%
-- · --	50–70%
-- --	20–50%
±	0–20%

trinucleotides had little or no detectable template activity for unfractionated *E. coli* AA-sRNA but served as active templates with *Xenopus* and guinea pig AA-sRNA: AGG, CGG, arginine; AUA, isoleucine; AAG, lysine; AGU, AGC, serine; and UGA, cysteine. Those trinucleotides with high template activity for *E. coli* AA-sRNA but low activity for *Xenopus* or guinea pig liver AA-sRNA were: UUG, N-formyl-methionine; GCG, alanine; and UCG, serine. Possible differences also were observed with ACG, threonine; AUC, isoleucine; CAC, histidine; GCC, alanine; and GUC, valine. No species dependent differences were found with Asp-, Gly-, Glu-, Phe-, Pro-, and Tyr-codons.

Thus, some degenerate trinucleotides were active templates with sRNA from each species studied, whereas others were active with sRNA from one species but not from another.

UAA and UAG do not appreciably stimulate binding of unfractionated *E. coli* AA-sRNA (AA-sRNA for each amino acid tested); *Xenopus* Arg-, Phe-, Ser-, or Tyr-sRNA; or guinea pig Ala-, Arg-, Asp-, His-, Ile-, Met-, Pro-, Ser-, or Thr-sRNA.

Nucleotide sequences recognized by *Xenopus* skeletal muscle Arg-, Lys-, Met-, and Ser-sRNA were determined and compared with sequences recognized by corresponding *Xenopus* liver AA-sRNA preparations. No differences between liver and muscle AA-sRNA were detected, either in nucleotide sequences recognized or in relative responses to synonym codons.

Fossil records of bacteria 3.1 billion years old have been reported (Barghoorn and Schopf, 1966). The first vertebrates appeared approximately 510 million years ago, and amphibians and mammals, 355 and 181 million years ago, respectively. The presence of bacteria 3 billion years ago may indicate the presence of a functional genetic code at that time. Almost surely the code has functioned for more than 500 million years. The remarkable similarity in codon base sequences recognized by bacterial, amphibian, and mammalian AA-sRNA suggest that most, if not all, forms of life on this planet use almost the same genetic language, and that the language has been used, possibly with few major changes, for at least 500 million years.

UNUSUAL ASPECTS OF CODON RECOGNITION AS POTENTIAL INDICATORS OF SPECIAL CODON FUNCTIONS

Most codons correspond to amino acids; however, some codons serve in other capacities, such as initiation, termination or regulation of protein synthesis. Although only a few codons have been assigned special functions thus far, we think it likely that many additional codons eventually may be found to serve special functions. Unusual properties of codon recognition sometimes may indicate special codon functions. For example, the properties of initiator and terminator codons, during codon recognition, are quite distinctive (see below). We find that approximately 20 codons have unusual properties related either to codon position, template activity, specificity, patterns of degeneracy, or stability of codon-ribosome-sRNA complexes. Until more information is available these observations will be considered as *possible* indicators of special codon functions.

Conclusions will be stated first to provide a frame of reference for discussion:

(1) A codon may have alternate meanings. (For example, UUG at or near the 5'-terminus of mRNA may correspond to N-formyl-methionine; whereas, an internal UUG codon may correspond to leucine.)

(2) A codon may serve multiple functions simultaneously. (For example, a codon may specify both initiation and an amino acid, perhaps via AA-sRNA with high affinity for peptidyl-sRNA sites on ribosomes.)

(3) Codon function sometimes is subject to modification.

(4) Degenerate codons for the same amino acid often differ markedly in template properties.

CODON FREQUENCY AND DISTRIBUTION

Often, multiple species of sRNA corresponding to the same amino acid recognize different synonym codons. Degenerate codon usage in mRNA sometimes is nonrandom (Garen, pers. comm.; also von Ehrenstein; Weigert et al., this volume). The possibility that different sets of sRNA may be required for the synthesis of two proteins with the same amino acid composition suggests that protein synthesis sometimes may be regulated by codon frequency and distribution coupled with differential recognition of degenerate codons. Possibly, the rates of synthesis of certain proteins may be regulated simultaneously by alterations which affect the apparatus recognizing one degeneracy but not another (see reviews by Ames and Hartman, 1963; and Stent, 1964).

CODON POSITION

As discussed in an earlier section, the template properties of 5'-terminal-, 3'-terminal-, and internal- codons may differ. Regulatory mechanisms based on such differences have been suggested. Reading of mRNA probably is initiated at or near the 5'-terminal codon and then proceeds toward the 3'-terminus of the RNA chain (Salas, Smith,

Stanley, Jr., Wahba, and Ochoa, 1965). It is not known whether mechanisms of 5′-terminal and internal initiation in polycistronic messages are similar. Also, internal- and 3′-terminal mechanisms of termination remain to be defined.

N-formyl-Met-sRNA may serve as an initiator of protein synthesis in *E. coli* (Clark and Marcker, 1966; Adams and Capecchi, 1966; Webster, Englehardt, and Zinder, 1966; Thach, Dewey, Brown, and Doty, 1966). Met-sRNA$_1$ can be converted enzymatically to N-formyl-Met-sRNA$_1$ and responds to UUG, CUG, AUG and, to a lesser extent, GUG. Met-sRNA$_2$ does not accept formyl-moieties and responds primarily to AUG (Clark and Marcker, 1966; Marcker et al., this volume; also Kellogg, Doctor, Loebel, and Nirenberg, 1966). In *E. coli* extracts protein synthesis is initiated in at least two ways: by initiator codons specifying N-formyl-Met-sRNA or, at somewhat higher Mg^{++} concentrations, by another means, probably not dependent upon N-formyl-Met-sRNA since many synthetic polynucleotides without known initiator codons direct cell-free protein synthesis (Nakamoto and Kolakofsky, 1966). Poly U, for example, directs di- as well as polyphenylalanine synthesis (Arlinghaus, Schaeffer, and Schweet, 1964). Probably codons for N-formyl-Met-sRNA initiate protein synthesis with greater accuracy than codons which serve as initiators only at relatively high Mg^{++} concentrations.

UAA and UAG may function as terminator codons (Brenner, Stretton, and Kaplan, 1965; Weigert and Garen, 1965). The trinucleotides UAA and UAG do not stimulate binding appreciably of *unfractionated E. coli* AA-sRNA to ribosomes. However, sRNA fraction(s) corresponding to UAA and/or UAG are not ruled out.

Extragenic suppressors may affect the specificity of UAA and/or UAG recognition (see review by Beckwith and Gorini, 1966). The efficiencies of ochre suppressors (UAA) are relatively low compared to that of amber suppressors (UAG). Since amber suppressors do not markedly affect the rate of cell growth, and ochre suppressors with high efficiency have not been found, UAA may specify chain termination in vivo more frequently than UAG. In a study of great interest, Newton, Beckwith, Zipser and Brenner (1965) have shown that the synthesis of protein (probably mRNA also) is regulated by the relative position in the RNA message of codons sensitive to amber suppressors. Therefore, a codon may perform a regulatory function at one position but not at another.

TEMPLATE ACTIVITY

Trinucleotides with little activity for AA-sRNA (in studies thus far) are: UAA, UAG, and UUA,

(perhaps CUA also). In addition, the following trinucleotides are active templates with AA-sRNA from one organism, but not from another: AGG, AGA, CGG, arginine; UUG, (N-formyl-)-methionine; GCG, alanine; AUA, isoleucine; AAG, lysine; UCG, AGU, AGC, serine; and UGA, cysteine (see Universality Section and Table 9). However, some inactive trinucleotides possibly function as active codons at internal positions. For example, the following codon base compositions were estimated with synthetic polynucleotides and a cell-free protein synthesizing system from *E. coli*; AUA, isoleucine; AGA, arginine; and AGC, serine (Nirenberg et al., 1963; Speyer et al., 1963; also see Jones, Nishimura, and Khorana, 1966, for results with AGA). Among the many possible explanations for low template activities of trinucleotides in binding assays are: special codon function; codon position; appropriate species of sRNA absent or in low concentration; competition for codons or for ribosomal sites by additional species of sRNA; high ratio of de-acylated to AA-sRNA; cryptic (non-acylatable) sRNA; reaction conditions, e.g., low concentration of Mg^{++} or other components, time or temperature of incubation.

CODON SPECIFICITY

Often synonym trinucleotides differ strikingly in template specificity. Such observations may indicate that template specificities of terminal- and internal-codons differ, or that special function codons or suppressors are present. At 0.010–0.015 M Mg^{++}, trinucleotide template specificity is high, in many cases higher than that of a polynucleotide; for example, poly U, but not UUU, stimulates binding of Ile-sRNA to ribosomes. However, at 0.03 M Mg^{++} ambiguous recognitions of tri- and polynucleotides are observed more frequently.

Relative template activities of synonym trinucleotides in reactions containing 0.01 or 0.03 M Mg^{++} are shown in Table 9. In some cases, only one or two trinucleotides in a synonym set are active templates at 0.01 M Mg^{++}; whereas all degeneracies are active at 0.03 M Mg^{++} (e.g., Glu, Lys, Ala, Thr). In other cases either all synonym trinucleotides are active at 0.01 M Mg^{++} as well as at 0.03 M Mg^{++} (e.g., Val), or none are active at the lower Mg^{++} concentration (e.g., Tyr, His, Asn). Such data suggest that codon-ribosome-AA-sRNA complexes formed with degenerate trinucleotides often differ in stability.

MODIFICATION OF CODON RECOGNITION DUE TO PHAGE INFECTION

N. and T. Sueoka (1964; also see Sueoka et al., this volume) have shown that infection of *E. coli*

TABLE 9. TEMPLATE ACTIVITY OF TRINUCLEOTIDES IN 0.01 OR 0.03 M Mg⁺⁺

	U	C	A	G	
U	PHE	[SER]	TYR	[CYS]	U
	PHE	SER	TYR	CYS	C
		(SER)			A
	[F-MET]	[SER]		(TRP)	G
C		PRO	HIS	[ARG]	U
		PRO	HIS	ARG	C
		(PRO)	GLN	[ARG]	A
	LEU	(PRO)	[GLN]	ARG	G
A	ILE	[THR]	ASN	SER, CYS	U
	ILE	THR	ASN	SER, CYS	C
		THR	[LYS]		A
	[MET]	[THR]	LYS		G
G	[VAL]	ALA	[ASP]	[GLY]	U
	[VAL]	ALA	ASP	[GLY]	C
	[VAL]	[ALA]	[GLU]	(GLY)	A
	[VAL]	[ALA]	GLU	(GLY)	G

Legend: 0.01 M Mg   0.03 M Mg

[box] =   +   ；

No Box =   −   ；

( ) = not tested

Relative template activities of trinucleotides in reactions containing 0.01 or 0.03 M Mg⁺⁺. A plus (+) sign in the legend means that the trinucleotide stimulates AA-sRNA binding to ribosomes at that magnesium concentration; a minus (−) sign means it is relatively inactive as a template. The results refer to AA-sRNA from E. coli strains B and/or W3100. The data are from Anderson, Nirenberg, Marshall, and Caskey (1966).

by T2 bacteriophage results, within one to three minutes, in the modification of one or more species of Leu-sRNA present in the E. coli host. Concomitantly, E. coli, but not viral protein synthesis is inhibited. Protein synthesis is required, however, for modification of Leu-sRNA.

In collaboration with N. and T. Sueoka, modification of Leu-sRNA has been correlated with codon recognition specificity. sRNA preparations were isolated from E. coli before phage infection and at one and eight min after infection. After acylation, Leu-sRNA preparations were purified by MAK column chromatography and the binding of each pooled Leu-sRNA fraction to ribosomes in response to templates was determined (Fig. 6). The profile of Leu-sRNA (eight min after infection) acylated with yeast, rather than E. coli.

Leu-sRNA synthetase is shown also (Fig. 6D); thus, both anticodon and enzyme recognition sites were monitored. In Fig. 7 the approximate chromatographic mobility on MAK columns of each Leu-sRNA fraction is shown diagrammatically, together with the relative response of each fraction to tri- and polynucleotide templates and acylation specificity of E. coli and yeast Leu-sRNA synthetase preparations.

Within one minute after infection, a marked decrease was observed in Leu-sRNA$_2$, responding to CUG, and a corresponding increase was seen in Leu-sRNA$_1$, responding to poly UG, but not to the trinucleotides, UUU, UUG, UGU, GUU, UGG, GUG, GGU, CUU, CUC, CUG, UAA, UAG, UGA, or to poly U or poly UC. However, Leu-sRNA$_1$ was not detected 8 min. after infection.

A marked increase in the response of Leu-sRNA$_5$ to UUG was observed one minute after infection, and an even greater increase was seen eight minutes after infection.

Greater responses of Leu-sRNA$_3$ and Leu-sRNA$_{4a,b}$ to poly UC also were observed eight minutes after phage infection. Leu-sRNA fractions 3 and 4 differ in chromatographic mobility and in acylation specificity by yeast and E. coli Leu-sRNA synthetase preparations. Thus, Leu-sRNA$_3$ and a component in fraction 4 differ, although both fractions 3 and 4 respond to poly UC. The multiple responses of Leu-sRNA$_{4a,b}$ to poly U, poly UC, and the trinucleotides, CUU and CUC, suggest that fraction 4 may contain two or more Leu-sRNA species. Striking increases in response of fraction 4 to poly U were observed one and eight minutes after infection.

Leu-sRNA fractions 1, 2, and 3 are related, for each is recognized by yeast as well as by E. coli Leu-sRNA synthetase preparations. In contrast, Leu-sRNA$_{4a,b}$ and Leu-sRNA$_5$ are recognized by E. coli, but not yeast Leu-sRNA synthetase; thus, fraction 4 is related to fraction 5. Two different cistrons of Leu-sRNA are predicted: Leu-sRNA fractions 1, 2, and 3 may be products of one cistron; whereas, fractions 4 and 5 may be products of a different cistron. In this regard, Berg, Lagerkvist, and Dieckman (1962) have shown that E. coli Leu-sRNA contains two base sequences at the 4th, 5th, and 6th base positions from the 3'-terminus of the sRNA.

The data suggest the following sRNA precursor-product relationships. Leu-sRNA$_2$ is a product of "cistron A"; the decrease in Leu-sRNA$_2$ and the simultaneous increase in Leu-sRNA$_1$ (within one minute after infection) suggests that Leu-sRNA$_2$ is the precursor of Leu-sRNA$_1$. The data also suggest that Leu-sRNA$_2$ is a precursor of Leu-sRNA$_3$. The following anticodons and mRNA

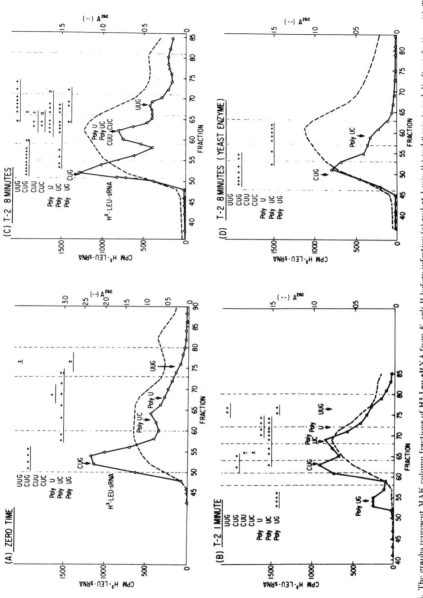

FIGURE 6. The graphs represent MAK column fractions of H³-Leu-sRNA from *E. coli* B before infection (a) and at 1 min (b) and 8 min (c and d) after infection with T2 phage. sRNA was acylated prior to chromatography with H³-leucine using *E. coli* (a, b, c) or yeast (d) synthetase preparations. Column eluates were pooled as indicated by the vertical broken lines; dialyzed against 5 × 10⁻⁴ M potassium cacodylate, pH 5.5, and lyophilized. Then binding of each fraction to ribosomes in response to tri- or polynucleotide templates was determined. At the top of each graph relative responses of Leu-sRNA fractions to templates are indicated as follows: No symbol, no detectable response of Leu-sRNA; ±, possible response; +, slight response; and —— to ——·, moderate to strong responses. Profiles represented by broken lines indicate A²⁶⁰ units; △—△ represent H³-Leu-sRNA. Data are from Kano-Sueoka, Nirenberg, and Sueoka (unpubl.). Also see Sueoka et al., this volume.

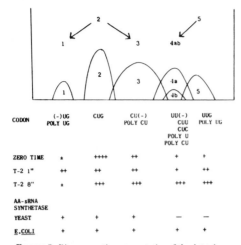

CODON	(-)UG POLY UG	CUG	CU(-) POLY CU	UU(-) CUU CUC POLY U POLY CU	UUG POLY UG
ZERO TIME	±	++++	++	+	+
T-2 1"	++	++	++	+	++
T-2 8"	±	+++	+++	+++	+++
AA-sRNA SYNTHETASE					
YEAST	+	+	+	—	—
E.COLI	+	+	+	+	+

FIGURE 7. Diagrammatic representation of the data shown in Fig. 6. The relative mobilities of multiple species of Leu-sRNA, before and after phage infection, fractionated by MAK column chromatography, are shown at the top. Leu-sRNA peaks are numbered. Arrows represent predicted Leu-sRNA precursor-product relationships (Fractions 2 and 5 possibly are products of different cistrons).
Tri- and polynucleotide codons recognized by each Leu-sRNA peak are shown below. Approximate relative responses of Leu-sRNA₁-₅ to codons are indicated as follows: ±, possible response, + to + + + +, slight to strong responses.
On the bottom are shown the specificities of *E. coli* (zero time, 1 and 8 min after infection) and yeast (8 min after infection only) Leu-sRNA synthetase preparations for sRNAᴸᵉᵘ fractions.

codons are suggested for Leu-sRNA fractions 2, 3, and 1, respectively (note: asterisks represent modifications of a nucleotide base; codon and anticodon sequences are written with 3′,5′-phosphodiester linkages; antiparallel hydrogen bonding between codon and anticodons is assumed): Leu-sRNA₂-product of "cistron A", CAG anticodon, [CUG codon]; Leu-sRNA₃- derived from fraction 2, C*AG anticodon, [CU(−) codon]; Leu-sRNA₁- derived from fraction 2, CAG** anticodon, [(−)UG codon].

Leu-sRNA₅ is a product of "cistron B", and differs from Leu-sRNA₂ in anticodon and Leu-sRNA synthetase recognition sites. The sequence, CAA, is suggested for the Leu-sRNA₅ anticodon, corresponding to a UUG mRNA codon. Leu-sRNA₄ₐ,ᵦ are derived from fraction 5. Possible anticodons and codons are: C*AA anticodon, [UU(−) codon]; C*IA anticodon, [UU(−), UC(−), UA(−) codons]; C*AI anticodon, [UU(−), CU(−), AU(−) codons].

Since modification of Leu-sRNA after phage infection is dependent upon protein synthesis,

enzyme(s) may be needed to modify bases in Leu-sRNA fractions.

The inhibition of host *E. coli*, but not viral protein synthesis following viral infection may result from modification of Leu-sRNA fractions. N-formyl-Met-sRNA₁ serves as an initiator of protein synthesis in *E. coli* and responds to two trinucleotides, UUG and CUG, which are also recognized by Leu-sRNA fractions (see previous discussion on special function codons). Possibly, initiation or termination of *E. coli*, but not viral protein synthesis is affected. Further studies are needed, however, to elucidate the mechanism of viral induced inhibition of host protein synthesis.

ACKNOWLEDGMENTS

It is a pleasure to thank Miss Norma Zabriskie, Mrs. Theresa Caryk, Mr. Taysir M. Jaouni, and Mr. Wayne Kemper for their invaluable assistance. D. Kellogg is a Postdoctoral fellow of the Helen Hay Whitney Foundation. J. Levin is supported by USPHS grant 1-F2-GM-6369-01. F. Rottman is supported by grant PF-244 from the American Cancer Society.

REFERENCES

ADAMS, J. M., and M. R. CAPECCHI. 1966. N-formyl-methionyl-sRNA as the initiator of protein synthesis. Proc. Natl. Acad. Sci. *55*: 147–155.

AMES, B. N., and P. E. HARTMAN. 1963. The histidine operon. Cold Spring Harbor Symp. Quant. Biol. *28*: 349–356.

ANDERSON, W. F., M. W. NIRENBERG, R. E. MARSHALL, and C. T. CASKEY. 1966. RNA codons and protein synthesis: Relative activity of synonym codons. Fed. Proc. *25*: 404.

ARLINGHAUS, R., J. SHAEFFER, and R. SCHWEET. 1964. Mechanism of peptide bond formation in polypeptide synthesis. Proc. Natl. Acad. Sci. *51*: 1291–1299.

BARGHOORN, E. S., and J. W. SCHOPF. 1966. Micro-organisms three billion years old from the precambrian of South Africa. Science *152*: 758–763.

BECKWITH, J. R., and L. GORINI. 1966. Suppression. Ann. Rev. Microbiol., in press.

BERG, P., U. LAGERKVIST, and M. DIECKMANN. 1962. The enzymic synthesis of amino acyl derivatives of ribonucleic acid. VI. Nucleotide sequences adjacent to the ...pCpCpA end groups of isoleucine- and leucine-specific chains. J. Mol. Biol. *5*: 159–171.

BERNFIELD, M. 1966. Ribonuclease and oligoribonucleotide synthesis. II. Synthesis of oligonucleotides of specific sequence. J. Biol. Chem. *241*: 2014–2023.

BRENNER, S., A. O. W. STRETTON, and S. KAPLAN. 1965. Genetic Code: the 'nonsense' triplets for chain termination and their suppression. Nature *206*: 994–998.

BRIMACOMBE, R., J. TRUPIN, M. NIRENBERG, P. LEDER, M. BERNFIELD, and T. JAOUNI. 1965. RNA codewords and protein synthesis. VIII. Nucleotide sequences of synonym codons for arginine, valine, cysteine and alanine. Proc. Natl. Acad. Sci. *54*: 954–960.

CERUTTI, P., H. T. MILES, and J. FRAZIER. 1966. Interaction of partially reduced polyuridylic acid with a polyadenylic acid. Biochem. Biophys. Res. Commun. *22*: 466–472.

CLARK, B., and K. MARCKER. 1966. The role of N-formyl-methionyl-sRNA in protein biosynthesis. J. Mol. Biol., 17: 394–406.

CONNELLY, C. M., and B. P. DOCTOR. 1966. Purification of two yeast serine transfer ribonucleic acids by counter-current distribution. J. Biol. Chem. 241: 715–719.

CRICK, F. H. C. 1966. Codon-Anticodon Pairing: The wobble hypothesis. J. Mol. Biol., 19: 548–555.

HEPPEL, L. A., P. R. WHITFIELD, and R. MARKHAM. 1955. Nucleotide exchange reactions catalyzed by ribonuclease and spleen phosphodiesterase. 2. Synthesis of polynucleotides. Biochem. J. 60: 8–15.

HOLLEY, R. W., J. APGAR, G. A. EVERETT, J. T. MADISON, M. MARQUISEE, S. H. MERRILL, J. R. PENSWICK, and A. ZAMIR. 1965. Structure of a ribonucleic acid. Science 147: 1462–1465.

INGRAM, V. M., and J. A. SJÖQUIST. 1963. Studies on the structure of purified alanine and valine transfer RNA from yeast. Cold Spring Harbor Symp. Quant. Biol. 28: 133–138.

JONES, D. S., S. NISHIMURA, and H. G. KHORANA. 1966. Studies on polynucleotides LVI. Further syntheses, in vitro, of copolypeptides containing two amino acids in alternating sequence dependent upon DNA-like polymers containing two nucleotides in alternating sequence. J. Mol. Biol. 16: 454–472.

KELLER, E. B., and M. F. FERGER. 1966. Alanyl-sRNA in the aminoacyl polymerase system of protein synthesis. Fed. Proc. 25: 215.

KELLOGG, D. A., B. P. DOCTOR, J. E. LOEBEL, and M. W. NIRENBERG. 1966. RNA codons and protein synthesis, IX. Synonym codon recognition by multiple species of valine-, alanine-, and methionine-sRNA. Proc. Natl. Acad. Sci. 55: 912–919.

LEDER, P., M. F. SINGER, and R. L. C. BRIMACOMBE. 1965. Synthesis of trinucleoside diphosphates with polynucleotide phosphorylase. Biochem. 4: 1561–1567.

MADISON, J. T., G. A. EVERETT, and H. KUNG. 1966. Nucleotide sequence of a yeast tyrosine transfer RNA. Science 153: 531–534.

NAKAMOTO, T., and D. KOLAKOFSKY. 1966. A possible mechanism for initiation of protein synthesis. Proc. Natl. Acad. Sci. 55: 606–613.

NEWTON, W. A., J. R. BECKWITH, D. ZIPSER, and S. BRENNER. 1965. Nonsense mutants and polarity in the Lac operon of Escherichia coli. J. Mol. Biol. 14: 290–296.

NIRENBERG, M. W., O. W. JONES, P. LEDER, B. F. C. CLARK, W. S. SLY, and S. PESTKA. 1963. On the coding of genetic information. Cold Spring Harbor Symp. Quant. Biol. 28: 549–557.

NIRENBERG, M., and P. LEDER. 1964. RNA codewords and protein synthesis. I. The effect of trinucleotides upon the binding of sRNA to ribosomes. Science 145: 1399–1407.

NIRENBERG, M., P. LEDER, M. BERNFIELD, R. BRIMACOMBE, J. TRUPIN, F. ROTTMAN, and C. O'NEAL. 1965. RNA codewords and protein synthesis, VII. On the general nature of the RNA code. Proc. Natl. Acad. Sci. 53: 1161–1168.

ROTTMAN, F., and P. CERUTTI. 1966. Template activity of uridylic acid-dihydrouridylic acid copolymers. Proc. Natl. Acad. Sci. 55: 960–966.

ROTTMAN, F., and M. NIRENBERG. 1966. Regulatory mechanisms and protein synthesis XI. Template activity of modified RNA codons. J. Mol. Biol., in press.

SALAS, M., M. A. SMITH, W. M. STANLEY, JR., A. J. WAHBA, and S. OCHOA. 1965. Direction of reading of the genetic message. J. Biol. Chem. 240: 3988–3995.

SÖLL, D., E. OHTSUKA, D. S. JONES, R. LOHRMANN, H. HAYATSU, S. NISHIMURA, and H. G. KHORANA. 1965. Studies on polynucleotides, XLIX. Stimulation of the binding of aminoacyl-sRNA's to ribosomes by ribotrinucleotides and a survey of codon assignments for 20 amino acids. Proc. Natl. Acad. Sci. 54: 1378–1385.

SONNEBORN, T. M. 1965. Degeneracy of the genetic code: Extent, nature and genetic implications. pp. 377–397. In: V. Bryson and H. J. Vogel (ed.) Evolving Genes and Proteins. Academic Press, New York.

SPEYER, J., P. LENGYEL, C. BASILIO, A. WAHBA, R. GARDNER, and S. OCHOA. 1963. Synthetic polynucleotides and the amino acid code. Cold Spring Harbor Symp. Quant. Biol. 28: 559–567.

STENT, G. S. 1964. The operon: On its third anniversary. Science 144: 816–820.

SUEOKA, N., and T. KANO-SUEOKA. 1964. A specific modification of Leucyl-sRNA of Escherichia coli after phage T2 infection. Proc. Natl. Acad. Sci. 52: 1535–1540.

THACH, R. E., K. F. DEWEY, J. C. BROWN, and P. DOTY. 1966. Formylmethionine codon AUG as an initiator of polypeptide synthesis. Science 153: 416–418.

THACH, R. E., and P. DOTY. 1965. Enzymatic synthesis of tri- and tetranucleotides of defined sequence. Science 148: 632–634.

WEBSTER, R. E., D. L. ENGELHARDT, and N. D. ZINDER. 1966. In vitro protein synthesis: Chain initiation. Proc. Natl. Acad. Sci. 55: 155–161.

WEIGERT, M., and A. GAREN. 1965. Base composition of nonsense codons in E. coli: Evidence from amino-acid substitutions at a tryptophan site in alkaline phosphatase. Nature 206: 992–994.

ZACHAU, H., D. DÜTTING, and H. FELDMANN. 1966. Nucleotide sequences of two serine-specific transfer ribonucleic acids (1). Angew. Chem. 5: 422, English Edition.

33

Reprinted from *J. Biol. Chem.*, **236**(6), 1726–1734 (1961)

The Enzymic Synthesis of Amino Acyl Derivatives of Ribonucleic Acid

I. THE MECHANISM OF LEUCYL-, VALYL-, ISOLEUCYL-, AND METHIONYL RIBONUCLEIC ACID FORMATION*

PAUL BERG,[†] FRED H. BERGMANN,[‡] E. J. OFENGAND,[§] AND M. DIECKMANN[†]

From the Department of Microbiology, Washington University School of Medicine, St. Louis 10, Missouri

(Received for publication, November 8, 1960)

The enzymic formation of enzyme-bound amino acyl adenylates from adenosine triphosphate and amino acid (Equation 1) has been recognized for several years (1–8) and enzymes specific for certain of the amino acids have been isolated in a number of laboratories (2, 9, 10). These same enzymes are now known (11–15) to catalyze a second reaction involving the transfer of the amino acyl moiety from the adenosine phosphate moiety to a specific type of ribonucleic acid (Equation 2). The over-all reaction catalyzed by such amino acyl ribonucleic acid synthetases[1] is summarized in Equation 3.

$$AMP\text{-}PP + RCHNH_2COOH + enzyme \xrightarrow{Mg^{++}}$$
$$\overset{O}{\underset{\|}{}}$$
$$enzyme\text{—}AMPCCHNH_2R + PP_i \qquad (1)$$

$$\overset{O}{\underset{\|}{}}$$
$$Enzyme\text{—}AMPCCHNH_2R + RNA\text{—}OH \rightharpoonup$$
$$\qquad\qquad (2)$$
$$\overset{O}{\underset{\|}{}}$$
$$RNA\text{—}OCCHNH_2R + AMP$$

$$AMP\text{-}PP + RCHNH_2COOH + RNA\text{—}OH \xrightarrow{Mg^{++}}$$
$$\overset{O}{\underset{\|}{}}$$
$$RNA\text{—}OCCHNH_2R + AMP + PP_i \qquad (3)$$

* This investigation was supported by grant funds from the National Institutes of Health of the United States Public Health Service.

† Present address, Department of Biochemistry, Stanford University School of Medicine, Palo Alto, California.

‡ Postdoctoral Research Fellow of the National Institutes of Health, United States Public Health Service; present address, Department of Biochemistry, Brandeis University, Waltham, Massachusetts.

§ Predoctoral Research Fellow of the National Science Foundation; present address, Medical Research Council Unit, Cavendish Laboratory, Cambridge University, Cambridge, England.

[1] Enzymes which catalyze an amino acid-dependent ATP-PP exchange and ATP-dependent amino acid hydroxamate formation have been referred to as amino acid-activating enzymes (1). Inasmuch as these activities are partial manifestations of the over-all reaction leading to amino acyl RNA formation (12, 16, 17) we propose to designate this class of enzymes as amino acyl RNA synthetases and the enzyme specific for a single amino acid, *e.g.* leucine, as leucyl RNA synthetase. This nomenclature, we feel, is consistent with the practice of including some indication of the nature of the product formed in the reaction. Moreover, it minimizes any ambiguity arising from situations in which amino acid activation occurs by reactions not involving amino acyl RNA

By the above reaction, the amino acids are bound to the acceptor ribonucleic acid through an ester linkage to the 2'- or 3'-hydroxyl group of the terminal nucleotidyl ribose moiety (23–25), and where this has been examined, each amino acid is linked to a terminal adenylic acid (16, 23–25).

The results of our investigations on the mechanism of amino acyl ribonucleic acid formation are reported in the present communication. The purification and characterization of the specific amino acyl ribonucleic acid synthetases and the amino acid-acceptor ribonucleic acid from *Escherichia coli* are presented in the following papers (26, 27). The fourth communication (28) describes the enzymic removal and resynthesis of the 3'-hydroxy-ended trinucleotide portion of the acceptor ribonucleic acid.

EXPERIMENTAL PROCEDURE

Materials

Enzymes—In some of our earlier studies and in several experiments reported here, extracts of *E. coli* and a mixture of uniformly C^{14}-labeled amino acids were used as a means of generating highly labeled amino acyl adenylates. Extracts were prepared from cells grown as described in Paper II (26) by treatment of a washed cell suspension (4 ml of 0.05 M glycylglycine buffer, pH 7.0, per g wet weight of cells) in a cooled Raytheon 10 kc sonic oscillator for 15 minutes or by disruption in a Waring Blendor with glass beads (26). Both types of extract were dialyzed for about 24 hours against 30 to 40 volumes of 0.01 M Tris buffer, pH 8.0.

The leucyl-, valyl-, isoleucyl-, and methionyl RNA synthetases from *E. coli* were prepared as described in Paper II of this series (20); the isolation of the methionyl RNA synthetase from yeast has been reported previously (2, 29).

Crystalline inorganic pyrophosphatase (30) was kindly supplied by Drs. G. Perlmann and M. Kunitz.

Amino Acid-Acceptor RNA Preparations—The acceptor RNA was isolated as described by Ofengand *et al.* (27), and in almost all cases, the material eluted from Ecteola (Brown Company) was used. The concentration of the acceptor RNA is expressed in terms of its nucleotide content and determined by its optical density at 260 mμ in 0.01 N KOH with a value of 10.0 as equal to 1 μmole of RNA nucleotide.

formation, *e.g.* S-adenosyl methionine (18), glutamine (19), glycineamide ribonucleotide (20, 21) formation, and very likely the formation of peptides (22).

C¹⁴-labeled Amino Acids—The uniformly C¹⁴-labeled amino acid mixture was obtained from the protein of *Chromatium* grown in the presence of NaHC¹⁴O₃ as carbon source (31). The protein was hydrolyzed in 6 N HCl at 110° for 18 hours. The specific activity of the amino acids was 2.5 to 3.0 × 10⁵ c.p.m. per μg atom of carbon. DL-Leucine-1-C¹⁴, DL-valine-1-C¹⁴, and L-methionine-CH₃-C¹⁴ were purchased from Isotope Specialties, Inc., and uniformly labeled L-isoleucine-C¹⁴ was obtained from Volk Radiochemical Company. The specific activities of the amino acids ranged from 3 to 17 × 10⁶ c.p.m. per μmole counted in a windowless gas flow counter.

Miscellaneous—PPᵢ³² was made as previously described (29). Nucleoside mono-, di-, and triphosphates were obtained from the Sigma Chemical Company, and unlabeled amino acids were purchased from the California Foundation for Biochemical Research or from Nutritional Biochemicals. As pointed out elsewhere (26), it was necessary in certain cases to use synthetic preparations of the amino acids to avoid trace contaminations by other amino acids.

Methods

Measurement of Amino Acyl RNA Formation—Depending upon the experiment, one of two assays for amino acyl RNA formation was carried out. The first determined the *yield* of amino acyl RNA formed when the enzyme, ATP, and amino acids were present in excess and the amount of acceptor RNA was limiting. The standard conditions for this measurement were as follows. The incubation mixture contained in a total volume of 0.5 ml, 50 μmoles of sodium cacodylate buffer, pH 7.0; 0.5 μmole of ATP; 1.0 μmole of MgCl₂ (for leucyl- and valyl RNA formation) or 5.0 μmoles of MgCl₂ (for isoleucyl- and methionyl RNA formation); either 0.3 μmole of DL-leucine-1-C¹⁴, 0.4 μmole of DL-valine-1-C¹⁴, 0.03 μmole of uniformly labeled L-isoleucine-C¹⁴ or 0.3 μmole of L-methionine-CH₃-C¹⁴; 0.2 to 1.0 μmole of acceptor RNA nucleotide; 100 μg of crystalline beef serum albumin; 2 μmoles of reduced glutathione; 5 μmoles of potassium chloride (for methionyl RNA formation); and either 0.9, 0.5, 7, or 3 μg of protein of the leucyl-, valyl-, isoleucyl-, or methionyl RNA synthetase preparations, respectively. The mixture was incubated at 30° for 20 minutes (a time which was sufficient for the reaction to come to completion) and the reaction was stopped by the addition of 0.5 to 1.5 mg of carrier yeast RNA and 3 ml of a cold solution containing 0.5 M NaCl and 67% ethanol. After 5 minutes at 0°, the precipitate was centrifuged and washed three times by resuspension in the ethanol-salt mixture. The precipitate was dissolved in 1 ml of 1.5 N NH₄OH, and a suitable aliquot was dried in small dishes and counted in a windowless gas flow counter. The results are expressed as millimicromoles of amino acid bound per μmole of acceptor RNA nucleotide. Data to be presented below (Fig. 5) show that, under these conditions, the amount of each of the amino acids bound is proportional to the amount of acceptor RNA added.

In contrast to the first assay, which determined the yield of product, the second assay measured the *rate* of amino acyl RNA formation and was carried out under the conditions described above, except with less enzyme and more acceptor RNA (1.0 to 2.0 μmoles of RNA nucleotide). The reaction rate was proportional to enzyme concentration over the range shown in Fig. 1.

Measurement of Amino Acyl Adenylate Formation—The capacity of each of the enzymes to form amino acyl adenylates was

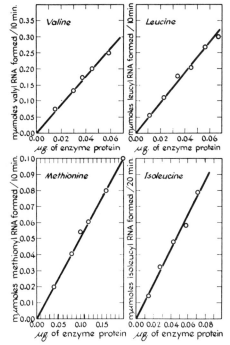

FIG. 1. Linear relationship of the rate of amino acyl RNA formation with enzyme concentration. See text for conditions.

measured by the amino acid-dependent exchange of ATP and PPᵢ³² (26). Inasmuch as the rate of amino acyl adenylate formation was the rate-determining step in the over-all exchange reaction (2, 29), the amino acid-dependent incorporation of PPᵢ³² into ATP actually measured the rate of amino acyl adenylate formation.

For comparisons of amino acyl adenylate and amino acyl RNA formation under the same conditions, the following assay was used. In a volume of 1.0 ml were 100 μmoles of sodium cacodylate buffer, pH 7.0, 5 μmoles of MgCl₂, 2 μmoles of ATP, 2 μmoles of PPᵢ³² (specific activity, 0.5 to 1.0 × 10⁵ c.p.m. per μmole), 2 μmoles of the L-form of leucine, valine, isoleucine, or methionine, 200 μg of serum albumin, 4 μmoles of reduced glutathione (where indicated above), 10 μmoles of KCl (where indicated above), and enough enzyme to give an incorporation of 0.01 to 0.3 μmole of PPᵢ³² into ATP. The mixture was incubated at 30° for 15 minutes and the ATP was isolated and counted as previously described (2). All values were corrected for any ATP³² formed in the absence of amino acid. This blank was always less than 5% of that observed with amino acid.

RESULTS

Required Components for Enzymatic Synthesis of Amino Acyl RNA Compounds—Formation of the amino acyl RNA derivatives was observed in the presence of ATP, Mg⁺⁺, a specific RNA

TABLE I

Requirements for amino acyl RNA formation by amino acyl RNA synthetases from E. coli

The incubation mixtures and conditions used for measuring the rate of formation of each amino acyl RNA derivative are described under "Methods." The column headings refer to the isolated enzymes which are relatively specific for the amino acids listed (26).

Components	Leucine	Valine	Isoleucine	Methionine
		μmoles/mg/hour		
Complete............	3.1	21.0	3.3	3.3
Minus ATP............	<0.03	<0.2	<0.02	0.1
Minus RNA............	<0.03	<0.2	<0.02	<0.1
Minus Mg++............	0.33	0.6	0.15	<0.1
Minus enzyme.........	<0.03	<0.2	<0.02	<0.1

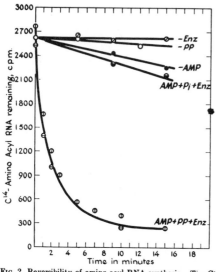

FIG. 2. Reversibility of amino acyl RNA synthesis. The C^{14} labeled mixed amino acyl RNA was prepared by incubating 2000 μmoles of cacodylate buffer, pH 7.0, 90 μmoles of MgCl₂, 8 μmoles of ATP, 1.17×10^7 c.p.m. of the C^{14}-amino acid mixture, 93 μmoles of acceptor RNA, and 8.0 mg of protein of a sonic extract of *E. coli* in a volume of 20 ml for 60 minutes at 30°. The product was isolated by the addition of NaCl to a concentration of 1.5 M followed by 2 volumes of cold ethanol. After chilling the mixture, the precipitated product was removed by centrifugation and extracted with 0.01 M cacodylate buffer, pH 7.0. Denatured protein was removed by centrifugation, and the process of precipitation, buffer extraction, and removal of denatured protein was repeated twice more. The final product was dissolved in 0.02 M succinate buffer, pH 6.

The amino acyl RNA was incubated with 20 μmoles of cacodylate buffer, pH 7.0, 15 μmoles of MgCl₂ and, where indicated, approximately 2 μmoles of C^{12}-amino acid mixture prepared from *Chromatium*, 2 μmoles of AMP, 2 μmoles of PP$_i$, 4 μmoles of P$_i$, and 0.2 mg of protein of a sonic extract of *E. coli* in a volume of 0.5 ml, for the indicated time at 30°. The amount of amino acid remaining bound to the RNA was determined by measuring the amount of C^{14}-amino acid still precipitable by 0.6 M perchloric acid. The abbreviation used is: Enz, enzyme.

TABLE II

Formation of ATP from valyl RNA, AMP, and PP

The complete system contained, per ml, 100 μmoles of sodium cacodylate buffer, pH 7.0, 2 μmoles of MgCl₂, 50 μmoles of potassium fluoride, 200 μg of serum albumin, 2.38 mμmoles of valine-C^{14} as valyl RNA, 106 μg of valyl RNA synthetase protein, 0.10 μmole of AMP, and 0.06 μmole of PP$_i$^{32} (3.4×10^6 c.p.m. per μmole). Valyl RNA was hydrolyzed to free acceptor RNA and valine by heating at 55° for 15 minutes at pH 9. The incubation was at 30° for 15 minutes and the reaction was stopped by boiling for 2 minutes. An aliquot of the reaction mixture was removed and the amount of valyl RNA remaining was determined by the amount of radioactivity still precipitable after the addition of the NaCl-ethanol mixture described under "Methods." After the addition of unlabeled ADP, ATP, and PP$_i$ to the remainder of the reaction mixture, the nucleotides were adsorbed on charcoal. After the charcoal was washed several times with 0.01 M PP$_i$, the nucleotides were eluted with 50% ethanol containing 0.3 M NH₄OH and chromatographed on a Dowex 1-Cl⁻ column (32). In control experiments in which PP$_i$ was omitted, there was no disappearance (<5%) of valyl RNA or ATP when added in the amounts obtained in the experiment. Over 90% of the radioactivity was eluted with the carrier ATP, but in the experiment with hydrolyzed valyl RNA, less than 5% of the P^{32} appeared with the ATP. In the former case, the specific activity of the ATP was essentially constant over the entire peak. The isolated material was further identified as ATP by the following two experiments. After reaction of the ATP^{32} with glucose and hexokinase, 45% of the P^{32} was isolated in the glucose 6-phosphate and 55% in the ADP. With an excess of valyl RNA synthetase, L-valine, and unlabeled PP$_i$, under conditions of the ATP-PP$_i$^{32} exchange reaction (26), 95% of the P^{32} in the ATP was found in the PP$_i$ fraction.

Conditions	Valyl RNA			ATP	
	Initial	Final	Δ	Final	Δ
		mμmoles			*mμmoles*
Complete system....	1.19	0.06	−1.13	1.11	+1.06
Complete system with hydrolyzed valyl RNA........	0.04			0.05	

fraction isolated from *E. coli*, a C^{14}-labeled amino acid, and the purified enzyme fraction capable of converting that amino acid to the corresponding amino acyl adenylate (Table I). In each case, omission of any one of the cited components resulted in a marked decrease in the rate of amino acyl RNA synthesis. Substitution of the ATP by UTP, GTP, CTP, dATP, or ADP lead to a decrease in the rate to less than 1%. Ribosomal RNA (27) from *E. coli*, or the equivalent fractions from animal and other bacterial sources, and synthetic polynucleotides prepared with polynucleotide phosphorylase (31), failed to function as amino acid acceptors under these conditions (32).

Reversibility of Amino Acyl RNA Formation—Equation 3 predicts that amino acyl RNA formation is reversible. Substantiation of this prediction was given by the following experiments. When amino acid-acceptor RNA, to which a mixture of C^{14}-labeled amino acids had been linked, was incubated with AMP, PP$_i$ and a dialyzed extract of *E. coli*, there was a rapid removal of the labeled amino acids from the RNA (Fig. 2). This occurred whether or not a pool of unlabeled amino acids was added. If either the AMP, PP$_i$, or the extract was omitted, or if UMP or CMP replaced AMP, or if P$_i$ was substituted for PP$_i$, the rate

TABLE III

Determination of equilibrium constant of L-valyl RNA formation

For each experiment, 350 μmoles of sodium cacodylate buffer, pH 7.0, 7 μmoles of MgCl₂, 0.7 mg of serum albumin, 175 μmoles of KF, and the reactants shown in the table were incubated in a volume of 3.5 ml. The valyl RNA was labeled with L-valine-1-C¹⁴ (6 × 10⁴ c.p.m. per μmole). Samples were removed at 2, 3, 4, 5, 10, 15, and 20 minutes and the valyl RNA formed or remaining was determined as described in the standard assay procedure. In each case, the reaction was followed until no further change in the amount of valyl RNA could be detected. Completion of the reaction occurred by 5 minutes, and no change was measurable up to 20 minutes. The concentration of RNA is expressed as millimicromoles of valine-specific acceptor sites. The concentration of the RNA was calculated by the difference between the amount of valine-specific acceptor RNA added and the amount of valyl RNA formed or from the amount of valyl RNA which disappeared. Inasmuch as the concentrations of each of the other components was large compared to the amount of reaction which had occurred, the initial concentrations of each were used in the calculation. In separate experiments, it was shown that under these conditions there was no detectable destruction of the valine acceptor RNA chains nor was there any disappearance of ATP, AMP, or PP₁ (<4%) when added separately.

Experiment No.	Initial concentrations						Valyl RNA synthetase	Final concentrations Valyl RNA	K*
	AMP	PP₁	ATP	Valine	Valyl RNA	RNA			
	μmoles/ml						*units/ml*	*μmoles/ml × 10⁴*	
1	0.40	1.02	0.40	0.52	1.36 × 10⁻³	0	2.3	2.1	0.36
2	0.40	0.51	0.80	0.52	1.36 × 10⁻³	0	1.2	5.4	0.32
3	0.40	0.51	0.80	0.52	0	1.36 × 10⁻³	0.3	4.8	0.28
4	0.40	0.51	0.80	0.52	1.36 × 10⁻³	0	0.3	5.1	0.30
Average									0.32

$$*K = \frac{(AMP)(PP_i)(Valyl\ RNA)}{(ATP)(Valine)(RNA)}.$$

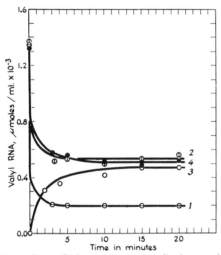

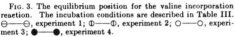

FIG. 3. The equilibrium position for the valine incorporation reaction. The incubation conditions are described in Table III. ⊖——⊖, experiment 1; ⊕——⊕, experiment 2; ○——○, experiment 3; ●——●, experiment 4.

of amino acid removal from the RNA occurred at less than 1% the rate. The failure to remove amino acids from the RNA in the presence of only AMP and the enzyme was also consistent with the formation of enzyme-bound amino acyl adenylates by the reverse as well as the forward reaction (7, 8).

That the previous observations do represent reversal of amino acyl RNA synthesis was established by the finding that incubation of C¹⁴-valyl RNA with AMP, PP₁³², and the specific valyl RNA synthetase resulted in essentially complete removal of the valine from the RNA and stoichiometric formation of ATP³² (Table II). No ATP³² formation is observed when an equivalent amount of RNA and valine is substituted for the valyl RNA.

Determination of the equilibrium constant for valyl RNA formation was made by measuring the steady state concentration of valyl RNA in the presence of the other components of the system (Table III and Fig. 3). The average K_{eq} value of 0.32² showed that there was little change in free energy resulting from the formation of valyl RNA at the expense of the cleavage of ATP.

Existence of Specific Acceptor RNA for Each Amino Acid—An examination of the kinetics of amino acyl RNA formation showed that the reaction proceeded linearly with time and then reached a limit (Fig. 4). This limit was not appreciably increased (less than 5%) by the addition of up to 5 times more enzyme, ATP, or amino acid, whether added initially or when the reaction had stopped. The addition of 2.6 μg of crystalline inorganic pyrophosphatase did not affect the extent of amino acyl RNA formation. However, the addition of acceptor RNA, either at the beginning of the reaction or at the time the reaction ceased, lead to an increased yield of amino acyl RNA. If in each case the reaction was allowed to proceed to completion in the presence of varying amounts of acceptor RNA, the amount of amino acyl RNA formed was a linear function of the amount of acceptor RNA added (Fig. 5). It should be noted, however, that the yield of each amino acyl RNA was different. Thus, although

² It should be pointed out that the calculation of the K_{eq} does not take into account the concentrations of possible complexes of the phosphorylated derivatives (*e.g.* ATP, RNA, etc.) with Mg⁺⁺.

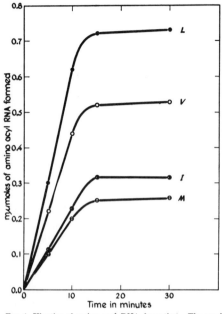

FIG. 4. Kinetics of amino acyl RNA formation. The conditions used were those described for the usual assay of amino acyl RNA formation except that 1.2, 0.1, 0.4, or 0.5 μg of leucyl-(*L*), valyl-(*V*), isoleucyl-(*I*), or methionyl-(*M*) RNA synthetase protein, respectively, and 1.0 μmole of acceptor RNA nucleotide were used.

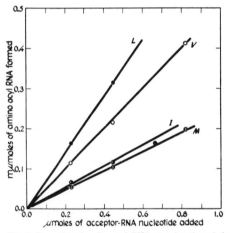

FIG. 5. Formation of amino acyl RNA as a function of the amount of acceptor RNA. The conditions used were those described under "Methods." Abbreviations are as in Fig. 4.

TABLE IV

Separate sites for linking amino acids to acceptor RNA

Experiment 1. The reaction mixture (0.5 ml) contained 20 μmoles of sodium cacodylate buffer, pH 7.0, 1 μmole of MgCl₂, 0.2 μmole of ATP, either 0.25 μmole of DL-leucine-1-C¹⁴ (5.1 × 10⁶ c.p.m. per μmole), 0.25 μmole of DL-valine-1-C¹⁴ (6.1 × 10⁶ c.p.m. per μmole) or 0.16 μmole of L-methionine-CH₃-C¹⁴ (3.0 × 10⁶ c.p.m. per μmole), 0.9 μmole of acceptor RNA, and 100 μg of a sonic extract of *E. coli*. The incubation was for 20 minutes at 30°. In the experiment with a mixture of the three amino acids, all were present initially.

Experiment 2. The incubation mixtures and conditions were as described under "Methods." When the incorporation of two amino acids was examined, the second amino acid and the appropriate enzyme were added after 20 minutes, and the incubation was continued for an additional 20 minutes.

Experiment No.	Amino acid added	Incorporation
		total c.p.m.
1	Leucine	3759
	Valine	1646
	Methionine	468
	Mixture of above	5832
	Calculated sum	5873
2	Valine	947
	Leucine	1038
	Valine, then leucine	2006
	Calculated sum	1985
	Valine	947
	Methionine	448
	Valine, then methionine	1434
	Calculated sum	1395

the RNA acted stoichiometrically with each amino acid, for a given amount of acceptor RNA the amount of amino acyl RNA formed varied with the amino acid.

Two possible interpretations of this result are that (*a*) there was a single binding site which reacted with each amino acid to a different extent, or (*b*) there existed different and specific sites for the individual amino acids. These alternatives were distinguished by the following experiments (Table IV). When leucine, valine, and methionine were present together, the total amount of amino acid linked to the acceptor RNA was equal to the sum of the amounts obtained when each amino acid was present by itself (Experiment 1). Moreover, saturation of the acceptor RNA with one amino acid (*e.g.* L-valine) did not affect the amount of any other amino acid (*e.g.* L-leucine or L-methionine) which could subsequently be linked to the acceptor RNA (Experiment 2). These data ruled out the common binding site hypothesis but were consistent with the existence of a limited and fixed number of binding sites, each specific for a particular amino acid.

Further support for this view has come from studies on the destruction of amino acid acceptor sites by periodate (28). This work showed that periodate oxidation of acceptor RNA destroys the ability to accept all amino acids. However, similar treatment of leucyl-, valyl-, or methionyl RNA followed by removal of the amino acid yielded preparations of RNA which could accept only that amino acid which was linked to the RNA during the exposure to periodate. Since, according to currently accepted ideas of RNA structure, polynucleotide chains are unbranched and therefore the only *cis*-hydroxyl configuration resides on the terminal nucleotide with a free 3'-hydroxyl group, it may be inferred that each amino acid is linked exclusively to either the 2'- or 3'-hydroxyl group of the terminal nucleotidyl ribose unit of individual RNA molecules. For the leucine- and valine-specific polynucleotide chains this terminal nucleotide is adenylic acid (28), although it is now clear that, in the acceptor RNA of *E. coli*, adenylic acid is the sole terminal nucleotide containing a free 3'-hydroxyl group on the ribose moiety (27).

Heterogeneity of Acceptor RNA Chains Reacting with Single Amino Acid—The conclusion stated above predicts that acceptor RNA represents a heterogeneous population of polynucleotide chains, each chain being specific for a particular amino acid. It is essential before considering any analysis of the chemical basis of the amino acid specificity of acceptor RNA to know whether there exists a second order of heterogeneity, namely, whether all the chains reacting with a particular amino acid are identical. The following experiments suggested that they are not.

Acceptor RNA from *E. coli* bound methionine to a different extent depending upon whether the methionyl RNA synthetase from *E. coli* or yeast was used (Fig. 6). Although the amount of methionine fixed was a direct function of the acceptor RNA added, the slopes of the two curves differed by a factor of about 2.5; that is, 2.5 times more methionine was bound per unit of RNA when the synthetase from *E. coli* was used as compared with the one from yeast. Although the addition of more yeast methionyl RNA synthetase, ATP, or methionine did not increase the yield of methionyl RNA, it was clear that there still were sites available to accept methionine. This is shown by the experiment (Table V) in which the *E. coli* synthetase was added when the reaction with the yeast enzyme had come to completion. The reciprocal experiment, in which the acceptor RNA was reacted to a limit with methionine with the *E. coli* enzyme and then exposed to the yeast enzyme, showed no additional formation of methionyl RNA. It may be inferred from this result that of the polynucleotide chains specific for methionine, 40% can function with either enzyme, whereas 60% of the chains are available only to the *E. coli* synthetase.

Support for this interpretation was obtained by the periodate oxidation technique for selectively inactivating those polynucleotide chains not linked to amino acids (28). Samples of methionyl RNA prepared with the *E. coli* or yeast synthetases were treated with periodate, reisolated, and then the amino acids were removed with alkali. The regenerated acceptor RNA preparations were retested for their capacity to accept methionine with each of the enzymes (Table VI). When the methionyl RNA was prepared with *E. coli* enzyme, all sites specific for methionine survived the periodate oxidation when tested with either enzyme. On the other hand, when methionyl RNA was prepared with the yeast enzyme, 60% of the chains which accept methionine were inactivated as judged by the test with the enzyme from *E. coli* but all were conserved when assayed with the yeast enzyme. These data show that within the population of acceptor RNA molecules there were at least two distinguishable classes of polynucleotide chains which could accept methionine.

Nature of Enzymes Catalyzing Amino Acyl RNA Formation—The enzyme preparations used in the present studies were purified on the basis of their activity for amino acyl adenylate formation (26). Although these same preparations catalyzed the formation of the amino acyl RNA derivatives, it was not clear which of the following hypotheses was operative.

1. The formation of the specific enzyme-amino acyl adenylate

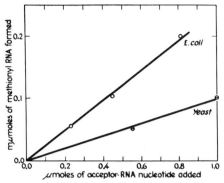

Fig. 6. Formation of methionyl RNA by methionyl RNA synthetases from yeast and *E. coli*. The conditions used are those described under "Methods."

TABLE V

Formation of methionyl RNA by methionyl RNA synthetases from E. coli and yeast

The reaction mixtures contained, in a volume of 0.5 ml, 50 μmoles of sodium cacodylate buffer, pH 7.0, 1 μmole of $MgCl_2$, 0.5 μmole of ATP, 0.06 μmole of L-methionine-CH_2-C^{14} (7.5 × 10⁶ c.p.m. per μmole), 5 μmoles of potassium chloride, 100 μg of serum albumin, 0.81 μmole of acceptor RNA, and either 30 μg of protein of methionyl RNA synthetase from *E. coli* or 125 μg of protein of the similar enzyme from yeast added as indicated in the table. The incubation was at 30°.

Enzyme additions	Total time of incubation	Methionyl RNA formation
	min	*mμmole/μmole nucleotide*
E. coli enzyme at time zero	20	0.28
	40	0.27
E. coli enzyme at time zero and again at 20 minutes	40	0.28
E. coli enzyme at time zero; yeast enzyme at 20 minutes	40	0.28
Yeast enzyme at time zero	20	0.12
	40	0.14
Yeast enzyme at time zero and again at 20 minutes	40	0.13
Yeast enzyme at time zero; *E. coli* enzyme at 20 minutes	40	0.28
Yeast and *E. coli* enzyme at time zero	20	0.27

complex may be followed by a nonenzymic transfer of the amino acid to the RNA.

2. There may be a single amino acid-specific enzyme which catalyzes the formation of an amino acyl adenylate and the transfer of that amino acid from the adenylate moiety to the acceptor RNA.

3. There may be required in addition to the enzyme forming each amino acyl adenylate either (*a*) separate specific amino acyl transferases for linking each amino acyl residue to the appropriate acceptor RNA, or (*b*) a single amino acyl transferase which

TABLE VI

Evidence for heterogeneity among acceptor RNA chains for L-methionine

Methionyl RNA was prepared under the usual conditions with either the methionyl RNA synthetase from *E. coli* or from yeast and was isolated as already described. Both samples were treated with sodium metaperiodate in 0.1 M sodium succinate buffer at pH 5.6 in the ratio of 1 µmole of periodate per 14 µmoles of acceptor RNA nucleotide. The RNA samples were reisolated and the bound methionine removed by 0.1 M glycine buffer, pH 10.2, at 30° for 60 minutes. The two RNA preparations were again recovered and then tested as acceptors of methionine with each of the methionyl RNA synthetase preparations as described in Table V.

Source of enzyme used to measure methionyl RNA formation	Methionyl RNA formation		
	Original RNA	Periodate-treated methionyl RNA prepared with	
		E. coli enzyme	Yeast enzyme
	mµmoles/µmole RNA nucleotide		
E. coli.............	0.25	0.23	0.10
Yeast.............	0.10	0.10	0.10

transfers any amino acyl group but only to the appropriate acceptor RNA chain.

Although nonenzymic acylation of RNA by amino acyl adenylates has been observed, the amino acids appeared to be bound to any RNA and in a variety of linkages (34). Such a mechanism would therefore not account for the fact that only a particular fraction of RNA functions as an amino acid acceptor (11, 27). It also seems unlikely that a nonenzymic mechanism would manifest the high degree of specificity inherent in linking each amino acid exclusively to the terminal nucleotide of a particular RNA chain. Furthermore, the different yields of methionyl RNA produced in the presence of two different methionyl RNA synthetases are inconsistent with a nonenzymic transfer reaction.

Since hypotheses 2 and 3 predict that the synthesis of a given amino acyl RNA derivative is preceded by the formation of the corresponding amino acyl adenylate, it is implicit in either alternative that the formation of each amino acyl RNA compound must be at least as specific with respect to the amino acid as is the synthesis of the amino acyl adenylate. Table VII shows that with four enzymes from *E. coli* and one from yeast only that amino acid which is converted to the adenylate is linked to the acceptor RNA. The only deviation from an exact correlation of the two specificities is the case of the isoleucyl RNA synthetase. Although this enzyme can form both isoleucyl- and valyl adenylates (26), it synthesizes only isoleucyl RNA. The reasons why the valyl moiety is not transferred to the acceptor RNA (which can accept valine from the valyl RNA synthetase preparation) remain to be determined.

Hypothesis 2, in contrast to 3, predicts that the ratio of the activity for amino acyl adenylate formation and amino acyl RNA synthesis must be constant throughout the purification of the enzymes. Any alteration in this ratio during purification would suggest the existence of separable activities. Table VIII shows that the ratio of activities for methionyl adenylate and methionyl RNA formation was constant during the course of an approximately 100-fold purification of the methionyl RNA synthetase from yeast. Similar findings have been made with the leucyl- and methionyl RNA synthetases of *E. coli*. The ratio of leucyl

TABLE VII

Specificity of enzyme preparations for amino acyl adenylate and amino acyl RNA formation

The rate of amino acyl adenylate formation was measured as described elsewhere (26), and the rate of amino acyl RNA synthesis was determined as described under "Methods."

Enzyme	Amino acid tested	Amino acyl adenylate formation	Amino acyl RNA formation
		µmoles/mg/hour	
Leucine	Leucine	358	3.2
	Valine	13.2	0.18
	Methionine	8.0	<0.01
	Isoleucine	<3.0	<0.01
Valine	Valine	560	25
	Leucine	<0.5	<0.01
	Isoleucine	<4.0	<0.01
	Methionine	2.0	<0.01
Isoleucine	Isoleucine	768	3.3
	Leucine	31	<0.07
	Valine	416	<0.03
	Methionine	41	0.07
Methionine (*E. coli*)	Methionine	356	3.5
	Leucine	4.0	<0.01
	Valine	<3.0	<0.01
	Isoleucine	<3.0	<0.01
Methionine (Yeast)	Methionine	44	0.016
	Leucine	<0.4	<0.001
	Valine	<0.4	<0.001
	Phenylalanine	<0.4	<0.001

TABLE VIII

Methionyl adenylate and methionyl RNA formation in various fractions obtained during purification of methionyl RNA synthetase from yeast

Enzyme fraction*	Methionyl adenylate formation (A)	Methionyl RNA formation (B)	A/B × 10⁻¹
	µmoles/ mg/hour	*mµmoles/ mg/hour*	
Crude extract..................	0.61	0.23	2.6
Alcohol Fraction 2..............	14.8	5.2	2.8
Ammonium sulfate Fraction 1......	23.2	8.0	2.9
Ammonium sulfate Fraction 2......	44.0	16.2	2.7
Alumina Cγ gel eluate...........	55.7	19.7	2.8

* The various enzyme fractions were prepared as previously described (35) and the specific enzyme activities for the formation of methionyl adenylate and methionyl RNA were measured as described under "Methods."

adenylate formation to that of leucyl RNA formation was 40 in crude extracts and 43 in the purified preparation (50-fold purified). The analogous ratios with methionine in the crude and purified preparations (75-fold purified) were 79 and 81, respectively.

The failure to observe separation of amino acyl adenylate and amino acyl RNA formation might also result (*a*) if the enzymes involved fractionated identically in the procedures we have used,

TABLE IX

Comparison of rate of amino acyl adenylate and amino acyl RNA formation under similar conditions

The rate of amino acyl RNA formation was measured as already described, and the synthesis of amino acyl adenylates was determined by the amino acid-dependent ATP-PP$_i$[32] exchange (26). To compare the two rates, however, the ATP-PP$_i$[32] exchange reaction was carried out under the same conditions used for amino acyl RNA formation, except that unlabeled amino acid and 0.002 M ATP and PP$_i$[32] were added.

Enzyme	Amino acyl adenylate formation (A)	Amino acyl RNA formation (B)	A/B
	μmoles/mg/hour		
Leucyl RNA synthetase	120	2.8	43
Valyl RNA synthetase	574	26.4	22
Isoleucyl RNA synthetase	460	3.3	140
Methionyl RNA synthetase (*E. coli*)	244	3.0	81
Methionyl RNA synthetase (yeast)	22.1	0.007	3200

and (b) if the hypothetical enzyme catalyzing the transfer of the amino acyl group to RNA were present in excess in both the crude and subsequent enzyme fractions. The first point can only be answered by more extensive purification studies. The second objection, however, is eliminated by the observation that with each enzyme preparation it is the formation of the amino acyl adenylate which is by far the faster reaction, *i.e.* the transfer of the amino acyl group to the RNA is the rate-limiting step (Table IX). Note that this difference in the rate of amino acyl adneylate- and amino acyl RNA formation is of the order of 20- to 140-fold and in one case is about 3000 times. These data imply that the transfer of the amino acyl moiety from the enzyme-amino acyl adenylate complex to the acceptor RNA is the rate-limiting reaction. It should be pointed out that the slow transfer of methionine by the yeast methionyl RNA synthetase may be due to the use of the acceptor RNA from *E. coli*.

At present, our data are consistent with the hypothesis that a single enzyme catalyzes both the formation of a specific enzyme-amino acyl adenylate complex and the transfer of the amino acyl group to the acceptor RNA.

DISCUSSION

The findings reported here and those presented recently by other workers (16, 17) indicate strongly that the so-called "amino acid-activating enzymes" are in essence amino acyl RNA synthetases. Whereas the initial reaction between ATP, amino acid, and a specific enzyme results in the formation of an enzyme-bound amino acyl adenylate, in the presence of the appropriate acceptor RNA chain, the amino acyl moiety is transferred to the RNA and more specifically to the 2'- or 3'-hydroxyl group of the terminal nucleotidyl ribose. A mechanism of this type not only minimizes spontaneous destruction of the highly unstable free amino acyl adenylate under physiological conditions (34, 36), but it also eliminates the requirement of additional specific enzymes to form each amino acyl RNA derivative. Indeed, it has recently been shown (37) that synthetic tryptophanyl adenylate, in the presence of purified tryptophanyl RNA synthetase, serves as tryptophan donor to amino acid acceptor RNA. From a mechanistic view, the amino acyl RNA synthetases are analogous to the enzymes which catalyze the formation of acyl-CoA derivatives (38–40), pantothenic acid (41), and carnosine (42) in that there is a primary formation of an enzyme-bound acyl adenylate and a subsequent transfer of the acyl moiety to an acceptor molecule.

Recently Zillig *et al.* (43) reported that the yield of amino acyl RNA is a function of the amount of amino acyl RNA synthetase added. We have not observed this phenomenon in our studies. Rather, only the initial rate of amino acyl RNA formation is influenced by the amount of enzyme present. The final yield of amino acyl RNA is, with sufficient time, independent of enzyme concentration and depends entirely on the amount of acceptor RNA present. In our early studies of valyl RNA synthesis, we observed that the yield of valyl RNA did vary with the amount of enzyme added. This, however, was found to be due to inactivation of the enzyme during the course of the reaction, and it could be circumvented by the addition of serum albumin to the incubation mixture. Under these latter conditions, the enzyme continues to act until the acceptor RNA is saturated with respect to valine.

The finding that amino acyl RNA synthesis is reversible is surprising in light of the ester linkage between amino acid and the acceptor RNA. The K_{eq} of 0.32 for valyl RNA synthesis and the values of 0.7 and 0.37 reported for threonyl RNA synthesis (16, 17) indicate that the amino acyl moiety is maintained at a high energy level. Whether this thermodynamic activation of the amino acid is a consequence of an adjacent hydroxyl group on the ribose or to some other structural feature of the combination remains to be determined.

An interesting aspect of the mechanism of amino acyl RNA synthesis concerns the basis of the specificity in linking each amino acid to the appropriate polynucleotide chain. This question may be considered on the basis of the two reactions catalyzed by the enzyme: in the first phase, the enzyme forms a specific enzyme-amino acyl adenylate complex and in the second this complex reacts with a specific acceptor RNA chain to form the appropriate amino acyl RNA derivative. With respect to the first phase of the reaction, it is clear from studies with the purified amino acyl RNA synthetases that they exhibit a relatively high degree of selectivity for a single naturally occurring amino acid. The significance of the slight activity sometimes noted with other amino acids is difficult to assess in the absence of more precise data concerning the purity of the enzyme preparations and the amino acid substrates (26). There are two exceptions, however, which should be noted. The purified isoleucyl RNA synthetase forms valyl adenylate as well as isoleucyl adenylate, and the valyl RNA synthetase forms threonyl adenylate (26). In both cases, the K_m for the "unnatural" substrate is about 100-fold higher than that for the "natural" one, so that with equal concentrations of the "natural" and "unnatural" amino acids, the enzyme reacts almost exclusively with the "natural" substrate.

In an analysis of the factors which control the transfer of the amino acyl moiety to its specific acceptor RNA chain, several aspects must be considered. First, we might ask, "What portions of the synthetase-amino acyl adenylate complex function in selecting the appropriate acceptor RNA chain?" Our data suggest that both the amino acid and protein moieties function in this selection. The fact that the enzyme-isoleucyl adenylate complex transfers isoleucine to the isoleucine-specific RNA chain but that the same enzyme in combination with valyl adenylate does not transfer the valine to any acceptor RNA chain emphasizes the role of the amino acid side chain. Similarly, the observation that different amounts of methionyl RNA are formed when methionyl adenylate is linked to two different proteins points to a specific function for the protein in the selection of the correct RNA chain. There is no information at present concerning the chemical structures of the RNA chains which allow for the "recognition" between a specific enzyme-amino acyl adenylate complex and its appropriate RNA chain. Clearly, the terminal nucleotide, to which the amino acid is bound, cannot account for this specificity since for each amino acid this unit is adenylic acid (16, 24, 25). Whether the differentiation between acceptor RNA chains relies on differences in nucleotide sequence, configuration, or to some unknown factors remains to be determined. The indications that there may be heterogeneity amongst RNA chains specific for a single amino acid may serve to complicate the analysis of this problem.

SUMMARY

Purified enzymes from *Escherichia coli* which form L-leucyl-, L-valyl-, L-isoleucyl-, or L-methionyl adenylates also catalyze the formation of the corresponding amino acyl ribonucleic acid derivatives. Each amino acid is bound through its carboxyl group to the terminal nucleotide (2'- or 3'-hydroxyl end) of specific polynucleotide chains. The synthesis of amino acyl ribonucleic acid derivatives is reversible, and in the case of L-valyl ribonucleic acid formation the equilibrium constant is 0.32. Indications were obtained that the polynucleotide chains specific for accepting L-methionine are heterogeneous.

REFERENCES

1. HOAGLAND, M. B., KELLER, E. B., AND ZAMECNIK, P. C., *J. Biol. Chem.*, **221**, 45 (1956).
2. BERG, P., *J. Biol. Chem.*, **222**, 1025 (1956).
3. DEMOSS, J. A., AND NOVELLI, G. D., *Biochim. et Biophys. Acta*, **22**, 49 (1956).
4. NISMANN, B., BERGMANN, F. H., AND BERG, P., *Biochim. et Biophys. Acta*, **26**, 639 (1957).
5. WEBSTER, G. C., *J. Biol. Chem.*, **229**, 535 (1957).
6. CLARK, J. M., JR., *J. Biol. Chem.*, **233**, 421 (1958).
7. KARASEK, M., CASTELFRANCO, P., KRISHNASWAMY, P. R., AND MEISTER, A., *J. Am. Chem. Soc.*, **80**, 2335 (1958).
8. KINGDON, H. S., WEBSTER, L. T., JR., AND DAVIE, E. W., *Proc. Natl. Acad. Sci. U. S.*, **44**, 757 (1958).
9. DAVIE, E. W., KONINGSBERGER, V. V., AND LIPMANN, F., *Arch. Biochem. Biophys.*, **65**, 21 (1956).
10. SCHWEET, R. S., AND ALLEN, E. H., *J. Biol. Chem.*, **233**, 1104 (1958).
11. HOAGLAND, M. B., STEPHENSON, M. L., SCOTT, J. F., HECHT, L. I., AND ZAMECNIK, P. C., *J. Biol. Chem.*, **231**, 241 (1958).
12. BERG, P., AND OFENGAND, E. J., *Proc. Natl. Acad. Sci. U. S.*, **44**, 78 (1958).
13. HOLLEY, R. W., *J. Am. Chem. Soc.*, **79**, 658 (1957).
14. WEISS, S. B., ACS, G., AND LIPMANN, F., *Proc. Natl. Acad. Sci. U. S.*, **44**, 189 (1958).
15. SCHWEET, R. S., BOVARD, F. C., ALLEN, E. H., AND GLASSMAN, E., *Proc. Natl. Acad. Sci. U. S.*, **44**, 173 (1958).
16. LIPMANN, F., HÜLSMANN, W. C., HARTMANN, G., BOMAN, H. G., AND ACS, G., *J. Cell. Comp. Physiol.*, **54**, 75 (1959).
17. ALLEN, E. H., GLASSMAN, E., AND SCHWEET, R. S., *J. Biol. Chem.*, **235**, 1061 (1960).
18. CANTONI, G. L., *J. Biol. Chem.*, **189**, 745 (1951).
19. KRISHNASWAMY, P. R., PAMILJANS, V., AND MEISTER, A., *J. Biol. Chem.*, **235**, PC 39 (1960).
20. GOLDTHWAIT, D. A., *J. Biol. Chem.*, **222**, 1051 (1956).
21. HARTMAN, S. C., AND BUCHANAN, J. M., *J. Biol. Chem.*, **233**, 456 (1958).
22. BELJANSKI, M., *Biochim. et Biophys. Acta*, **41**, 111 (1960); *Compt. rend. Acad. Sci.*, **250**, 624 (1960).
23. ZACHAU, H. G., ACS, G., AND LIPMANN, F., *Proc. Natl. Acad. Sci. U. S.*, **44**, 885 (1958).
24. PREISS, J., BERG, P., OFENGAND, E. J., BERGMANN, F. H., AND DIECKMANN, M., *Proc. Natl. Acad. Sci. U. S.*, **45**, 319 (1959).
25. HECHT, L. I., STEPHENSON, M. L., AND ZAMECNIK, P. C., *Proc. Natl. Acad. Sci. U. S.*, **45**, 505 (1959).
26. BERGMANN, F. H., BERG, P., AND DIECKMANN, M., *J. Biol. Chem.*, **236**, 1735 (1961).
27. OFENGAND, E. J., DIECKMANN, M., AND BERG, P., *J. Biol. Chem.*, **236**, 1741 (1961).
28. PREISS, J., DIECKMANN, M., AND BERG, P., *J. Biol. Chem.*, **236**, 1748 (1961).
29. BERG, P., *J. Biol. Chem.*, **233**, 601 (1958).
30. KUNITZ, M., *J. Gen. Physiol.*, **35**, 423 (1952).
31. GRUNBERG-MANAGO, M., ORTIZ, P. J., AND OCHOA, S., *Biochim. et Biophys. Acta*, **20**, 269 (1956).
32. OFENGAND, E. J., Ph.D. Thesis, Washington University, St. Louis, Mo., 1959.
33. COHN, W. E., AND CARTER, C. E., *J. Am. Chem. Soc.*, **72**, 4273 (1950).
34. MOLDAVE, K., CASTELFRANCO, P., AND MEISTER, A., *J. Biol. Chem.*, **234**, 841 (1959).
35. BERG, P., *J. Biol. Chem.*, **233**, 601 (1958).
36. DEMOSS, J. A., GENUTH, S. M., AND NOVELLI, G. D., *Proc. Natl. Acad. Sci. U. S.*, **42**, 325 (1956).
37. WONG, G. K., AND MOLDAVE, K., *J. Biol. Chem.*, **235**, 694 (1960).
38. BERG, P., *J. Biol. Chem.*, **222**, 991 (1956).
39. JENCKS, W. P., AND LIPMANN, F., *J. Biol. Chem.*, **225**, 207 (1957).
40. KELLERMAN, G. M., *J. Biol. Chem.*, **231**, 427 (1958).
41. MAAS, W. K., *Federation Proc.*, **12**, 241 (1953).
42. KALYANKAR, G. D., AND MEISTER, A., *J. Biol. Chem.*, **234**, 3210 (1959).
43. ZILLIG, W., SCHACTSCHABEL, D., AND KRONE, W., *Z. physiol. Chem.*, **318**, 100 (1960).

34

Reprinted from *Science,* **147**(3664), 1462–1465 (1965)

STRUCTURE OF A RIBONUCLEIC ACID

Robert W. Holley et al.

Abstract. *The complete nucleotide sequence of an alanine transfer RNA, isolated from yeast, has been determined. This is the first nucleic acid for which the structure is known.*

Transfer RNA's are the smallest biologically active nucleic acids known. Their function is to carry activated amino acids to the site of protein synthesis. During protein synthesis, the amino acid sequence of the polypeptide chain is determined by the interaction of a messenger RNA with transfer RNA's specific for a given amino acid. The structures of transfer RNA's are crucial in this process.

Three transfer RNA's, obtained from yeast, and specific for alanine, tyrosine, and valine, respectively, have been purified in our laboratories (*1*). Studies of the compositions of pancreatic ribonuclease digests have established that the three RNA's have quite different structures, as indicated by oligonucleotide sequences (*2*). The identification of a number of oligonucleotides obtained from the three RNA's has been described (*3*). We now summarize the determination of the first complete nucleotide sequence, that of the alanine RNA.

The structure determination involved the identification of small fragments formed by complete digestion of the RNA with pancreatic ribonuclease and takadiastase ribonuclease T1, followed by the determination of the structures of successively larger fragments, until the complete sequence of the RNA was established.

Complete digestion of the alanine RNA with pancreatic ribonuclease, an enzyme that cleaves the RNA chain next to pyrimidine nucleotides, for example, C- and U- (*4*), gives the 19 products listed in Table 1 (*5*). Complete digestion of the RNA with takadiastase ribonuclease T1, a highly specific enzyme (*6*) that cleaves the RNA chain next to G- and I-, gives the 17 products listed in Table 2 (*5*). Proof of the structures of these products required the use of both classical and new methods of sequence determination, as well as the identification of certain new nucleotides (*5, 7*).

Combination of the results summarized in Tables 1 and 2 permits description of the structure of the alanine RNA in terms of 16 oligonucleotide sequences, shown in Table 3 (*5*). Except for the positions of the two end sequences, the arrangement of the 16 oligonucleotide sequences is not established by these data. The presence of a 5'-phosphate on the pG-G-G-C- sequence establishes that this is the left end of the RNA molecule as conventionally written, and the 3'-hydroxyl on the U-C-C-A-C-C-Aoн sequence establishes that this is the right end of the molecule. The 16 oligonucleotide sequences account for a total of 77 nucleotide residues and give a calculated molecular weight for the RNA of 26,600 as the sodium salt.

Takadiastase ribonuclease T1 cleaves the alanine RNA selectively, under controlled conditions, and gives a number of large oligonucleotide fragments (*8, 9*). Analysis of these large fragments, in combination with the data in Tables 1 to 3, has furnished sufficient information to establish the complete nucleotide sequence of the RNA.

The structure of the alanine RNA is shown at the top of Fig. 1. Large fragments that were crucial in the proof of structure are shown in the lower part of Fig. 1.

Summary of proof of structure. To determine the structures of the large oligonucleotide fragments *a* to *k* (Fig. 1), a large fragment was digested with ribonuclease T1 giving certain of the previously identified fragments listed in Table 2, and additional information was then used to establish the arrangement of these complete digest fragments.

The isolation and proof of structure of fragments *a*, *b*, *c*, and *d* (Fig. 1) have been described (*8*). These frag-

Table 1. List of fragments obtained by complete digestion of the alanine RNA with pancreatic ribonuclease.

Coнt	MeG-G-C-
13 C-	A-G-C-
ψ-	A-G-DiHU-
6 U-	G-A-U-
A-C-	I-G-C-
MeI-ψ-	G-G-T-
DiMeG-C-	G-G-DiHU-
2 G-C-	G-G-A-C-
4 G-U-	pG-G-G-C-
G-G-G-A-G-A-G-U*-	

† The presence of a free 3'-hydroxyl group on this fragment indicates that cytidine occupies the terminal position in the purified alanine RNA. This establishes that the terminal adenylic acid residue is missing, as it is from most transfer RNA's isolated from commercial baker's yeast. A terminal adenylic acid residue is replaced under assay conditions before the amino acid is attached.

ments were obtained from a limited ribonuclease T1 digest (1 hour at 0°C) of the RNA. Fragments *e* to *i* were isolated by rechromatography (at 55°C) of the largest fragments obtained from the same limited digest. The two halves of the RNA molecule, *j* and *k*, were obtained from a much more limited ribonuclease T1 digest (9).

Analyses of complete ribonuclease T1 digests of *j* and *k* indicated that the fragments listed in Table 2 fell into two groups, corresponding to the two halves of the molecule (9). Determination of the complete structure of *k*, the right half of the molecule, was simpler and will be considered first.

The analyses established that *k* contained the ribonuclease T1 fragments present in *c* and *d* plus three oligonucleotides, U*-C-U-C-C-G-, T-ψ-C-G, and A-U-U-C-C-G-, as well as additional G-'s. Fragment *g* gave, on ribonuclease T1 digestion, the components of *d* plus one of these oligonucleotides, A-U-U-C-C-G-, and one G-. The sequence shown for *g* in Fig. 1 is the

Table 2. List of fragments obtained by complete digestion of the alanine RNA with takadiastase ribonuclease Tl.

9 G-	DiHU-C-G-
pG-	DiHU-A-G-
C-DiMeGp!	C-MeI-ψ-G-
U-MeGp!	T-ψ-C-G-
4 C-G-	A-C-U-C-G-
2 A-G-	U-C-C-A-C-C₀ₕ†
U-G-	U*-C-U-C-C-G-
U-A-G-	A-U-U-C-C-G-
C-U-C-C-C-U-U-I-	

† See Table 1.

only arrangement of these fragments that is consistent with the presence of a G-G-A-C- sequence in the RNA (Table 1).

Fragment *f* contained all the ribonuclease T1 fragments present in *k* that were not already accounted for in *g*. The structure of *f* was established by two pieces of information. First, the presence of G-G-G-A-G-A-G-U*- in a pancreatic ribonuclease digest of the RNA (Table 1) places the sequence U*-C-U-C-C-G- at the right end of *c* (Table 3). Second, pancreatic

ribonuclease digestion of *f* gave G-G-T-, which could be obtained only if the sequence G-T-ψ-C-G is at the right end of the U*-C-U-C-C-G- sequence. The sequence of *f* must therefore be that shown in Fig. 1.

Fragment *i* confirmed the nature of the attachment of *f* to *g*. The sequence of *k*, the right half of the molecule, is therefore that shown in Fig. 1.

Determination of the structure of the left half of the molecule was more complicated, with the proof of structure of *e* being most difficult. Essential information was obtained from partial and complete enzymatic digestion of *e*, as well as from limitations imposed by the sequences shown in Table 1. Complete degradation of *e* with ribonuclease T1 gave four fragments, DiHU-C-G, G-, DiHU-A-G-, and C-G-, in addition to the components of *b*. Digestion of *e* with pancreatic ribonuclease gave Ip!, establishing that Ip! must be at the right end of *e*. All four of the small ribonuclease T1 fragments must therefore be to the left of *b*. Five

STRUCTURE OF AN ALANINE RNA

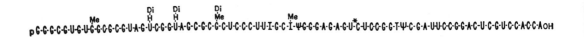

LARGE OLIGONUCLEOTIDE FRAGMENTS

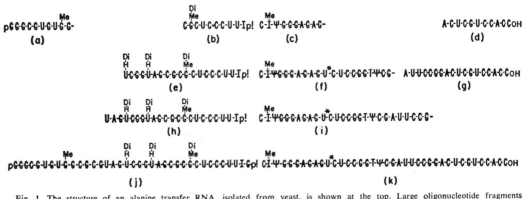

Fig. 1. The structure of an alanine transfer RNA, isolated from yeast, is shown at the top. Large oligonucleotide fragments that were crucial in the proof of structure are shown below.

arrangements of the four small fragments were consistent with the data of Table 1: DiHU-C-G-G-DiHU-A-G-C-G, DiHU-C-G-C-G-G-DiHU-A-G-, DiHU-A-G-C-G-G-DiHU-C-G-, C-G-G-DiHU-A-G-DiHU-C-G-, and G-DiHU-A-G-DiHU-C-G-C-G-. Since a pancreatic ribonuclease digest of *e* contained A-G-C-, the last two arrangements were eliminated. The third arrangement was eliminated because partial ribonuclease T1 digestion (½

hour at 0°C) removed a trinucleotide, DiHU-C-G, and left the remainder of *e* intact. Pancreatic ribonuclease digestion of the remainder of *e* gave only one trinucleotide, A-G-C, and a mixture of dinucleotides, eliminating the second arrangement. Therefore, the sequence DiHU-C-G-G-DiHU-A-G-C-G-must be attached at the left end of *b*; and fragment *e* has the structure shown in Fig. 1.

Fragment *h* differs from *e* only in

the presence of an additional U-A-G-sequence. The presence of A-G-DiHU-in a pancreatic ribonuclease digest of the RNA (Table 1) is sufficient to establish the structure of *h* as that shown in Fig. 1.

Finally, the structure of the left half of the molecule, *j*, was established by the isolation of I-Gp! from a pancreatic ribonuclease digest of this fragment, an indication that -I-Gp! is the right end of *j*. With this terminal Gp! plus all the sequences known to be present in *a* and *h*, everything in *j* is accounted for except two C-G- sequences. Only one arrangement of these is possible, placing them between *a* and *h*. This gives the structure of *j* shown in Fig. 1. This sequence is consistent with the presence of MeG-G-C- and two G-C-sequences in Table 1.

Joining the two halves of the molecule, *j* and *k*, gives the I-G-C- sequence that is found in the pancreatic ribonuclease digest of the RNA (Table 1).

The entire RNA molecule is now accounted for except for one nucleotide. The purified alanine RNA, in common with most of the transfer RNA's isolated from commercial baker's yeast, lacks a 3'-terminal pA residue. Since this terminal pA residue is replaced before the amino acid becomes attached, the complete structure of the yeast alanine RNA is that shown at the top of Fig. 1.

Discussion. The structure shown in Fig. 1 is of interest in several respects:

There is no obvious pattern in the distribution of the "minor" or "unusual" nucleotide residues. They are scattered throughout most of the molecule, though none is present near the amino acid–acceptor end.

The pentanucleotide sequence G-T-ψ-C-G, which is believed to be a common feature of transfer RNA's (*10*), is located approximately 20 nucleotides from the amino acid–acceptor end of the molecule. It seems likely that this sequence is present at the same position in other transfer RNA's.

There are several possible trinucleotide sequences that might represent the coding triplet or "anticodon" for the transfer of alanine (*11*). If it is assumed that the coding triplet contains two G's and one C, though this has not been established for this alanine RNA, there are two particularly intriguing possibilities. One is that the coding triplet is between the two dihydrouridylic acid residues in the sequence DiHU-C-G-G-DiHU. The func-

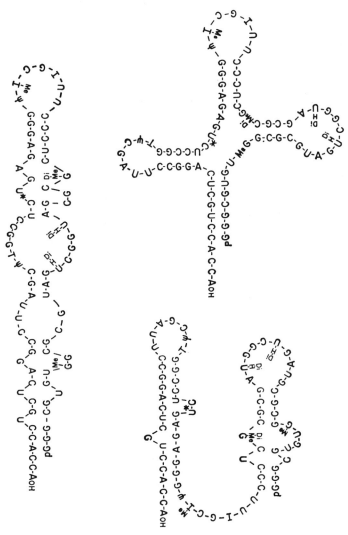

Fig. 2. Schematic representation of three conformations of the alanine RNA with short, double-stranded regions.

Table 3. Sequences that account for all of the nucleotide residues in the alanine RNA.

pG-G-G-C-, G-U-G-, U-MeG-G-C-, G-C-, G-U-A-G-, DiHU-C-G-, G-DiHU-A-G-, C-G-, C-DiMeG-, C-U-C-C-C-U-U-I-G-C-, MeI-ψ-, G-G-G-A-G-A-G-U*-C-U-C-C-G-, G-T-ψ-C-G-, A-U-U-C-C-G-, G-A-C-U-C-G-, U-C-C-A-C-C-A_{OH}

tion of the dihydrouridylic acid residues may be to "insulate" the coding triplet from the influence of neighboring nucleotides in the polynucleotide chain. Another possibility for the coding triplet is the I-G-C sequence, which is in the middle of the chain and is particularly sensitive to ribonuclease T1. This sequence would be expected to be equivalent to G-G-C in coding properties. The sequence that is the actual coding triplet is still unknown.

It is of interest to consider the structure from the standpoint of the conformation of the RNA in solution. Examination of the nucleotide sequence shows that there are no long complementary sequences that could give G paired with C and A paired with U. In fact, the longest complementary sequences contain only five nucleotides. As a consequence, double-stranded regions must be relatively short or must contain many imperfections in base pairing. Three conformations that utilize different regions of the RNA chain in Watson-Crick type base pairing are shown schematically in Fig. 2. These should be considered only speculative, and in any case it is likely that the actual conformation of the RNA in solution varies with the conditions. The factors that might be expected to influence the conformation of an RNA have been discussed (*12*).

The determination of this structure of a nucleic acid represents the successful conclusion of a major undertaking— an attempt to isolate a biologically active nucleic acid and establish its structure. At the time that the isolation of the alanine RNA was undertaken, it was not known whether an individual nucleic acid could be isolated from a complex mixture of nucleic acids. Once individual transfer RNA's were isolated, it was not known whether a nucleic acid structure could be determined. In retrospect, the part of the structure determination that was completely uncharted, namely the determination of long nucleotide sequences, turned out to be easier than was anticipated. The time-consuming part of our work was the identification of new nucleotides and the proof of structure of the small fragments obtained by complete digestion of the RNA with pancreatic ribonuclease and ribonuclease T1. Once the analyses of the two digests were finished, the elucidation of the structure was greatly facilitated by two experimental developments. The first was the discovery that ribonuclease T1 shows a high selectivity of action at 0°C, making it possible to isolate large fragments from the RNA molecule. The second was the development of highly efficient chromatographic methods which make use of $7M$ urea solutions (*13*) with long, narrow DEAE-cellulose columns and which provide the resolution required to separate small amounts of many different oligonucleotides (*8, 9*). Together, these developments made it possible to isolate and analyze large fragments of the RNA, and the results furnished sufficient information to establish the sequence.

Determination of the structure of the alanine RNA indicates that the structures of other nucleic acids can also be determined and provides a basis for attempts to synthesize a biologically active nucleic acid.

ROBERT W. HOLLEY, JEAN APGAR
GEORGE A. EVERETT
JAMES T. MADISON
MARK MARQUISEE, SUSAN H. MERRILL
JOHN ROBERT PENSWICK, ADA ZAMIR

U.S. Plant, Soil, and Nutrition Laboratory, U.S. Department of Agriculture, and *Department of Biochemistry, Cornell University, Ithaca, New York*

References and Notes

1. J. Apgar, R. W. Holley, S. H. Merrill, *J. Biol. Chem.* **237**, 796 (1962).
2. R. W. Holley, J. Apgar, S. H. Merrill, P. L. Zubkoff, *J. Am. Chem. Soc.* **83**, 4861 (1961).
3. R. W. Holley, J. Apgar, G. A. Everett, J. T. Madison, S. H. Merrill, A. Zamir, *Cold Spring Harbor Symp. Quant. Biol.* **28**, 117 (1963); B. P. Doctor, C. M. Connelly, G. W. Rushizky, H. A. Sober, *J. Biol. Chem.* **238**, 3985 (1963); V. M. Ingram and J. A. Sjöquist, *Cold Spring Harbor Symp. Quant. Biol.* **28**, 133 (1963); A. Armstrong, H. Hagopian, V. M. Ingram, I. Sjöquist, J. Sjöquist, *Biochemistry* **3**, 1194 (1964).
4. Abbreviations: p and - are used interchangeably to represent a phosphate residue; A-, adenosine 3'-phosphate; C-, cytidine 3'-phosphate; C_{OH}, cytidine (with the free 3'-hydroxyl group emphasized); DiHU-, 5,6-dihydrouridine 3'-phosphate; DiMeG-, N²-dimethylguanosine 3'-phosphate; I-, inosine 3'-phosphate; MeG-, 1-methylguanosine 3'-phosphate; MeI-, 1-methylinosin 3'-phosphate; ψ-, pseudouridine 3'-phosphate; T-, ribothymidine 3'-phosphate; U-, uridine 3'-phosphate; U*-, a mixture of U-, and DiHU-; p!, 2',3'-cyclic phosphate, for example: Ip!, inosine 2',3'-cyclic phosphate; DEAE, diethylaminoethyl.
5. R. W. Holley, G. A. Everett, J. T. Madison, A. Zamir, *J. Biol. Chem.*, in press.
6. K. Sato-Asano and F. Egami, *Nature* **185**, 462 (1960).
7. R. W. Holley, J. T. Madison, A. Zamir, *Biochem. Biophys. Research Commun.* **17**, 389 (1964); J. T. Madison and R. W. Holley, *ibid.* **18**, 153 (1965).
8. J. Apgar, G. A. Everett, R. W. Holley, *Proc. Natl. Acad. Sci. U.S.*, in press.
9. J. R. Penswick and R. W. Holley, *ibid.*, in press.
10. A. Zamir, R. W. Holley, M. Marquisee, *J. Biol. Chem.*, in press.
11. M. R. Bernfield and M. W. Nirenberg, *Science* **147**, 479 (1965).
12. A. S. Spirin, *Progr. Nucleic Acid Res.* **1**, 301 (1963); M. Spencer, *Cold Spring Harbor Symp. Quant. Biol.* **28**, 77 (1963); J. R. Fresco, L. C. Klotz, E. G. Richards, *ibid.* **28**, 83 (1963); J. R. Fresco and B. M. Alberts, *Proc. Natl. Acad. Sci. U.S.* **46**, 311 (1960); J. R. Fresco, B. M. Alberts, P. Doty, *Nature* **188**, 98 (1960); G. L. Brown and G. Zubay, *J. Mol. Biol.* **2**, 287 (1960); K. S. McCully and G. L. Cantoni, *ibid.* **5**, 497 (1962).
13. R. V. Tomlinson and G. M. Tener, *J. Am. Chem. Soc.* **84**, 2644 (1962).
14. Supported in part by the NSF and the NIH. John Scott Poucher contributed invaluable technical assistance. We thank Dr. Elizabeth B. Keller for many helpful suggestions.

8 January 1965

35

Reprinted from *J. Mol. Biol.*, 19(2), 548–555 (1966)

Codon—Anticodon Pairing:

The Wobble Hypothesis

F. H. C. CRICK

Medical Research Council, Laboratory of Molecular Biology

Hills Road, Cambridge, England

(*Received 14 February 1966*)

It is suggested that while the standard base pairs may be used rather strictly in the first two positions of the triplet, there may be some wobble in the pairing of the third base. This hypothesis is explored systematically, and it is shown that such a wobble could explain the general nature of the degeneracy of the genetic code.

Now that most of the genetic code is known and the base-sequences of sRNA molecules are coming out, it seems a proper time to consider the possible base-pairing between codons on mRNA and the presumed anticodons on the sRNA.

The obvious assumption to adopt is that sRNA molecules will have certain common features, and that the ribosome will ensure that all sRNA molecules are presented to the mRNA in the same way. In short, that the pairing between one codon–anticodon matching pair will to a first approximation be "equivalent" to that between any other matching pair.

As far as I know, if this condition has to be obeyed, and if all four bases must be distinguished in any one position in the codon, then the pairing in this position is *highly likely* to be the standard one; that is:[†]

$$G ==== C$$
$$\text{and} \quad A ==== U$$

or some equivalent ones such as, for example,

$$I ==== C$$
$$\text{and} \quad A ==== T$$

since this is the only type of pairing which allows all four bases to be distinguished in a strictly equivalent way.

We now know enough of the genetic code to say that in the *first two* positions of the codon the four bases are clearly distinguished; certainly in many cases, and probably in all of them. I thus deduce that the pairings in the first two positions are likely to be the standard ones.

[†] Throughout this paper the sign $====$ is used to mean "pairs with". If two bases are equivalent in their coding properties, this is written $\genfrac{}{}{0pt}{}{U}{C}$ or $\left.\genfrac{}{}{0pt}{}{U}{C}\right\}$

However, what we know about the code has already suggested two generalizations about the third place of the codon. These are:

(1) U⎱† this already appears true in about a dozen cases out of the possible 16,
    C⎰ and there are no data to suggest any exceptions.

(2) A⎱ probably true in about half of the possible 16 cases, but the evidence
    G⎰ suggests it may perhaps be incorrect in several other cases.

The detailed experimental evidence is rather complicated and will not be discussed here. (For details of the code see, for example, Nirenberg *et al.*, 1965; and Söll *et al.*, 1965.) It suffices that these rules *may* be true, as suggested by Eck (1963) a little time ago. Alternatively, only the first one may be true.

This naturally raises the question: Does *one* sRNA molecule recognize more than one codon, e.g. both UUU *and* UUC. Some evidence for this was first presented by Bernfield & Nirenberg (1965). They showed that *all* the sRNA for phenylalanine can be bound by poly U, although this sRNA also recognizes the triplet UUC, at least in part. More recent evidence along these lines is presented in Söll *et al.* (1966) and Kellogg *et al.* (1966). Again I do not wish to discuss here the evidence in detail, but simply to ask: If one sRNA codes both XYU and XYC, how is this done?

Now if we do not know anything about the geometry of the situation, it might be thought that almost any base pairs might be used, since it is well known that the bases can be paired (i.e. form at least two hydrogen bonds) in many different ways. However, it occurred to me that if the first two bases in the codon paired in the standard way, the pairing in the third position might be *close* to the standard ones.

We therefore ask: How many base pairs are there in which the glycosidic bonds occur in a position close to the standard one? Possible pairs are:

$$G ==== A \tag{1}$$

In my opinion this will not occur, because the NH_2 group of guanine cannot make one of its hydrogen bonds, even to water (see Fig. 1).

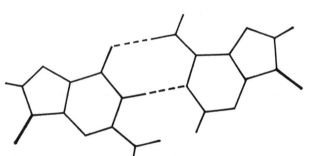

Fig. 1. The unlikely pair guanine–adenine.

$$U ==== C \tag{2}$$

This brings the two keto groups rather close together and also the two glycosidic bonds, but it may be possible (see Fig. 2).

† This symbol implies that both U and C code the same amino acid.

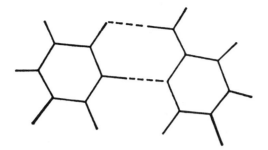

FIG. 2. The close pair uracil–cytosine.

$$U ==== U \qquad\qquad (3)$$

Again rather close together (see Fig. 3).

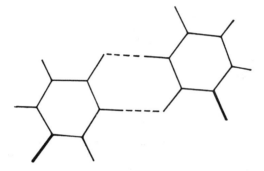

FIG. 3. The close pair uracil–uracil.

$$G ==== U \qquad\qquad (4)$$
$$\text{or } I ==== U$$

These only require the bond to move about 2·5 Å from the standard position (see Fig. 4).

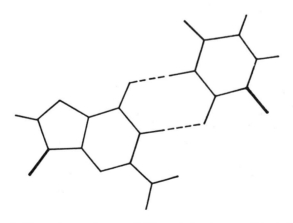

FIG. 4. The pair guanine–uracil (the pair inosine–uracil is similar).

323

$$I ==== A \qquad (5)$$

This is perfectly possible. Poly I and poly A will form a double helix. The distance between the glycosidic bonds is increased (see Fig. 5).

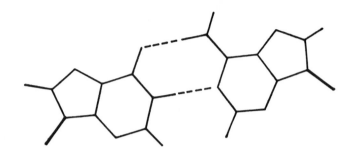

FIG. 5. The pair inosine–adenine.

As far as I know, these are all the possible solutions if it is assumed that the bases are in their usual tautomeric forms.

I now postulate that in the base-pairing of the third base of the codon there is a certain amount of play, or wobble, such that more than one position of pairing is possible.

As can be seen from Fig. 6, there are seven possible positions which might be reached by wobbling. However, it by no means follows that all seven are accessible, since the molecular structure is very likely to impose limits to the wobble. We should there-fore strictly consider all possible *combinations of allowed positions*. There are 127 of these, but most of them are trivial. If we adopt the rule that *all four bases* on the codon (in the third position) must be recognized (that is, paired with) we are left with 51 different combinations. This is too many for easy consideration, but fortunately we can eliminate most of them by only accepting combinations which do not violate the broad features of the code. If we assume:

(a) that all four bases must be recognizable;

(b) that the code must *in some cases* distinguish between

$\left.\begin{matrix} U \\ C \end{matrix}\right\}$ and $\left.\begin{matrix} A \\ G \end{matrix}\right\}$ as it appears to do for the **pairs**

| Phe | Tyr | His | Asn | Asp |
| Leu | C.T.† | Gln | Lys | Glu |

(not all of which are likely to be wrong)

then by strictly logical argument it can be shown both that the standard position must be used, and that the three positions on the left of Fig. 6 cannot be used.

This leaves us with only four possible sites to consider one of which—the standard one—must be included. There are therefore only seven possible combinations. I have examined all these, but I shall restrict myself here to the case in which all four posi-tions are used, as this is structurally the most likely and also seems to give the code (called code 4 in the note privately circulated) which best fits the experimental data.

† C.T., Chain termination.

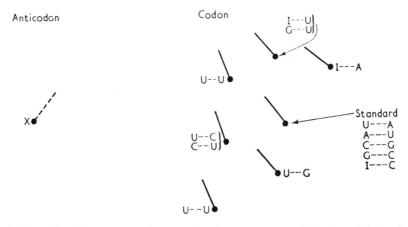

FIG. 6. The point X represents the position of the C_1' atom of the glycosidic bond (shown dotted) in the anticodon. The other points show where the C_1' atom and the glycosidic bond fall for the various base pairs. (Pairs with inosine in the codon have been omitted for simplicity.) The wobble code suggested uses the four positions to the right of the diagram, but not the three close positions.

The rules for pairing between the third base on the codon and the corresponding base on the anticodon are set out in Table 1. It can be seen that these rules make several strong predictions:

(1) it is not possible to code for either C alone, or for A alone.

For example, at the moment the codon UGA has not been decisively allocated. Wobble theory states that UGA might either:

    (a) code for cysteine, which has UGU and UGC; or

    (b) code for tryptophan, which has UGG; or

    (c) not be recognized.

TABLE 1

Pairing at the third position of the codon

Base on the anticodon	Bases recognized on the codon
U	A G
C	G
A†	U
G	U C
I	U C A

† It seems likely that inosine will be formed enzymically from an adenine in the nascent sRNA. This may mean that A in this position will be rare or absent, depending upon the exact specificity of the enzyme(s) involved.

However it does *not* permit UGA to code for any amino acid other than cysteine or tryptophan. This rule could also explain why no suppressor has yet been found which suppresses only *ochre* mutants (UAA), although suppressors exist which suppress both *ochre* and *amber* mutants (UA$_G^A$).

(2) If an sRNA has inosine in the place at the relevant position on the anticodon (i.e. enabling it to pair with the third base of the codon), then it must recognize U, C and A in the third place of the codon. Conversely, those amino acids coded only by XY$_C^U$ (such as Phe, Tyr, His, etc.) cannot have inosine in that place on their sRNA.

(3) Wobble theory does not state exactly how many different types of sRNA will actually be found for any amino acid. However if an amino acid is coded for by all four bases in the third position (as are Pro, Thr, Val, etc.), then wobble theory predicts that there will be at least two sRNA's. These can have the recognition pattern:

$$\left.\begin{matrix} U \\ C \end{matrix}\right\} \text{ plus } \left.\begin{matrix} A \\ G \end{matrix}\right\}$$

$$\text{or} \qquad \left.\begin{matrix} U \\ C \\ A \end{matrix}\right\} \text{ plus } G$$

Note that the sets actually used for any amino acid may well vary from species to species.

The Anticodons

At this point it is useful to examine the experimental evidence for the anticodon. In the sRNA for alanine from yeast, Holley *et al.* (1965) have the following sequences:

$$--- \text{pUpUpIp Gp CpMeIp}\Psi\text{p} ---$$

$$\text{position} --- 36 \ 37 \ 38 ---.$$

Zachau and his colleagues (Dütting, Karan, Melchers & Zachau, 1965) have for one of the serine sRNA's from yeast:

$$--- \text{p}\Psi\text{pUpIpGpApA}^+\text{p}\Psi\text{p} ---$$

(A$^+$ stands for a modified A)

For the valine sRNA from yeast, Ingram & Sjöquist (1963) have shown that the only inosine occurs in the sequence:

$$--- \text{pIpApCp} ---$$

Holley *et al.* (1965) have already pointed out that IGC is a possible anticodon for alanine, and the additional evidence makes it almost certain to my mind that this is correct, and that the anticodons are as given in the Table below†:

† *Note added 26 April 1966.* Drs J. T. Madison, G. A. Everett and H. Kung (personal communication) have completed the sequence of the tyrosine sRNA from yeast. The sequence strongly suggests that the anticodon in this case is GΨA, corresponding to the known codons UA^U. Since Ψ can form the same base pairs as U, this is in excellent agreement with the previous data.

Yeast sRNA

	Anticodon	Codon
Ala	I G C	G C ?
Ser	I G A	U C ?
Val	I A C	G U ?

remembering that the pairing proposed between codon and anticodon is *anti*-parallel. Thus I confidently predict: the anticodon is a triplet at (or very near) positions 36–37–38 on every sRNA, and that the *first two bases* in the codon pair with this (in an anti-parallel manner) *using the standard base pairs.*

However, inosine does not occur in every sRNA. In particular Holley *et al.* (1963) (and personal communication) have reported that the tyrosine sRNA has two peaks, neither of which contains inosine. Moreover, Sanger (personal communication) tells me that there is rather little inosine in the total sRNA from *E. coli.*

Testing the Theory

Two obvious tests present themselves:

(1) To find which triplets are bound by any one type of sRNA. This is being done by Khorana and his colleagues (Söll *et al.*, 1966), and also by Nirenberg's group (Kellogg, Doctor, Loebel & Nirenberg, 1966). The difficulty here is to be sure that the sRNA used is pure, and not a mixture.

(2) To discover unambiguously the position of the anticodon on sRNA, and to find further anticodons. This will certainly happen as our knowledge of the base sequence of sRNA molecules develops. The absence of inosine from any anticodon is obviously of special interest.

In conclusion it seems to me that the preliminary evidence seems rather favourable to the theory. I shall not be surprised if it proves correct.

I thank my colleagues for many useful discussions and the following for sending me material in advance of publication: Dr M. W. Nirenberg, Dr H. G. Khorana, Dr G. Streisinger, Dr W. Holley, Dr J. Fresco, Dr H. G. Zachau, Dr C. Yanofsky, Dr H. G. Wittmann, Dr H. Lehmann and Dr J. D. Watson.

REFERENCES

Bernfield, M. R. & Nirenberg, M. W. (1965). *Science,* **147,** 479.
Dütting, D., Karan, W., Melchers, F. & Zachau, H. G. (1965). *Biochim. biophys. Acta,* **108,** 194.
Eck, R. V. (1963). *Science,* **140,** 477.
Holley, R. W., Apgar, J., Everett, G. A., Madison, J. T., Marquisee, M., Merrill, S. H., Penswick, J. R. & Zamir, A. (1965). *Science,* **147,** 1462.
Holley, R. W., Apgar, J. Everett, G. A., Madison, J. T., Merrill, S. H. & Zamir, A. (1963). *Cold Spr. Harb. Symp. Quant. Biol.* **28,** 117.
Ingram, V. M. & Sjöquist, J. A. (1963). *Cold. Spr. Harb. Symp. Quant. Biol.* **28,** 133.

Kellogg, D. A., Doctor, B. P., Loebel, J. E. & Nirenberg, M. W. (1966). *Proc. Nat. Acad. Sci., Wash.* **55**, 912.

Nirenberg, M., Leder, P., Bernfield, M., Brimacombe, R., Trupin, J., Rottman, F. & O'Neal, C. (1965). *Proc. Nat. Acad. Sci., Wash.* **53**, 1161.

Söll, D., Jones, D. S., Ohtsuka, E., Faulkner, R. D., Lohrmann, R., Hayatsu, H., Khorana, H. G., Cherayil, J. D., Hampel, A. & Bock, R. M. (1966). *J. Mol. Biol.* **19**, 556.

Söll, D., Ohtsuka, E., Jones, D. S., Lohrmann, R., Hayatsu, H., Nishimura, S. & Khorana, H. G. (1965). *Proc. Nat. Acad. Sci., Wash.* **54**, 1378.

36

Conformation of the Anticodon Loop in tRNA

by
W. FULLER
A. HODGSON
Biophysics Department and
MRC Biophysics Research Unit,
King's College, University of London

A molecular model for the anticodon arm is proposed which is compatible with chemical, X-ray and genetic evidence. It provides a stereochemical basis for Crick's "wobble" hypothesis.

NUCLEOTIDE sequences determined for a number of amino-acid specific tRNA molecules[1-4] have led to the suggestion that these molecules have a "clover leaf" structure (Fig. 1). This was because, despite their different nucleotide sequences, there are striking structural homologies when the tRNA molecules are folded so that the number of intramolecular Watson–Crick base-pairs is a maximum (Fig. 1). Diagrams like Fig. 1, however, indicate little of the three-dimensional appearance of such structures and their implications. Therefore we have constructed three-dimensional models, and here describe a molecular model-building study of the anticodon arm.

Using chemical information about nucleotide sequence and X-ray evidence on the conformation of base-paired regions in the tRNA, the maintenance of reasonable stereochemical constraints leads to a model for the anticodon arm. This model accounts for the observed

degeneracy in the reading of the third position of the codon and also makes a prediction about the site of the distortion required to accommodate this degeneracy.

Model Building Technique

We used Corey, Pauling and Koltun[5] spacefilling models and also skeletal models with a scale of 4 cm to 1 Å (ref. 6). The former ensure that short van der Waals contacts are avoided during preliminary investigations. Because, however, the atomic centres in them are inaccessible, we used skeletal models when preliminary study suggested that a particular conformation merited detailed analysis. Lengths and angles of covalent bonds and short van der Waals contacts were calculated from atomic co-ordinates measured on skeletal models and the co-ordinates were adjusted until acceptable stereochemistry was obtained, that is lengths of covalent bonds within 0·05 Å of accepted values, covalent angles within 6° and no non-covalently bonded contacts more than 0·4 Å short of the sum of the atomic van der Waals radii. We do not necessarily believe that our models describe the actual molecular conformations to an accuracy of a few hundredths of an angstrom, but the analysis shows that a model with the general characteristics we propose can be built with acceptable stereochemistry. Only if model building is treated as a rigid discipline with strict attention paid to detailed stereochemistry can the results of a study such as this be considered reliable and meaningful.

Conformation of the Anticodon Arm

X-ray diffraction suggests that the molecules of tRNA (ref. 7), in common with all RNA molecules so far studied by this method, contain helical regions with a conformation similar to that determined for two-stranded reovirus RNA (ref. 8). We have assumed that the Watson–Crick base-paired regions in the clover leaf structure have a conformation like the eleven-fold double-helical structure of reovirus RNA (rather than the less favoured ten-fold possibility). In the anticodon arm there is a loop of seven nucleotides at the end of the helical region. From considerations of biological function, the structural homologies in the different tRNA species might be expected to extend to the conformation of this loop.

The characteristic features of polynucleotide secondary structure are provided by interbase hydrogen bonding and base-stacking. The tRNA nucleotide sequences so far determined do not suggest an intramolecular base-pairing scheme which would give a similar structure for all the anticodon loops (Fig. 2). Therefore we searched for conformations of this loop which maximized single-

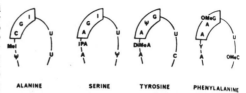

Fig. 2. The nucleotide sequences in yeast tRNA anticodon loops. The anticodon is shown boxed. The symbols ψ, MeI, C, G, I, U, A, IPA, DiMeA, Y, OMeG, OMeC stand for respectively: pseudouracil, 1-methyl-inosine, cytosine, guanine, inosine, uracil, adenine, isopentenyl-adenine, dimethyl-adenine, a so-far unidentified purine, O-methyl guanine, O-methyl cytosine. (The references from which the sequences were taken are in the caption to Fig. 1.)

stranded base-stacking. In doing this we also attempted to: (a) avoid negatively charged phosphate groups coming closer to each other than in accurately determined crystalline fibrous structures; (b) ensure that hydrogen bond donor groups on unpaired bases and ribose sugars were not buried in the structure, so they were unavailable for hydrogen bonding; (c) maintain single bond orientations in the polynucleotide chain (for example, the conformation at the glycosidic link) within the limits of values observed in model compounds and other polynucleotides.

When these stereochemical constraints are maintained, model-building studies suggest that the polynucleotide chain has surprisingly little conformational freedom. Furthermore, orientation of the single bonds in the only two polyribonucleotides whose structures have been determined in detail by X-ray analysis (two-stranded helical RNA (ref. 8), and two-stranded polyadenylic acid[9]) are rather similar. Therefore if stacking is to be maintained, the polynucleotide chain might be expected to have a conformation similar to that in one of the structures described for ribopolynucleotides. Stacking as much as possible of the anticodon loop on top of the double helical region of the anticodon arm might be expected to "nucleate" the structure of the single-stranded region so that its nucleotide conformation is similar to that in the double helical region, that is that of the eleven-fold model for reovirus RNA. (We refer to this conformation as standard.) There is some support from physical studies on solutions of polynucleotides and dinucleotides for postulating that the conformation of a single-stranded polynucleotide with base-stacking is similar to the conformation it would have as one of the strands in a two-stranded structure[10,11].

Figs. 3 and 4 illustrate the structure which stacks the greatest number of the nucleotides in the anticodon loop. Five nucleotides are stacked in the standard conformation so that they lie on the same helix as that chain in the anticodon arm double-helix nearer the tRNA 3' end. This structure represents a unique solution to the problem of maximizing base-stacking in the anticodon loop. Conformations with slightly different base tilt and rotation and translation of each nucleotide (for example if the standard nucleotide conformation was that of the ten-fold rather than eleven-fold RNA model) could of course give a similar degree of base-stacking. Stacking combinations of nucleotides other than those stacked in this structure, however, result in less than five nucleotides being stacked. In particular five nucleotides cannot be arranged so that they lie on the same helix as that chain of the anticodon arm double-helix nearer the tRNA 5' end. This is shown in Fig. 4 where A and B are closer together than they would be for a structure with bases perpendicular to the helix axis. If, however, the five bases were stacked on the chain of the two-stranded helix nearer the tRNA 5' end, the base tilt would make the distance spanned by the two non-standard nucleotides greater than for a structure with bases perpendicular to the helix axis. In addition to this increased distance, the two nucleotides would have to span the RNA groove containing the 2-keto

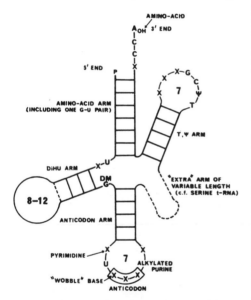

Fig. 1. Generalized clover-leaf structure for yeast tRNA based on sequences determined for tRNAs specific for alanine[1], serine[2], tyrosine[3] and phenylalanine[4]. A base-pair is indicated by a line linking two parts of the RNA chain. A base-pair indicated by a dashed line indicates a base-pair occurring in some tRNAs but not in others); X indicates a nucleotide which varies with the tRNA species; nucleotides which occur at an equivalent position in all sequences are denoted as follows—uracil (U), dimethylguanine (DMG), adenine (A), cytosine (C), thymine (T), pseudouracil (ψ). The number at the centre of each loop indicates the number of nucleotides in the loop.

groups rather than that containing the 6-keto groups which is spanned in the model illustrated in Figs. 3 and 4. The RNA conformation is such that spanning the groove containing the 2-keto groups requires a much longer poly-nucleotide chain.

In the structure illustrated in Figs. 3 and 4 the two nucleotides of the anticodon loop not in the standard conformation have the planes of their bases approximately parallel, with some overlap of their hydrophobic surfaces. There is, however, some flexibility in this region of the structure and this conformation should be regarded as typical of a number of related possibilities. All the tRNA nucleotide sequences so far determined are compatible with the poorer stacking of these two nucleotides as compared with that in the standard helix because these bases are always pyrimidines (of which at least one is uracil). These bases are generally thought to stack least well. In all these sequences, the second of the five nucleotides in the single stranded helix (that is 7 in Fig. 4) has a chemically modified hydrogen bonding donor group on the base. In addition to inhibiting base-pairing which might favour alternative structures to that in Figs. 3 and 4, this group may increase hydrophobic stabilization of this stacked conformation.

The structure illustrated in Figs. 3 and 4 could describe the conformation of the loop of 7 unpaired bases in the Tψ loop (Fig. 1). The occurrence of uracil and cytosine, however, at what would be positions 6 and 7 in the single-stranded helix (Fig. 4) make it a rather less attractive solution than it is for the anticodon loop.

A loop at the end of an RNA double helix need be no longer than three nucleotides[11]. The base-stacking in such a structure, however, is probably much less than that in the standard conformation and such a loop would only be expected to occur if it contained nucleotides with poor stacking interactions, for example the loop with UUU and UCU at the end of the extra arm in the two serine tRNAs (Fig. 1).

Codon–Anticodon Interactions

From consideration of the likely anticodon in a number of tRNAs and a knowledge of the different codons which will

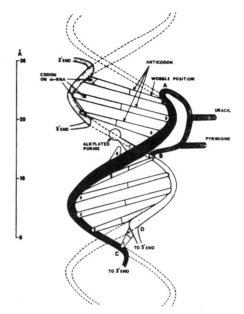

Fig. 4. Schematic diagram of the tRNA anticodon loop illustrating its relationship to the codon and the helical character of the structure. The letters A, B, C and D identify the same points on the structure as in Fig. 3. The bases in nucleotides 1 to 10 are stacked on one another and follow the regular helix which is shown black. The chain of the anticodon double helix between D and B is shaded like the codon to indicate that they follow the same helix. This helix is complementary to the black one. The two nucleotides not in the standard conformation are represented by dark line shading. The representation of their conformation is very schematic because they lie behind nucleotides 8, 9 and 10 in the black chain. The dotted lines indicate the generic helix from which the structure can be imagined to be derived.

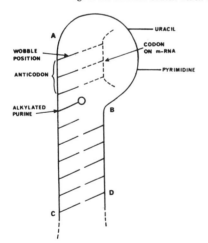

Fig. 3. Schematic diagram of the model for the tRNA anticodon arm illustrating its relationship to the codon. The helical regions are shown as straight in this diagram. CD is the first base-pair in the double helical region of the anticodon arm and all the bases between A and C are stacked on one another and follow a regular helix. The companion set between B and D and the set of three bases in the codon follow the complementary helix. In space (see Fig. 4) A and B are quite close together because five nucleotide pairs is about half a turn of the helix.

recognize a particular tRNA species, Crick[13] has proposed a hypothesis for codon–anticodon recognition. This involves standard Watson–Crick base-pairing between the bases in the codon and anticodon triplets while allowing the possibility of "wobble" or limited alternative pairing in the third position. When codon and anticodon are paired in this way, the atomic sequence in one chain is the reverse of that in the other and the two triplets can be arranged as the two strands in a regular RNA double helix. In our model the anticodon triplet occupies positions 8, 9 and 10 of the anticodon helix (Figs. 3 and 4) and a messenger RNA (mRNA) codon can be base-paired to it without steric hindrance between the rest of the anticodon arm and adjacent mRNA codons. In fact simultaneous recognition of two anticodon arms by adjacent codons is stereochemically possible (Fig. 5). There is not much flexibility in the relative position and orientation of two anticodon arms when they are interacting simultaneously with adjacent mRNA codons. They can interact, however, with each other in a way similar to neighbouring helices in crystalline fibres of reovirus RNA (ref. 8) (see caption to Fig. 5). The possible biological significance of intermolecular hydrogen bonds between the sugar hydroxyl of one helix and the phosphate oxygen of another has been noted[8]. One such hydrogen bond is formed between the two anticodon arms when arranged as in Fig. 5. Preliminary studies suggest that it is possible to arrange the clover leaf arms of the generalized tRNA molecule (Fig. 1) so that there is no steric hindrance between two tRNA molecules whose anticodon arms are interacting in this way.

It might be asked if mistakes in translation could occur by codon "recognition" of bases at positions 7, 8 and 9

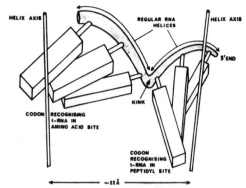

Fig. 5. Schematic diagram of successive codons in mRNA simultaneously recognizing anticodon arms. Each codon of the mRNA has the conformation illustrated in Figs. 3 and 4. The operation required to move the anticodon arm from the amino-acid site to the peptidyl site involves a rotation θ about and a translation t along the anticodon helix axis and a translation d perpendicular to the helix axis along a line joining the two helix axes illustrated in this figure. If the anticodon helices are linked by a hydrogen bond and have a stereochemical relationship like two reovirus RNA helices in the crystalline fibre then the symmetry operation can be defined precisely as follows (otherwise it is an approximate description). $\theta = 87 \cdot 3^\circ$ (that is $120-32 \cdot 7$), $t = -2 \cdot 73$ Å, $d = 22$ Å. Values of θ and t which would move nucleotide 9 into the position occupied by nucleotide 10 (Fig. 4) are taken as positive.

(or even 6, 7 and 8) rather than 8, 9 and 10 of the anticodon helix. Studies with spacefilling models suggest that recognition of both 7, 8 and 9 and 6, 7 and 8 could not be excluded even if it is insisted that adjacent codons recognize anticodon arms simultaneously. Recognition of 6, 7 and 8 is the less plausible stereochemically. If the geometry of the tRNA–mRNA–ribosome interaction is inadequate to prevent mis-reading of this kind, it may be that it is prevented by the chemical modification which occurs at nucleotide 7 (Fig. 4) in all sequences determined so far[1-4]

Stereochemical Apects of the "Wobble" Hypothesis

We have considered possible distortions of our model for codon–anticodon interaction which would accommodate the alternative base-pairings described by Crick in his "wobble" hypothesis[13] (Table 1). The separation and relative orientation of the glycosidic links in these alternative base-pairs differ from that in the standard Watson–Crick pairs. The position of the wobble base-pair in our model is illustrated in Figs. 3 and 4. Accommodation of adenine-inosine in this position requires extension of the sugar–phosphate chain linking the second and third bases of the anticodon (at positions 9 and 10) or compression of the chain joining the second and third bases of the codon. The chain in the RNA helix is already rather compressed (about 5·6 Å between successive phosphates as compared to about 7 Å in a completely extended chain) and further compression results in steric hindrance between the 2′ hydroxyl (and the sugar carbon to which it is attached) and the base of the previous nucleotide. Therefore the principal distortion involved in accommodating adenine-inosine (and any other alternative pairs with an interglycosidic link separation larger than the standard pair) must occur at the anticodon.

In contrast accommodation of an alternative pair with an interglycosidic link separation smaller than the standard pair would require either extension of the sugar-phosphate chain between the second and third bases of the codon or compression of the chain between the second and third bases of the anticodon. Significant compression of the chain can be excluded, and so it appears that the principal distortion involved in accommodating pairs with

an interglycosidic link separation significantly shorter than standard must occur at the codon.

Site of Distortion in "Wobble" Pairing

Using skeletal models we have found that all the alternative pairing required to account for the genetic evidence on degeneracy in the third position of the codon-anticodon interaction can be accommodated in our model by distortion of the anticodon alone (Table 1). Distortion of the codon conformation is not required. Further, our model building studies indicate that a uracil–uracil pairing can be accommodated if distortion is allowed at the codon. Therefore, because the genetic evidence excludes such pairing in this position, we can conclude that it does not occur because the codon conformation cannot be significantly distorted. It should be noted, however, that our criteria for an acceptable pairing relate to the geometry of the interbase hydrogen bonds and the stereochemistry of the sugar–phosphate chain. While these are clearly necessary requirements, other considerations may also be relevant to the occurrence of a particular alternative pair, for example interbase dipole–dipole interactions. (It may be that non-occurrence of the uracil–uracil pair is the result of such effects rather than of the codon being rigidly held.)

The assignment of "wobble" distortion to the anticodon rather than the codon seems reasonable from general considerations because one might expect each codon to be held on the 30S ribosome in a way which is independent of its position in the mRNA and therefore through bonds involving groups near to or part of the sugar–phosphate chain of the codon currently being read. Such bonds would be expected to limit the conformational flexibility of the codon as compared with the anticodon (which is a relatively small part of the tRNA molecule and not necessarily close to the ribosomal binding site on the tRNA) in a way which is compatible with the above assignment of the distortion

Table 1. THE ACCOMMODATION OF ALTERNATIVE BASE-PAIRS AT THE "WOBBLE" POSITION IN THE CODON-ANTICODON COMPLEX AS A FUNCTION OF WHETHER THE DISTORTION IS ALLOWED IN THE CODON OR ANTICODON CONFORMATION

Alternative codon-anticodon pairs (groups involved in interbase hydrogen bonding)	Genetic evidence for its occurrence (— denotes occurrence) (X denotes non-occurrence)	Stereochemistry of the polynucleotide chain according to the site of distortion (— denotes acceptable stereochemistry) (X denotes unacceptable stereochemistry)		
		Distortion at anticodon only	Distortion at codon only	Distortion at codon and anticodon
Adenine-inosine (6-amino to 6-keto and N1 to N1)	—	(The torsion angle of the inosine glycosidic link is about 5° outside the acceptable range)	X	
Guanine-uracil (6-keto to N1 and N1 to 2-keto)	—	(There is a hydrogen-oxygen non-bonded contact of about 2·2 Å, i.e., about 0·3 less than the sum of the van der Waals radii of these atoms)	X	
Uracil-guanine (N1 to 6-keto and 2-keto to N1)	—	(There is a hydrogen-oxygen non-bonded contact of about 2·2 Å, i.e., about 0·3 less than the sum of the van der Waals radii of these atoms)	X	
Uracil-uracil (6-keto to N1 and N1 to 2-keto)	X	X	X	X
Uracil-uracil (N1 to 6-keto and 2-keto to N1)	X	X	—	—
Uracil-cytosine (6-keto to 6-amino and N1 to N1)	X	X	X	X

For none of the pairings denoted as "stereochemically acceptable" is the stereochemistry quite as satisfactory as that in the undistorted standard conformation. The departures from acceptable stereochemistry are noted and are small enough for it to be concluded that the codon–anticodon complex could be distorted to accommodate these pairs. In contrast the stereochemistry of the pairs denoted "stereochemically unacceptable" is quite unacceptable with non-covalently bonded contacts 1 or 2 Å less than normal values and with torsion angles 40 to 50° outside the range of observed values.

associated with "wobble" pairing to the anticodon. Further, the occurrence of the "wobble" base at the top of the single strand anticodon helix allows distortion in the part of the sugar–phosphate chain to which it is attached to be absorbed in the conformational flexibility of the two unstacked pyrimidines next to it.

The model we propose for the anticodon arm of tRNA allows codon–anticodon interaction through Watson–Crick base-pairing. The codon and anticodon nucleotide triplets have the conformation of the two strands in a regular RNA double helix. The alternative or "wobble" pairings suggested for the third base of the anticodon can be accommodated in this model by distortion of the anticodon conformation. It is not necessary to postulate distortion of the anticodon conformation. The observation that uracil–uracil is not a wobble pairing suggests that codon conformation distortion does not take place. This model of the anticodon arm allows adjacent mRNA codons to simultaneously recognize anticodon arms.

In a study such as this it is important to identify clearly the principal assumptions on which the model building is based. Essentially the only assumption we make is that the number of stacked bases in the anticodon loop should be a maximum; this leads to a unique solution for the conformation of the loop. The model receives support from the base sequences which have been determined for the anticodon loop in a number of tRNAs (Fig. 2): the pyrimidines (mainly uracil) are in the irregular part of the loop, the wobble base is at that position in the stacked part of the structure which has the most conformational flexibility, and the modified purine is at a position which could prevent a wrong set of three nucleotides in the anticodon being recognized by the codon. There is no structural or genetic evidence in conflict with this model and, while the model building study does not prove it to be correct, its stereochemical neatness and the manner in which it accounts for what is known about the anticodon region of tRNA suggest that it is essentially correct.

One of us (A. H.) is the holder of a Medical Research Council award. We thank Professor Sir John Randall, Professor M. H. F. Wilkins, and Dr F. H. C. Crick for their interest, Miss A. Kernaghan for preparing the figures and Mr Z. Gabor for carrying out photographic work.

Received June 16; revised August 3, 1967.

[1] Holley, R. W., Apgar. J., Everett, G. A., Madison, J. T., Marquisee, M., Merrill, S. H., Penswick, J. R., and Zamir, A., Science, 147, 1462 (1965).
[2] Zachau, H., Dütting, D., and Feldman, M., Angew Chemie, 78, 393 (1966).
[3] Madison, J. T., Everett, G. A., and Kung, H., Science, 153, 531 (1966).
[4] RajBhandary, U. L., Chang, S. M., Stuart, A., Faulkner, R. D., Hoskinson, R. H., and Khorana, H. G., Proc. US Nat. Acad. Sci., 57, 751 (1967).
[5] Koltun, W. L., Biopolymers, 3, 665 (1965).
[6] Langridge, R., Marvin, D. A., Seeds, W. E., Wilson, H. R., Hooper, C. W., Wilkins, M. H. F., and Hamilton, L. D., J. Mol. Biol., 2, 38 (1960).
[7] Dover, S. D., Spencer, M., Wilkins, M. H. F., and Fuller, W. (in preparation).
[8] Arnott, S., Hutchinson, F., Spencer, M., Wilkins, M. H. F., Fuller, W., and Langridge, R., Nature, 211, 227 (1966).
[9] Rich, A., Davies, D. R., Crick, F. H. C., and Watson, J. D., J. Mol. Biol., 3, 71 (1961).
[10] McDonald, C. C., Phillips, W. D., and Lazar, J. (in the press).
[11] Buch, C. A., and Tinoco, jun., I., J. Mol. Biol., 23, 601 (1967).
[12] Spencer, M., Fuller, W., Wilkins, M. H. F., and Brown, G. L., Nature, 194, 1014 (1962).
[13] Crick, F. H. C., J. Mol. Biol., 19, 548 (1966).

37

Reprinted from *Science*, **179**, 285–288 (Jan. 1973)

Three-Dimensional Structure of Yeast Phenylalanine Transfer RNA: Folding of the Polynucleotide Chain

S. H. Kim, G. J. Quigley, F. L. Suddath, A. McPherson, D. Sneden,
J. J Kim, J. Weinzierl and Alexander Rich

Abstract. *At 4 Å resolution the polynucleotides in yeast phenylalanine transfer RNA are seen in a series of electron dense masses about 5.8 Å apart. These peaks are probably associated with the phosphate groups, while lower levels of electron density between segments of adjacent polynucleotide chains are interpreted as arising from hydrogen-bonded purine-pyrimidine base pairs. It is possible to trace the entire polynucleotide chain with only two minor regions of ambiguity. The polynucleotide chain has a secondary structure consistent with the cloverleaf conformation; however, its folding is different from that proposed in any model. The molecule is made of two double-stranded helical regions oriented at right angles to each other in the shape of an L. One end of the L has the CCA acceptor; the anticodon loop is at the other end, and the dihydrouridine and TΨC loops form the corner.*

Transfer RNA (tRNA) has a key role in the translation of the polynucleotide sequences of messenger RNA into the polypeptide sequences of protein. A considerable body of information has accumulated regarding these molecules but up to the present time the three-dimensional folding of the polynucleotide chain was unknown. Eight years ago Holley and his collaborators sequenced alanine tRNA from yeast and pointed out that the sequence could be folded into a cloverleaf conformation in which there are four base paired stem regions connected with loops (*1*). Approximately 40 tRNA molecules from various sources have now been sequenced, and all of them can be arranged in a similar cloverleaf arrangement. We have been carrying out an x-ray diffraction analysis of yeast phenylalanine tRNA (*2*) whose sequence is known (*3*). Recently we described the heavy atom derivatives which allowed us to calculate a three-dimensional electron density map at 5.5 Å resolution (*4*). That map allowed us to discern the external shape of portions of the molecule and to trace short segments of the polynucleotide chain. We have continued this work and now report our interpretation of the electron density map at 4.0 Å resolution which allows us to determine the positions of most of the phosphate groups in the nucleotides of yeast phenylalanine tRNA. The polynucleo-

tide chain has been traced and its three-dimensional folding is presented.

Yeast phenylalanine tRNA crystallizes in an orthorhombic unit cell, space group $P2_12_22_1$, $a = 33$ Å, $b = 56$ Å, and $c = 161$ Å, with four molecules in the unit cell (*2*). The methods used in preparing crystals of yeast phenylalanine tRNA, and the chemistry of the isomorphous heavy atom replacements have been described (*4*). Three types of heavy atom derivatives containing platinum, osmium, or samarium have been used. The 4 Å data including anomalous pairs were collected for the osmium and samarium derivatives and 5.5 Å data were collected for the platinum derivative. The positions of these heavy atoms have been reported (*4*). The overall figure of merit for the 2806 reflections collected is 0.70, and the R factors (modulus) are 0.58 for osmium 5.5 Å data; 1.56 for the 5.5 Å to 4.0 Å data; 0.47 for the 4.0 Å samarium data; and 1.05 for the 5.5 Å platinum data (*4*).

The electron density map reported at 5.5 Å resolution (*4*) had a number of intense peaks 5 to 7 Å apart, which were interpreted as arising from the phosphate groups of the tRNA polynucleotide chain. Although portions of the polynucleotide chain could be traced at 5.5 Å resolution, there were too many ambiguities to trace the entire chain. However, at 4.0 Å resolution the individual peaks of electron

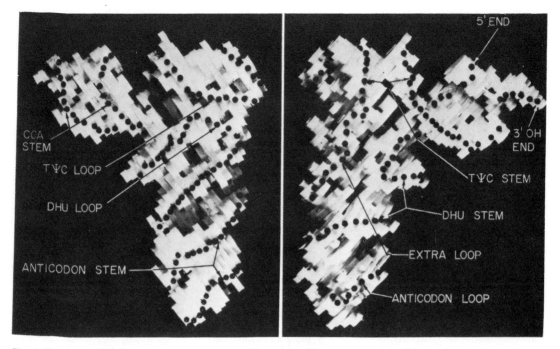

Fig. 1. Two views from opposite sides of a solid molecular model of yeast phenylalanine tRNA as seen at 4.0 Å resolution. The molecule is approximately 20 Å thick in a direction perpendicular to the page. The vertical distance in the molecule is 77 Å. In order to make the tracing of the chains more visible, a series of round-headed pins have been inserted into the molecule. These do *not* represent atoms, but are designed to show the folding of the polynucleotide chain. Hydrogen-bonded base paired stem regions can be readily identified because the adjacent polynucleotide chains are connected to each other.

density are largely resolved. We can observe approximately 80 substantial peaks of electron density in the asymmetric unit. Since yeast phenylalanine tRNA contains 76 nucleotides as well as a number of tightly bound cations, this number was considered adequate for the interpretation of the map.

As was pointed out in the 5.5 Å analysis, large portions of the tRNA molecule can be visualized directly because the tRNA molecules are clearly separated from neighboring molecules by large regions of mother liquor. This includes the unusual 20 to 30 Å separations between the molecules along the *c* axis (*4*). However, portions of the molecule along the *a* and *b* axes are closely packed next to neighboring molecules. The shape of the molecule as described at 5.5 Å resolution was largely correct, except for one region of electron density which appeared to be attached at the end of the elongated molecule. In the 4 Å map, it is apparent that this region is attached to the side of an adjacent molecule and thus the molecule is L-shaped. A solid three-dimensional model of the molecule has been made at 4.0 Å resolution, and two views are shown in Fig. 1. The

sequence of yeast phenylalanine tRNA is shown in Fig. 2.

The peaks of electron density in the map occur with an average separation of 5.8 Å. This spacing is close to that expected from adjacent phosphate groups on a polynucleotide chain. However, individual spacings vary. This would be anticipated in a molecule that had not only a regular polynucleotide conformation in helical regions but also less regular conformations that occur in other parts of the molecule. Interpretation of the electron density map was aided considerably by the observation that there are four regions in the map in which two adjacent chains of electron dense peaks are connected through regions of lowered electron density. The lower electron density regions typically connect two chains as shown in the electron density maps in Fig. 3. In Fig. 3a there are two polynucleotide chains perpendicular to the page, and the phosphate peaks are designated by *X*. It can be seen that the two high density chains are connected by a lower density region which lies largely on one side of both chains. Because of the asymmetry in the geometry of polynucleotide

chains, this suggests that we are observing pairs of hydrogen-bonded bases connecting the two chains, each of which has opposite polarity and is running in opposite directions.

Another region connecting two polynucleotide chains is shown in Fig. 3b. Here we can see the trace of two polynucleotide chains rising toward the reader and then dipping down. Peaks representing the phosphate groups on the polynucleotide chain are marked, and there is a region of weaker electron density connecting the two chains. Four such regions are found in the map and are interpreted to be right-handed, antiparallel double-stranded stems.

The 3'-OH terminus of the molecule is single-stranded and has the only ribose which contains free *cis*-OH groups. Careful inspection of the electron density map revealed that a segment of extended chain containing four peaks of electron density went from one molecule to an adjoining molecule approximately 22 Å away. Since there was no other polynucleotide chain nearby, it seemed to be a suitable candidate for the CCA end (*5*). In addition, the terminal phosphate peak was 7.0 Å away from the heavy atom osmium position. Osmium forms complexes with ribonucleotides through the 2'- and 3'-OH groups, and it seems likely that the osmium is complexed in tRNA to the terminal adenosine. Through a study of related osmium complexes we estimate the distance between osmium and the next phosphate group on the chain could be between 4.0 and 7.6 Å, depending on the orientation of the phosphate. The observed position of the osmium atom next to the phosphate group at the end of this single chain of four residues led us to infer that this was the 3'-OH end of the molecule.

Initial attempts at tracing the poly-nucleotide chain involved a simple in-spection of the electron density map to determine which peaks were within a reasonable distance to be considered as adjoining phosphates on a polynucleo-tide chain. When a polynucleotide chain is fully extended, it has a phos-phate-to-phosphate distance of approxi-mately 7.5 Å; however, the chain can be folded so that in some cases the distance between the phosphates can be less than 5 Å. Accordingly, we carried out the chain tracing by looking for the nearest neighbors within that distance range. Study of the three-dimensional electron density map revealed that only a small number of chain tracings were possible. In most regions the chain can be traced unambiguously; however, in a few places in the map the electron density of a peak was somewhat de-creased so that we could not be certain that it was a nucleotide phosphate group. Alternatively, in two places too many peaks were clustered together, so that it was not easy to make an unam-biguous assignment of correct neigh-bors. However, it quickly became appar-ent that one of the chain tracings not only provided a reasonable assignment of nearest neighbor relations but also demonstrated a direct and simple physi-cal arrangement which embodied the secondary structure implicit in the cloverleaf model. The four double helical regions described above only oc-curred along portions of the chain in which base pairing is expected in the cloverleaf conformation. This correla-tion between the hydrogen-bonded base pairs in the cloverleaf and the low elec-tron density regions between chains in the map gave us more confidence that this was the correct chain tracing. In this regard it should be pointed out that the evidence for the existence of the base pairing in the cloverleaf stems is based not only on the sequence of a large number of tRNA molecules but also on recent high-resolution nuclear

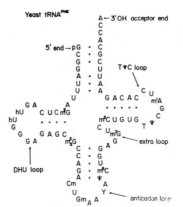

Fig. 2. The sequence of nucleotides in yeast phenylalanine tRNA shown in the conventional cloverleaf diagram (3).

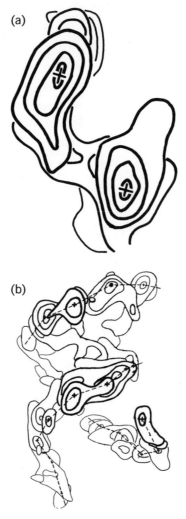

magnetic resonance studies of yeast phenylalanine tRNA in solution (6). Accordingly, the chain tracing found in the three-dimensional electron density map which appeared to utilize the base pairing in the stems of the cloverleaf was selected as the correct tracing. However, it should be noted that one or two alternative tracings are still con-sidered possible at 4.0 Å resolution if one ignores the evidence for the clover-leaf secondary structure. At the pres-ent time we are collecting 3 Å data, and it is anticipated that it will allow us to remove any remaining uncertain-ties regarding the course of the poly-nucleotide chain.

The folding of the molecule is illus-trated in the photographs of the solid model shown in Fig. 1 as well as in the perspective drawing of Fig. 4. The molecule contains two segments of double helix, each about one turn in length. These are oriented at approxi-mately right angles to each other to form an L. The CCA stem is connected to the TΨC stem in a continuous double helix in which the two helical axes of the stem regions are nearly colinear. The TΨC loop occurs at the corner of the L-shaped molecule. Immediately adjoining it is the DHU loop, whose stem is connected to the CCA stem through a short seg-ment of chain containing two phos-phate groups. Both the DHU stem and the anticodon stem form the other continuous double helix around an axis that is oriented approximately at right angles to the axis containing the CCA and TΨC stems. The anti-codon loop is located at the very end of the molecule while the DHU loop is in a position immediately adjoining the TΨC loop at the corner of the mol-ecule. The arrangement of these chains is shown diagrammatically in Fig. 5.

Fig. 3. Sections of the electron density map which illustrates the connections be-tween adjacent segments of polynucleotide chain. (a) Two polynucleotide chains are shown at several levels of the electron density map. The ribose phosphate chains run perpendicular to the page. X marks the position of the phosphate peaks, and lower levels of electron density are seen to connect these chains (scale 1 cm = 2.5 Å). (b) A superposition of segments of the electron density map showing por-tions of three polynucleotide chains. The two chains on the left are joined by regions of lower electron density. The chain on the lower right, although passing nearby, is not joined. X marks the position of the phosphate peaks, and the dotted line shows the continuity of the ribose phosphate polynucleotide chain (scale, 1 cm = 4.7 Å).

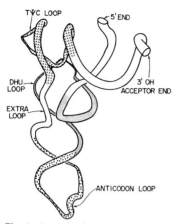

Fig. 4. A perspective diagram in which the polynucleotide chain is represented as a continuous coiled tube. The different shading represents the various loops and stem regions of the tRNA molecule. It can be seen that the TΨC and DHU loops come in very close contact.

The major part of the molecule containing the anticodon loop at one end and the TΨC loop at the other end is 77 Å long. The distance from the anticodon to the end of the CCA stem is 82 Å. An unusual feature of the molecule that was not immediately apparent is the fact that almost the entire structure appears to be 20 Å thick, which is the thickness of an RNA double helix.

It is interesting to note that the major framework of the molecule is due to two segments of helical RNA, each of which is approximately one turn in length. These two helical segments are joined by additional nucleotides in two places. A shorter stretch involves the nucleotides in positions 8 and 9, which connect the CCA stem with the DHU stem. The longer stretch contains five nucleotides in the extra loop and connects the TΨC stem with the anticodon stem. In this regard it is interesting that the extra loops in other tRNA molecules have lengths ranging from 4 to 21 nucleotides (7). We estimate that it would be possible to connect the two helical segments by using as few as four nucleotides, and it is clear from the position of the extra loop on the outside of the molecule that there is ample room there to accommodate a much larger number of nucleotides, even a group containing another elongated stem region, such as is found in some tRNA's with longer extra loop regions.

A number of attempts have been made to predict the three-dimensional

structure of tRNA by constructing models (7). Almost all of the proposed molecular models contain some of the features which are found in three-dimensional structure of tRNA. This is related to the fact that all of them use the secondary structure of the cloverleaf as the basis for their models; however, none of them describe the structure as we now see it. In several models (7) the stems of the various arms are joined to make a continuous double helix, in some cases involving a coaxial arrangement of the CCA stem and the TΨC stem as well as the DHU and anticodon stems; however, none of them illustrate the arrangement of these two helical regions at right angles to each other as is observed in the electron density map.

It is worth drawing attention to some features in the molecule. It has been reported that photoactivation of *Escherichia coli* tRNA$^{Val}_1$ results in the formation of a photodimer involving the 4-thioU residue found in position 8 and the cytosine in position 13 (8). This suggests that these two bases are in close proximity, and this feature has been incorporated in some models. In the electron density map the distance between the phosphates of residues 8 and 13 is near 10 Å, a distance close enough to allow the formation of a photodimer. In all tRNA's so far sequenced the base at position 15 was found to be complementary to the base between the TΨC stem and the extra arm (base 48 in yeast phenylalanine tRNA) (9). The phosphates of these residues are 16 Å apart in our present map, a distance which is compatible with the existence of a base pair. However, the visualization of the purine and pyrimidine bases is necessarily crude at 4 Å resolution, and we will await the analysis of the 3 Å map before citing in detail which interactions stabilize the three-dimensional conformation of tRNA.

A number of enzymatic cleavages, base specific chemical modifications, and oligonucleotide binding studies on native tRNA have been described (7), and most of these seem compatible with the structure we see in the map. A thorough discussion of these reactions together with a discussion of the possible mechanism of tRNA denaturation will await further analysis. However, one feature is worthy of note. The CCA stem projects out and appears free from contacts with the rest of the molecule. It seems to us entirely possible that this stem may be capable of changing

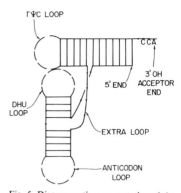

Fig. 5. Diagrammatic representation of the yeast phenylalanine tRNA. This shows the way in which the cloverleaf representation must be transformed in order to show the physical connections between various parts of the molecule. The orientation of the anticodon loop is the same in this diagram as it is in the conventional cloverleaf representation.

its orientation somewhat either in this molecule or in other tRNA molecules. If an effect of this type does occur it could be of considerable importance in understanding various mechanisms that exist in charging tRNA. This might also be of importance during protein synthesis in the ribosome.

S. H. Kim*, G. J. Quigley
F. L. Suddath, A. McPherson
D. Sneden, J. J. Kim
J. Weinzierl, Alexander Rich
*Department of Biology,
Massachusetts Institute of Technology,
Cambridge, 02139*

References and Notes

1. R. W. Holley, J. Apgar, G. A. Everett, J. T. Madison, M. Marquisse, S. H. Merrill, J. R. Penwick, A. Zamir, *Science* 147, 1462 (1965).
2. S. H. Kim, G. J. Quigley, F. L. Suddath, A. Rich, *Proc. Nat. Acad. Sci. U.S.A.* 68, 841 (1971).
3. U. L. RajBhandary and S. H. Chang, *J. Biol. Chem.* 243, 598 (1968).
4. S. H. Kim, G. J. Quigley, F. L. Suddath, A. McPherson, D. Sneden, J. J. Kim, J. Weinzierl, P. Blattman, A. Rich, *Proc. Nat. Acad. Sci. U.S.A.* 69, 3746 (1972).
5. Abbreviations: C, cytidine; Ψ, pseudouridine; T, ribothymidine; U, uridine; DHU, dihydrouridine; tRNA$^{Val}_1$, valine tRNA, species 1.
6. Y. P. Wong, D. R. Kearns, B. R. Reid, R. G. Shulman, *J. Mol. Biol.*, in press.
7. F. Cramer, *Progr. Nucleic Acid Res. Mol. Biol.* 11, 391 (1971).
8. M. Yaniv, A. Favre, B. G. Barrell, *Nature* 223, 1331 (1969).
9. D. Hirsh, *ibid.* 228, 57 (1970).
10. Supported by grants from NIH (CA 04186-15 at M.I.T. and GM 15000-04 at the Biochemistry Department, Duke University), NSF, NASA, the American Cancer Society; and by an NIH postdoctoral fellowship to G.J.Q.; American Cancer Society postdoctoral fellowships to A.M. and F.L.S.; and NIH training grants to D.S. and J.W.
* Present address. Department of Biochemistry, Duke University, Durham, N.C. 27707

19 December 1972

AUTHOR CITATION INDEX

SUBJECT INDEX

About the Editor

ROBERT A. NIEDERMAN is Assistant Professor of Microbiology at Rutgers University, where he teaches courses in molecular biology and bacterial physiology. He received his B.S. and M.S. degrees in bacteriology from the University of Connecticut and his D.V.M. and Ph.D. degrees from the University of Illinois.

Professor Niederman served as a U.S. Atomic Energy Commission Postdoctoral Fellow in Biochemistry at Michigan State University in 1967–1968 and as a Postdoctoral Fellow at the Roche Institute of Molecular Biology in 1968–1970. He is currently studying molecular aspects of bacterial membrane development.